Microbiology Monographs

Volume 9

Series Editor
Alexander Steinbüchel
Münster, Germany

More information about this series at http://www.springer.com/series/7171

Jan Tachezy
Editor

Hydrogenosomes and Mitosomes: Mitochondria of Anaerobic Eukaryotes

Second Edition

Springer

Editor
Jan Tachezy
Faculty of Science
Department of Parasitology
BIOCEV
Charles University
Vestec, Czech Republic

ISSN 1862-5576 ISSN 1862-5584 (electronic)
Microbiology Monographs
ISBN 978-3-030-17940-3 ISBN 978-3-030-17941-0 (eBook)
https://doi.org/10.1007/978-3-030-17941-0

This Springer imprint is published by the registered company Springer Nature Switzerland AG.
The registered company address is: Gewerbestrasse 11, 6330 Cham, Switzerland

Preface

We were very pleased by the invitation of Springer-Verlag to prepare the second edition of the *Microbiology Monographs volume Hydrogenosomes and Mitosomes: Mitochondria of Anaerobic Eukaryotes*. First, it was a signal that the first edition was successful and interesting for readers in cell biology, parasitology, evolutionary biology, and other fields of biology as we hoped. Second, it provides us the opportunity to upgrade the first edition from 2008; there was a tremendous increase in knowledge about evolution, biogenesis, and function of mitochondria and their various anaerobic forms. New technologies, particularly in genomics and proteomics, allow a much more detailed analysis of hydrogenosomes and mitosomes to learn about their functions. For example, initial proteomic study identified only 61 proteins in hydrogenosomes of *Trichomonas vaginalis* in 2007, while about tenfold more proteins are known today. Moreover, various mitochondria that are adapted to anaerobiosis to a great extent have been described in disparate protists across almost all eukaryotic lineages. In this context, hydrogenosomes and mitosomes appeared to be only two most extreme forms of mitochondrial continuum, functioning as aerobic, facultatively anaerobic, or anaerobic organelles. The unexpected versatility of mitochondria was found not only in energy metabolism that ranges from classical Krebs cycle and oxidative phosphorylation, via extended glycolysis to the complete loss of ATP synthesis, but also in the most conserved pathways mediating assembly of FeS (ISC). Although ISC pathway that was inherited from proteobacterial ancestor is present in the vast majority of mitochondria, there are organisms that were able to replace ISC with alternative nitrogen fixation system (NIF) or sulphur-mobilization proteins (SUF) during the course of adaptation to anaerobic environments. NIF system can operate in mitochondria and in the cytosol, which is a case of *Mastigamoeba balamuthi* or exclusively in the cytosol in related *Entamoeba histolytica*. Similarly, SUF proteins were suggested to reside in mitochondria and the cytosol of *Pygsuia biforma* and in the cytosol of *Monocercomonoides* sp. The cytosolic localization of the SUF pathway possibly enabled even the entire loss of mitochondria in the latter organism. In addition to hydrogenosomes and mitosomes, different names for mitochondria with peculiar biochemical properties were defined such as mitochondrion-related organelles in broad sense for mitochondria in anaerobic protists, anaerobic mitochondria that do not produce hydrogen, and hydrogen-producing mitochondria that retain membrane-associated electron transport chain. However, additional studies revealed more subtypes of mitochondria and a more complex mosaic of specific functions. For example, how to call mitochondria in *Dysnectes* that most likely produce hydrogen as hydrogenosomes but do not synthesize ATP as mitosomes? These discoveries gradually erode lines between defined mitochondrion subtypes, and perhaps it is time to call all these organelles simply mitochondria based on their common evolutionary origin.

We highly appreciate the efforts of all the authors who enthusiastically upgraded their original chapters or contributed with new chapters to this book. We believe that the second edition will provide comprehensive contemporary information to scientists with broad interest in biology as well as specialists in cell biology, protistology, and the evolution of eukaryotic cell.

Vestec, Czech Republic — Jan Tachezy
Münster, Germany — Alexander Steinbüchel
June 2019

Contents

The Hydrogenosome Is Born: In Memoriam Donald G. Lindmark 1
Miklós Müller

The Evolution of Oxygen-Independent Energy Metabolism in Eukaryotes with Hydrogenosomes and Mitosomes 7
Verena Zimorski and William F. Martin

Protein Import into Hydrogenosomes and Mitosomes 31
Pavel Dolezal, Abhijith Makki, and Sabrina D. Dyall

Structure of the Hydrogenosome 85
Marlene Benchimol and Wanderley de Souza

Hydrogenosomes of Anaerobic Ciliates 111
Johannes H. P. Hackstein, Rob M. de Graaf, Jaap J. van Hellemond, and Aloysius G. M. Tielens

Metabolism of Trichomonad Hydrogenosomes 127
Ivan Hrdý, Jan Tachezy, and Miklós Müller

Hydrogenosomes of Anaerobic Fungi: An Alternative Way to Adapt to Anaerobic Environments 159
Johannes H. P. Hackstein, Scott E. Baker, Jaap J. van Hellemond, and Aloysius G. M. Tielens

The Proteome of *T. vaginalis* Hydrogenosomes 177
Petr Rada and Jan Tachezy

Mitosomes in Parasitic Protists 205
Jan Tachezy and Ondřej Šmíd

The Mitochondrion-Related Organelles of *Cryptosporidium* Species 243
Anastasios D. Tsaousis and Janet S. Keithly

The Mitochondrion-Related Organelles of *Blastocystis* 267
Anastasios D. Tsaousis, Nigel Yarlett, and Kevin S. W. Tan

Mitochondrion-Related Organelles in Free-Living Protists 287
Michelle M. Leger, Martin Kolísko, Courtney W. Stairs,
and Alastair G. B. Simpson

Organisms Without Mitochondria, How It May Happen? 309
Vladimír Hampl

Index 319

The Hydrogenosome Is Born: In Memoriam Donald G. Lindmark

Miklós Müller

Es ist schlimm genug, rief Eduard,
dass man jetzt nicht mehr für sein ganzes Leben lernen kann.
Unsre Vorfahren hielten sich an den Unterricht,
den sie in ihrer Jugend empfangen;
wir aber müssen jetzt alle fünf Jahre umlernen,
wenn wir nicht ganz aus der Mode kommen wollen.
Johann Wolfgang Goethe, Die Wahlverwandtschaften, 1807
It is bad enough, said Eduard,
that our learning does not last for our whole life.
Our ancestors could hold onto the learning they acquired in their youth;
We have to relearn everything every five years, just to remain fashionable.
(Author's translation)

Contents

References 4

The first version of this multi-author work devoted to hydrogenosomes and mitosomes, younger cousins of the well-known mitochondria, was published 10 years ago (Tachezy 2008). Although Eduard, the hero of Goethe's novel, lamented already in 1807 that one "had to relearn everything every five years," the publication of a new edition, although not 5 but only 10 years later, still indicates a significant increase of our knowledge and understanding of these organelles.

Needless to say that the appearance of this new volume gives a great personal satisfaction to this author who, together with his late coworker, Donald G. Lindmark,

M. Müller (✉)
The Rockefeller University, New York, NY, USA
e-mail: mmuller@mail.rockefeller.edu

J. Tachezy (ed.), *Hydrogenosomes and Mitosomes: Mitochondria of Anaerobic Eukaryotes*, Microbiology Monographs 9,
https://doi.org/10.1007/978-3-030-17941-0_1

was responsible almost half a century ago, for adding to the biological nomenclature (or rather cluttering it with) a new term, hydrogenosome, which now appears on the title page of this volume (Lindmark and Müller 1973). The other term on the title page, mitosome, appeared in the literature later, introduced by Tovar et al. (1999). *Both terms are fully incorporated in the literature and are used extensively.*

As an octogenarian, I hope I will be excused if I briefly describe my own road to the hydrogenosome, a story I told before (e.g., Müller 2007, 2012). I think, however, that it might be a fitting introduction to this volume. I dedicate this brief writing to the memory of my late friend, colleague, and co-discoverer of the hydrogenosome, Donald Gustav Lindmark (1942–2013) (Jaroll 2014).

Protists were always close to my heart. As a high school student, I had a decent microscope and liked to look at free-living organisms in various water samples. As medical student I joined a student group in the Department of Histology and Embryology of the Budapest Medical School and began my apprenticeship in cell biology research, primarily exploring the mechanisms of intracellular digestion in protists with histochemical methods. From this time on, I also participated in teaching medical biology to first-year students. To give relevance to medical students, some laboratory sessions were devoted to human parasites. This is where I first met the human genitourinary parasite, *Trichomonas vaginalis*. Lacking continuous cultures, with some colleagues, we took tubes with culture medium to the Ob/Gyn clinic and obtained positive samples. Of such cultures we made permanent slides, but the students also got to observe the live organisms as it is done in routine clinical examinations for trichomoniasis. Little did I realize then that later in my life I shall devote years and years to studies of these organisms.

My work on protist intracellular digestion in Budapest suggested the relationship of food vacuoles to lysosomes, recently described by Christian de Duve (see Müller 1967). From 1964 I was fortunate to continue my studies on protist cell organelles in the Laboratory of Biochemical Cytology of de Duve at the Rockefeller University in New York, but now with the use of modern methods of biochemical cytology (Müller 1985).

The serial endosymbiosis theory of the origin of the complex eukaryotic cell emerging in the 1950s–1960s solved many puzzles; among them it provided a plausible hypothesis for the origin of the mitochondria. The origin of the peroxisomes, found in cells of mammals but also in aerobic protists, presented a puzzle less easily solved. De Duve proposed that peroxisomes might be a more ancestral respiratory organelle preceding the mitochondria (de Duve 1969) and suggested a search for protist lineages that already acquired peroxisomes but are premitochondrial.

I selected as candidate for this search trichomonad flagellates. These organisms do not have mitochondrial-type respiration but contain subcellular organelles of the size of mitochondria and peroxisomes without cristae, called chromatic granules. In their morphology, these granules resemble peroxisomes. Several species of trichomonads were available in bacterium-free cultures, making biochemical studies possible. Following cell fractionation schemes developed by de Duve's group, I was able to study the distribution of various enzymes in fractions enriched in diverse cell components. The fractions enriched in chromatic granules did not contain catalase, a marker enzyme for peroxisomes, although it is present in some of the trichomonad species explored. It was not localized in the granules but was cytosolic. No other enzymes characteristic of peroxisomes were present in the organism (Müller 1973). Briefly, trichomonad flagellates were not de Duve's hypothetical premitochondrial organism. But the nature of the chromatic granules remained an enigma. Trichomonad flagellates were known to have unusual energy metabolism. Notably, it was known that they can use protons instead of oxygen as terminal electron acceptors, thus forming molecular hydrogen as a metabolic end product. This showed the presence of a metabolic system more akin to that of obligate anaerobic prokaryotes than to typical aerobic eukaryotes. It seemed to me an interesting possibility that the chromatic granules might harbor the enzymatic pathway of hydrogen production.

Not being adept in the study of oxygen-sensitive enzymes of anaerobic metabolism, I needed the participation of a coworker familiar with the applicable techniques. I fortunately found a young microbiologist who could contribute knowledge and experimental expertise to our quest. Donald G. Lindmark obtained his PhD with work on sensitive anaerobic enzymes of bacteria and was ready to join our effort. With him, we continued our cell fractionation experiments and soon could demonstrate the localization of two key enzymes of hydrogen formation (pyruvate:ferredoxin oxidoreductase and hydrogenase) in the chromatic granules. Based on this finding, we proposed a hypothetical metabolic map for the granules and concluded that they are the intracellular site of hydrogen formation in trichomonad flagellates. This part of our studies actually took less than 1 year, thanks to the expertise of Lindmark with anaerobic enzymes and our previous results in cell fractionation of trichomonad flagellates and other protists. Thus the concept of hydrogenosome was born (Lindmark and Müller 1973).

The biochemical nature of these organelles did intrigue also other scientists who departed from the idea of their putative relationship with mitochondria. Extensive studies indicated similarities to mitochondrial metabolism, e.g., the presence of respiratory control (Čerkasov et al. 1978) and of some mitochondrial enzymes (Brugerolle and Metenier 1973). While the marked difference from mitochondria of their main metabolic pathway argued for an independent nature of the hydrogenosomes, subsequent studies demonstrating a common origin with mitochondria revealed the prophetic nature of these studies.

The results of subsequent work of many investigators on hydrogenosomes, and later also on mitosomes, are amply discussed in various chapters of this volume. There is no need and place in this short introduction to give a parallel survey of these. I just mention the main directions our work and work in other laboratories took.

It was immediately asked whether hydrogenosomes are only present in trichomonad flagellates or can be found in other organisms. The highly satisfying answer was that both hydrogenosomes and mitosomes are broadly distributed in symbiotic and parasitic protists that live in microaerophilic or anaerobic habitats.

Subsequent work revealed the presence of a number of additional enzymes in hydrogenosomes and mitosomes of different protists, enabling the establishment of more detailed metabolic maps, than proposed in our initial proposal. These results revealed much about the function and diversity of these organelles and contributed interesting insights into the diversity of eukaryotic organisms.

A vexing question was whether hydrogenosomes represent an independent organelle type or are they related to other cell organelles. Again the answer was clear cut. These organelles turned out to be homologous with aerobic mitochondria. All share a common ancestor. Since these organelles are present in independent protist lineages, their evolutionary history is still incompletely understood and is actively explored.

A still much discussed question is the nature of the common ancestor of these organelles and the way it became part of the eukaryotic cell.

Of course none of this would have remained hidden and obviously would have emerged in later studies by others, probably starting with other organisms and following very different lines of research, but it is still greatly satisfying to see how our biochemical description of a "novel" cell organelle of a specialized parasite entered into debates on several aspects of cell biology and led to ever-expanding insights into our understanding of the diversity of eukaryotic cell life.

Acknowledgments My research could not have unfolded without the generous support of those who accommodated me in their Departments and laboratories: Imre Törő and Béla Párducz in Budapest, Christian de Duve and William Trager in New York, Heinz Holter and Cicily Chapman-Andresen in Copenhagen, André Adoutte and Hervé le Guyader in Orsay, and William Martin in Düsseldorf. I am indebted to all. My wife and parasitologist colleague Jan S. Keithly stood by me for long, long years. Many colleagues participated in our work, to many to be listed here. Their contributions are acknowledged in the individual papers, which emerged from this work. Much of the work was supported with past grants from NIH and NSF.

References

Brugerolle G, Metenier G (1973) Localisaion intracellulaire et characterization des deux types de malate déshydrogénase chez *Trichomonas vaginalis* Donné. J Protozool 20:320–327

Čerkasov J, Čerkasovová A, Kulda J, Vilhelmová D (1978) Respiration of hydrogenosomes of *Tritrichomonas foetus*. I. J Biol Chem 253:1207–1214

de Duve C (1969) Evolution of the peroxisome. Ann N Y Acad Sci 168:369–381

Jaroll EL (2014) In memoriam: Donald G. Lindmark (1942-2013). J Eukaryot Microbiol 61:446–447

Lindmark DG, Müller M (1973) Hydrogenosome, a cytoplasmic organelle of the anaerobic flagellate *Tritrichomonas foetus*, and its role in pyruvate metabolism. J Biol Chem 248:7724–7728

Müller M (1967) Digestion. In: Florkin M, Scheer BT, Kidder GW (eds) Chemical zoology, vol 1 Protozoa. Academic, New York, pp 351–380

Müller M (1973) Cytology of trichomonad flagellates. I. Subcellular localization of hydrolases, dehydrogenases, and catalase in *Tritrichomonas foetus*. J Cell Biol 57:453–474

Müller M (1985) Search for cell organelles in protozoa. J Protozool 32:59–563

Müller M (2007) The road to hydrogenosomes. In: Martin WF, Müller M (eds) Origin of mitochondria and hydrogenosomes. Springer, Berlin, pp 1–11

Müller M (2012) Discovering the hydrogenosome. In: Moberg CL (ed) Entering an unseen world. A founding laboratory and origins of modern cell biology 1910-1914. The Rockefeller University Press, New York, pp 305–309

Tachezy J (ed) (2008) Hydrogenosomes and mitosomes: mitochondria of anaerobic eukaryotes. Springer, Berlin

Tovar J, Fischer A, Clark CG (1999) The mitosome, a novel organelle related to mitochondria in the amitochondrial parasite *Entamoeba histolytica*. Mol Microbiol 32:1013–1021

The Evolution of Oxygen-Independent Energy Metabolism in Eukaryotes with Hydrogenosomes and Mitosomes

Verena Zimorski and William F. Martin

Contents

1 Introduction 8
2 Motivation and a Thirteenfold Insight About O_2 10
3 Older Views of How and Why the Mitochondrion Become Established 11
4 Anoxic and Sulfidic Oceans up to ~450 Million Years Ago 13
5 Anoxic Oceans Give Anaerobic Eukaryotes Room to Breathe 15
6 Conclusion 19
References 21

Abstract Hydrogenosomes and mitosomes are mitochondria in the evolutionary sense because they descend from one and the same bacterial endosymbiont. But the evolutionary significance of eukaryotic anaerobes that possess hydrogenosomes, mitosomes, and anaerobically functioning mitochondria is still an issue of some contention. This chapter serves to further revise the role of oxygen in eukaryote evolution in light of surprising findings about O_2 and energetics and in light of what geologists have discovered regarding the timing of oxygen accumulation in Earth history. According to substantial newer findings that geologists have been reporting for about 20 years, the Proterozoic witnessed a protracted period of Earth ocean history, during which the oceans were mostly anoxic and often sulfidic. That period started about 2.3 billion years ago and only came to an end about 580 million years ago, perhaps as recently as 450 million years ago. That was the time during which eukaryotes arose and diversified into their major lineages. In light of that, anaerobic eukaryotes with mitochondria are not, in an evolutionary sense, strange, obscure, unexpected, or otherwise out of the ordinary; hence no special or unusual mechanisms are required to explain their origin. They are normal in every respect and so are their mitochondria. The surprise from energetics is not that O_2 increases ATP yield from glucose but that the energetic costs for synthesis of cellular building blocks in

V. Zimorski · W. F. Martin (✉)
Institute of Molecular Evolution, Heinrich-Heine-University, 40225 Düsseldorf, Germany
e-mail: bill@hhu.de

J. Tachezy (ed.), *Hydrogenosomes and Mitosomes: Mitochondria of Anaerobic Eukaryotes*, Microbiology Monographs 9,
https://doi.org/10.1007/978-3-030-17941-0_2

oxic environments are 13 times greater than in anaerobic environments. Is oxygen overrated? The benefits of O_2 are offset by its costs.

1 Introduction

Hydrogenosomes and mitosomes are mitochondria in the evolutionary sense that they descend from one and the same eubacterial endosymbiont. But they differ from the mitochondria that most people know from textbooks in that neither hydrogenosomes nor mitosomes studied to date have been found to possess mechanisms of oxidative phosphorylation. Hydrogenosomes and mitosomes occur among eukaryotes that have oxygen-independent ATP synthesis. Accordingly, they occur among various protists that inhabit anaerobic environments and among various protists that have a parasitic lifestyle. The discovery and study of both organelle types have had substantial impact on our understanding of eukaryotic evolution. The role of mitochondria in eukaryote evolution has come more clearly into focus in recent years. Mitochondria were causal to the unique bioenergetic configuration of eukaryotic cells (Lane and Martin 2010), they contributed the lipid vesicles that formed the seed of the eukaryotic endomembrane system (Gould et al. 2016), they contributed three times as many genes to the eukaryotic lineage than the archaeal host did (Thiergart et al. 2012; Ku et al. 2015), and they were the physiological enabler for the origin of phagocytosis in eukaryotes (Martin et al. 2017). Hydrogenosomes and mitosomes carry the evolutionary heritage of mitochondria in eukaryotic anaerobes.

Hydrogenosomes were discovered in 1973 (Lindmark and Müller 1973) and turned out to be anaerobic forms of mitochondria from eukaryotes that inhabit anoxic environments (van der Giezen 2009). Mitosomes were discovered independently by Tovar et al. (1999) and Mai et al. (1999) in the human intestinal parasite *Entamoeba histolytica*. Mai et al. (1999) called the organelle a "crypton," but the name "mitosome" suggested by Tovar et al. (1999) has stuck. Both organelles are mitochondria in the evolutionary sense. The list of those who also agree on that issue would surely include those who discovered hydrogenosomes (Lindmark and Müller 1973), those who discovered mitosomes (Mai et al. 1999; Tovar et al. 1999), many authors of the first edition of this book, and authors of other publications on the topic (Müller 2007; Lane 2007; Allen et al. 2007; Sapp 2007; Tielens and van Hellemond 2007; Tachezy and Dolezal 2007; Hackstein et al. 2007; Cavalier-Smith 2007; Emelyanov 2007; Barberà et al. 2007; Tovar 2007). The evidence to support that view has been the subject of numerous topical reviews and minireviews (Embley et al. 1997; Biagini et al. 1997; van der Giezen et al. 2005; van der Giezen and Tovar 2005; Hackstein et al. 2006; Embley and Martin 2006; Mentel and Martin 2008; Tachezy 2008; van der Giezen 2009; Shiflett and Johnson 2010; Müller et al. 2012; Zimorski et al. 2014; Makiuchi and Nozaki 2014). The evidence from the genome

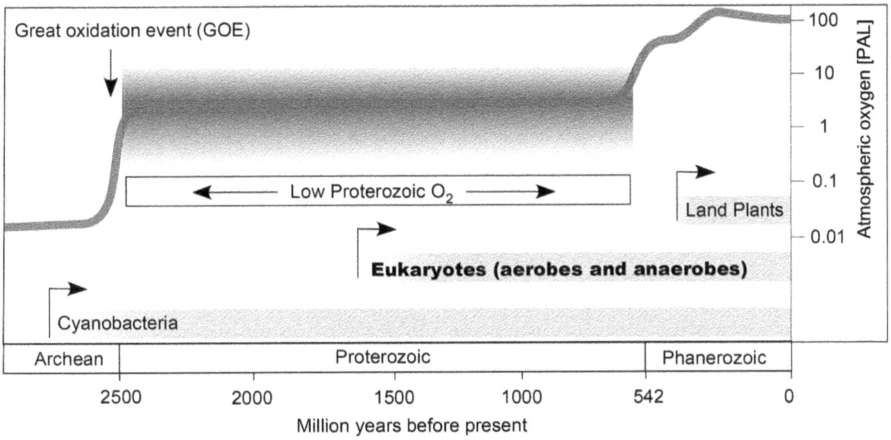

Fig. 1 Summary of oxygen accumulation of earth history. Data from Fischer et al. (2016), Lyons et al. (2014), Javaux and Lepot (2018), Lenton et al. (2016), and Stolper and Keller (2018). *PAL* percent of present atmospheric level

sequences of organisms that possess hydrogenosomes, *Trichomonas* (Carlton et al. 2007), or mitosomes—*Encephalitozoon* (Katinka et al. 2001), *Entamoeba* (Loftus et al. 2005), and *Giardia* (Morrison et al. 2007)—is also fully consistent with the view that both organelles share a common ancestry with mitochondria. Hence there is no need to recite here once more the evidence in support of that view that is based in gene sequence comparisons (van der Giezen et al. 2005; Embley and Martin 2006), shared mechanisms and components of protein import (Dolezal et al. 2006), and shared enzymes (Tielens et al. 2002; Müller 2003; Müller et al. 2012).

There have been outspoken critics of the view of common ancestry for mitochondria and hydrogenosomes. Critics tend to call evidence indicating the common ancestry for these organelles into question, to dismiss it as equivocal, or to marginalize its possible evolutionary significance (Margulis et al. 2006; de Duve 2007). Other critics of common origin favor unspecified lateral gene transfer (LGT) mechanisms to explain hydrogenosome enzyme origin (Andersson and Kurland 1999; Andersson et al. 2003, 2007; Barberà et al. 2007; Hug et al. 2010; Hampl et al. 2011; Stairs et al. 2011, 2014, 2015; Leger et al. 2013, 2016). The LGT theory has it that the earliest eukaryotes were strict aerobes (see however Fig. 1) that acquired the ability to survive low O_2 by means of interdomain and intradomain LGT of genes for proteins germane to anaerobe physiology. Yet a different camp disputes the endosymbiotic origin of mitochondria and hydrogenosomes (Harish and Kurland 2017). Most current evidence for LGT comes from the study of trees and branch lengths, and at present nobody knows which, what kind of, or how many phylogenetic trees we can really trust (see Dyall et al. 2004 vs. Hrdy et al. 2004 or Leger et al. 2016 vs. Martin 2017a).

There are thus two kinds of issues to consider: (1) the evolutionary relationship of hydrogenosomes and mitosomes to mitochondria as membrane-bounded compartments, on which everyone basically agrees and hence will not be the main focus here, and (2) the phylogenetic status and evolutionary significance of eukaryotic anaerobes that possess hydrogenosomes, mitosomes, and anaerobically functioning mitochondria in the geological context of life's history on Earth. The latter issue is of some interest. We will not deal with the LGT theory for eukaryote anaerobe origin here in any detail because that has been dealt with recently in considerable detail elsewhere. The problems with LGT theories for eukaryote anaerobe origin are numerous and severe. LGT theories for eukaryote anaerobe origin are Lamarckian in nature (Martin 2018), but that is the least of their problems (Martin 2017a). The main problem with LGT theories is that the genes for anaerobic energy metabolism germane to mitochondria and hydrogenosomes (and the cytosol in eukaryotes that have mitosomes) trace to the eukaryote common ancestor, as Martin and Müller (1998) predicted, and that eukaryotes arose and diversified in low oxygen and anaerobic environments (Müller et al. 2012), as shown in Fig. 1. Eukaryotes required genes for anaerobic energy metabolism from the onset of the lineage, not as late lineage-specific lateral acquisitions.

2 Motivation and a Thirteenfold Insight About O_2

This chapter serves to further develop the thesis presented in earlier papers (Martin et al. 2003; Martin 2007, 2017a, b, c, 2018; Mentel and Martin 2008; Müller et al. 2012), that (1) the role of oxygen in eukaryote evolution needs to be viewed in light of what geologists are telling us about oxygen in Earth history and that (2) mitochondriate anaerobes are altogether normal eukaryotes with no special kind of evolutionary rank or status. It will be argued that the common ancestry of ATP-synthesizing forms of mitochondria among eukaryotic aerobes and anaerobes and the presence of mitochondria in eukaryotes that synthesize their ATP in the cytosol are most readily interpreted in light of evolutionary specializations of the mitochondrion from a facultatively anaerobic ancestral eukaryotic state. Conversely, the strictly oxygen-dependent forms of the organelle that we know from organisms that live above the soil line on land are also most easily interpreted as evolutionary specializations of the organelle from a more generalized ancestral state—facultatively anaerobic—that existed during a protracted, anaerobic period of Earth ocean history.

Geochemists have been reporting that the oceans had very low oxygen levels and were subject to widespread anoxia until about 580 million years ago (Lyons et al. 2014; Lenton et al. 2016; Reinhard et al. 2016; Fischer et al. 2016; Brocks et al. 2017; van de Velde et al. 2018) with evidence indicating that the rise to roughly our current atmospheric levels of O_2 was generated by early land plants only some 400 million years ago (Lenton et al. 2016), reaching present levels only about 450 million years ago (Lenton et al. 2016; Stolper and Keller 2018). The rise of land plants appears to have been essential for the permanent transition to fully oxygenated oceans (Wallace et al. 2017). Thus, the consensus view on oxygen in

Earth history that was current in the 1990s, namely, that O_2 rose to modern levels 2.4 billion years ago (Kasting 1993), has changed: O_2 rose late, about 500 million years ago, aerobic metabolisms evolved late in Earth history, and anaerobes and aerobes have always been around. Eukaryotes have been around for 1.6 billion years; the first billion years of their lineage diversification took place in anoxic or low oxygen environments. This is sketched in Fig. 1.

A main goal of this chapter will be to convey to the readers the new view of oxygen accumulation during Earth history that has emerged over the last years. The newer view of oxygen in Earth history (oxic habitats late) would render anaerobic eukaryotes with mitochondria a natural result of evolution to be expected in many disparate lineages. Under that view, anaerobic eukaryotes with mitochondria are not, in an evolutionary sense, strange, obscure, or otherwise out of the ordinary; hence no special or unusual mechanisms are required to explain their origin. They are normal in every respect and so are their mitochondria.

What is the 13-fold insight about oxygen? It is this. We know that oxidative phosphorylation using O_2 as the terminal acceptor generates about 30 ATP per glucose (Rich and Maréchal 2010), which corresponds to a sevenfold higher ATP yield than the anaerobic fermentations typical of hydrogenosomes (4 ATP per glucose) or a sixfold higher ATP yield compared to anaerobic mitochondria of the type found in liver fluke (5 ATP per glucose) (Müller et al. 2012). That sounds like a big energetic benefit conferred by O_2. A similar calculation has been made for the ATP yield from amino acid oxidation (Martin et al. 2017) which is relevant because predatory eukaryotes eat mainly cells, and cells are mainly protein. But that is only half the calculation. Less well known is the circumstance that cells living in the presence of O_2 require 13 times more energy to synthesize their cell mass and building blocks for replication than cells that live in the complete absence of O_2 (Lever et al. 2015). Eukaryotes that use O_2 as the terminal acceptor obtain, if we are generous, 10 times more energy per glucose, yes, but they have to spend 13 times more energy per glucose to use it, because they live in an oxic environment. Where is the energetic benefit? It seems that if we look at the matter openly, there is none. The reason is that O_2 is a strong oxidant. It can react with lipids, proteins, nucleic acids, and biosynthetic intermediates in such a way as to oxidize them. Living in an O_2 atmosphere is costly because cells require energy input just to synthesize and maintain the reduced organic compounds that are the building blocks of life in an environment that constantly tends to oxidize them.

3 Older Views of How and Why the Mitochondrion Become Established

Current concepts relating to the origin of mitochondria are most readily understood before the backdrop of how those concepts arose. Since about 1967, mainstream reasoning on the rationale behind mitochondrial origins has focused on oxygen,

ATP, and improved energy yield from glucose breakdown through oxidative phosphorylation. There are exceptions to that rule (Blackstone 1995), but the exceptions prove the rule, as set forth recently in a contribution (Martin 2017c) to a special issue of *J Theoret Biol* devoted to the 50th anniversary of Lynn Margulis's 1967 paper (Sagan 1967) that revived endosymbiotic theory. In order for there to be a biochemical rationale behind mitochondrial origins, there has to be a null hypothesis about the nature of the host, so that one can explicate the possible nature of host-symbiont interactions that lead to a stable symbiosis and hence, in turn, to specify the selective advantages for either partner during the transition from endosymbiont to organelle. The host has always been the weak partner in endosymbiotic theory as it concerns the origin of mitochondria.

Following the rise and fall of endosymbiotic theory in the first half of the twentieth century (Mereschkowsky 1905; Sapp 1994), Margulis (named Sagan in 1967) suggested that the host that acquired the mitochondrial endosymbiont (hereafter just called "the host") was an anaerobic, heterotrophic, fermenting, cell wall-lacking (amoeboid, in the broad sense) prokaryote perhaps similar to modern *Mycoplasma* (Sagan 1967; Margulis 1970). This host corresponded, in terms of cell topology, to the nucleocytoplasmic component of eukaryotes. Margulis opined that the selective advantage that the mitochondrion conferred upon its assumedly fermenting host was improved ATP yield from glucose breakdown by virtue of oxygen respiration and oxidative phosphorylation. Sagan (1967, p. 229) wrote: "The anaerobic breakdown of glucose to pyruvate along the Embden-Meyerhof pathway occurred in the soluble cytoplasm under the direction of the host genome. [. . .] The greater amounts of energy available after the incorporation of the mitochondrion resulted in large cells with amoeboid and cyclotic movement."

This idea was highly compatible with another view that emerged at the time, namely, that the origin of eukaryotes (and their mitochondria) corresponded temporally and causally to the global rise in atmospheric oxygen levels ~2 billion years ago. For example, Sagan (1967, p. 225) wrote: "The subsequent evolution of aerobic metabolism in prokaryotes to form aerobic bacteria (protoflagella and protomitochondria) presumably occurred during the transition to the oxidizing atmosphere." A main point of this chapter, to which will return shortly, concerns the last word of that quote, because microbial evolution did not take place in the atmosphere.

That dual assumption of Margulis's original versions of mitochondrial origins, which one can call the "oxygen/ATP" argument (Martin 2007), was accepted by many subsequent authors, most notably in the present context by those who contest the common ancestry of mitochondria, hydrogenosomes, and mitosomes. However, most people have always rejected the part of Margulis's theory suggesting that flagella (and in later formulations, also the nucleus) stem from spirochaete symbionts. Similarly, most people have always rejected the part of de Duve's (1969, 2007) modification of Margulis's (Sagan 1967; Margulis et al. 2006) theory in which peroxisomes, instead of spirochaetes, are the additional endosymbiont. Possibly as a consequence of that and other early controversies about endosymbiosis in cell evolution (Cavalier-Smith 1975), a new view emerged about 1980 (van Valen and

Maiorana 1980; Doolittle 1980) that placed the oxygen/ATP advantage in the context of a modified form of Margulis's theory that (1) lacked the spirochaete endosymbiont and that (2) assumed the newly discovered archaebacteria to be relatives of the host lineage, flanked by the assumption that the host was a phagocytotic, anaerobic, fermenting eukaryote (possessing a nucleus and other salient eukaryotic features).

That 1980 view assumed that the prokaryote-to-eukaryote transition occurred via gradualist mechanisms such as point mutation and hence did not involve symbiosis at all (van Valen and Maiorana 1980; Doolittle 1980) and culminated with a cell that possessed a nucleus but lacked mitochondria. This is what Doolittle (1998) has called the "standard model". In that view, mitochondria are interpreted as a small tack-on to, and mechanistically unrelated to, the process that made eukaryotic cells nucleated and complex (Cavalier-Smith 2002). In the standard model, mitochondria (and chloroplasts) are descended from endosymbionts, but the nuts-and-bolts of the prokaryote-to-eukaryote transition (the origin of eukaryote-specific traits) was seen as having occurred independently from, and prior to, the origin of mitochondria. The paper by van Valen and Maiorana (1980) expresses that view in clear physiological terms: The host was assumed to be an amoeboid, anaerobic, fermenting cell related to archaebacteria; the advantage of the mitochondrial endosymbiont was to supply ATP.

The most important sentence of this section probably got lost in the foregoing; hence it is repeated here: The 1960s idea that oxygen and ATP drove the origin of mitochondria was highly compatible with another 1960s view, namely, that the origin of eukaryotes (and their mitochondria) corresponded temporally and causally to the global rise in atmospheric oxygen levels ~2 billion years ago. In that 1960s view, there was no room for anaerobic eukaryotes and especially no room for their organelles. The evolutionary significance of hydrogenosomes and mitosomes is, today, still marginalized by proponents of theories that have one assumption in common: The origin of eukaryotes and their mitochondria was mechanistically related to global oxygen change. But that idea might be wrong. So, it is worthwhile to see if geochemical evidence tells a different story about the global history of oxygen in 2018 than in 1970. That is the subject of the next section. We will see that eukaryotic anaerobes fit much better in modern views of global oxygen change.

4 Anoxic and Sulfidic Oceans up to ~450 Million Years Ago

The older, traditional model of global oxygen history arguably starts off with a paper (Cloud 1968) that appeared at the time when Margulis was reviving endosymbiotic theory. Further developed and supported by much evidence into the 1990s, the model went more or less like this: The early Earth was devoid of O_2; O_2 in the atmosphere stems from photosynthesis; cyanobacteria produced several millions of billions of tons of O_2, much of which served to oxidize Fe(II) in the early oceans; and once that titration process was complete, the planet was oxidized and O_2 started accumulating in the atmosphere about 2–2.3 billion years ago as evidenced by the disappearance of particular uranium minerals and the appearance of redbeds (iron oxidized on continents) at that time (Holland and Beukes 1990; Kasting 1993). In that model, the transition from anoxic oceans to oxic oceans occurred in a very narrow window of time (<100 million years ago) and coincided with the appearance of atmospheric O_2 (Kasting 1993). Under that view, the oceans were chemically much like today's (oxic through and through) for about the last 2 billion years. Because the oceans were where life began (Weiss et al. 2016) and eukaryotic evolution was taking place, and because mitochondria were thought to be all about oxygen, anaerobes only fit on the evolutionary map if they lacked mitochondria, and the anaerobic mitochondria of the animals made no evolutionary sense whatsoever, other than as some late and evolutionarily insignificant adaptation to anaerobic habitats. That is the evolutionary position of eukaryotic anaerobes under the traditional view of ocean oxygen history. Thus, eukaryotes were seen as a very ancient lineage; the most primitive lineages branched off early and lack mitochondria; and with the advent of O_2, mitochondria and the lineage leading to higher eukaryotes arose (Knoll 1992). This is also the essence of the recent contributions by Margulis et al. (2006) and de Duve (2007), who cast doubt upon the common ancestry of mitochondria and hydrogenosomes, who disregard mitosomes, and who derive eukaryotes via their own versions of endosymbiotic theory that entail hypothetical symbioses (flagella vs. peroxisomes, respectively) preceding that which gave rise to the mitochondrion. The main biological interpretation of those models and other models that have no room for hydrogenosomes, mitosomes, or anaerobes among evolutionarily advanced eukaryotic lineages is as follows: Anaerobic eukaryotes without mitochondria arose early (and are somehow poorly understood or mysterious), aerobic eukaryotes with mitochondria arose late, and the border between early and late coincides with O_2 accumulation. First chinks in the old geological model came from unexpectedly late occurrences and reinterpretation of particular deposits called banded iron formations (BIFs) (Holland 1999), but the model still stood.

Over the last years, new models of ocean geochemistry arose. The new models of global oxygen history start off like the old one: The ancient Earth was devoid of O_2, and O_2 stems from photosynthesis at the beginning of the early Phanerozoic Eon. The increase of oxygen in the Earth's atmosphere was caused by photosynthesis by cyanobacteria (ca. 3 Ga) and by eukaryotic algae (ca. 1.5 Ga) (Rensing 2018). The oxygen content leading to 100% of present atmospheric level may have originated from the advent of terrestrial photosynthesis and biomass production accompanied by carbon burial (Lenton et al. 2016; Daines et al. 2017). Early land plants, which evolved from charophytic algae (Nishiyama et al. 2018) and came onto land approximately 500 million years ago (Morris et al. 2018; Reski 2018), generated large amounts of oxygen (Rensing 2018) and O_2 reached present levels only about 450 million years ago (Lenton et al. 2016; Stolper and Keller 2018). The accumulation of oxygen in the oceans was a much slower process. Oxygenation of the upper ocean to relatively sustainable near-modern conditions happened at about 200 million years ago (Lu et al. 2018). Many lines of evidence indicate that the oxygenation of the oceans came to a completion less than about 600 million years ago (Sahoo et al. 2012; Lyons et al. 2014).

First traces of microbial life are found in rocks that are 3.95–3.8 billion years (Ga) of age (Tashiro et al. 2017, Arndt and Nisbet 2012, Ueno et al. 2006). At that time the early Earth was a highly reducing environment; thus life on earth originated and evolved in anoxic environments (Fischer et al. 2016). Thus, "facultative anaerobes early" fits seamlessly with the newer findings that oceans were oxygen-poor and largely anoxic until about 580 million years ago (Poulton et al. 2004; Lyons et al. 2014; Fischer et al. 2016; Planavsky et al. 2016; Lenton et al. 2016; Reinhard et al. 2016, Laakso and Schrag 2017). In the view of facultative anaerobes early, no convoluted LGT narratives are required to deliver the same genes for anaerobic survival to many different eukaryotic lineages, rather the genes required to underpin anaerobic growth were present in the eukaryote common ancestor, whereby most symbiont genes were transferred to the host's chromosomes (Martin and Müller 1998; Timmis et al. 2004; Ku et al. 2015), a select few for components of the electron transport chain were retained in organelles (Allen 2015), and the symbiotic origin of mitochondria, not oxygen, was the decisive factor in eukaryote origin (Lane and Martin 2010).

Early eukaryotes arose about 1.65 billion years ago and had diversified by 0.8 billion years ago (Porter et al. 2018). That occurred in anoxic oceans; hence the widespread occurrence of anaerobic energy metabolism among eukaryotes no longer is a curious puzzle, nor does it require some kind of special explanation. The broad phylogenetic distribution of a uniform and recurrently conserved energy metabolic repertoire among anaerobic eukaryotes no longer has to be explained away as some adaptation to anaerobic niches from an assumed aerobic-specialized ancestral state. In many different lineages, the anaerobic lifestyle in eukaryotes can be readily understood as a direct holdover among diversified descendants of a kingdom's anaerobic youth. The new models of "oxygen late" and models for eukaryote origins that accommodate the ubiquity of mitochondriate eukaryotes in anaerobic habitats fit perfectly together.

5 Anoxic Oceans Give Anaerobic Eukaryotes Room to Breathe

Under the new model of Earth's oxygen, hydrogenosomes, mitosomes, and anaerobic eukaryotes in general fit quite naturally into the overall scheme of Earth and life history, without any kinds of special explanation required to account for anaerobic lifestyles. Many protists today still inhabit environments where virtually no free O_2 is available, such as anaerobic sediments (Fenchel and Finlay 1995; Bernhard et al. 2000; Stoeck et al. 2009; Beinart et al. 2018; Porter et al. 2018), and new species are still discovered (Hanousková et al. 2018). In addition, numerous species, in particular among the ciliates, harbor methanogenic archaebacteria that live as endosymbionts within their hosts (Embley and Finlay 1994; Edgcomb et al. 2011), whereby methanogens are among the strictest anaerobes known (Welte and Deppenmeier 2014). Thus, the terms microaerophilic and anaerobic can be equally applicable for many protists. Few eukaryotes studied in detail so far are truly strict anaerobes, in that most of them regularly encounter a bit of O_2 in their natural habitats. Accordingly, they have biochemical means for dealing with O_2 and can readily tolerate it in small amounts.

Globally widespread oxic (aerobic) marine environments are a very late arrival on the evolutionary scene, almost as late as life on land. In that sense, the ecological specialization to strictly oxic environments is a comparatively late evolutionary development. That statement stands in direct opposition to much traditional argumentation regarding the origin and evolution of mitochondria (de Duve 2007; Margulis et al. 2006; Cavalier-Smith 2004), so let's contrast it to some readily available observations to see how it fares.

Let's move back in evolution just one single step more, into the invertebrates. Early eukaryotes appeared ca. 1.65 Ga ago and diversified ca. 800 million years ago (Porter et al. 2018). Vertebrate life on land only goes back about 350 million years (Long and Gordon 2004), when the Earth's atmosphere was enriched with oxygen. Thus, all large animals living on land are specialized to life with O_2 in the atmosphere, specialized to oxic environments, because their mitochondria are specialized to oxic environments. They arose after the oceans were oxic and moved onto land at a time when they had evolved lungs and were able to crawl. Most marine invertebrates (worms, molluscs, and crustaceans) can also use oxygen as terminal electron acceptor in their mitochondria but can also endure sustained anoxia without any problems whatsoever, during which their mitochondria function anaerobically, excreting acetate, propionate, succinate, and CO_2 as end products of mitochondrial ATP synthesis, without the participation of oxygen. Their energy metabolism has been summarized in various extensive reviews (de Zwaan 1991; Schöttler and Bennet 1991; Zebe 1991; Grieshaber et al. 1994; Müller et al. 2012). The biochemical steps involved in the synthesis of the main end products of mitochondrial ATP synthesis—succinate, propionate, acetate, and alanine (Schöttler and Bennet 1991; de Zwaan 1991; Grieshaber et al. 1994)—are the same in various free-living marine invertebrates as those found in various parasitic worms that afflict humans and livestock (Tielens 1994; Tielens et al. 2002; van Hellemond et al. 2003; Bringaud et al. 2010; Harder 2016). The steps involve fumarate reductase (FRD), rhodoquinone instead of ubiquinone at the FRD step (Tielens and van Hellemond 1998; Tielens et al. 2002; Müller et al. 2012), and an acetate/succinate CoA transferase (ASCT) as occurs in hydrogenosomes (Müller 2003). The methylmalonyl-CoA route to propionyl-CoA in anaerobic mitochondrial energy metabolism of invertebrates is nothing special, it is simply a reversal of the pathway used by the mammalian digestive tract to incorporate propionate into glycogen (Schöttler and Bennet 1991), and it is also used by *Euglena* mitochondria (Schneider and Betz 1985), which contain rhodoquinone as required for the FRD step (Hoffmeister et al. 2004), during anaerobic respiration leading to synthesis of odd chain fatty acids and alcohols during *Euglena*'s wax ester fermentation (Tucci et al.

2007). Succinate production in anaerobic metazoan mitochondria is not a process of substrate-level phosphorylation; it involves rhodoquinone-dependent proton pumping at the FRD step and chemiosmotic ATP synthesis (Tielens et al. 2002; van Hellemond et al. 2003), and it could thus be described as fumarate respiration, the term commonly used when discussing energy metabolism in prokaryotes (Fenchel and Finlay 1995), although the term malate dismutation is also used (Tielens and van Hellemond 2007). The typical method for cultivating the marine invertebrates to characterize their anaerobic mitochondrial metabolism is to incubate them in N_2-saturated water over longer periods of time (Grieshaber et al. 1994). For very small metazoans (less than ~2 mm in size; meiofauna), biochemical data about anaerobic metabolism are scarce at best, but it is known that many of them live in the anoxic layer of marine sediment (Nicholas 1991).

The circumstance that mitochondria of marine invertebrates (metazoans, higher eukaryotes in any sense of the word) can readily do their job of synthesizing ATP for the animal without the help of oxygen is not very well known or well appreciated in the evolutionary community (see Martin 2017b), even though the biochemistry and physiology of invertebrate anaerobic mitochondria have been studied in considerable detail. Grieshaber et al. (1994, p. 64) write: "Meanwhile, details of the anaerobic energy metabolism have been elucidated in all major phyla of invertebrates and even in some insect species (Zebe 1991) which are usually considered as completely oxygen dependent." The widespread occurrence of anaerobic mitochondria among marine invertebrates might seem like an odd adaptation of some sort if we think that mitochondria arose because of oxygen, but it makes sense if we think that mitochondria arose at a time where only little amounts of oxygen were available and that the low availability of oxygen persisted throughout most of eukaryotic history in the oceans.

If we dig a bit deeper into eukaryotic phylogeny and look at the fungi, then we already arrive at the anaerobic chytridiomycetes, a basal group of true fungi that inhabit anaerobic habitats and have fully fledged hydrogenosomes (Yarlett et al. 1986; Yarlett 1994; Biagini et al. 1994; Boxma et al. 2004). The chytrid hydrogenosomes are anaerobic mitochondria that produce ATP without an electron transport chain (without chemisomosis) and that produce H_2. The main end products of ATP synthesis in hydrogenosomes of *Piromyces* sp. E2 are formate and acetate, with a comparatively small amount of H_2; additional end products of energy metabolism are succinate, lactate, formate, and ethanol produced in the cytosol (Boxma et al. 2004). The production of formate is the main difference to trichomonad hydrogenosomes (Mountfort et al. 1994); it entails the activity of pyruvate/formate lyase (PFL), which is a comparatively rare metabolic enzyme among most eukaryotes studied so far, but the same PFL enzyme occurs in other chytridiomycetes and in the green alga *Chlamydomonas reinhardtii* (Gelius-Dietrich and Henze 2004; Atteia et al. 2006), suggesting the presence of PFL already in the eukaryote common ancestor. The ethanol produced by *Piromyces* is generated from acetyl-CoA by the action of bifunctional aldehyde/alcohol dehydrogenase (ADHE), which is found in a variety of anaerobic protists (Boxma et al. 2004), in particular many possessing mitosomes (Müller 2003), and which is also found in

Chlamydomonas (Atteia et al. 2006, 2013; Mus et al. 2007) and its colorless chlorophyte relative, *Polytomella* sp. (Atteia et al. 2003). The presence of ADHE in the eukaryote common ancestor also seems likely (Atteia et al. 2003). The hydrogenase of chytrid hydrogenosomes is an iron-only hydrogenase ([Fe]-Hyd), the same type as occurs in hydrogenosomes of other anaerobic eukaryotes and also in *Chlamydomonas* (Horner et al. 2002; Embley et al. 2003). The ATP-synthesizing enzyme of the chytrid hydrogenosome, succinyl-CoA synthase (also called succinate thiokinase), is an enzyme of the citric acid cycle and is the same one (Brondijk et al. 1996; Dacks et al. 2006) as is found in other hydrogenosomes and in mitochondria that have an electron transport chain (Lahti et al. 1994).

Thus, the small repertoire of a few enzymes involved in anaerobic energy metabolism in chytridiomycete hydrogenosomes and their cytosol reveals nothing unique to the group; it entails the same enzymes as are found in other unicellular eukaryotes that deal with anaerobiosis permanently or on a regular basis and as also are found in eukaryotes that produce oxygen. In light of a predominantly anaerobic ocean past, that should not be surprising in the slightest. The reader might wonder how *Chlamydomonas* could encounter anaerobiosis, because it is often regarded as an oxygenic producer, but it is a generalist that commonly inhabits soil (Sack et al. 1994; Merchant et al. 2006), that can regularly encounter anaerobic conditions, and that avidly grows heterotrophically on acetate (Heifetz et al. 2000), a common currency of metabolic end product and carbon source among microbes (Wolfe 2005).

The example of the chytrid hydrogenosomes uncovers many links of shared common ancestry for enzymes regarded as "characteristic" for eukaryotic anaerobes, but the same enzymes also occur among oxygen-adapted forms like *Chlamydomonas*. That makes sense in the context of oxygen late in Earth history. So, if we look around in other eukaryotic groups, do we see anything really special among the anaerobes?

In the trichomonads, the group where hydrogenosomes were discovered, the map of metabolism is slightly different from the chytrids, because the trichomonad hydrogenosomes studied so far possess pyruvate/ferredoxin oxidoreductase (PFO), an enzyme far more widespread among eukaryotic anaerobes (Müller 2003) than PFL. The end products of energy metabolism in trichomonad hydrogenosomes are equimolar amounts of H_2, CO_2, and acetate per mol of pyruvate that enters the organelle (Steinbüchel and Müller 1986). The H_2 is generated by the same type of typical [Fe]-HYD as found in chytrids and *Chlamydomonas* (Horner et al. 2002). The same kind of PFO is found in some ciliate hydrogenosomes (Ellis et al. 1994) but also in the mitochondria of the photosynthetic protist *Euglena* (Rotte et al. 2001) and in *Chlamydomonas* (Atteia et al. 2006), where it is located in the chloroplast (van Lis et al. 2013).

In anaerobic eukaryotes that perform all of their ATP synthesis in the cytosol, like *Entamoeba* or *Giardia*, both of which have mitosomes (Tovar et al. 1999, 2003; Mai et al. 1999), several of the enzymes otherwise found in mitochondria or hydrogenosomes, for example, PFO and [Fe]-HYD (Müller 2003), have been relocalized to the cytosol. *Entamoeba* or *Giardia* tend to produce acetate and ethanol, or a mixture of both, as their main metabolic end products (Müller 2003), although *Giardia* can also produce hydrogen under highly anoxic conditions (Lloyd et al. 2002). The enzymes involved are PFO (Townson et al. 1996), ADHE (Sanchez 1998), and acetyl-CoA synthase (ADP-forming) (Sanchez and Müller 1996; Sanchez et al. 2000), that, like PFL, seems to be a rare enzyme in eukaryote energy metabolism, but the corresponding gene also occurs in the malaria parasite *Plasmodium falciparum* (Sanchez et al. 2000), which uses the mitochondrial electron transport chain in some stages of the life cycle (Painter et al. 2007). So again, there is no case to be made that the enzyme is genuinely specific to anaerobes. In some highly reduced parasitic (but not anaerobic) eukaryotes that possess mitosomes (Williams et al. 2002), it is not clear that the parasite actually produces its own ATP, so particularly among energy parasites, the presence of highly reduced mitochondria need not correlate with the anaerobic lifestyle.

Although hydrogenosomes and mitosomes are both mitochondrion-derived organelles (Tachezy 2008; van der Giezen 2009; Shiflett and Johnson 2010; Zimorski et al. 2014), there is one important difference between them—mitosomes do not produce ATP. The only function discovered so far that links mitosomes to mitochondria and hydrogenosomes is iron-sulfur cluster assembly (Tovar et al. 2003; Goldberg et al. 2008), which seems to be seen as evolutionarily significant. Iron sulfur cluster assembly is a notoriously O_2-sensitive process, so it is not surprising that in oxygen-addicted vertebrates like humans, the process occurs in the mitochondrial matrix, which is generally oxygen-poor, similar to the situation in *Bradyrhizobium*, where N_2 fixation in the cytosol takes place in the presence of active O_2 respiration at the surrounding plasma membrane (Preisig et al. 1996).

To sum up this section, if we look around among the eukaryotic anaerobes that possess hydrogenosomes and mitosomes in search of enzymes or other biochemical attributes that set them distinctly apart from eukaryotes with mitochondria specialized to intermittently oxic or to permanently oxic environments, we come up with an empty set, hence the older ideas of that there are some eukaryotic lineages that diverged after a first endosymbiosis entailing spirochaetes (flagella) or some unspecified bacterium (peroxisomes) but before the origin of mitochondria should probably be abandoned; in 50 years no molecular evidence has come forth in support of either idea (Jekely and Arendt 2006; Gabaldon et al. 2006; Martin 2017c).

6 Conclusion

The common ancestry of mitochondria, hydrogenosomes, and mitosomes does not seem to be an issue anymore. But views concerning the origin, age, and polarity of the ability of eukaryotes to thrive in anaerobic habitats remain an issue of some debate. There are some who would possibly prefer the idea that the enzymes of anaerobic energy metabolism in eukaryotes are acquisitions via LGT in response to the adaptation to anaerobic habitats (Andersson and Kurland 1999; Andersson et al. 2003, 2007; Barberà et al. 2007; Hug et al. 2010; Hampl et al. 2011; Stairs et al.

2011, 2014, 2015; Leger et al. 2013, 2016), that the ancestral state of the eukaryotic lifestyle is aerobic (Cavalier-Smith 2004), or that eukaryotic anaerobes do not have mitochondria (Margulis et al. 2006; de Duve 2007). But if we actually look at the anaerobes, they are altogether normal eukaryotes, possessing no special sorts of attributes that would set them apart from the aerobes. If we had all learned the biochemistry of mitochondria in college using the example of marine invertebrates, then we would see mitochondria as organelles that produce ATP under aerobic and anaerobic conditions. But we learned mitochondrial biochemistry with the example of rat liver mitochondria. Rats do not survive without oxygen; marine invertebrates do.

A dozen or so genes for anaerobic fermentations also entered the eukaryotic lineage at mitochondrial origin, such that aerobic respiration, anaerobic respiration, and hydrogen-producing fermentations, in addition to heterotrophy in general, entered the eukaryotic lineage at mitochondrial origin, as a single inheritance from the facultatively anaerobic metabolism of the mitochondrial endosymbiont, followed by ecological specialization and differential loss in independent mitochondrion bearing eukaryotic lineages (Ku et al. 2015; Martin 2017a). While aerobic eukaryotes lost the ability to survive under anoxic conditions, anaerobic eukaryotes kept the anaerobic metabolism for specialization to anaerobic niches.

The anaerobic energy metabolism in eukaryotes carries the rather unmistakable imprint of a single acquisition via endosymbiosis in that it was acquired once, it represents a very narrow sample of preexisting prokaryotic metabolic diversity, and, at least with regard to an important quinone (rhodoquinone), it is traced physiologically to a particular group (facultatively anaerobic α-proteobacteria). This situation is similar to the cyanobacterial origin of plastids, which also entailed the single acquisition of a small sample of preexisting photosynthetic diversity via a plastid endosymbiont (Gould et al. 2008), followed by specialization and differential loss (e.g., of phycobilisomes in some lineages). Beyond the mitochondrion, there is no evidence for the participation of any other endosymbionts in the origin of the primarily heterotrophic eukaryotic groups, involving microbodies (Gabaldon et al. 2006), flagella (Jekely and Arendt 2006), or otherwise (Martin et al. 2001, 2015).

More robust characters, like the presence or absence of genes and enzymes, the presence of organelles, or the intermingling of aerobic and anaerobic lineages, as, for example, among the ciliates (Embley et al. 1995) or higher eukaryotic groups (Embley and Martin 2006; Müller et al. 2012), suggest that there is no phylogenetic distinction at all between eukaryotic aerobes and eukaryotic anaerobes. Similarly, there is no phylogenetic distinction between eukaryotes that possess or lack flagella (missing among many amoebae, all red algae, and most fungi) and no phylogenetic distinction between eukaryotes that possess or lack peroxisomes (typically lacking in anaerobes). That is, however, not to say that there are no phylogenetic distinctions among eukaryotes at all, for there clearly is a phylogenetic distinction between eukaryotes that possess or lack primary plastids, and the same is true for many other characters as well (see Adl et al. 2005 for a compilation). In light of the late rise of oxygen on Earth, eukaryotic anaerobes should hardly be surprising. The brunt of eukaryote evolution appears now to have occurred in anaerobic marine environments (Martin 2017d).

That is the reason that many eukaryotes have preserved enzymes of anaerobic energy metabolism even though they are aerobes. Many O_2-producing algae have preserved enzymes of anaerobic energy metabolism (Atteia et al. 2013), for example. A more surprising case is perhaps the amoeba *Naegleria gruberi*. Its genome sequence revealed the presence of many enzymes involved in energy metabolism of hydrogenosomes, suggesting that it might use them for anaerobic growth (Fritz-Laylin et al. 2010). However, recent work showed that *N. gruberi* is an obligate aerobe that gains its ATP from O_2-dependent lipid oxidation and that shuns both glucose and amino acids and will not grow in the absence of oxygen (Bexkens et al. 2018). *Naegleria*'s unique O_2-dependent specialization toward lipid oxidizing energy metabolism (Bexkens et al. 2018) underscores the message of Müller et al. (2012) that studies of physiology and growth are essential, because investigations of genes alone can be positively misleading, if we are unlucky. Indeed, the lipid- and O_2-dependent metabolism of *N. gruberi* appears to be fully conserved in *N. fowleri*, the deadly "brain-eating amoeba," explaining why it avidly consumes nerve tissue en route to a lipid-rich and the most O_2-rich human organ—the human brain, which consumes 20% of our O_2 intake but accounts for only 2% of our body weight (Erecińska and Silver 1989). Nonetheless, the retention of genes for anaerobic metabolism in the strict aerobe *Naegleria* is readily understood as a relic from the anaerobic past of eukaryotes and as an evolutionarily recent ecological specialization to late arising oxic environments.

References

Adl SM, Simpson AG, Farmer MA, Andersen RA et al (2005) The new higher level classification of eukaryotes with emphasis on the taxonomy of protists. J Eukaryot Microbiol 52:399–451. 10.1111/j.1550-7408.2005.00053.x

Allen JF (2015) Why chloroplasts and mitochondria retain their own genomes and genetic systems: colocation for redox regulation of gene expression. Proc Natl Acad Sci U S A 112:10231–10238. 10.1073/pnas.1500012112

Allen CA, van der Giezen M, Allen JF (2007) Origin, function, and transmission of mitochondria. In: Martin W, Müller M (eds) Origin of mitochondria and hydrogenosomes. Springer, Berlin, pp 39–56. 10.1007/978-3-540-38502-8_3

Andersson SGE, Kurland CG (1999) Origins of mitochondria and hydrogenosomes. Curr Opin Microbiol 2:535–541. 10.1016/S1369-5274(99)00013-2

Andersson SGE, Karlberg O, Canback B, Kurland CG (2003) On the origin of mitochondria: a genomics perspective. Philos Trans R Soc Lond B 358:165–177. 10.1098/rstb.2002.1193

Andersson JO, Sjogren AM, Horner DS, Murphy CA et al (2007) A genomic survey of the fish parasite *Spironucleus salmonicida* indicates genomic plasticity among diplomonads and significant lateral gene transfer in eukaryote genome evolution. BMC Genomics 8:51. 10.1186/1471-2164-8-51

Arndt N, Nisbet E (2012) Processes on the young Earth and the habitats of early life. Annu Rev Earth Planet Sci 40:521–549. 10.1146/annurev-earth-042711-105316

Atteia A, van Lis R, Mendoza-Hernández G, Henze K et al (2003) Bifunctional aldehyde/alcohol dehydrogenase (ADHE) in chlorophyte algal mitochondria. Plant Mol Biol 53:175–188. 10.1023/B:PLAN.0000009274.19340.36

Atteia A, van Lis R, Gelius-Dietrich G, Adrait A et al (2006) Pyruvate:formate lyase and a novel route of eukaryotic ATP-synthesis in anaerobic *Chlamydomonas* mitochondria. J Biol Chem 281:9909–9918. 10.1074/jbc.M507862200

Atteia A, van Lis R, Tielens AGM, Martin WF (2013) Anaerobic energy metabolism in unicellular photosynthetic eukaryotes. Biochim Biophys Acta 1827:210–223. 10.1016/j.bbabio.2012.08.002

Barberà MJ, Ruiz-Trillo I, Leigh J, Hug LA et al (2007) The diversity of mitochondrion-related organelles amongst eukaryotic microbes. In: Martin W, Müller M (eds) Origin of mitochondria and hydrogenosomes. Springer, Berlin, pp 239–276. 10.1007/978-3-540-38502-8_10

Beinart RA, Beaudoin DJ, Bernhard LM, Edgcomb VP (2018) Insights into the metabolic functioning of a multipartner ciliate symbiosis from oxygen-depleted sediments. Mol Ecol 27:1794–1807. 10.1111/mec.14465

Bernhard JM, Buck KR, Farmer MA, Bowser SS (2000) The Santa Barbara Basin is a symbiosis oasis. Nature 403:77–80. 10.1038/47476

Bexkens ML, Zimorski V, Sarink MJ, Wienk H et al (2018) Lipids are the preferred substrate of the protist *Naegleria gruberi*, relative of a human brain pathogen. Cell Rep 25:537–543.e3. 10.1016/j.celrep.2018.09.055

Biagini GA, van der Giezen M, Hill B, Winters C et al (1994) Ca^{2+}-accumulation in the hydrogenosome of *Neocallimastix frontalis*: a mitochondrial-like physiological role. FEMS Microbiol Lett 149:227–232. 10.1111/j.1574-6968.1997.tb10333.x

Biagini GA, Finlay BJ, Lloyd D (1997) The evolution of the hydrogenosome. FEMS Microbiol Lett 155:133–140. 10.1111/j.1574-6968.1997.tb13869.x

Blackstone NW (1995) Perspective a units-of-evolution perspective on the endosymbiont theory of the origin of the mitochondrion. Evolution 49:785–796. 10.1111/j.1558-5646.1995.tb02315.x

Boxma B, Voncken F, Jannink S, van Alen T et al (2004) The anaerobic chytridiomycete fungus *Piromyces* spE2 produces ethanol via pyruvate:formate lyase and an alcohol dehydrogenase E. Mol Microbiol 51:1389–1399. 10.1046/j.1365-2958.2003.03912.x

Bringaud F, Ebikeme C, Boshart M (2010) Acetate and succinate production in amoebae, helminths, diplomonads and trypanosomatids: common and diverse metabolic strategies used by parasitic lower eukaryotes. Parasitology 137:1315–1331. 10.1017/S0031182009991843

Brocks JJ, Jarrett AJM, Sirantoine E, Hallmann C et al (2017) The rise of algae in Cryogenian oceans and the emergence of animals. Nature 548:578–581. 10.1038/nature23457

Brondijk THC, Durand R, van der Giezen M, Gottschal JC et al (1996) *scsB*, a cDNA encoding the hydrogenosomal beta subunit of succinyl-CoA synthetase from the anaerobic fungus *Neocallimastix frontalis*. Mol Gen Genet 253:315–323. 10.1007/PL00008598

Carlton JM, Hirt RP, Silva JC, Delcher AL et al (2007) Draft genome sequence of the sexually transmitted pathogen *Trichomonas vaginalis*. Science 315:207–212. 10.1126/science.1132894

Cavalier-Smith T (1975) The origin of nuclei and of eukaryotic cells. Nature 256:463–468. 10.1038/256463a0

Cavalier-Smith T (2002) The phagotrophic origin of eukaryotes and phylogenetic classification of Protozoa. Int J Syst Evol Microbiol 52:297–354. 10.1099/00207713-52-2-297

Cavalier-Smith T (2004) Only six kingdoms of life. Proc Biol Sci 271:1251–1262. 10.1098/rspb.2004.2705

Cavalier-Smith T (2007) The chimaeric origin of mitochondria: photosynthetic cell enslavement, gene-transfer pressure, and compartmentation efficiency. In: Martin W, Müller M (eds) Origin of mitochondria and hydrogenosomes. Springer, Berlin, pp 161–200. 10.1007/978-3-540-38502-8_8

Cloud PE Jr (1968) Atmospheric and hydrospheric evolution on the primitive earth. Science 160:729–736. 10.1126/science.160.3829.729

Dacks JB, Dyal PL, Embley TM, van der Giezen M (2006) Hydrogenosomal succinyl-CoA synthetase from the rumen-dwelling fungus *Neocallimastix patriciarum*; an energy-producing enzyme of mitochondrial origin. Gene 373:75–82. 10.1016/j.gene.2006.01.012

Daines SJ, Mills BJW, Lenton TM (2017) Atmospheric oxygen regulation at low Proterozoic levels by incomplete oxidative weathering of sedimentary organic carbon. Nat Commun 8:14379. 10.1038/ncomms14379

de Duve C (1969) Evolution of the peroxisome. Ann N Y Acad Sci 168:369–381. 10.1111/j.1749-6632.1969.tb43124.x

de Duve C (2007) The origin of eukaryotes: a reappraisal. Nat Rev Genet 8:395–403. 10.1038/nrg2071

de Zwaan A (1991) Molluscs. In: Bryant C (ed) Metazoan life without oxygen. Chapman and Hall, London, pp 186–217

Dolezal P, Likic V, Tachezy J, Lithgow T (2006) Evolution of the molecular machines for protein import into mitochondria. Science 313:314–318. 10.1126/science.1127895

Doolittle WF (1980) Revolutionary concepts in evolutionary cell biology. Trends Biochem Sci 5:146–149. 10.1016/0968-0004(80)90010-9

Doolittle WF (1998) A paradigm gets shifty. Nature 392:15–16. 10.1038/32033

Dyall SD, Yan WH, Delgadillo-Correa MG, Lunceford A et al (2004) Non-mitochondrial complex I proteins in a hydrogenosomal oxidoreductase complex. Nature 431:1103–1107. 10.1038/nature02990

Edgcomb VP, Leadbetter ER, Bourland W, Beaudoin D et al (2011) Structured multiple endosymbiosis of bacteria and archaea in a ciliate from marine sulfidic sediments: a survival mechanism in low oxygen, sulfidic sediments? Front Microbiol 2:1–16. 10.3389/fmicb.2011.00055

Ellis JE, Lindmark DG, Williams AG, Lloyd D (1994) Polypeptides of hydrogenosome-enriched fractions from rumen ciliate protozoa and trichomonads: Immunological studies. FEMS Microbiol Lett 117:211–216. 10.1016/0378-1097(94)90197-X

Embley TM, Finlay BJ (1994) The use of small subunit rRNA sequences to unravel the relationships between anaerobic ciliates and their methanogen endosymbionts. Microbiology 140:225–235. 10.1099/13500872-140-2-225

Embley TM, Martin W (2006) Eukaryotic evolution, changes and challenges. Nature 440:623–630. 10.1038/nature04546

Embley TM, Finlay BJ, Dyal PL, Hirt RP et al (1995) Multiple origins of anaerobic ciliates with hydrogenosomes within the radiation of aerobic ciliates. Proc R Soc Lond B 262:87–93. 10.1098/rspb.1995.0180

Embley TM, Horner DA, Hirt RP (1997) Anaerobic eukaryote evolution: hydrogenosomes as biochemically modified mitochondria? Trends Ecol Evol 12:437–441. 10.1016/S0169-5347.(97)01208-1

Embley TM, van der Giezen M, Horner DS, Dyal PL et al (2003) Hydrogenosomes, mitochondria and early eukaryotic evolution. IUBMB Life 55:387–395. 10.1080/15216540310001592834

Emelyanov VV (2007) Constantin Merezhkowsky and the endokaryotic hypothesis. In: Martin W, Müller M (eds) Origin of mitochondria and hydrogenosomes. Springer, Berlin, pp 201–238. 10.1007/978-3-540-38502-8_9

Erecińska M, Silver IA (1989) ATP and brain function. J Cereb Blood Flow Metab 9:2–19. 10.1038/jcbfm.1989.2

Fenchel T, Finlay BJ (eds) (1995) Ecology and evolution in anoxic worlds. Oxford University Press, Oxford

Fischer WW, Hemp J, Valentine JS (2016) How did life survive Earth's great oxygenation? Curr Opin Chem Biol 31:166–178. 10.1016/j.cbpa.2016.03.013

Fritz-Laylin LK, Prochnik SE, Ginger ML, Dacks JB et al (2010) The genome of *Naegleria gruberi* illuminates early eukaryotic versatility. Cell 140:631–642. 10.1016/j.cell.2010.01.032

Gabaldon T, Snel B, van Zimmeren F, Hemrika W et al (2006) Origin and evolution of the peroxisomal proteome. Biol Direct 1:8. 10.1186/1745-6150-1-8

Gelius-Dietrich G, Henze K (2004) Pyruvate formate lyase (PFL) and PFL activating enzyme in the chytrid fungus *Neocallimastix frontalis*: a free-radical enzyme system conserved across divergent eukaryotic lineages. J Eukaryot Microbiol 51:456–463. 10.1111/j.1550-7408.2004.tb00394.x
Goldberg AV, Molik S, Tsaousis AD, Neumann K et al (2008) Localization and functionality of microsporidian iron-sulphur cluster assembly proteins. Nature 452:624–628. 10.1038/nature06606
Gould SB, Waller RR, McFadden GI (2008) Plastid evolution. Annu Rev Plant Biol 59:491–517. 10.1146/annurev.arplant.59.032607.092915
Gould SB, Garg S, Martin WF (2016) Bacterial vesicle secretion and the evolutionary origin of the eukaryotic endomembrane system. Trends Microbiol 24:525–534. 10.1016/j.tim.2016.03.005
Grieshaber M, Hardewig I, Kreutzer U, Pörtner H-O (1994) Physiological and metabolic responses to hypoxia in invertebrates. Rev Physiol Biochem Pharmacol 125:43–147
Hackstein JH, Tjaden J, Huynen M (2006) Mitochondria, hydrogenosomes and mitosomes: products of evolutionary tinkering! Curr Genet 50:225–245. 10.1007/s00294-006-0088-8
Hackstein JHP, Tjaden J, Koopman W (2007) Hydrogenosomes (and related organelles, either) are not the same. In: Martin W, Müller M (eds) Origin of mitochondria and hydrogenosomes. Springer, Berlin, pp 135–160. 10.1007/978-3-540-38502-8_7
Hampl V, Stairs CW, Roger AJ (2011) The tangled past of eukaryotic enzymes involved in anaerobic metabolism. Mob Genet Elem 1:71–74. 10.4161/mge.1.1.15588
Hanousková P, Táborsky P, Cepicka I (2018) *Dactylomonas* gen. nov., a novel lineage of heterolobosean flagellates with unique ultrastructure, closely related to the amoeba *Selenaion koniopes* Park, De Jockheere, Simpson, 2012. J Eukaryot Microbiol 66(1):120–139. 10.1111/jeu.12637
Harder A (2016) The biochemistry of *Haemonchus contortus* and other parasitic nematodes. Adv Parasitol 93:69–94. 10.1016/bs.apar.2016.02.010
Harish A, Kurland CG (2017) Mitochondria are not captive bacteria. J Theoret Biol 434:88–98. 10.1016/j.biochi.2017.04.013
Heifetz PB, Förster B, Osmond CB, Giles LJ et al (2000) Effects of acetate on facultative autotrophy in *Chlamydomonas reinhardtii* assessed by photosynthetic measurements and stable isotope analyses. Plant Physiol 122:1439–1446. 10.1104/pp.122.4.1439
Hoffmeister M, van der Klei A, Rotte C, van Grinsven KWA et al (2004) *Euglena gracilis* rhodoquinone:ubiquinone ratio and mitochondrial proteome differ under aerobic and anaerobic conditions. J Biol Chem 279:22422–22429. 10.1074/jbc.M400913200
Holland HD (1999) When did the Earth's atmosphere become oxic? A reply. Geochem News 100:20–22
Holland HD, Beukes N (1990) A paleoweathering profile from Griqualand West, South Africa: evidence for a dramatic rise in atmospheric oxygen between 2.2 and 1.9 bybp. Am J Sci 290:1–34
Horner DS, Heil B, Happe T, Embley TM (2002) Iron hydrogenases – ancient enzymes in modern eukaryotes. Trends Biochem Sci 27:148–153. 10.1016/S0968-0004(01)02053-9
Hrdy I, Hirt RP, Dolezal P, Bardonova L et al (2004) *Trichomonas* hydrogenosomes contain the NADH dehydrogenase module of mitochondrial complex I. Nature 432:618–622. 10.1038/nature03149
Hug LA, Stechmann A, Roger AJ (2010) Phylogenetic distributions and histories of proteins involved in anaerobic pyruvate metabolism in eukaryotes. Mol Biol Evol 27:311–324. 10.1093/molbev/msp237
Javaux EJ, Lepot K (2018) The Paleoproterozoic fossil record: implications for the evolution of the biosphere during Earth's middle-age. Earth-Sci Rev 176:68–86. 10.1016/j.earscirev.2017.10.001
Jekely G, Arendt D (2006) Evolution of intraflagellar transport from coated vesicles and autogenous origin of the eukaryotic cilium. Bioessays 28:191–198. 10.1002/bies.20369
Kasting JF (1993) Earth's early atmosphere. Science 259:920–926. 10.1126/science.1079894

Katinka MD, Duprat S, Cornillot E, Méténier G et al (2001) The genome of the intracellular parasite, *Encephalitozoon cuniculi*. Nature 414:450–453. 10.1038/35106579

Knoll AH (1992) The early evolution of eukaryotes: a geological perspective. Science 256:622–627. 10.1126/science.1585174

Ku C, Nelson-Sathi S, Roettger M, Sousa FL et al (2015) Endosymbiotic origin and differential loss of eukaryotic genes. Nature 524:427–432. 10.1038/nature14963

Laakso TA, Schrag DP (2017) A theory of atmospheric oxygen. Geobiology 15:366–384. 10.1111/gbi.12230

Lahti CJ, Bradley PJ, Johnson PJ (1994) Molecular characterization of the alpha-subunit of *Trichomonas vaginalis* hydrogenosomal succinyl CoA synthetase. Mol Biochem Parasitol 66:309–318. 10.1016/0166-6851(94)90157-0

Lane N (2007) Mitochondria: key to complexity. In: Martin W, Müller M (eds) Origin of mitochondria and hydrogenosomes. Springer, Berlin, pp 13–38. 10.1007/978-3-540-38502-8_2

Lane N, Martin W (2010) The energetics of genome complexity. Nature 467:929–934. 10.1038/nature09486

Leger MM, Gawryluk RM, Gray MW, Roger AJ (2013) Evidence for a hydrogenosomal-type anaerobic ATP generation pathway in *Acanthamoeba castellanii*. PLoS One 8:e69532. 10.1371/journal.pone.0069532

Leger MM, Eme L, Hug LA, Roger AJ (2016) Novel hydrogenosomes in the microaerophilic jakobid *Stygiella incarcerata*. Mol Biol Evol 33:2318–2336. 10.1093/molbev/msw103

Lenton TM, Dahl TW, Daines SJ, Mills BJW et al (2016) Earliest land plants created modern levels of atmospheric oxygen. Proc Natl Acad Sci U S A 113:9704–9709. 10.1073/pnas.1604787113

Lever MA, Rogers KL, Lloyd KG, Overmann J et al (2015) Life under extreme energy limitation: a synthesis of laboratory- and field-based investigations. FEMS Microbiol Rev 39:688–728. 10.1093/femsre/fuv020

Lindmark DG, Müller M (1973) Hydrogenosome, a cytoplasmic organelle of the anaerobic flagellate, *Tritrichomonas foetus*, and its role in pyruvate metabolism. J Biol Chem 248:7724–7728

Lloyd D, Ralphs JR, Harris JC (2002) *Giardia intestinalis*, a eukaryote without hydrogenosomes, produces hydrogen. Microbiology 148:727–733. 10.1099/00221287-148-3-727

Loftus B, Anderson I, Davies R, Alsmark UC et al (2005) The genome of the protist parasite *Entamoeba histolytica*. Nature 433:865–868. 10.1038/nature03291

Long JA, Gordon MS (2004) The greatest step in vertebrate history: a paleobiological review of the fish-tetrapod transition. Physiol Biochem Zool 77:700–719. 10.1086/425183

Lu W, Ridgwell A, Thomas E, Hardisty DS et al (2018) Late inception of a resiliently oxygenated upper ocean. Science 361:174–177. 10.1126/science.aar5372

Lyons TW, Reinhard CT, Planavsky NJ (2014) The rise of oxygen in Earth's early ocean and atmosphere. Nature 506:307–315. 10.1038/nature13068

Mai ZM, Ghosh S, Frisardi M, Rosenthal B et al (1999) Hsp60 is targeted to a cryptic mitochondrion-derived organelle ("crypton") in the microaerophilic protozoan parasite *Entamoeba histolytica*. Mol Cell Biol 19:2198–2205. 10.1128/MCB.19.3.2198

Makiuchi T, Nozaki T (2014) Highly divergent mitochondrion-related organelles in anaerobic parasitic protozoa. Biochimie 100:3–17. 10.1016/j.biochi.2013.11.018

Margulis L (1970) Origin of eukaryotic cells. Yale University Press, New Haven, CT

Margulis L, Chapman M, Guerrero R, Hall J (2006) The last eukaryotic common ancestor (LECA): acquisition of cytoskeletal motility from aerotolerant spirochetes in the Proterozoic Eon. Proc Natl Acad Sci U S A 103:13080–13085. 10.1073/pnas.0604985103

Martin W (2007) Eukaryote and mitochondrial origins: two sides of the same coin and too much ado about oxygen. In: Falkowski P, Knoll AH (eds) primary producers of the sea. Elsevier, Amsterdam, pp 53–73. 10.1016/B978-012370518-1/50006-0

Martin WF (2017a) Too much eukaryote LGT. Bioessays 39. 10.1002/bies.201700115

Martin WF (2017b) Unmiraculous facultative anaerobes (comment on DOI 10.1002/bies.201600174). Bioessays 39. 10.1002/bies.201700041

Martin WF (2017c) Physiology, anaerobes, and the origin of mitosing cells 50 years on. J Theor Biol 434:2–10. 10.1016/j.jtbi.2017.01.004
Martin WF (2017d) Symbiogenesis, gradualism and mitochondrial energy in eukaryote evolution. Period Biol 119:141–158. 10.18054/pb.v119i3.5694
Martin WF (2018) Eukaryote lateral gene transfer is Lamarckian. Nat Ecol Evol 2:754. 10.1038/s41559-018-0521-7
Martin W, Müller M (1998) The hydrogen hypothesis for the first eukaryote. Nature 392:37–41. 10.1038/32096
Martin W, Hoffmeister M, Rotte C, Henze K (2001) An overview of endosymbiotic models for the origins of eukaryotes, their ATP-producing organelles (mitochondria and hydrogenosomes), and their heterotrophic lifestyle. Biol Chem 382:1521–1539. 10.1515/BC.2001.187
Martin W, Rotte C, Hoffmeister M, Theissen U et al (2003) Early cell evolution, eukaryotes, anoxia, sulfide, oxygen, fungi first (?), and a tree of genomes revisited. IUBMB Life 55:193–204. 10.1080/1521654031000141231
Martin W, Garg S, Zimorski V (2015) Endosymbiotic theories for eukaryote origin. Philos Trans R Soc Lond B 370:20140330. 10.1098/rstb.2014.0330
Martin WF, Tielens AGM, Mentel M, Garg SG et al (2017) The physiology of phagocytosis in the context of mitochondrial origin. Microbiol Mol Biol Rev 81:e00008–e00017. 10.1128/MMBR.00008-17
Mentel M, Martin W (2008) Energy metabolism among eukaryotic anaerobes in light of Proterozoic ocean chemistry. Philos Trans R Soc Lond Ser B Biol Sci 363:2717–2729. 10.1098/rstb.2008.0031
Merchant SS, Allen MD, Kropat J, Moseley JL et al (2006) Between a rock and a hard place: trace element nutrition in *Chlamydomonas*. Biochim Biophys Acta 1763:578–594. 10.1016/j.bbamcr.2006.04.007
Mereschkowsky C (1905) Über Natur und Ursprung der Chromatophoren im Pflanzenreiche. Biol Centralbl 25:593–604 [English translation in Eur J Phycol 34:287–295 (1999) doi:10.1080/09670269910001736342]
Morris JL, Puttick MN, Clark JW, Edwards D et al (2018) The timescale of early land plant evolution. Proc Natl Acad Sci U S A 115:E2274–E2283. 10.1073/pnas.1719588115
Morrison HG, McArthur AG, Gillin FD, Aley SB et al (2007) Genomic minimalism in the early diverging intestinal parasite *Giardia lamblia*. Science 317:1921–1926. 10.1126/science.1143837
Mountfort DO, Orpin CG, Dekker M (1994) Anaerobic fungi: biology, ecology and function. Marcel Dekker, New York, NY
Müller M (2003) Energy metabolism. Part I: anaerobic protozoa. In: Marr J (ed) Molecular medical parasitology. Academic, London, pp 125–139. 10.1016/B978-012473346-6/50008-9
Müller M (2007) The road to hydrogenosomes. In: Martin W, Müller M (eds) Origin of mitochondria and hydrogenosomes. Springer, Berlin, pp 1–12. 10.1007/978-3-540-38502-8_1
Müller M, Mentel M, van Hellemond JJ, Henze K et al (2012) Biochemistry and evolution of anaerobic energy metabolism in eukaryotes. Microbiol Mol Biol Rev 76:444–495. 10.1128/MMBR.05024-11
Mus F, Dubini A, Seibert M, Posewitz MC et al (2007) Anaerobic acclimation in *Chlamydomonas reinhardtii* – anoxic gene expression, hydrogenase induction, and metabolic pathways. J Biol Chem 282:25475–25486. 10.1074/jbc.M701415200
Nicholas WL (1991) Interstitial meifauna. In: Bryant C (ed) Metazoan life without oxygen. Chapman and Hall, London, pp 129–145
Nishiyama T, Hidetoshi S, de Vries J, Buschmann H et al (2018) The *Chara* genome: secondary complexity and implications for plant terrestrialization. Cell 174:448–464. 10.1016/j.cell.2018.06.033
Painter HJ, Morrisey JM, Mather MW, Vaidya AB (2007) Specific role of mitochondrial electron transport in blood-stage *Plasmodium falciparum*. Nature 446:88–91. 10.1038/nature05572

Planavsky NJ, Cole DB, Reinhard CT, Diamond C et al (2016) No evidence for high atmospheric oxygen levels 1,400 million years ago. Proc Natl Acad Sci U S A 113:E2550–E2551. 10.1073/pnas.1601925113

Porter SM, Agic H, Riedman LA (2018) Anoxic ecosystems and early eukaryotes. Emerg Top Life Sci 2:299–309. 10.1042/ETLS20170162

Poulton SW, Fralick PW, Canfield DE (2004) The transition to a sulphidic ocean ~1.84 billion years ago. Nature 431:173–177. 10.1038/nature02912

Preisig O, Zufferey R, Thöny-Meyer L, Appleby CA et al (1996) A high-affinity *cbb3*-type cytochrome oxidase terminates the symbiosis-specific respiratory chain of *Bradyrhizobium japonicum*. J Bacteriol 178:1532–1538. 10.1128/jb.178.6.1532-1538.1996

Reinhard CT, Planavsky NJ, Olson SL, Lyons TW et al (2016) Earth's oxygen cycle and the evolution of animal life. Proc Natl Acad Sci U S A 113:8933–8938. 10.1073/pnas.1521544113

Rensing SA (2018) Great moments in evolution: the conquest of land by plants. Curr Opin Plant Biol 42:49–54. 10.1016/j.pbi.2018.02.006

Reski R (2018) Enabling the water-to-land transition. Nat Plants 4:67–68. 10.1038/s41477-018-0101-5

Rich PR, Maréchal A (2010) The mitochondrial respiratory chain. Essays Biochem 47:1–23. 10.1042/bse0470001

Rotte C, Stejskal F, Zhu G, Keithly JS et al (2001) Pyruvate:$NADP^+$ oxidoreductase from the mitochondrion of *Euglena gracilis* and from the apicomplexan *Cryptosporidium parvum*: a fusion of pyruvate:ferredoxin oxidoreductase and NADPH-cytochrome P450 reductase. Mol Biol Evol 18:710–720. 10.1093/oxfordjournals.molbev.a003853

Sack L, Zeyl C, Bell G, Sharbel T et al (1994) Isolation of four new strains of *Chlamydomonas reinhardtii* (chlorophyta) from soil samples. J Phycol 30:770–773. 10.1111/j.0022-3646.1994.00770.x

Sagan L (1967) On the origin of mitosing cells. J Theoret Biol 14:225–274. 10.1016/0022-5193(67)90079-3

Sahoo SK, Planavsky NJ, Kendall B, Wang X et al (2012) Ocean oxygenation in the wake of the Marinoan glaciation. Nature 489:546–549. 10.1038/nature11445

Sanchez LB (1998) Aldehyde dehydrogenase (CoA-acetylating) and the mechanism of ethanol formation in the amitochondriate protist, *Giardia lamblia*. Arch Biochem Biophys 354:57–64. 10.1006/abbi.1998.0664

Sanchez LB, Müller M (1996) Purification and characterization of the acetate forming enzyme, acetyl-CoA synthetase (ADP-forming) from the amitochondriate prostist, *Giardia lamblia*. FEBS Lett 378:240–244. 10.1016/0014-5793(95)01463-2

Sanchez LB, Galperin MY, Müller M (2000) Acetyl-CoA synthetase from the amitochondriate eukaryote *Giardia lamblia* belongs to the newly recognized superfamily of acyl-CoA synthetases (nucleoside diphosphate-forming). J Biol Chem 275:5794–5803. 10.1074/jbc.275.8.5794

Sapp J (1994) Evolution by association: a history of symbiosis. Oxford University Press, New York, p 255

Sapp J (2007) Mitochondria and their host: morphology to molecular phylogeny. In: Martin W, Müller M (eds) Origin of mitochondria and hydrogenosomes. Springer, Berlin, pp 57–84. 10.1007/978-3-540-38502-8_4

Schneider T, Betz A (1985) Wax ester fermentation in *Euglena gracilis* T. Factors favouring the synthesis of odd-numbered fatty acids and alcohols. Planta 166:67–73. 10.1007/BF00397387

Schöttler U, Bennet EM (1991) Annelids. In: Bryant C (ed) Metazoan life without oxygen. Chapman and Hall, London, pp 165–185

Shiflett AM, Johnson PJ (2010) Mitochondrion-related organelles in eukaryotic protists. Annu Rev Microbiol 64:409–429. 10.1146/annurev.micro.62.081307.162826

Stairs CW, Roger AJ, Hampl V (2011) Eukaryotic pyruvate formate lyase and its activating enzyme were acquired laterally from a firmicute. Mol Biol Evol 28:2087–2099. 10.1093/molbev/msr032

Stairs CW, Eme L, Brown MW, Mutsaers C et al (2014) A SUF Fe-S cluster biogenesis system in the mitochondrion-related organelles of the anaerobic protist *Pygsuia*. Curr Biol 24:1176–1186. 10.1016/j.cub.2014.04.033

Stairs CW, Leger MM, Roger AJ (2015) Diversity and origins of anaerobic metabolism in mitochondria and related organelles. Philos Trans Soc Lond B Biol Sci 370:20140326. 10.1098/rstb.2014.0326

Steinbüchel A, Müller M (1986) Anaerobic pyruvate metabolism of *Tritrichomonas foetus* and *Trichomonas vaginalis* hydrogenosomes. Mol Biochem Parasitol 20:57–60. 10.1016/0166-6851(86)90142-8

Stoeck T, Behnke A, Christen R, Amaral-Zettler L et al (2009) Massively parallel tag sequencing reveals the complexity of anaerobic marine protistan communities. BMC Biol 7:72. 10.1186/1741-7007-7-72

Stolper DA, Keller B (2018) A record of deep-ocean dissolved O_2 from the oxidation state of iron in submarine basalts. Nature 553:323–327. 10.1038/nature25009

Tachezy J (ed) (2008) Hydrogenosomes and mitosomes: mitochondria of anaerobic eukaryotes. Springer, Berlin. 10.1007/978-3-540-76733-6

Tachezy J, Dolezal P (2007) Iron–sulfur proteins and iron–sulfur cluster assembly in organisms with hydrogenosomes and mitosomes. In: Martin W, Müller M (eds) Origin of mitochondria and hydrogenosomes. Springer, Berlin, pp 105–134. 10.1007/978-3-540-38502-8_6

Tashiro T, Ishida A, Hori M, Igisu M et al (2017) Early trace of life from 3.95 Ga sedimentary rocks in Labrador, Canada. Nature 549:516–518. 10.1038/nature24019

Thiergart T, Landan G, Schenk M, Dagan T et al (2012) An evolutionary network of genes present in the eukaryote common ancestor polls genomes on eukaryotic and mitochondrial origin. Genome Biol Evol 4:466–485. 10.1093/gbe/evs018

Tielens AGM (1994) Energy generation in parasitic helminths. Parasitol Today 10:346–352. 10.1016/0169-4758(94)90245-3

Tielens AGM, van Hellemond JJ (1998) The electron transport chain in anaerobically functioning eukaryotes. Biochim Biophys Acta 1365:71–78. 10.1016/S0005-2728(98)00045-0

Tielens AGM, van Hellemond JJ (2007) Anaerobic mitochondria: properties and origins. In: Martin W, Müller M (eds) Origin of mitochondria and hydrogenosomes. Springer, Berlin, pp 85–104. 10.1007/978-3-540-38502-8_5

Tielens GM, Rotte C, van Hellemond J, Martin W (2002) Mitochondria as we don't know them. Trends Biochem Sci 27:564–572. 10.1016/S0968-0004(02)02193-X

Timmis JN, Ayliffe MA, Huang CY, Martin W (2004) Endosymbiotic gene transfer: organelle genomes forge eukaryotic chromosomes. Nat Rev Genet 5:123–135. 10.1038/nrg1271

Tovar J (2007) Mitosomes of parasitic protozoa: biology and evolutionary significance. In: Martin W, Müller M (eds) Origin of mitochondria and hydrogenosomes. Springer, Berlin, pp 277–300. 10.1007/978-3-540-38502-8_11

Tovar J, Fischer A, Clark CG (1999) The mitosome, a novel organelle related to mitochondria in the amitochondrial parasite *Entamoeba histolytica*. Mol Microbiol 32:1013–1021. 10.1046/j.1365-2958.1999.01414.x

Tovar J, León-Avila G, Sánchez LB, Sutak R et al (2003) Mitochondrial remnant organelles of *Giardia* function in iron-sulphur protein maturation. Nature 426:172–176. 10.1038/nature01945

Townson SM, Upcroft JA, Upcroft P (1996) Characterization and purification of pyruvate:ferredoxin oxidoreductase from *Giardia duodenalis*. Mol Biochem Parasitol 79:183–193. 10.1016/0166-6851(96)02661-8

Tucci S, Proksch P, Martin W (2007) Fatty acid biosynthesis in mitochondria of *Euglena gracilis*. In: Benning C, Ohlrogge J (eds) Current advances in the biochemistry and cell biology of plant lipids. Aardvark Global Publishing Company, LLC, Salt Lake City, UT, pp 133–136

Ueno Y, Yamada K, Yoshida N, Maruyama S et al (2006) Evidence from fluid inclusions for microbial methanogenesis in the early Archaean era. Nature 440:516–519. 10.1038/nature04584 van de Velde S, Mills BJW, Meysman FJR, Lenton TM et al (2018) Early Palaeozoic ocean anoxia and global warming driven by the evolution of shallow burrowing. Nat Commun 9:2554. 10.1038/s41467-018-04973-4

van der Giezen M (2009) Hydrogenosomes and mitosomes: conservation and evolution of functions. J Eukaryot Microbiol 56:221–231. 10.1111/j.1550-7408.2009.00407.x

van der Giezen M, Tovar J (2005) Degenerate mitochondria. EMBO Rep 6:525–530. 10.1038/sj.embor.7400440

van der Giezen M, Tovar J, Clark CG (2005) Mitochondrion-derived organelles in protists and fungi. Int Rev Cytol 244:175–225. 10.1016/S0074-7696(05)44005-X

van Hellemond JJ, van der Klei A, van Weelden SW, Tielens AG (2003) Biochemical and evolutionary aspects of anaerobically functioning mitochondria. Philos Trans R Soc Lond B 358:205–213. 10.1098/rstb.2002.1182

van Lis R, Baffert C, Couté Y, Nitschke W et al (2013) *Chlamydomonas reinhardtii* chloroplasts contain a homodimeric pyruvate:ferredoxin oxidoreductase that functions with FDX1. Plant Physiol 161:57–71. 10.1104/pp.112.208181

van Valen LM, Maiorana VC (1980) The archaebacteria and eukaryotic origins. Nature 287:248–250. 10.1038/287248a0

Wallace MW, Hood AS, Shuster A, Greig A et al (2017) Oxygenation history of the Neoproterozoic to early Phanerozoic and the rise of land plants. Earth Planet Sci Lett 466:12–19. 10.1016/j.epsl.2017.02.046

Weiss MC, Sousa FL, Mrnjavac N, Neukirchen S et al (2016) The physiology and habitat of the last universal common ancestor. Nat Microbiol 1:16116. 10.1038/nmicrobiol.2016.116

Welte C, Deppenmeier U (2014) Bioenergetics and anaerobic respiratory chains of aceticlastic methanogens. Biochim Biophys Acta 1837:1130–1147. 10.1016/j.bbabio.2013.12.002

Williams BA, Hirt RP, Lucocq JM, Embley TM (2002) A mitochondrial remnant in the microsporidian *Trachipleistophora hominis*. Nature 418:865–869. 10.1093/gbe/evu031

Wolfe AJ (2005) The acetate switch. Microbiol Mol Biol Rev 69:12–50. 10.1128/MMBR.69.1.12-50.2005

Yarlett N (1994) Fermentation product formation. In: Mountfort DO, Orpin CG (eds) Anaerobic fungi: biology, ecology and function. Marcel Dekker, New York, pp 129–146

Yarlett N, Orpin CG, Munn EA, Yarlett NC et al (1986) Hydrogenosomes in the rumen fungus *Neocallimastix patriciarum*. Biochem J 236:729–739. 10.1042/bj2360729

Zebe E (1991) Arthropods. In: Bryant C (ed) Metazoan life without oxygen. Chapman and Hall, London, pp 218–237

Zimorski V, Ku C, Martin WF, Gould SB (2014) Endosymbiotic theory for organelle origins. Curr Opin Microbiol 22:38–48. 10.1098/rstb.2014.0330

Protein Import into Hydrogenosomes and Mitosomes

Pavel Dolezal, Abhijith Makki, and Sabrina D. Dyall

Contents

1 Introduction 32
2 Protein Trafficking in Eukaryotes 33
2.1 The Nucleus 33
2.2 The Endoplasmic Reticulum 34
2.3 The Mitochondrion 34
3 The Evolution of the Mitochondrial Protein Import Machinery 38
4 Studying Hydrogenosomal and Mitosomal Protein Import 40
4.1 Laboratory Techniques and Tools 40
4.2 Mining Genome Sequence Data 42
5 Organellar Targeting Signals 44
5.1 Mitochondrial Targeting Signals 44
5.2 Signals on Precursors of Soluble Hydrogenosomal and Mitosomal Proteins 45
5.3 Signals on Hydrogenosomal and Mitosomal Membrane Proteins 54
6 Crossing the Organellar Membranes 56
6.1 The Outer Membrane 56
6.2 The Intermembrane Space Chaperones 60
6.3 The Inner Membrane 61
7 The Protein Import Motor 63
8 Preprotein Processing Peptidases 64
8.1 The Mitochondrial Processing Peptidase (MPP) 65
8.2 The Inner Membrane Protease 66
9 Folding Newly Imported Soluble Proteins 67
10 Perspectives 69
References 70

P. Dolezal (✉) · A. Makki
Faculty of Science, Department of Parasitology, Charles University, Vestec, Czech Republic
e-mail: pavel.dolezal@natur.cuni.cz; abhijith.makki@natur.cuni.cz

S. D. Dyall
Department of Biosciences and Ocean Studies, University of Mauritius, Réduit, Mauritius
e-mail: s.dyall@uom.ac.mu

J. Tachezy (ed.), *Hydrogenosomes and Mitosomes: Mitochondria of Anaerobic Eukaryotes*, Microbiology Monographs 9,
https://doi.org/10.1007/978-3-030-17941-0_3

Abstract In the past decade, studies on protein targeting to hydrogenosomes and mitosomes have revealed several characteristics in common with mitochondrial protein targeting. Proteins from one system can readily be imported into another, strongly suggesting that targeting signals on hydrogenosomal, mitosomal and mitochondrial preproteins are conserved. By extension, these observations, together with the proposed common origin of hydrogenosomes, mitosomes and mitochondria, led to the proposition that components of the respective protein import machineries for these organelles are conserved. With the advent of complete genome sequence databases for diverse eukaryotes, we are now in a better position to examine this proposition. In this review, we report and integrate the latest experimental and bioinformatics data on the state of protein import in hydrogenosomes, mitosomes and mitochondria.

1 Introduction

Eukaryotic cells have internal membranes defining subcellular compartments, each of which has discrete metabolic and/or biosynthetic functions. To fulfil such functions, specific sets of proteins must be precisely targeted to, and quantitatively imported and localized within, the compartment in a timely fashion. Protein trafficking has been extensively studied in fungi and mammals, and a number of elaborate machines have been described that specifically import certain proteins into mitochondria. Recent studies have demonstrated that there are several mechanistic and structural features in common between import into hydrogenosomes, mitosomes and mitochondria (Makiuchi and Nozaki 2014; Pyrihová et al. 2018; Makki et al. 2019). Despite their different morphologies and non-mitochondrial functional pathways, hydrogenosomes and mitosomes are thought to be related descendants of the endosymbiont that gave rise to mitochondria (Roger et al. 2017). Due to the limited number of the experimental models, we focus the current review to *Trichomonas vaginalis, Neocallimastix* sp. and *Nyctotherus ovalis* hydrogenosomes and to *Entamoeba histolytica*, *Giardia intestinalis*, *Cryptosporidium* sp. and microsporidian mitosomes. We also include additional information inferred from the recent genomic and transcriptomic data. We shall examine what is known about hydrogenosomal and mitosomal biogenesis in those species and shall discuss the evolution of protein import in relation to bacterial, mitochondrial and other organellar systems.

2 Protein Trafficking in Eukaryotes

Of over 6500 proteins encoded in the nuclear genome of *Saccharomyces cerevisiae*, about 1000 proteins are targeted to the endoplasmic reticulum (ER) and from there are subsequently localized throughout the endomembrane system, and a similar number of proteins are targeted to mitochondria. Nearly 50% of the proteins are folded and localized in the cytoplasm, while around 25% of them are retargeted to the nucleus, and a smaller portion is distributed in other compartments such as peroxisomes (Kumar et al. 2002; Picotti et al. 2013). In general, a sophisticated system of membrane translocases with associated propelling machines recognizes the address on individual protein molecules. In most cases, protein trafficking is fuelled by the hydrolysis of either ATP or GTP. Additionally, proteins travelling through or into mitochondrial inner membranes require an electrochemical membrane potential generated by the mitochondrial electron transport chain (Wickner and Schekman 2005).

Despite these general similarities, the respective mechanisms responsible for protein import into the nucleus, the ER and the mitochondrion differ fundamentally, with each employing distinct molecular machine. While the molecular machines operating in the ER were recruited during the evolution of the eukaryotic cell from an ancestral prokaryote, and adapted towards current needs, nuclear and mitochondrial protein import systems seem to be almost entirely created de novo by the eukaryotic cell (Dolezal et al. 2006; Lithgow and Schneider 2010; Cautain et al. 2015; Mani et al. 2016; Fukasawa et al. 2017).

2.1 *The Nucleus*

The nuclear envelope is perforated with huge macromolecular assemblies of ~30 different proteins that form nuclear pore complexes with a central channel of 25–30 nm in diameter. This channel fuses the inner and the outer nuclear membranes and allows proteins smaller than 40 kDa to passively traverse. Larger proteins are actively transported across the nuclear envelope and contain nuclear localization signal (NLS) sequence motifs. These signals consist of one or two clusters of four or five basic residues localized usually within the polypeptide chain. The import of proteins with NLS through the channel is facilitated by the carrier heterodimer of importin-α/β (Pemberton and Paschal 2005; Lange et al. 2007). Upon passing through the nuclear pore, the interaction of the complex with RanGTP initiates the release of cargo protein from the importins. The whole process of translocation is regulated by the nucleotide state of Ran, which accordingly cycles between the nucleus and the cytoplasm (Stewart 2007).

2.2 The Endoplasmic Reticulum

In contrast to cytosolic, nuclear and most mitochondrial proteins that are synthesized on free ribosomes, mRNA transcripts encoding ER-destined proteins are translated on ribosomes tightly bound to the ER membrane. Nascent luminal proteins are equipped with an N-terminal signal sequence that consists of a basic amino-terminus followed by a stretch of 8–14 non-polar residues and a cleavage motif for the signal peptidase (Blobel and Dobberstein 1975a, b; von Heijne 1990) (Blobel and Dobberstein 1975a, b). Membrane proteins usually contain internal topological signals instead of the N-terminal signal peptide. Translation and translocation are coordinated by the signal recognition particle (SRP), a complex of 7S RNA and six protein subunits (Nyathi et al. 2013). Initially, SRP binds the signal peptide emerging from the ribosome. Translation slows down until SRP is recognized by its ER-bound receptor, whereupon translocation can resume following the binding of GTP to both SRP and its receptor. The passage through the membrane is formed by the Sec61 translocon consisting of a Sec61α channel and two accessory subunits β and γ (Park and Rapoport 2012). The co-translational transport of substrate protein through the channel is driven by the elongation of the polypeptide by the ribosome (Connolly and Gilmore 1986; Ménétret et al. 2007).

In case of post-translational transport, the chaperones protecting the translated polypeptides are released upon the contact with Sec61 translocon, which is accompanied by the additional Sec62/Sec63 complex of so far unclear role. The translocation is then driven by the action of lumenal Hsp70 (Osborne et al. 2005). Analogously, in bacteria, the SecYEG translocon is used for secretion of proteins across the plasma membrane, and the signal peptides of secreted proteins share similar characteristics with ER proteins (Park and Rapoport 2012). While protein import into ER requires nucleotide triphosphates, bacteria need additional membrane electrochemical gradient to export proteins across the plasma membrane. After translocation into the ER, the signal peptide is cleaved from the precursor polypeptide by the signal peptidase. This step is necessary for releasing the protein from the membrane lipid bilayer to which it is bound via the hydrophobic signal peptide. The ER signal peptidase is a membrane protein that shares ancestry with both the bacterial signal peptidase and the inner membrane protease complex in mitochondria (Dalbey et al. 1997).

2.3 The Mitochondrion

About 99% of the mitochondrial proteins are nuclear-encoded and are synthesized in the cytosol from where they are imported into mitochondria (Wiedemann and Pfanner 2017). Protein import into mitochondria is mostly post-translational, although recent data suggest that mRNA subpopulations are organized in the proximity of the mitochondrial outer membrane (García-Rodríguez et al. 2007).

Besides having to be targeted to the organelle, these proteins have to be internally sorted to either of four distinct sub-compartments: the outer membrane, the intermembrane space (IMS), the inner membrane or the mitochondrial matrix. Proteins destined to mitochondria possess N-terminal and/or internal targeting signals which ensure their correct delivery to the organelle. A large majority of mitochondrial proteins are synthesized with N-terminal cleavable presequences of variable length, ranging from 10 to 80 amino acids (Wiedemann and Pfanner 2017). The presequence is not generally conserved at the primary sequence level among the different preproteins, but it is rather the α-helical structure engaged by the presequence upon interaction with the outer membrane receptor Tom20 that appears to be the common factor (Schatz and Dobberstein 1996; Abe et al. 2000). This α-helix is amphipathic, containing patches of positively charged, and hydrophobic amino acids, respectively, on opposite surfaces of the theoretical cylinder. The presequence is usually processed by the mitochondrial processing peptidase (MPP), and the mature protein is sorted to either the matrix or to the inner membrane if it bears a hydrophobic stop-transfer sequence. Some mitochondrial proteins, mostly destined to the membranes, do not have cleavable N-terminal presequences but have internal targeting signals that are not well characterized (Wiedemann and Pfanner 2017).

Translocation across mitochondrial membranes is carried out by several molecular machines. These machines consist of core transmembrane translocases complemented by additional components that provide specificity. Such a modular structure enables the independent evolution and function of each molecular machine (Dolezal et al. 2006). The translocase of the outer membrane (TOM) complex constitutes the central recognition point and gate for all nuclear-encoded mitochondrial proteins (Fig. 1). Tom70 and Tom20 are receptor subunits that recognize the precursor proteins and release them subsequently into the translocation channel. This transfer is assisted by Tom22, which, together with Tom40 and Tom5, represents the core and essential part of the TOM complex (Meisinger et al. 2001). Two other small proteins, Tom6 and Tom7, participate in the maintenance of the complex. The translocation pore is formed by several subunits of Tom40, which is most likely a β-barrel protein (Hill et al. 1998; Bausewein et al. 2017). After passing through the TOM complex, preproteins may interact with either of three distinct molecular machines (Fig. 1), depending on their final destination.

The insertion and assembly of β-barrel outer membrane proteins, including Tom40, are assisted by the sorting and assembly machinery (SAM) complex (Fig. 1). The SAM complex consists of four subunits, the core translocase Sam50, which is itself a β-barrel protein (Kozjak et al. 2003; Paschen et al. 2003; Gentle et al. 2004), and the additional proteins Sam35, Sam37 and Mdm10 (Bohnert et al. 2007). In addition to SAM complex Mdm10 takes part in the ER-mitochondria tethering complex known as ERMES (Kornmann et al. 2009). The structure of β-barrel precursors does not allow their lateral insertion into the lipid bilayer directly from the TOM channel, and the precursors must first be released into the IMS. The passage of β-barrel precursors from the TOM complex to the SAM complex is assisted by the so-called small translocase of the inner membrane (TIM) chaperone

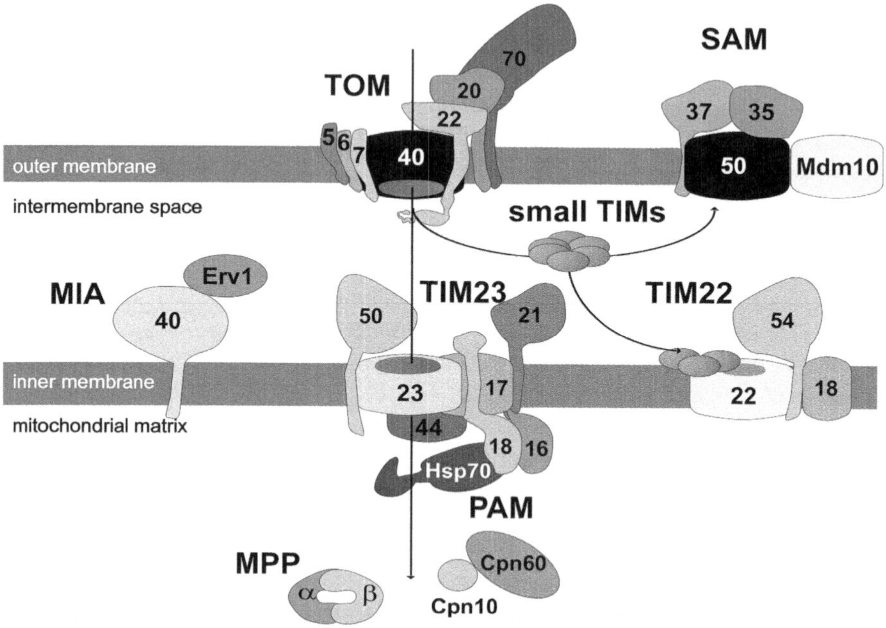

Fig. 1 The mitochondrial protein import machinery as defined in *S. cerevisiae*. *TOM* translocase of the outer mitochondrial membrane, *SAM* sorting and assembly machinery, *TIM* translocase of the inner mitochondrial membrane, *MIA* mitochondrial IMS assembly machinery, *PAM* presequence-associated motor, *IMP* inner membrane protease, *MPP* mitochondrial processing peptidase. The numbers on the individual TOM, SAM, TIM or PAM components represent their approximate molecular masses in kDa. See text for mechanistic details

complexes (small Tims). These soluble complexes are trimeric assemblies of either Tim9/Tim10 heterodimers or of Tim8/Tim13 heterodimers and protect the exposed hydrophobic epitopes of some membrane proteins from the aqueous environment of the IMS (Hoppins and Nargang 2004; Wiedemann et al. 2004; Koehler 2004). Specific class of outer membrane proteins with a-helical transmembrane segment are inserted into the membrane via distinct mitochondrial import (MIM) complex (Becker et al. 2008; Hulett et al. 2008).

Some of the precursors destined for the IMS have a sorting signal at the N-terminus. The inner membrane protease (IMP) complex is responsible for the maturation of these proteins. IMP comprises the two proteases Imp1 and Imp2 and a regulatory subunit Som1. Some of the precursors contain bipartite presequences consisting of a matrix-targeting signal followed by an IMS-sorting signal (Gakh et al. 2002). However, majority of the IMS proteins lack any N-terminal presequences and instead contain internal CX3C or CX9C motives (Stojanovski et al. 2012), which are recognized by the mitochondrial IMS import and assembly (MIA) machinery (Mesecke et al. 2005). The machinery consists of IMS components Mia40, Erv1 and Hot13p (Fig. 1). According to the proposed model, Mia40 binds the precursors in the IMS via disulphide bridges, thereby trapping them after their entrance through the Tom40 pore. Further isomerization of disulphide bridges releases the precursors from Mia40, which is subsequently oxidized by Erv1 (Mesecke et al. 2005). In this process, Hot13p might perform a reducing action on the precursors (Curran et al. 2004).

Some precursors that are to be integrated in the inner membrane, such as the mitochondrial carrier family (MCF) proteins, are inserted by the TIM22 machine (Fig. 1), which is built around a Tim22 subunit and contains another two membrane-integral subunits Tim54 and Tim18 (Rehling et al. 2003). The precursors are shuttled from the TOM to the TIM22 complex by the Tim9/Tim10 chaperone through the IMS. Tim12, peripherally associated with TIM22, serves as a docking site for the Tim9/Tim10 complex that detaches from the precursor upon contact with the TIM22 complex (Koehler 2004).

Soluble matrix-destined preproteins, usually synthesized with a cleavable N-terminal presequence, are passed from the TOM complex to the distinct TIM23 machine through interaction with the IMS domain of Tim50 (Fig. 1). In yeast, TIM23 consists of Tim23, Tim17, Tim50 and Tim21 (Bohnert et al. 2007). Tim23 forms a protein-conducting channel that is regulated by the action of Tim50 and Tim17 (Meinecke et al. 2006; Martinez-Caballero et al. 2007). The TIM23 complex is also capable of inserting preproteins into the inner mitochondrial membrane (Koehler 2004). These inner membrane preproteins are synthesized with a short hydrophobic sorting sequence downstream of the N-terminal presequence (Glick et al. 1992; Stuart 2002). To discriminate between and to coordinate the dual translocation and insertion activities, TIM23 either interacts with Tim21 or associates with the presequence translocase-associated motor (PAM) complex (Fig. 1). This cycle is regulated by Tim17, which recruits the PAM complex to the TIM23 complex (Chacinska et al. 2005). The PAM complex, together with the membrane potential, drives translocation of the precursor through the TIM23 complex (Rehling et al. 2003). First, the negatively charged matrix face of the inner membrane generates an electrophoretic force on the predominantly positively charged presequence and mediates preprotein insertion into the Tim23 channel. Consequently, the PAM complex completes the translocation of the bulk polypeptide in an ATP-dependent manner (Voos and Röttgers 2002).

The central component of the PAM complex is the molecular chaperone Hsp70 (Fig. 1). The activity of Hsp70 is regulated by two inner membranes, J-domain-containing proteins Pam18 and Pam16, and the soluble nucleotide exchange factor Mge1. The PAM complex is tethered to the translocation channel probably by the peripheral membrane protein Tim44. Upon translocation into the matrix, the N-terminal presequence of precursors is processed by MPP, and the mature protein is thereafter folded into its native conformation. Some preproteins with an octapeptide-containing presequence require sequential processing of the targeting presequence by MPP followed by the mitochondrial intermediate peptidase (MIP), which removes the octapeptide sequence (Gakh et al. 2002).

After being driven in by the PAM complex and having their presequence processed, some newly imported matrix proteins require further assistance from molecular chaperones. First, soluble Hsp70 and its co-chaperone Mdj1 accost the substrate proteins to partially fold them. Two populations of Hsp70 thus exist in mitochondria: (1) a Tim44-bound membrane-associated form serving as the protein import driver and (2) a matrix-soluble chaperone that assists protein folding (Horst et al. 1997). After release from Hsp70, the protein is passed along an Hsp60/Hsp10 system, which is the major chaperone system for protein folding in the matrix (Manning-Krieg et al. 1991). A homo-oligomer of Hsp60 provides a protected cavity for protein folding, while Hsp10 regulates the ATPase cycle of Hsp60 and the behaviour of individual subunits (Martin et al. 1991).

3 The Evolution of the Mitochondrial Protein Import Machinery

Mitochondria are of endosymbiotic origin and have descended from bacteria. Extensive sequence analyses have shown that mitochondria form a monophyletic group and have demonstrated strong affinities between mitochondrial genomes and present-day alphaproteobacteria-related organisms (Andersson et al. 1998; Gray et al. 1999; Martijn et al. 2018). Consequently, the endosymbiotic theory for the origin of mitochondria purports that the mitochondrion originates from a single endosymbiont, which formed a symbiotic relationship with a pre-eukaryotic or a primitive eukaryotic cell around two billion years ago. Over time, the endosymbiont lost its capacity to function and reproduce as an independent organism, and its fate was sealed within the host as it transferred the bulk of its genome to the host nucleus or simply discarded some of it. The possible reasons or driving forces behind the symbiosis, and the subsequent loss of the endosymbiont genome and why and how that happened, are beyond the scope of this chapter and are comprehensively covered in Chap. 2 and in reviews, e.g. (Martin et al. 2015; Roger et al. 2017). The outcome of, or perhaps the support for, the endosymbiont transferring its genes to the nucleus was the evolution of new machinery in the eukaryotic cell to send the nuclear-encoded proteins back to the degenerate endosymbiont to allow the latter to function. Moreover, it is of note that the large majority of extant mitochondrial proteins are not of endosymbiotic or α-proteobacterial origin. These proteins have either been recruited from other bacterial sources or have been invented de novo by the evolving eukaryote (Andersson et al. 2003; Gabaldón and Huynen 2003; Gray 2015). All these proteins would have had to develop targeting signals, while the eukaryote was inventing a new machine, either from scratch, or by tinkering existing protein targeting components, to intake the nuclear-encoded precursors into the proto-mitochondrion.

Of the six protein import machines characterized to date in yeast mitochondria, namely, TOM, SAM, MIM, TIM22, TIM23 and MIA, only the SAM complex bears a component, Sam50, which is clearly related to a bacterial translocase (Gentle et al. 2004). Several import components such as Pam18 and Tim44 contain domains found in bacteria (Clements et al. 2009). Thus, the majority of translocase components are a product of eukaryotic invention. Much can be inferred about the evolution of mitochondrial biogenesis by examining these translocases and their features, as discussed later in this chapter. It is of particular interest whether hydrogenosomes and mitosomes use phylogenetically similar translocases as mitochondria.

The majority of mitochondrial proteins have an N-terminal presequence that is both necessary and sufficient to target a passenger protein to mitochondria (von Heijne et al. 1989). How and when these presequences were initially acquired and how they have been distributed to genes on different loci are intriguing questions. It has been demonstrated that synthetic mitochondrial presequences can penetrate either artificial or bacterial lipid bilayers (Roise et al. 1986; Maduke and Roise 1993; Neupert 1997). Of the presequence properties, it is mainly their amphiphilic character and the net charge which have a decisive role in targeting proteins to mitochondria and other organelles. It is therefore plausible that presequences were developed prior to the existence of specific receptors or pores on the outer surface of the proto-mitochondrion and that the latter were later evolved to enhance the efficiency of translocation. Conversely, it was hypothesized that in order to avoid crosstalk between targeting to host cell translocases at the plasma membrane and the newly developing organelle, the early mitochondrial targeting might not have relied on the presequence at all (Garg and Gould 2016). It was proposed that a primitive PAM might have driven the presequence and the mature polypeptide into the proto-mitochondrion in the absence of any TOM or TIM component (Herrmann 2003). Sequences with presequence-like features are commonly found in genomes, and the odds of a transferred gene landing in such a locus may have been quite good (Baker and Schatz 1987; Lucattini et al. 2004). Since some presequences are distributed over several exons in some nuclear-encoded mitochondrial genes, exon shuffling and alternative splicing have been proposed as mechanisms for presequence generation (McFadden 1999). Much insight on these aspects has been gained by studying mitochondrial gene transfer processes in flowering plants (Adams et al. 2000; Adams and Palmer 2003). In those species, mitochondrial gene transfer is an ongoing process, such that functional copies of some genes for mitochondrial proteins can be found in (1) both the nucleus and the mitochondrion of one species or (2) in the nucleus of one species, but only in the mitochondrion of a sister species. Thus, the changes required to target the newly transferred genes to the mitochondrion can be examined. Productive gene transfer not only involves the evolution of a targeting signal but is primarily dependent on the acquisition of gene expression signals. It was shown that mitochondrial copies of freshly transferred genes remain active for some indeterminate period and do not immediately get shut off (Choi et al. 2006). Transferred genes can acquire presequences by simply integrating into the locus for a duplicated nuclear-encoded mitochondrial gene (Sandoval et al. 2004; Murcha et al. 2005b; Choi et al. 2006). Curiously, some mitochondrial presequences have independently been acquired from the same donor gene, e.g. from *mt-hsp70* (Adams et al. 2000; Choi et al. 2006). Genes for non-mitochondrial proteins have also acted as donors for gene control regions and for fortuitous presequence-like stretches (Murcha et al. 2005b). Or, sometimes, the transferred gene does not gain a presequence but uses internal signals from the mature polypeptide (Murcha et al. 2005b; Choi et al. 2006). Additionally, changes in local hydrophobicity in the protein sequence of certain transferred genes have been implied in enhancing protein import (Daley et al. 2002). Therefore, a plethora of tricks exists to append presequences to, or to create internal signals within, mitochondrial proteins, and though the precise mechanisms are shady, it appears that these tricks occur repeatedly and independently.

It has been hypothesized that the process of inventing a protein import machine for mitochondria would have been so intricate and critical that it is unlikely to have occurred more than once (Cavalier-Smith 1987). By extension, similarities found between mitosomal and hydrogenosomal and mitochondrial protein import have been presented as a strong support that all these organelles use common components for import and are therefore one and the same. These observations have prompted a number of studies to shed light on the constitution and the evolution of the protein import machineries of hydrogenosomes and mitosomes.

4 Studying Hydrogenosomal and Mitosomal Protein Import

4.1 Laboratory Techniques and Tools

Unfortunately, a limited set of tools is available to study the hydrogenosomal or mitosomal species. In yeast, much has been deduced about mitochondrial biogenesis through extensive genetic manipulation, and a variety of mutants can readily be obtained that can be used to assess the function of individual components in the biogenetic pathway (Bonnefoy et al. 2007). These studies are complemented with a wealth of highly honed biochemical techniques such as in organello import, membrane separation, creation of mitoplasts and generation of protein import intermediates, to list a few (Stojanovski et al. 2007). Nonetheless, these techniques provide a basis for developing new methods to study protein import in hydrogenosomal or mitosomal species.

Among the species under study, *T. vaginalis* is currently one of the most experimentally tractable. It can be genetically transformed to express endogenous or exogenous proteins (Delgadillo et al. 1997) or to delete genes by homologous recombination (Land et al. 2003; Brás et al. 2013). More recently, a CRISPR/Cas9-mediated gene modification and gene knockout methods were developed for *T. vaginalis* that are based on techniques originally developed for *Cryptosporidium parvum* (Vinayak et al. 2015; Janssen et al. 2018). Establishment of a gene knockout technique can significantly advance the field and can help the biologists to prove the gene function. Of all the mitochondria-related organelles, only *T. vaginalis* hydrogenosomes (~0.8 μm in diameter) have been isolated to high purity, on a Percoll gradient, and have been demonstrated to be protein-import competent

(Bradley et al. 1997). The development of an assay for importing precursor proteins into isolated *T. vaginalis* hydrogenosomes has revealed several requirements that appear to be in common with mitochondrial protein import (Bradley et al. 1997). For this assay, recombinant precursor proteins are either metabolically radiolabelled in *Escherichia coli* for detection by autoradiography or purified with a C-terminal hexahistidine (His6) tag for detection by western analysis. The isolated organelles and the precursor are incubated in an isotonic import buffer supplemented with ATP and crude *T. vaginalis* cytosol. In mitochondrial import systems, the precursor protein is synthesized and radiolabelled in rabbit reticulocyte lysate or in wheat germ extract; these extracts contain cytosolic chaperones that support mitochondrial protein import. However, hydrogenosomal protein import is absolutely dependent on *T. vaginalis* crude cytosol that cannot be substituted for by either rabbit reticulocyte lysate or wheat germ extract (Bradley et al. 1997), emphasizing the specificity of the cytosolic factor(s). Successful import is measured as follows: resistance of the imported protein to externally added protease and presequence cleavage, as detected by faster electrophoretic migration on SDS-PAGE. The import of a matrix precursor protein, ferredoxin, has been shown to be linear and saturable, and dependent on a protease-sensitive component(s) on the outer hydrogenosomal surface, indicating the presence of a specific receptor (Plümper et al. 2000). Precursor ferredoxin import has been shown to be dependent on the presence of a specific presequence and, on ATP, weak electrochemical potential and temperature (Bradley et al. 1997), which are all requirements for import into mitochondria (Schleyer et al. 1982). Import studies have also been carried out to explore the functional conservation of import pathways between hydrogenosomes and mitochondria (Dyall et al. 2000, 2003) or between hydrogenosomes and mitosomes (Dolezal et al. 2005).

The second most studied hydrogenosomal species is *Neocallimastix* sp., although not many techniques are available for thorough studies. An enriched hydrogenosomal fraction can be prepared by differential centrifugation of disrupted cells of this species and can be used for rudimentary sub-organellar fractionation studies (Marvin-Sikkema et al. 1993). However, most of the studies on *Neocallimastix* sp. hydrogenosomal proteins have been done in heterologous fungal systems (van der Giezen et al. 1998, 2002, 2003).

Giardia and *Entamoeba* represent the best cellular models for studying the mitosomal biology. Genetic transformation techniques are used for both organisms and have been applied to localize mitosomal proteins and to investigate putative targeting signals, e.g. (Mai et al. 1999; Tovar et al. 2003; Regoes et al. 2005; Dolezal et al. 2005; Mi-ichi et al. 2009). Transcriptional silencing techniques exist for *Entamoeba*, e.g. (Linford et al. 2009), including elusive G3 strain-specific silencing (Bracha et al. 2003), although none of them represent a robust and reliable genetic tool. Similarly, ribozyme- (Dan et al. 2000) and morpholino-based (Carpenter and Cande 2009). RNA silencing has been developed for *G. intestinalis*, but the techniques have not been widely applied in the field. Recently, CRISPRi has been introduced to *Giardia*, opening new possibilities in manipulating expression of even multiple genes at once (McInally et al. 2018). Due to polyploid nature of both organisms, gene knockouts are not very feasible method, although complete knockout of *Giardia* cyst wall protein 1 was recently accomplished by employing Cre-Lox recombination to recycle selection marker for consecutive elimination of the four alleles (Ebneter et al. 2016).

A high-speed differential centrifugal fraction for *Giardia* mitosomes has been generated that has been successfully used to reconstitute Fe-S cluster formation (Tovar et al. 2003). Further enrichment of these fractions was obtained on sucrose gradients that yielded organelles of ~150 nm in diameter (Regoes et al. 2005; Dolezal et al. 2005). These organelles have been used for localization studies, but so far only limited protein import assays have been performed yet (Dagley et al. 2009). The proteome of mitosomes is far from complete, and only a few proteins have been physically localized in the mitosomes by specific polyclonal antibodies or by detection of their tagged recombinant versions. In vivo enzymatic tagging and immunoprecipitation techniques increased the number of newly identified mitosomal proteins, most of which remain of unknown function (Martincová et al. 2015; Rout et al. 2016). Gradient centrifugation of *E. histolytica* cellular lysates led to the purification of the mitosomes and the identification of unique sulphate-activation pathway within the organelles (Mi-ichi et al. 2009). While genetic manipulation of *Microsporidia* remains to be established, CRISPR/Cas9-based strategy was successfully introduced to *Cryptosporidium* parasites (Vinayak et al. 2015), although the experimental system still depends on a cumbersome propagation of the parasites. The likelihood of mitosomes being ever purified from *Microsporidia* or *Cryptosporidium* is not very high as the organelles are extremely small, ranging from 70 nm for the microsporidian *Trachipleistophora hominis* (Williams et al. 2002) to between 150 and 300 nm for the *C. parvum* mitosome (Riordan et al. 2003; Putignani et al. 2004). Moreover, the single *C. parvum* mitosome is entangled by the rough ER (Riordan et al. 2003; Putignani et al. 2004), which will render any disruption technique quite tricky. Despite these limitations, the studies of protein import into mitosomes of *G. intestinalis* (Regoes et al. 2005; Dolezal et al. 2005; Pyrihová et al. 2018), *E. histolytica* (Mai et al. 1999; Tovar et al. 1999) and of the two microsporidian species *Encephalitozoon cuniculi* and *Antonospora* (*Nosema*) *locustae* (Burri et al. 2006) brought exciting insight into the degree of functional similarity between mitosomes, hydrogenosomes and mitochondria, and also into the degree of adaptation of mitosomes within the microsporidia (Burri and Keeling 2007). Although both mitosomes and hydrogenosomes have arisen independently and repeatedly, the molecular basis of the reduced protein import machinery may offer clues as to the composition of the original sets of translocases installed in the membranes of proto-mitochondria.

4.2 Mining Genome Sequence Data

Thanks to the completed genome sequencing projects for hydrogenosomal and mitosomal species, e.g.(McArthur et al. 2000; Katinka et al. 2001; Abrahamsen et al. 2004; Xu et al. 2004; Loftus et al. 2005; Carlton et al. 2007), we have a tremendous amount of information about the biology and evolution of these organisms at hand. Having all these data available, we are presented with the difficult issue of efficient data mining. Our attempts to identify possible homologous sequences in the genomes of evolutionary diverse species are very often faced with the danger of false-negative results and therefore of incorrect conclusions. The widely used approach is BLAST based on pairwise sequence analyses (Altschul et al. 1990, 1997). BLAST searches are sometimes inefficient simply because a particular query may be too divergent to pick the target sequence from genome databases. In the field of mitochondrial protein import, queries originate primarily from *S. cerevisiae* or other fungal sequences. While pairwise sequence analyses were sufficient to identify equivalent components in animals, they work less well on plants and often fail to identify homologous sequences in other phylogenetic groups, especially protists (Hoogenraad et al. 2002). If the whole family of proteins instead of a single sequence is available, a search based on the hidden Markov model (HMM) offers a significantly more sensitive mining method when compared with BLAST (Eddy 1998; Finn et al. 2011; Alva et al. 2016). In practice, analyses based on HMMs represent a reversed search of the protein family (PFAM) database (Bateman et al. 2004). Instead of comparing one query sequence with all the available HMMs in PFAM, a single HMM is used to search the genome database. Although HMMs were first designed for speech recognition, they can be applied to a variety of problems, where hidden parameters need to be determined from obvious parameters, such as sequence alignment of homologous proteins. The parameters that are extracted from the sequence alignment, for instance, the probability of occurrence of certain amino acids in a particular position, can be then used for mining data from the conceptual translation of a genome sequence. Depending on the selection of sequences for the alignment, HMMs can even pick structural information otherwise hidden in the primary sequence (Dolezal et al. 2006; Likic et al. 2010). For example, the alignment of some homologous β-barrel proteins might provide enough information to find any β-barrel protein in the examined genome. Importantly, newly identified homologous sequences can be included into the alignment used for the building of new HMM, thus providing a refined and more sensitive tool for the next round of searches. Usually several cycles of refinement are used to craft a reliable HMM that is powerful enough to pick very divergent homologous sequences, but that is insensitive to unrelated sequences. The freely available HMMER software (https://www.ebi.ac.uk/Tools/hmmer/) enable the building of tailored HMMs based on the user' s protein sequence alignment. A more recent addition to the list of search tools is HHpred (https://toolkit.tuebingen.mpg.de/#/tools/hhpred). HHpred is a sensitive server for detecting remote homologues based on protein function and protein structure (Söding et al. 2005). Thus, the efficient searches using single or several HMMs in a large database can now be performed by the typical bench-work biologist. Collectively, in silico studies and laboratory studies have begun to reveal much about hydrogenosomal and mitosomal targeting signals, translocases, chaperones and processing peptidases.

5 Organellar Targeting Signals

Targeting signals contain the minimal information necessary for a protein precursor to be recognized by targeting machinery and to be directed to the correct compartment in a cell. These sequences are both "necessary and sufficient" to target a passenger protein to a given organelle. Other sequence stretches within the precursor may be necessary to target the protein to the correct sub-organellar location (Neupert 1997). Several categories of targeting signals have been defined for mitochondrial precursors and, to a lesser extent, for hydrogenosomal and mitosomal precursors.

5.1 Mitochondrial Targeting Signals

The majority of mitochondrial precursors are synthesized with N-terminal cleavable presequences. Typical presequences contain 10–80-amino acid residues, many of which are positively charged, hydrophobic and hydroxylated (Pfanner and Geissler 2001). Negatively charged amino acid residues are notoriously absent in most presequences (von Heijne et al. 1989). Generally, the amino acid residues on presequences are disposed in an amphipathic α-helix, which has one hydrophobic surface opposed by a positively charged surface (Roise et al. 1986; Abe et al. 2000). These contrasting surfaces make contact sequentially with Tom and Tim translocases as the presequence-containing precursor traverses both in mitochondrial membranes (Pfanner and Geissler 2001). For instance, structural studies have shown that the hydrophobic surface of the α-helical presequence makes contact with the Tom20 binding groove (Abe et al. 2000). Subsequently, the precursor is passed over to Tom22 through interaction of the positive surface of the α-helix with negative charges on Tom22 (Brix et al. 1997) and continues through a "binding chain" by contacting the various translocases (Pfanner and Geissler 2001). Eventually, as the precursor reaches TIM23, the positive charges on the presequence are acted upon by the membrane potential that draws the precursor into the channel (Martin et al. 1991). Upon translocation through TIM23, the presequence is generally cleaved by MPP from the majority of proteins (Gakh et al. 2002), but there are exceptions where presequences remain an integral part of the mature protein (Rospert et al. 1993). Curiously, one case of a cleavable presequence at the C-terminus of a matrix protein has been reported, where the precursor is translocated in a C- to N-terminal rather than the common N- to C-terminal orientation (Lee et al. 1999). Following cleavage, if any, the precursors are then either imported for further folding into the matrix or released by TIM23 into the inner membrane if they additionally possess a hydrophobic stop-transfer signal (Glick et al. 1992; Beasley et al. 1993; Bömer et al. 1997). Some inner membrane and IMS preproteins contain a bipartite presequence that comprises an N-terminal positively charged matrix-targeting sequence and a downstream sorting signal that is similar to sorting signals found on bacterial and ER secretory proteins. These preproteins are first processed by MPP and then undergo a second cleavage by IMP (Schneider et al. 1991; Nunnari et al. 1993). Other types of inner membrane proteins have internal targeting signals as a combination of transmembrane hydrophobic segments together with positively charged loops (Folsch et al. 1996; Davis et al. 1998). Multiple internal targeting signals that act cooperatively have been characterized for the inner membrane ADP/ATP carrier (AAC). No consensus sequence has been computed for these signals, but each segment of about 10-amino acid residues can be recognized individually by the Tom70 receptor (Brix et al. 2000; Wiedemann et al. 2001). In general, internal targeting signals for hydrophobic proteins are poorly characterized. Some outer membrane, monotopic proteins, like Tom70, have a non-cleavable presequence that directs the precursor to mitochondria and drives the insertion of a downstream hydrophobic stretch that acts as a membrane anchor (Hahne et al. 1994). Recent investigations have shown that the targeting signal for outer membrane β-barrel proteins resides in the β-hairpin motif positioned between any two β-strands of the protein. The signal seems to be recognized by Tom20 and in part by Tom70 on the mitochondrial surface.

5.2 *Signals on Precursors of Soluble Hydrogenosomal and Mitosomal Proteins*

Characterization of targeting signals within the hydrogenosomal and mitosomal proteins has mainly focused on putative N-terminal cleavable presequences. These are relatively straightforward to detect, by experimentally determining the N-terminal sequences on isolated endogenous organellar proteins and comparing those with the conceptual translation of the corresponding genes. Most of the putative N-terminal presequences on hydrogenosomal and mitosomal proteins have been either predicted by programmes that have been devised to search for mitochondrial presequences or by sequence comparison with eubacterial homologues. In many cases, the ability of these putative presequences to function as genuine targeting signals has been tested by assessing their efficiency in conducting passenger proteins to mitochondria or to the relevant organelle.

5.2.1 *Trichomonas* Hydrogenosomes

A hydrogenosomal N-terminal cleavable presequence was first noted in *T. vaginalis* ferredoxin, a matrix protein, when purified endogenous ferredoxin was found to lack 8-amino acid residues at the N-terminus, relative to the conceptual gene translation (Johnson et al. 1990). This presequence (Table 1) has an overall positive charge and is significantly shorter than typical mitochondrial targeting sequences which range from 10 to 80-amino acid residues (Pfanner and Geissler 2001). Using an in organello import assay, Bradley et al. (1997) demonstrated that deletion of this eight-aa sequence abolishes binding to, and thus translocation of the protein into,

Table 1 Presequences of precursors to hydrogenosomal and mitosomal proteins[a]

	N-terminal presequence	References
Trichomonas vaginalis, hydrogenosomal matrix		
Ferredoxin[b]	**MLSQVCRF**^GTITAVKGGVKK	Johnson et al. (1990)
Succinyl coA synthetase, α-subunit 1	**MLAGDFSRN**^LHKPLLFIDKD	Lahti et al. (1994)
Succinyl coA synthetase, α-subunit 2	MLSSSFERN^LHQPLLFIDKD	Lahti et al. (1994)
Succinyl coA synthetase, α-subunit 3	MLSSSFERN^LHQPLLFIDKD	Lahti et al. (1994)
Succinyl coA synthetase, β-subunit	**MLSSSFARN**^FNILEWQSKEI	Lahti et al. (1992)
Adenylate kinase	**MLSTLAKRF**^ASGKKDRMVVF	Länge et al. (1994)
Chaperonin60	MSLIEAAKHFTRAF^AKARDL	Bui et al. (1996)
Pyruvate:ferredoxin oxidoreductase subunit A, PFORA	**MLRSF**^GKRIPGDGNTAATSV	Hrdy and Müller (1995a)
PFORB	MLRNF^SKRVPGDGNTAATSV	Hrdy and Müller (1995a)
Malic enzyme A	MLTSSVSVPVRN^ICRAKVPT	Hrdy and Müller (1995b)
Malic enzyme B	**MLTSSVNFPARE**^LSRNVRPT	Hrdy and Müller (1995b)
Malic enzyme C	**MLTSVSYPVRN**^ICRSKLPLA	Hrdy and Müller (1995b)
Malic enzyme D	**MLTSVSLPVRN**^ICRSKLPVA	Hrdy and Müller (1995b)
Isd11[c]	MLSSFLSRTF>ANESVMANLRES	Richards and van der Giezen (2006)
[Fe]-hydrogenase maturase HydG[d]	MLASLSRS[]YGKLRADVSKTL	Pütz et al. (2006)
[Fe]-hydrogenase maturase HydF	MLTSIGRY[]FAKKGNDLPRTH	Pütz et al. (2006); our unpublished data
[Fe]-hydrogenase maturase HydE	MSHDHIVRL[]LNPRTKDEIDA	Pütz et al. (2006); our unpublished data
Ferredoxin[e]	MLSQCSPLRF()GSVTVTKGGA	Dolezal et al. (2005)
Iron-sulfur cluster assembly protein IscS-1	MLTNLYNKA()FHGHYLDAQATSI	Tachezy et al. (2001), Sutak et al. (2004)
IscS-2	MLGSVSRS()YFKGHYLDTQATSV	Tachezy et al. (2001), Sutak et al. (2004)
Iron-sulfur cluster assembly protein IscU	MLAAVSRS()SALNMMKPLGIM	Dolezal et al. (2005)
Complex I protein Ndh51/NuoF	MLAAYGHRF()QTKFLDPKDRI	Dyall et al. (2004), Hrdy et al. (2004)

Complex I protein Ndh24/NuoE	MLASVNTSRF ( ) FARLNKKS>VL	Dyall et al. (2004), Hrdy et al. (2004)
Glycine cleavage H protein H1[f]	MITSCFTRA*AKQYSKDHLWF	Mukherjee et al. (2006a)
Glycine cleavage H protein H2	MISTLCNCSRNF*TKLYAKT>H	Mukherjee et al. (2006a)
Serine hydroxymethyl transferase SHMT	MLKNVFHRF*SSSWILSEKVL	Mukherjee et al. (2006b)
Malic enzyme G	MLTSVSLPVRN*ICRSKLPVA	Dyall et al. (2004)
Malic enzyme H	MLTSSVSLPARE*LSRKVLPT	Dyall et al. (2004)
[Fe]-hydrogenase A	MLASSATAMKGFANSLRM*KD	Bui and Johnson (1996)
[Fe]-hydrogenase B	MLASSSRAAANIRW ( ) VDTSHN	Bui and Johnson (1996); our unpublished data
DnaK/Hsp70	MLSSVARSTSSLFSRG ( ) FAAG	Bui et al. (1996), Germot et al. (1996), Dyall et al. (2003)
Chaperonin 10	MLATFARNF*AAKKVTIKPLG	Bui et al. (1996); our unpublished data
Pam18	MSIVNKF*VEKALSLPTYAKA	Dolezal et al. (2005)
Mitochondrial processing peptidase β-MPP	MSIISRY*AVPQISKLSNGVRV ( )	Dolezal et al. (2005)
Trichomonas vaginalis, hydrogenosomal membrane		
Mitochondrial carrier family protein Hmp31	**MAQPAEQILIAT**^SPKPSLSP	Dyall et al. (2000)
Hmp31-a[g]	MATEADKVLIAT*SPNGALPT	Our unpublished data; Acc EAX95270
Hmp31-b	MKPADKILIAT*SPSDAKLKP	Our unpublished data; Acc EAY15971
Hmp31-d	MKIKFSFGQKQKKDNL*SPVQ	Our unpublished data; Acc EAY00062
Neocallimastix frontalis		
Malic enzyme	**MLAPIQTIARPVSSILPATGALAAKRT**^FFA	van der Giezen et al. (1997)
Succinyl coA synthetase, β-subunit	MLANVTRSTSKAAPALASIAQTAQKRF*LSV	Brondijk et al. (1996)
Neocallimastix sp L2		
[Fe] hydrogenase	MSMLSSVLNKAVVNPKLTRSLATAAEKMVNISINGRKF ( ) QV	Voncken et al. (2002)

(continued)

Table 1 (continued)

	N-terminal presequence	References
Neocallimastix patriciarum		
Chaperonin 60	MLSARSLICKSMIKSGFRRAVAPSVAMAASSMTLTARRNY*SSKY	van der Giezen et al. (2003)
DnaK/Hsp70	MFLSTLAKKSTTFGVSNVVKNALSSKVMRTTPRMFQRF*ESSK	van der Giezen et al. (2003)
Giardia intestinalis		
IscU	**MTSLQLSSTSLLQSVARF**^LTKKTSSDEVYSE	Tovar et al. (2003), Dolezal et al. (2005)
Ferredoxin	MSLLSSIRRFITFRV()VQQ>	Nixon et al. (2002), Dolezal et al. (2005)
Pam18	MLRVLSENRF*PLSLVAGVVA	Dolezal et al. (2005)
β-MPP	MLIRD()IVPGALPS>ATVVFSG	Dolezal et al. (2005)
Entamoeba histolytica		
Chaperonin 60	**MLSSSSHYNGKLLSLNIDCRE**^NVL	Clark and Roger (1995), Mai et al. (1999), Tovar et al. (1999)
Antonospora locustae putative *mitosome*		
mtG3PDH	MINKRTYTYAFAAIGTGVLGYVGHRYYRHRK*	Burri et al. (2006)
Cryptosporidium parvum		
Chaperonin 60	MLLRSGINLYKSVEGSIGLRSAAIRFGMRYISSGKE>()LSF	Riordan et al. (2003)
DnaK/Hsp70	MSMIINSSFNGVVNSSGIAARILKRSLPLVFSRY[]MSSK	Slapeta and Keithly (2004)
IscS	MIVHRYCRQWAPSVVRGISKLAFFSSMSSIAKKRPAY>[]FDY	LaGier et al. (2003)
IscU	MLQLRQLIDKRILIKKCVPICQRLFYS>DTV	LaGier et al. (2003)

[a]This survey only shows proteins that have been shown to localise to hydrogenosomes, mitosomes or heterologous mitochondria, as reported in the literature, or from unpublished proteomics data

[b]Experimentally determined presequences are shown in bold with the cleavage site marked with ^. Arg residues at the -2 or -3 position relative to the cleavage site are underlined

[c]Cleavage sites predicted by MitoProtII (Claros and Vincens 1996) are indicated by >

[d]Cleavage sites predicted by iPSORT (Bannai et al. 2002) are indicated by [ ]

[e]Cleavage sites predicted by PSORTII (Nakai and Horton 1999) are indicated by ()

[f]Cleavage sites suggested by authors or suggested here are indicated by *

[g]Acc = GenBank accession number

the hydrogenosome (Bradley et al. 1997). This result has been confirmed in vivo, when ferredoxin that lacked residues 2–8 was expressed in *T. vaginalis*, and was found to reside exclusively in the crude cytosolic fraction (Dyall et al. 2000). This suggests an important role for the presequence in binding to any receptor and/or pore that promotes entry into the hydrogenosome. In the last 10 years, dozens of *T. vaginalis* hydrogenosomal presequences have been characterized or predicted for proteins involved in various pathways (Table 1). The emerging picture is that, unlike mitochondrial and plastidic presequences, the *T. vaginalis* hydrogenosomal presequences are highly conserved at primary sequence level. This is even more striking upon examination of over a hundred soluble protein sequences identified during proteomic studies, where about 75% of the translated gene sequences have N-terminal sequences that closely resemble those shown in Table 1 (our unpublished data). How these presequences have been appended and are so well conserved remain a mystery, but it is possible that, as is the case for newly transferred angiosperm mitochondrial genes (Adams et al. 2000; Choi et al. 2006), a small subset of hydrogenosomal protein genes has preferentially been used as presequence donor.

The *T. vaginalis* hydrogenosomal presequences are generally short, ranging from 5 to 14-amino acid residues for those that have been proven experimentally and up to 17 residues for the predicted presequences (Table 1). The presequences are enriched in the amino acid residues Ser (20%), Leu (14%), Arg (11%), Ala (8%), Phe (7%), Val (6%), Thr (6%) and Asn (5%). The other amino acids are significantly under-represented. Incidentally, or accidentally, the three amino acids most commonly found in these presequences, Ser, Leu and Arg, are the ones that are each encoded by six codons. This may have been relevant in the evolution of these presequences. The mitochondrial matrix N-terminal presequences are enriched in Arg (14%), Leu (12%), Ser (11%) and Ala (14%). On the other hand, chloroplast leader peptides have a different amino acid composition with 19% Ser and 9% Thr (von Heijne et al. 1989). Markedly underrepresented in hydrogenosomal presequences are the acidic residues, as in the case of both mitochondrial and plastidic presequences (von Heijne et al. 1989).

Three of the frequently occurring amino acid residues in hydrogenosomal presequences are positionally conserved as well. Of the 13 hydrogenosomal matrix preproteins for which presequence cleavage sites have been experimentally determined, 12 have Leu at position 2 of the presequence, and the exception has a Leu residue at position 3 (Table 1). Thus, not only the presence but also the position of the Leu residue is conserved. This is even more striking when we examine predicted N-terminal presequences for a further 20 proteins that have been localized to hydrogenosomes, where 85% have Leu at position 2. Mutation of the Leu residue at position 2 in the ferredoxin presequence disrupted binding of the protein precursor to hydrogenosomes (Bradley et al. 1997), suggesting that this particular residue plays a critical role in binding. The Arg residue occurs at the -2 or -3 position relative to the cleavage site in all the experimentally determined presequences, with 77% at the -2 position. Phe residues can be frequently found in the vicinity of the Arg residue. Interestingly, many, but not all, mitochondrial N-terminal presequences likewise contain Arg at -2 or -3 from the cleavage site (Gavel and von Heijne 1990), but the role of the residue in specifying the cleavage site for MPP is unclear (Gakh et al. 2002). The exact role of these conserved residues, i.e. whether they are important for binding, for translocation or for cleavage, is not known. Nonetheless, some of these conserved features were applied to devise consensus sequences that were used to screen the *T. vaginalis* genome sequence database. A genome-wide search using the consensus sequences M-L-(S/T/A)-$x_{(1-15)}$-R-(N/F/E/xF), M-S-L-$x_{(1-15)}$-R-(N/F/xF) or M-L-R-(S/N)-F picked out 138 sequences with 67% showing similarity to known proteins involved in metabolic pathways, electron transport, protein import, protein folding and oxygen scavenging pathways (Carlton et al. 2007). There are undoubtedly variations on these consensuses, as have been found during proteomic studies (our unpublished data).

Apart from a similar amino acid enrichment, a common feature of these hydrogenosomal presequences and of the mitochondrial N-terminal presequences is their ability to form amphipathic α-helices (Johnson et al. 1990; Lahti et al. 1992; Dolezal et al. 2005). The amphipathic α-helical structure within mitochondrial N-terminal presequences has been shown to be critically important for sequential electrostatic or hydrophobic interaction with various translocases (Pfanner and Geissler 2001). The hydrogenosomal presequence may be interacting with hydrogenosomal translocases using a similar "binding chain" mechanism. To date, it has not been demonstrated that the typical *T. vaginalis*-soluble preprotein presequence is sufficient for translocating the protein into hydrogenosomes, although it has been shown to be necessary (Bradley et al. 1997; Dyall et al. 2000). It is possible that there are additional downstream signals in the mature part of the protein that participate in translocation at stages beyond binding. It has been shown, however, that hydrogenosomal presequences can target a passenger protein into *Trypanosoma brucei* and *S. cerevisiae* mitochondria, but at very low efficiency (Hausler et al. 1997). Some hydrogenosomal matrix proteins, α-subunit of succinyl-CoA synthetase (SCS), Fdx1, malic enzyme and IscA without their predicted N-terminal hydrogenosomal targeting sequence (HTS) were shown to be targeted to the hydrogenosomes suggesting that HTSs are non-essential for targeting (Zimorski et al. 2013). Further, the same set of proteins were found to be localized in mitochondria when they were expressed in yeast without their HTS (Garg et al. 2015). An unusual case is that of ATP-dependent PFK1, which is primarily a glycolytic enzyme that was found to be present in the proteome of hydrogenosomes. Further investigations revealed that it localizes to hydrogenosomes, although it has no predictable HTS, and when expressed in yeast, the protein was targeted to mitochondria (Rada et al. 2011, 2015). Such cases point out that the N-terminal presequence-independent pathway exists for the import of some hydrogenosomal matrix proteins in *T. vaginalis*, and this feature seems to be conserved in yeast as well. In hydrogenosomes, the loss of the respiratory chain complexes and the membrane potential led to the loss of positive charge on the HTS, and subsequently, the HTS might have become dispensable or, in certain cases, the HTS itself was lost (Garg et al. 2015; Rada et al. 2015). It has been hypothesized that this could represent a "primitive" or an ancestral route of protein import into mitochondria.

However, the presence of cryptic or internal targeting sequence in these matrix proteins cannot be ruled out.

5.2.2 *Neocallimastix* Hydrogenosomes

Although there have been several reports of putative *Neocallimastix* sp. hydrogenosomal proteins in silico, few have actually been localized to hydrogenosomes or to heterologous mitochondria (Brondijk et al. 1996; van der Giezen et al. 1997; Voncken et al. 2002; van der Giezen et al. 2003). These proteins have quite similar N-terminal extensions (Table 1), but only one of them has been experimentally confirmed (van der Giezen et al. 1997). The extensions range from 27 to 37-amino acid residues, and are within the range for typical mitochondrial N-terminal presequences, with similar amino acid enrichment and characteristics. Indeed, when expressed in yeast, the hydrogenosomal malic enzyme was targeted to mitochondria in a presequence-dependent fashion (van der Giezen et al. 1998). The predicted N-terminal presequences on *N. patriciarum* Cpn60 and Hsp70 were sufficient to target the green fluorescent protein (GFP) to mammalian mitochondria, although some non-specific targeting was observed for the Cpn60 presequence, suggesting that additional signals are present in the mature part of the protein (van der Giezen et al. 2003).

5.2.3 *Cryptosporidium* Mitosomes

The complexity of mitochondria-related organelles differs significantly among different *Cryptosporidium* species with the most reduced mitosomes found in *C. parvum*, *C. hominis* and *C. ubiquitum* (Liu et al. 2016). Only four proteins have been experimentally localized to mitosome of *C. parvum*. Cpn60 has a putative 38-aa N-terminal presequence (Table 1) which does not follow the Arg 2 rule, but the N-terminal 57-amino acid portion of Cpn60 was necessary and sufficient to target GFP into yeast mitochondria (Riordan et al. 2003). Likewise, the predicted N-terminal extensions on mitosomal IscU and IscS (Table 1) were both sufficient to target GFP to yeast mitochondria (LaGier et al. 2003). The predicted 34-amino acid presequence of Hsp70 closely resembles typical mitochondrial presequences with a predicted amphipathic α-helical domain and similar enrichment in amino acids and has an Arg-2 cleavage site motif (Gavel and von Heijne 1990; Slapeta and Keithly 2004). This predicted presequence could specifically deliver GFP into yeast and *Toxoplasma gondii* mitochondria, and it was shown that the specific presequence region critical for targeting included the predicted amphipathic α-helical domain (Slapeta and Keithly 2004).

5.2.4 *Entamoeba* Mitosomes

Not many *E. histolytica* mitosomal proteins have been identified, leaving us with very little information on protein targeting signals. The only experimental data come from the analysis of Cpn60 that has an N-terminal extension of 15 amino acids (Table 1) shown to be important for mitosomal targeting (Mai et al. 1999; Tovar et al. 1999). This presequence, like most of the *T. vaginalis* presequences, has a Leu residue at position 2 and is highly enriched in Ser residues. While deletion of the extension leads to the accumulation of Cpn60 in the cytosol, the swapping of the extension with the N-terminal presequence from mitochondrial Hsp70 of *Trypanosoma cruzi* delivers the protein back into the enriched mitosomal fraction (Tovar et al. 1999). On the other hand, the components of mitosomal sulphate activation pathway (Mi-ichi et al. 2009) as well the orthologue of Cpn10 (Chan et al. 2005) lack any recognizable N-terminal extension, which thus indicates existence of so far unknown internal targeting signals.

No further functional data are currently available on the processing of targeting presequences in *E. histolytica.*

5.2.5 *Giardia* Mitosomes

A number of proteins have been successfully localized to *G. intestinalis* mitosomes (Tovar et al. 2003; Regoes et al. 2005; Dolezal et al. 2005; Rada et al. 2011; Rout et al. 2016; Pyrihová et al. 2018). The import of giardial homologues of IscU and [2Fe-2S] ferredoxin (Table 1) was shown to be dependent on the N-terminal targeting sequence (Regoes et al. 2005; Dolezal et al. 2005) as their truncated versions were mislocalized and/or degraded in the cytosol. These two presequences are enriched in Ser, Thr, Leu and Arg which are very similar to the *T. vaginalis* presequences. The N-terminal sequences of IscU and [2Fe-2S] ferredoxin, extending beyond the respective predicted presequence cleavage sites, were sufficient to target GFP into mitosomes. The increased electrophoretic mobility of the fusion protein in organellar fractions suggested that the N-terminal presequences were removed upon targeting (Regoes et al. 2005). These two targeting sequences consist of 15–18-amino acid residues that can be projected to form amphipathic α-helices (Dolezal et al. 2005). Interestingly, the gene coding for [2Fe-2S] ferredoxin was demonstrated to contain a spliceable intron just between the exons coding for the N-terminal targeting sequence and the mature ferredoxin (Nixon et al. 2002).

Other soluble proteins that have been localized in mitosomes have recognizable N-terminal presequences (Tovar et al. 2003; Regoes et al. 2005; Dolezal et al. 2005; Rada et al. 2011; Rout et al. 2016; Pyrihová et al. 2018). However, some proteins like IscS and Cpn60 seem to rely on internal signals, which is quite unusual since these mitochondrial proteins typically contain cleavable presequences in other studied organisms. This was experimentally demonstrated when *G. intestinalis* IscS was expressed as a 202-aa N-terminal polypeptide and a 232-aa C-terminal polypeptide, and both truncated proteins could be successfully delivered to organelles (as tested on *T. vaginalis* hydrogenosomes) showing that targeting information is found in multiple loci within the protein (Dolezal et al. 2005). Deletion of the first five amino acids on *G. intestinalis* Cpn60 did not affect the targeting of the protein to mitosomes (Regoes et al. 2005). Thus, *G. intestinalis* mitosomes display both presequence-dependent and presequence-independent targeting for soluble preproteins (Regoes et al. 2005; Dolezal et al. 2005).

The targeting information on mitosomal proteins can be recognized and processed by the heterologous systems of human and yeast mitochondria, as well as *T. vaginalis* hydrogenosomes (Regoes et al. 2005; Dolezal et al. 2005). The [2Fe-2S] ferredoxin N-terminal presequence was sufficient to deliver a passenger protein into human mitochondria (Regoes et al. 2005), and *T. vaginalis* hydrogenosomes can specifically import *G. intestinalis* [2Fe-2S] ferredoxin, IscU, IscS, Pam18 and β-MPP (Dolezal et al. 2005). Furthermore, the presequence on IscU was sufficient to efficiently target a passenger protein into *T. vaginalis* hydrogenosomes. The N-terminal presequence of IscU can be processed by *S. cerevisiae* mitochondrial extract and also by purified recombinant rat MPP (Dolezal et al. 2005). Altogether, these results strongly suggest that targeting information on *G. intestinalis* mitosomal proteins can be cross-recognized by the respective protein import machineries of mitochondria and hydrogenosomes.

5.2.6 Microsporidia Mitosomes

So far, multiple proteins predominantly of ISC pathway have been localized in situ in *T. hominis* and *E. cuniculi* mitosomes (Williams et al. 2002; Goldberg et al. 2008; Freibert et al. 2017). The presence of N-terminal targeting sequences was not thoroughly investigated in these two species, but at least some *E. cuniculi* protein have predicted cleavable N-terminal presequence (Katinka et al. 2001).

Instead, analysis of *A. locustae* genome has provided interesting insight and surprising differences in mitosomal protein import mechanisms (Slamovits ct al. 2004). Of the identified mitosomal proteins, only a handful has amphipathic N-terminal presequences, and others do not appear to have any extensions nor have many characteristics in common (Burri et al. 2006). As no genetic transformation technique has been developed as yet for microsporidia, the targeting information on these proteins was investigated by expressing the full-length and truncated versions of these proteins in *S. cerevisiae* as fusions with GFP (Burri et al. 2006). Of the 16 proteins under investigation, only 6, most from *A. locustae*, could direct GFP to mitochondria. Deletion of the N-terminal predicted extensions from four of these fusion proteins disrupted targeting to mitochondria, showing that the extensions are necessary for cross-organellar targeting. The other two proteins, including mitochondrial glycerol-3-phosphate dehydrogenase (mtG3PDH), could still be delivered to mitochondria, suggesting that internal targeting signals are sufficient for targeting. However, the N-terminal sequence of mtG3PDH (Table 1) was also found to be sufficient to deliver GFP to yeast mitochondria. The N-terminal extensions from the other proteins were not sufficient to target GFP to mitochondria. This finding undermines the exclusive role of the N-terminal sequence in organellar protein targeting. It is apparent that a combination of N-terminal and mainly internal signals seems to fulfil the targeting role in microsporidian mitosomes (Burri et al. 2006).

5.3 Signals on Hydrogenosomal and Mitosomal Membrane Proteins

Targeting signals on most membrane proteins are generally poorly characterized. Not only do membrane protein precursors require targeting, membrane sorting and insertion signals, but they also require a means of protection against premature folding or aggregation in the hydrophilic environments they encounter during transport to the organelle membrane. A variety of membrane proteins are targeted to mitochondria: β-barrels, tail-anchored and α-helical polytopic and monotopic proteins have been characterized. Given this diversity in structure, specific but sometimes overlapping pathways are utilized for their insertion (Rehling et al. 2003; Koehler 2004; Bohnert et al. 2007). Most of the data available on membrane protein insertion has been generated for members of the mitochondrial carrier family (MCF), particularly for AAC, the model precursor.

The *T. vaginalis* hydrogenosomal Hmp31 precursor protein, a member of the MCF, was found to have a cleavable 12-amino acid N-terminal presequence. Although this sequence is predicted to be mostly α-helical, it does not have an amphipathic disposition but has an overall negative charge. This presequence was found not to be necessary for targeting and integration of mature Hmp31 in the membrane, suggesting that Hmp31 utilizes internal targeting signals, like virtually all MCF proteins. However, the presequence was necessary, and sufficient, to target a passenger protein to the soluble hydrogenosomal fraction and as such acted as a targeting signal. Thus, the Hmp31 precursor has internal targeting signals and a functional N-terminal targeting signal (Dyall et al. 2000). Four more Hmp31 orthologues were found during proteomic analyses (our unpublished data), and two of those, Hmp31-a and Hmp31-b, were each found to have a similar N-terminal extension (Table 1). Another orthologue, Hmp31-d, however, had a putative N-terminal extension that resembles the matrix-targeting N-terminal presequence, with an overall positive charge, and the fourth one does not appear to have an N-terminal extension. However, none of these Hmp31 orthologues have had their N-termini experimentally determined as yet. No such presequence has been found on either the *Neocallimastix* hydrogenosomal AAC (van der Giezen et al. 2002) or on the *Entamoeba* mitosomal AAC (Chan et al. 2005). Although most MCF proteins are synthesized without N-terminal extensions, a subset of precursors has cleavable presequences. Plant mitochondrial AACs are synthesized with long N-terminal presequences, but these are both not necessary and not sufficient to target passenger proteins to mitochondria and are therefore not acting as targeting signals (Glaser et al. 1998; Murcha et al. 2005a). Another MCF protein, the mammalian phosphate carrier, bears a presequence that may act as an enhancer for translocation but is not strictly necessary, though it was marginally sufficient to target a passenger protein to mitochondria (Zara et al. 2007). It has been suggested that the dispensable N-terminal presequences of mammalian and fish citrate carriers may in fact act as chaperones to increase the solubility of the preprotein in the cytosol through electrostatic interaction (Zara et al. 2007).

All MCF members that have been characterized in hydrogenosomes or mitosomes have been successfully imported into yeast mitochondria (Dyall et al. 2000; van der Giezen et al. 2002; Chan et al. 2005). Therefore, all three precursors must have targeting signals that are compatible with the specific mitochondrial pathway used for mitochondrial carriers (Rehling et al. 2003). Indeed, *T. vaginalis* Hmp31 imported into mitochondria was found not only to be dependent on membrane potential but also on the presence of the small TIM chaperone complex (Fig. 1) that is essential for proper mitochondrial AAC translocation (Dyall et al. 2000). Conversely, mitochondrial AAC was efficiently targeted to *T. vaginalis* hydrogenosomes, showing that targeting signals are compatible between the two systems (Dyall et al. 2000). The targeting of β-barrel membrane proteins is conserved between hydrogenosomes and mitochondria, as a unique hydrogenosomal β-barrel protein, Hmp35, could be targeted to mitochondrial membranes where it associated with, or assembled into, a high molecular weight complex (Dyall et al. 2003). It is notable that β-barrel precursors from eubacteria and plastids can be successfully imported and assembled into mitochondria as well (Röhl et al. 1999; Müller et al. 2002). Thus, targeting and insertion pathways for β-barrel proteins appear to be conserved between eubacteria, mitochondria, plastids and hydrogenosomes. *T. vaginalis* hydrogenosomes may be using a conserved SAM-like pathway (Fig. 1) for insertion of β-barrel proteins, as a homologue of Sam50 has been discovered in the *T. vaginalis* genome (Dolezal et al. 2006). Proteomic analysis of *T. vaginalis* hydrogenosomes had shown the presence of 12 tail-anchored (TA) proteins (Rada et al. 2011). Mitochondrial TA proteins carry a single transmembrane domain at their C-terminus, and their targeting signal often reside in the TMD and its flanking region (Horie et al. 2002). In yeast, the overall charge of the flanking regions or the hydrophobicity determines the destination of the protein as TA proteins are present in ER and peroxisomes as well. However, in *T. vaginalis* since peroxisomes are absent, the TA proteins are targeted either to the hydrogenosomes or the ER. A global search for *T. vaginalis* TA proteins and localization experiments have shown that the upstream regions of the TMD of hydrogenosomal TA proteins have low hydrophobicity compared to those of ER TA proteins (Rada et al. 2019). The TMD flanking regions of TA proteins contain lysine and arginine residues and, thus, have an overall positive charge. Replacement of lysines with serines or extension of the upstream region of the TMD by multiple valines mistargeted the hydrogenosomal TA protein to the ER (Rada et al. 2019).

6 Crossing the Organellar Membranes

All hydrogenosomes and mitosomes examined to date appear to have double membranes, which implies the presence of an intermembrane space. Given that hydrogenosomal and mitosomal preproteins bear signals that are recognized by the mitochondrial protein import machinery, it is likely that some components are phylogenetically and/or functionally conserved between these organelles. In hydrogenosomes and mitosomes, only the core subunits that are conserved in most eukaryotes are readily identifiable, while some subunits that are found in animals and fungi are either absent or too diverged to be identified through sequence-based searches. To date, a very limited number of hydrogenosomal and mitosomal protein import components have been functionally characterized, and some putative players have been identified in the genomes of the protists through sequence comparison with mitochondrial translocases from various species. More insight into potential import processes can be gained by examining in greater detail how mitochondrial preproteins interact with translocases to cross organellar membranes.

6.1 *The Outer Membrane*

Two major protein import machineries have been characterized to date in the mitochondrial outer membrane: the TOM and the SAM complexes (Fig. 1).

6.1.1 Translocase of the Outer Membrane (TOM Complex)

Most mitochondrial proteins enter mitochondria through a general import pore, the TOM complex. In yeast mitochondria, this complex consists of a pore-forming β-barrel Tom40 and six α-helical proteins: two primary receptors Tom70 and Tom20, core receptor Tom22 and three small Toms Tom5, Tom6 and Tom7 (Pfanner and Geissler 2001; van der Laan et al. 2006a; Bohnert et al. 2007).

Tom70 is the preferred receptor for hydrophobic preproteins with or without presequences (Wiedemann et al. 2001; Chan et al. 2006), although Tom20 also participates in binding (Brix et al. 1997). A typical substrate for Tom70 is the precursor to AAC. AAC has multiple internal targeting signals that are recognized by several Tom70 dimers, which probably act to prevent aggregation of these hydrophobic precursors (Brix et al. 2000; Wiedemann et al. 2001). Preproteins with N-terminal presequences initially make contact with Tom20 (Söllner et al. 1989). This interaction occurs through the hydrophobic surface of the amphipathic helix formed by the presequence, as demonstrated by structural studies (Abe et al. 2000). Thereafter, the two surfaces of the presequence are differentially recognized by binding domains of increasing affinity within the downstream translocases (Pfanner and Geissler 2001). Preproteins from both Tom20 and Tom70 are subsequently transferred to Tom22, which acts both as a docking point for Tom20 and Tom70 and as a central receptor for preproteins within the TOM complex (Hönlinger et al. 1995; Bolliger et al. 1995; Brix et al. 1997; van Wilpe et al. 1999).

The cytosolic domain of Tom22 interacts with the positively charged surface of the amphipathic helix formed by N-terminal presequences (Brix et al. 1997). Next, the small protein Tom5 transfers preproteins from Tom22 to the Tom40 channel for translocation across the outer membrane (Dietmeier et al. 1997; Künkele et al. 1998; Hill et al. 1998). Besides making up the channel, Tom40 also has a binding site for presequences (Hill et al. 1998). After they pass through the channel, presequence-containing precursors bind to the IMS domain of Tom22 through the positive surface of the presequence and are subsequently sorted to the TIM23 complex. Therefore, a typical N-terminal presequence is recognized at least five times by Tom proteins, through either hydrophobic or ionic interactions (Pfanner and Geissler 2001; Bohnert et al. 2007). Following passage through the Tom channel, other types of preproteins are sorted into their respective specialized biogenesis pathways. Precursors to outer membrane β-barrel and inner membrane carrier proteins are guided by the small TIM chaperone complexes to their respective SAM or TIM22 pathway. Precursors to the small Tims and to other IMS proteins are taken up into the MIA pathway for further processing (Bohnert et al. 2007). Small Toms, Tom5, Tom6 and Tom7, are involved in the assembly and disassembly of the complex (Model et al. 2001).

Structural studies for TOM have been conducted so far only in two organisms, *S. cerevisiae* and *Neurospora crassa*. As observed via cryo-electron microscopy, the ~550 kDa holo TOM complex (trimeric) measuring around 140 Å has three protein translocation channels with each pore measuring around 20 Å, while the core TOM complex (dimeric) measuring 120 Å has two channels (Model et al. 2008; Bausewein et al. 2017). It has been speculated that the trimeric structure seems to be generic for the TOM complex in all forms of mitochondria (Fukasawa et al. 2017). However, this hypothesis needs to be tested in representative organisms that bear divergent and reduced forms of mitochondria. The TOM complex is a highly dynamic structure with the trimeric and dimeric forms switching alternately during the biogenesis of a new Tom40 (Shiota et al. 2015).

Given the intricacy and specificity displayed by the yeast mitochondrial protein import machinery, one might expect that the outer membrane translocases, or Tom proteins, would be conserved across species. Moreover, the demonstrated ability to successfully and specifically import mitosomal and hydrogenosomal preproteins into mitochondria led many to infer that similar and phylogenetically related receptors were present in hydrogenosomes and mitosomes as in mitochondria. This inference has in turn been used as supporting evidence for a common origin for mitochondria and related organelles. However, recent sequence surveys of complete genome databases have taught us that to start with, not all Tom proteins are conserved across all mitochondrial species (Maćasev et al. 2004; Likić et al. 2005; Chan et al. 2006; Perry et al. 2006), let alone mitosomal or hydrogenosomal species. Indeed, a comprehensive survey of available completed eukaryotic genomes revealed that only Tom7, Tom22 and Tom40 sequences are conserved among the majority of eukaryotes, including animals, plants, fungi and some protists (Maćasev et al. 2004). Other components such as Tom20 and Tom70 have only been found in the genomes of animals and fungi so far, although a functional homologue of Tom70 was found in *Blastocystis* sp. and its homologues were found in other Stramenopiles (Likić et al. 2005; Chan et al. 2006; Tsaousis et al. 2011; Fukasawa et al. 2017). Although "Tom20" has been named and functionally characterized in plants (Heins and Schmitz 1996; Werhahn et al. 2001), it bears no primary sequence similarity to the fungal and animal Tom20 sequences and is likely to be of independent origin (Likić et al. 2005; Perry et al. 2006). Strikingly, though, the plant Tom20 has similar but oppositely orientated structural domains to the fungal Tom20, which appear to fulfil similar functions (Abe et al. 2000; Likić et al. 2005; Perry et al. 2006). These observations, taken together, have led to the hypothesis that the mitochondrial ancestor to eukaryotes had invented a core TOM complex consisting of Tom40, Tom22 and Tom7 and that other components subsequently evolved independently in the descendants as they progressively tweaked their respective mitochondrial protein import apparatuses (Maćasev et al. 2004; Dolezal et al. 2006). Of note is the TOM complex in the excavate *Trypanosoma brucei* termed archaic translocase of the outer membrane (ATOM) that is composed of highly diverged Tom40 (ATOM40) and a partially conserved Tom22-like protein (ATOM14) and four other subunits—two receptors, ATOM69 and ATOM46, ATOM11 and ATOM12 (Mani et al. 2015).

A proteomic study of the *Trichomonas* hydrogenosomes reported the presence of seven Tom40-like proteins that belong to the mitochondrial porin superfamily (Rada et al. 2011). A highly divergent TvTom40 (Isoform-2), present in a high molecular weight complex of ~570 kDa in the hydrogenosomal outer membrane, mediates the translocation of proteins across the OM (Makki et al. 2019). TvTom40-2 is associated with other isoforms of TvTom40, four tail-anchored proteins and Sam50, the core protein involved in β-barrel biogenesis (Makki et al. 2019). Two of the TA proteins, namely, Tom36 and Tom46, carrying an N-terminal Hsp20 chaperone-like and three TPR-like domains were shown to bind two hydrogenosomal preproteins, Fdx1 and α-SCS, and, hence, can function as TOM receptors. The other two TA proteins include a Tom22-like protein with a predicted molecular weight of 6.4 kDa which is present in the TvTOM that has short *cis* domain (cytosolic) and a conserved Tom22 TMD but lacks a *trans* domain (intermembrane space) and Homp19, which has no homology (Makki et al. 2019). Visualization of TvTOM via electron microscopy revealed a triplet-pore structure with an unconventional skull-like shape. EM and biochemical data suggest that TvTOM can associate with Sam50 (Makki et al. 2019).

Entamoeba histolytica mitosomes have a conserved Tom40 and a receptor protein named Tom60 that are part of a ~600 kDa TOM complex (Makiuchi et al. 2013). In the case of mitosomes in *Giardia intestinalis*, a Tom40 protein was identified using an HMM-based search that migrated in ~200 kDa complex, and more recently, GiMOMP35, a mitosomal outer membrane protein, was observed to be enriched when Tom40 was pulled down (Dagley et al. 2009; Martincová et al. 2015). However, it is not known whether GiMOMP35 is present in the GiTOM complex or if it plays any role in the mitosomal protein import. Studies over the years have shown that no two major eukaryotic groups share the same set of TOM receptors indicating that these proteins were gained after the separation of the groups.

Tom40 homologues were identified in the genome sequences of *Microsporidia* and *Cryptosporidium* species (Abrahamsen et al. 2004; Xu et al. 2004; Heinz and Lithgow 2013; Abrahamsen et al. 2004; Xu et al. 2004). Microsporidian *Nosema bombycis* Tom40 was shown to be targeted to mitochondria when expressed in *S. cerevisiae* (Lin et al. 2012). Moreover, *Microsporidia* contain clear homologue of Tom70 receptor (Waller et al. 2009).

6.1.2 Sorting and Assembling β-Barrel Proteins: The SAM Complex

Upon entering mitochondria through the TOM channel, precursors to β-barrel proteins such as Tom40, porin and Mdm10 are directed by the small TIM chaperone complexes to the SAM pathway for correct sorting and insertion into the outer membrane (Bohnert et al. 2007). It was deduced that the β-hairpin present in the β-barrel proteins act as the targeting signal that is recognized by Tom20 and partly by Tom70 (Jores et al. 2016). Recently, an in-depth crosslinking study dissected some of the crucial steps of β-barrel assembly. The β-signal at the carboxy-terminal of the precursor initiates an opening of the Sam50 between the 1st and the 16th strands, and the β-barrel precursor is assembled in the lumen of Sam50, perhaps using Sam50 itself as the template, and further the newly folded β-barrel protein is released laterally into the membrane (Höhr et al. 2018). In yeast, SAM is composed of Sam50, Sam35, Sam37 and under certain conditions, Mdm10. Out of those, only Sam50 (Kozjak et al. 2003; Paschen et al. 2003; Gentle et al. 2004) and Sam35 (Milenkovic et al. 2004; Waizenegger et al. 2004) are essential for cell viability, whereas Mdm10 (Meisinger et al. 2004) and Sam37 (Wiedemann et al. 2003), though involved in β-barrel biogenesis, are not essential components. In yeast, the TOM complex forms a labile supercomplex with SAM for the efficient translocation and assembly of the OM proteins, and Tom22-Sam37 interaction seems to play a crucial role in the formation of the supercomplex (Qiu et al. 2013; Wenz et al. 2015). More players such as Mim1 and Mdm12/Mmm1 have been characterized that act downstream of the core SAM complex (Ishikawa et al. 2004; Waizenegger et al. 2005; Meisinger et al. 2007). Some components appear to be important only for the biogenesis of subcategories of β-barrels, such that further specific pathways may be uncovered in the near future.

The insertion of β-barrel precursors is one of the two translocation processes, besides the sorting of inner membrane and IMS proteins, which are clearly derived from a eubacterial translocation system. β-Barrel proteins are exclusively found in the outer membranes of Gram-negative bacteria and in the endosymbiotic organelles such as mitochondria and plastids (Wimley 2003). The discovery that Sam50, a protein of eubacterial ancestry, played a critical role in the insertion of mitochondrial β-barrel proteins allowed several parallels to be drawn between the eubacterial and mitochondrial β-barrel biogenesis pathways (Paschen et al. 2005; Dolezal et al.

2006). Sam50 is itself a β-barrel protein that is homologous to the β-barrel bacterial protein Omp85, which is found in all bacteria that have an outer membrane. Omp85 is essential for bacterial viability and has been shown to be involved in the insertion of β-barrel protein precursors into the outer membrane of *Neisseria* (Voulhoux et al. 2003). Phylogenetic analyses revealed that the *sam50* gene is widely distributed among eukaryotes and probably derived from an α-proteobacterial-like bacterium, possibly the mitochondrial endosymbiont (Gentle et al. 2004). Another parallel crops up between the small TIM chaperone complexes and the chaperones Skp and SurA that assist β-barrel precursors as they navigate through the bacterial periplasmic space. In effect, the mitochondrial IMS represents the periplasmic space of the mitochondrial endosymbiont. Though the two chaperone systems are phylogenetically unrelated, they presumably function to prevent aggregation of the substrates according to similar principles (Paschen et al. 2005; Dolezal et al. 2006).

In *T. vaginalis* hydrogenosomes, Sam50 was found to form a stable association with the TOM complex (Makki et al. 2019). With the help of HMM analyses, homologous sequences to Sam50 have been found in the genomes of virtually all eukaryotes with complete genome sequences except *Giardia* and related metamonds (Leger et al. 2017). These putative translocases all have features common to mitochondrial Sam50 and possibly share a common ancestor though no phylogenetic analyses have been performed on the more recently discovered sequences (Dolezal et al. 2006). The distribution of other components of the SAM complex has not yet been thoroughly investigated among mitochondrial eukaryotes, but some components are limited to fungi. No convincing homologues to Sam35, Sam37, Mdm10, Mdm12, Mmm1 or Mim1 have been found by BLAST searches of any of the complete genomes of hydrogenosomal or mitosomal species.

6.2 *The Intermembrane Space Chaperones*

The small TIM chaperones have been shown to convey "complicated" substrates like β-barrel and polytopic hydrophobic membrane protein precursors across the hydrophilic environment of the IMS (Koehler 2004; Bohnert et al. 2007). The small Tims—Tim8, Tim9, Tim10, Tim12 and Tim13—are about 10 kDa in size and are characterized by a $C\text{-}x_3\text{-}C\text{-}x_{11\text{-}16}\text{-}C\text{-}x_3\text{-}C$ motif. Of the small Tims, only Tim9, Tim10 and Tim12 are essential, and Tim8 and Tim13 appear to be dispensable for yeast (Koehler 2004). However, both Tim9-Tim10 and Tim8-Tim13 complexes can bind AAC or β-barrel precursors (Gentle et al. 2007). Recent examination of the distribution of these small proteins using HMM analyses revealed that the small Tims have no prokaryotic homologues and may be eukaryotic inventions devised to assist membrane protein import. One or more small Tim proteins are diversely distributed among eukaryotes, but the only occurrence of small Tim-like homologues in the hydrogenosomal or mitosomal species occurs in *T. vaginalis* and *C. parvum* (Rada et al. 2011; Alcock et al. 2012). The occurrence of the *small tim* genes in diverse eukaryotes suggests an early origin for these genes (Gentle et al.

2007). It is conceivable that some other hydrogenosomal and mitosomal species developed or acquired independent chaperones. As we saw earlier, there are other types of proteins such as Skp and SurA that undertake similar chaperoning functions in bacteria. The dependence of the hydrogenosomal inner membrane protein Hmp31 on the small TIM chaperone complex, particularly Tim10, when imported into mitochondria suggests that this precursor is likely to be sensitive to the IMS, just like its mitochondrial counterparts (Dyall et al. 2000). In support of this, two paralogues of Tim9-Tim10 (A and B) were identified in the proteome of *T. vaginalis* hydrogenosomes (Rada et al. 2011).

6.3 The Inner Membrane

The two complexes that import cytosolic proteins through the mitochondrial inner membrane, TIM22 and TIM23, split the import pathways of hydrophobic inner membrane proteins from that of presequence-containing preproteins (Fig. 1).

6.3.1 The TIM22 Complex

The Tim9-Tim10 chaperone complex delivers MCF proteins such as AAC from the TOM to the TIM22 complex. Tim12, which is peripherally associated with TIM22, acts as a docking receptor for the chaperone complex. The TIM22 complex contains twin pores built from Tim22 that form a voltage-activated channel that is sensitive to synthetic peptides bearing AAC internal targeting signals, but insensitive to synthetic N-terminal presequences. The passage of the substrate through the channel is entirely dependent on the membrane potential and not on ATP hydrolysis (Kovermann et al. 2002; Rehling et al. 2003; Koehler 2004). Both Tim18 and Tim54 do not seem to mediate protein import but act as a scaffold for the TIM22 complex (Koehler et al. 2000; Hwang et al. 2007). However, like Tim22 (Sirrenberg et al. 1996) and Tim12 (Jarosch et al. 1997), Tim54 is essential (Kerscher et al. 1997), whereas Tim18 is not (Kerscher et al. 2000; Koehler et al. 2000). Recently, a metazoan-specific Tim29 was reported that is required for the stability of the TIM22 complex and for forming a contact site with TOM complex for the efficient transfer of hydrophobic proteins in the aqueous intermembrane space (Kang et al. 2016).

All components of the yeast TIM22 complex appear to be restricted to fungi, except for Tim22 which is widely distributed among eukaryotes (Rassow et al. 1999; Dolezal et al. 2006). Among the mitosomal and hydrogenosomal species, only a putative Tim22 sequence was reported in the genome of *E. cuniculi* and other *Microsporidia* (Katinka et al. 2001; Žárský and Doležal 2016; Pyrihová et al. 2018).

6.3.2 The TIM23 Complex

The TIM23 complex, which intakes presequence-containing precursors, is better characterized than the TIM22 complex and exhibits the most intricate import mechanisms (Bohnert et al. 2007). Within the complex, Tim23 forms a cation-selective, voltage-gated, protein-conducting, possibly twin-pore, channel that is specifically sensitive to synthetic presequence peptides (Truscott et al. 2001; Martinez-Caballero et al. 2007). Tim17, though homologous in sequence and a secondary structure to Tim23, does not form part of the channel but modulates its activity (Meier et al. 2005; Martinez-Caballero et al. 2007). Tim21 makes direct contact with the TOM complex by interacting with the IMS domain of Tom22, where it promotes precursor release by competing with presequence binding (Chacinska et al. 2005). Tim50 has a dual role, i.e. acting as a receptor for presequences and regulating the closure of the TIM23 channel (Geissler et al. 2002; Yamamoto et al. 2002; Meinecke et al. 2006). Tim21 also regulates the interaction between PAM and TIM23 by associating with TIM23. This complicated interaction serves to generate two types of TIM23 complexes: one that is matrix-import competent and the other that is competent to sort and insert the presequence-carrying inner membrane proteins (Chacinska et al. 2005; van der Laan et al. 2006a, b). All components of TIM23, except for Tim21 (Chacinska et al. 2005), are essential (Dekker et al. 1993; Emtage and Jensen 1993; Ryan et al. 1994; Maarse et al. 1994; Geissler et al. 2002; Yamamoto et al. 2002; Mokranjac et al. 2003a). This is quite surprising, given the central role played by Tim21 at various levels. Again, this attests to the flexibility of the mitochondrial protein import machinery.

Genes homologous to *tim23* and *tim17* have been found in most mitochondrial eukaryotes (Rassow et al. 1999; Dolezal et al. 2006), and *tim21* homologues can be found in animal, plant and fungal genomes (Chacinska et al. 2005) but not in protists (our unpublished observations). Tim50 contains a LIM domain commonly found in proteins of diverse functions, and no profound sequence analyses have yet been performed to assess its distribution among various species. Sequences related to *tim17* and *tim23* have been detected in the respective genomes of *T. vaginalis* and *C. parvum* (Abrahamsen et al. 2004; Henriquez et al. 2005). Four paralogues of Tim17/22/23 family (A-D), a Tim17-like protein and Tim44 were reported to be present in the hydrogenosomes of *T. vaginalis* (Rada et al. 2011). Thus, a core TIM23 complex could exist in the organelles of these two organisms.

Recently, single Tim17 family protein has been identified in the genome of *G. intestinalis* and localized to its mitosomes (Pyrihová et al. 2018). The protein forms disulphide bond-mediated dimers in the inner mitosomal membrane, where it seems to interact with Tim44 and other components of mitosomal protein import machinery. No clear homologues to any TIM23 component could be detected in the complete genome sequences of *E. histolytica*.

7 The Protein Import Motor

The final step of the journey of the mitochondrial matrix-targeted preprotein across the membranes involves the participation of an ATP-driven protein import motor, PAM, which pulls the preprotein from the Tim23 channel into the matrix (Fig. 1). In yeast, the core component of the PAM complex is Ssc1, or mt-Hsp70, which is assisted by Mge1, Pam18, Pam16, Pam17 and Tim44 (van der Laan et al. 2006a; Bohnert et al. 2007). With the exception of Pam17 (van der Laan et al. 2005), all PAM components identified so far are essential for yeast viability (Craig et al. 1987; Maarse et al. 1992; Bolliger et al. 1994; Mokranjac et al. 2003b; D'Silva et al. 2003; Truscott et al. 2003; Frazier et al. 2004; Kozany et al. 2004; Li et al. 2004).

Mt-Hsp70 is a member of the Hsp70 chaperone family that is distributed in all domains of life. The bacterial cytoplasmic homologues are called DnaK, and various types of *hsp70* genes are found in eukaryotes, with the products localizing to the cytosol, the ER, the mitochondrion or a plastid compartment (Gupta and Singh 1994; Bukau and Horwich 1998). Phylogenetic analyses show strong affinity and conserved signature sequences between mt-Hsp70 and α-proteobacterial DnaK, supporting the endosymbiotic origin of mitochondria from an α-proteobacterial-like ancestor (Boorstein et al. 1994; Falah and Gupta 1994; Gupta 2006). The Hsp70 proteins are the central part of protein folding machines that are powered by ATP. Generally, Hsp70 molecules have a highly conserved amino-terminal region with an ATPase domain and a carboxy-terminal region with a peptide-binding domain. The extensively studied chaperone system in *E. coli* revealed much about the mechanism of action of DnaK, which is assisted in its function by the nucleotide exchange factor GrpE, and the J-protein DnaJ that enhances ATPase activity (Bukau and Horwich 1998). A similar system can be found operating with a likewise mechanism at the matrix side of the TIM23 complex. In this situation, however, mt-Hsp70 is not involved in protein folding per se, but its properties are put to use to bind a largely unfolded incoming preprotein and to drive it completely into the mitochondrial matrix in an action regulated by ATP hydrolysis and co-chaperones. A fraction of mt-Hsp70 docks onto the TIM23 complex through the essential peripheral membrane protein Tim44 (Voos and Röttgers 2002; van der Laan et al. 2006a). Genes homologous to *tim44* have been found in all the completed genome sequences of mitochondrial eukaryotes, and also of α-proteobacteria, where the putative functions of the homologues are unknown (Dolezal et al. 2006; Clements et al. 2009). As the freshly imported preprotein enters the mitochondrial matrix, it is bound and/or pulled in by mt-Hsp70, which is assisted by the soluble matrix protein Mge1 (Bolliger et al. 1994; Voos and Röttgers 2002) and the inner membrane protein Pam18 (Mokranjac et al. 2003b; D'Silva et al. 2003; Truscott et al. 2003). These are the respective homologues of bacterial GrpE and DnaJ. Pam18 has a matrix-oriented J-domain with which it can stimulate mt-Hsp70 ATPase activity (Truscott et al. 2003). Pam18 is tightly bound to Pam16, which contains a degenerate J-domain and acts as a regulatory peripheral inner membrane protein within the motor (Frazier et al. 2004; Kozany et al. 2004; Li et al. 2004). The role of the final and non-essential component of PAM, Pam17, is unclear, but it is necessary for the stable modular association of Pam16 and Pam18 with TIM23 (van der Laan et al. 2005). This particular component appears to be fungi-specific as convincing homologues could not be found in other mitochondrial species or in eubacteria.

Mitochondrial-type Hsp70 is the only PAM component that has homologues in all mitosome- or hydrogenosome-containing species examined to date, namely, *T. vaginalis* (Bui et al. 1996; Germot et al. 1996), *G. intestinalis* (Morrison et al. 2001; Arisue et al. 2002), *E. histolytica* (Arisue et al. 2002), *C. parvum* (Slapeta and Keithly 2004), *E. cuniculi* (Peyretaillade et al. 1998), *A. locustae* (Germot et al. 1997), *T. hominis* (Williams et al. 2002), *N. ovalis* (Boxma et al. 2005) and *N. patriciarum* (van der Giezen et al. 2003). The complete set of PAM components (mt-Hsp70, Tim44, Pam18 and Pam16) has been identified in *T. vaginalis* hydrogenosomes (Rada et al. 2011; Schneider et al. 2011) and *G. intestinalis* mitosomes (Dolezal et al. 2005; Rada et al. 2011; Martincová et al. 2015) suggesting that a PAM system functions in the organelle. Mt-Hsp70, Pam18 and Tim44 have been found in the genome sequence of *C. parvum* (Abrahamsen et al. 2004), but only mt-Hsp70 has been localized to its mitosome so far (Slapeta and Keithly 2004). In *Microsporidia*, only genes for mt-Hsp70 and Pam18 were identified (Katinka et al. 2001; Waller et al. 2009). However, the most derived and reduced motor complex seems to exist in *E. histolytica* where only homologue of mt-Hsp70 was found (Arisue et al. 2002). The putative origins of most of the mitosomal and hydrogenosomal mt-Hsp70 homologues have been thoroughly pursued through phylogenetic analyses where most sequences group with the mitochondrial homologues with fairly strong support, except in the case of *G. intestinalis* mt-Hsp70, which is divergent (Morrison et al. 2001; Arisue et al. 2002). Generally, it is assumed that the mt-Hsp70 homologues originate from the α-proteobacterial-like endosymbiont that gave rise to the mitochondrion.

8 Preprotein Processing Peptidases

Upon translocation into the matrix, the N-terminal presequence of preproteins is processed by MPP, and the mature protein is thereafter folded into its native conformation (Fig. 1). Some preproteins have a bipartite presequence that is processed in two steps, the first part by MPP and the second part, which includes an octapeptide motif, by the mitochondrial intermediate peptidase (MIP). Precursors destined for the IMS have an IMS-sorting signal at the N-terminus. The IMP complex is responsible for the maturation of these proteins. Some of the precursors contain bipartite presequences consisting of a matrix-targeting signal followed by an intermembrane space-sorting signal (Gakh et al. 2002).

8.1 The Mitochondrial Processing Peptidase (MPP)

The mitochondrial processing peptidase is an essential zinc-dependent metallopeptidase (Yaffe et al. 1985; Luciano and Géli 1996; Gakh et al. 2002; Nomura et al. 2006). It cleaves the N-terminal presequence from precursors to matrix-targeted proteins and from precursors destined for the inner membrane or the IMS. Through sequence comparisons, Gavel and von Heijne (1990) defined four cleavage site motifs for MPP and MIP:

(a) The R-2 motif: x-R-x^x-(S/x)
(b) The R-3 motif: x-R-x-(Y/x)^(S/A/x)-x
(c) The R-10 motif: x-R-x^(F/L/I)-x-x-(S/T/G)-xxxx^, where the second cleavage site is for MIP
(d) The R-none motif: x-x^x-(S/x)

Surveys of mitochondrial presequences showed that, though quite common, these above motifs are not found in all of them and that the primary sequence for the cleavage site is quite degenerate.

The role of the Arg at the -2 or -3 position is unclear and may be presequence-dependent as studies on a variety of precursors revealed that mutating the Arg results in cleavage inhibition or modification in some cases, but not in others. It may be that the structure, rather than the primary sequence composition of the presequence and perhaps of the mature protein, determines the MPP cleavage site (Gakh et al. 2002).

Generally, the enzyme consists of two core subunits, α-MPP and β-MPP, each of about 50 kDa in size, which are widely distributed among mitochondrial eukaryotes. α-MPP and β-MPP are homologous to each other with up to about 30% identical residues in some species. The catalytic unit is β-MPP, which contains the conserved and critical zinc-binding motif H-x-x-H-x_{76}-E. This motif is characteristic of the pitrilysin protease family that includes bacterial proteases (Rawlings and Barrett 1995). The α-subunit is not involved in processing but may be involved in substrate recognition and interaction through a highly conserved glycine-rich loop. However, both subunits are required for processing the presequence in mitochondria (Geli et al. 1990).

MPP has long been thought to have evolved from a bacterial protease of the pitrilysin family (Gakh et al. 2002). Recently, a putative peptidase has been characterized from the α-proteobacterial parasitic bacterium *Rickettsia prowazekii* and related species and was found to have domains typical of both subunits of MPP (Kitada et al. 2007). Strikingly, the N-terminal domain of this rickettsial putative peptidase (RPP) resembles the N-terminal region β-MPP with an H-x-x-H-x_{76}-E motif, and the C-terminal domain of RPP resembles the C-terminal region of α-MPP, minus the glycine-rich loop. Unlike β-MPP, monomeric recombinant RPP was shown to have proteolytic activity on its own, cleaving basic synthetic peptides preferentially. RPP was able to cleave mitochondrial presequence peptides at specific sites in some cases, albeit at reduced efficiency compared with MPP. However, when tested on mitochondrial preproteins with short and long presequences, respectively, RPP was inactive on its own. Processing of the short presequence only occurred when RPP was stoichiometrically mixed with yeast β-MPP, and it was demonstrated that β-MPP was involved in the catalytic activity and not RPP. Thus, RPP behaved like α-MPP as an activator of β-MPP. The long presequence was not processed by either RPP/β-MPP or RPP/α-MPP, and mutational studies on MPP indicated that this could be due to the absence of a glycine-rich loop on RPP (Kitada et al. 2007). Given the close relationship between mitochondria and *Rickettsia* (Andersson et al. 1998), these findings indicate that RPP may represent an ancestral form of both α-MPP and β-MPP, derived from the α-proteobacterial-like mitochondrial endosymbiont.

Homologues to β-MPP, both with conserved catalytic motifs, were discovered recently in the genomes of *T. vaginalis* and *G. intestinalis*. The *Giardia* β-MPP homologue was localized to mitosomes, and N-terminal sequencing of mitosomal IscU confirmed the cleavage site of its presequence at the position suggested by the PSORT prediction programme (Table 1) (Dolezal et al. 2005). The biochemical characterization of *Giardia* β-MPP subunit has showed that the protein functions as a monomer without the assistance of the α-MPP subunit (Šmíd et al. 2008). The proteins seem to have co-evolved with the shorter mitosomal presequences and are not able to process presequences on hydrogenosomal or mitochondrial precursors (Šmíd et al. 2008). Similarly in *C. parvum*, a homologue to only β-MPP, but not α-MPP, has been reported (Abrahamsen et al. 2004; Henriquez et al. 2005).

In contrast, the hydrogenosomes of *T. vaginalis* contain typical dimeric MPP (Šmíd et al. 2008), which exhibits broader specificity as demonstrated on the efficient processing of presequences derived from mitosomal (*Giardia*) and mitochondrial (*S. cerevisiae*) precursor proteins (Šmíd et al. 2008). In *E. histolytica*, one presequence has been shown to be cleaved at a site predicted for MPP (Mai et al. 1999; Tovar et al. 1999), but no enzyme responsible for the processing has been identified yet. Similarly, no MPP homologue was found in *E. cuniculi* and other *Microsporidia*, but given the occurrence of presequence-independent protein import in microsporidia (Burri et al. 2006), they may have dispensed with processing peptidases during their reductive evolution. In ciliate *N. ovalis*, both MPP subunits have been identified suggesting conserved processing of precursor proteins in its hydrogenosomes (Boxma et al. 2005). In general, there have been no reports of MIP-like proteins nor of any R-10 motif on protein precursors in any of the hydrogenosomal or mitosomal species.

8.2 *The Inner Membrane Protease*

Anchored on the outer face of the inner membrane, the mitochondrial IMP complex consists of two proteases Imp1 and Imp2 and a regulatory subunit Som1 (Fig. 1). The two proteases have distinct specificities for IMS protein precursors. Some of the precursors contain bipartite presequences consisting of a matrix-targeting signal followed by an intermembrane space-sorting signal for sequential cleavage by

MPP and IMP. Imp1 and Imp2 show significant similarity to bacterial type I leader peptidases that cleave the N-terminal signal of precursors that traverse the bacterial membrane (Gakh et al. 2002).

The import route of mtG3PDH into microsporidian mitosomes seems to follow the stop-transfer pathway in *S. cerevisiae*, during which the translocation of mtG3PDH is stopped at the TIM23 complex, where the precursor remains in the membrane without release into the matrix (Esser et al. 2004). However, the processing step is different in *A. locustae* and *E. cuniculi* preproteins. In *A. locustae*, as in *S. cerevisiae,* the precursor seems to be processed by IMP that cleaves off the presequence at the position following the first transmembrane segment (Esser et al. 2004; Burri et al. 2006). In contrast, in *E. cuniculi*, the N-terminal domain is retained within the mature protein. *S. cerevisiae* IMP could process the *A. locustae* mtG3PDH precursor, and an IMP2 homologue is present in the *A. locustae* genome. Together, these data suggest that *A. locustae* has retained an IMP proteolytic processing pathway but that the related microsporidian species *E. cuniculi* may have discarded both MPP and IMP processing (Burri et al. 2006; Burri and Keeling 2007). Currently, there is no evidence for IMP processing in any of the other mitosomal or hydrogenosomal species.

9 Folding Newly Imported Soluble Proteins

Newly imported proteins enter mitochondria in an extended or only partly folded conformation. Two main chaperone systems have been characterized in mitochondria that fold these incoming proteins into a native state that permits them to perform their function. Mitochondria have inherited these efficient and intricate folding systems from their bacterial progenitor(s): one involving mt-Hsp70 and the other with Cpn60/Cpn10 or Hsp60/Hsp10 (Neupert 1997; Voos and Röttgers 2002).

Besides its role in preprotein translocation across the inner membrane through TIM23 and PAM, mt Hsp70 can also act as a protein folding chaperone. Indeed, mt-Hsp70 in yeast mitochondria is either found in a membrane-associated complex with Tim44 and PAM or in a soluble state in association with co-chaperones Mdj1 and Mge1. Mdj1 is a highly conserved non-essential mitochondrial homologue of bacterial DnaJ and was shown not to be involved in translocation but to be important for protein folding in association with the homologues of GrpE and DnaK (Neupert 1997; Voos and Röttgers 2002). The manner in which the mt-Hsp70 chaperone functions is very similar to that of bacterial DnaK, and the system is likely to have been inherited from the bacterial progenitor of mitochondria (Hartl et al. 1994; Stuart et al. 1994; Szabo et al. 1994). As we have reported in Sect. 7, homologues to mt-Hsp70/DnaK have been found in the genomes of all mitosomal or hydrogenosomal species examined to date, and homologues to Mge1/GrpE have been found in *T. vaginalis* (Carlton et al. 2007), *G. intestinalis* (Martincová et al. 2015) and *C. parvum* (Abrahamsen et al. 2004). Homologues to Mdj1/DnaJ have been reported in *T. vaginalis*, *E. cuniculi* and *N. ovalis* (Katinka et al. 2001; Boxma et al. 2005; Carlton et al. 2007). All the components of the DnaK-type machinery have been localized to *T. vaginalis* hydrogenosomes (Bozner 1997; Dyall et al. 2003; Rada et al. 2011; Schneider et al. 2011), suggesting that a similar protein folding mechanism occurs in these organelles.

The mitochondrial Cpn60/Cpn10 or Hsp60/Hsp10 chaperone system participates in the folding of the majority of newly imported matrix proteins (Neupert 1997; Voos and Röttgers 2002). This system functions downstream of the mt-Hsp70 system, but both systems are likely to cooperate in protein folding (Manning-Krieg et al. 1991). Cpn60 and Cpn10 derive from bacterial homologues GroEL and GroES, respectively, and phylogenetic and comparative analyses of both protein sequences show a robust relationship between the respective monophyletic mitochondrial groups and α-proteobacteria. Since the progenitor of mitochondria is likely to have been an ancestor of extant α-proteobacteria, these findings support the notion that Cpn60 and Cpn10 originate from the endosymbiont that gave rise to mitochondria (Gupta 2018). In bacteria, including α-proteobacteria, *groel* and *groes* genes are found on a single operon, such that the eukaryotic genes are likely to have a common origin (Gupta 2018). Much has been learnt about the mechanism of protein folding in bacteria through the structure of the bacterial GroEL/GroES complex. In *E. coli*, the GroEL proteins form a double-ring structure comprising two apposed heptameric rings that form a central cavity that binds protein folding intermediates of up to 50 kDa and facilitates folding to the native state. The chaperonin cavity switches from a binding to a folding state through conformational changes induced by ATP. This action is regulated by a saucer-shaped heptameric complex of GroES which modulates both the ATPase cycle and the conformation of GroEL monomers (Rye et al. 1997; Xu et al. 1997; Bukau and Horwich 1998). Both Cpn60 and Cpn10 are encoded by essential genes in yeast and are likely to function similarly to their bacterial homologues (Cheng et al. 1989; Rospert et al. 1993), but not all mitochondrial proteins require Cpn60 for folding (Rospert et al. 1996). Homologues to Cpn60 that show high affinity to mitochondrial Cpn60 have been found in *E. histolytica* (Clark and Roger 1995), *T. vaginalis* (Horner et al. 1996; Bui et al. 1996; Roger et al. 1996), *G. intestinalis* (Roger et al. 1998), *C. parvum* (Riordan et al. 2003; Putignani et al. 2004), *N. patriciarum* (van der Giezen et al. 2003) and *N. ovalis* (Boxma et al. 2005). These putative chaperones have been localized to either hydrogenosomes or mitosomes in *T. vaginalis* (Bui et al. 1996; Bozner 1997), *E. histolytica* (Mai et al. 1999; Tovar et al. 1999) , *G. intestinalis* (Regoes et al. 2005) (Regoes et al. 2005), *C. parvum* (Riordan et al. 2003; Putignani et al. 2004) and *N. patriciarum* (van der Giezen et al. 2003). So far, homologues to Cpn10 have been reported in the genomes of *T. vaginalis, G. intestinalis, E. histolytica and Cryptosporidium* species, but phylogenetic relationships with either mitochondrial or α-proteobacterial sequences could not be convincingly inferred (Bui et al. 1996; van der Giezen et al. 2005). Surprisingly, no homologue to either *cpn60* or *cpn10* was found in the genomes of microsporidia available to date. It may be that microsporidian mitosomal proteins do not require Cpn60/Cpn10 for protein folding, as has been noted for a subset of mitochondrial matrix proteins (Rospert et al. 1996). It is plausible that, for the sake of economy, the highly reduced microsporidian mitosomes utilize homologues to the multifunctional mt-Hsp70 protein to both translocate and fold newly imported proteins and have dispensed with the energetically expensive Cpn60 machinery.

10 Perspectives

It is evident that protein import mechanisms are conserved between hydrogenosomes, mitosomes and mitochondria. Although no protein import pathway has been functionally deciphered for hydrogenosomes and mitosomes, we have started to get a glimpse of some putative mitochondrial-like components that may be involved in importing, processing and folding preproteins during biogenesis. The species that have been shown to harbour either mitosomes or hydrogenosomes and have had their genomes completely sequenced offer us an opportunity to examine their putative mitochondrial protein import complement. Mitosomes of *C. parvum* and hydrogenosomes of *T. vaginalis* potentially house the most mitochondrial-like components, though many of them have not as yet been localized. Their organelles could have mitochondrial-like SAM, TIM23 and PAM machines and mitochondrial-like preprotein processing and folding. On the contrary, the mitosomes of *E. histolytica* and *G. intestinalis* lack Tim17 family protein and Sam50, respectively, which are the key components present in all other eukaryotes. Does it mean *that G. intestinalis* develop an alternate mechanism to assemble β-barrel proteins in the outer membrane? Or did their β-barrel proteins evolve such way that they do not require SAM complex anymore? Did *E. histolytica* build the mitochondrial inner membrane translocase around different channel subunits?

We seem to be reaching the limits of how much we can assimilate and conclude from genome sequence analyses. These have been invaluable in identifying some putative protein translocases and chaperones. More sensitive searches like HMM may indeed deliver further putative candidates for mitochondrial-type translocases from the genome sequence databases of the hydrogenosomal and mitosomal species. However, we shall need to go back to the bench to demonstrate their localization and investigate their involvement in organellar protein trafficking. In addition to the above-mentioned unknowns, more general questions appear in the light of newly discovered protein transport and biogenesis pathways in model organisms. Specifically, what is the relationship of the mitochondria-related organelles to the endoplasmic reticulum and other endomembrane system organelles in terms of their biogenesis and dynamics? How do these protists control the number and the metabolic capacity of their organelles? Some of these questions can be answered with the limited set of tools we have at hand, but it is imperative to develop new techniques if we want to dissect these pathways.

Once these questions are answered, we shall be in a better position to formulate hypotheses on how these fantastic protein transport machines have evolved. By comparing protein import mechanisms and examining the structure of translocases between hydrogenosomal, mitosomal, mitochondrial, plastidic and bacterial systems, it is likely that we discover common principles for protein targeting. We can ask further and broader impact questions. For instance, how do the intricacy of the protein targeting machines correlates with proteome size? How do targeting signals and translocases co-evolve? How hard is it for an endosymbiont to build a protein import machine?

References

Abe Y, Shodai T, Muto T, Mihara K, Torii H, Nishikawa S, Endo T, Kohda D (2000) Structural basis of presequence recognition by the mitochondrial protein import receptor Tom20. Cell 100:551–560

Abrahamsen MS, Templeton TJ, Enomoto S, Abrahante JE, Zhu G, Lancto CA, Deng M, Liu C, Widmer G, Tzipori S, Buck GA, Xu P, Bankier AT, Dear PH, Konfortov BA, Spriggs HF, Iyer L, Anantharaman V, Aravind L, Kapur V (2004) Complete genome sequence of the apicomplexan, *Cryptosporidium parvum*. Science 304:441–445

Adams KL, Palmer JD (2003) Evolution of mitochondrial gene content: gene loss and transfer to the nucleus. Mol Phylogenet Evol 29:380–395

Adams KL, Daley DO, Qiu Y-L, Whelan J, Palmer JD (2000) Repeated, recent and diverse transfers of a mitochondrial gene to the nucleus in flowering plants. Nature 408:354

Alcock F, Webb CT, Dolezal P, Hewitt V, Shingu-Vasquez M, Likić VA, Traven A, Lithgow T (2012) A small Tim homohexamer in the relict mitochondrion of *Cryptosporidium*. Mol Biol Evol 29:113–122

Altschul SF, Gish W, Miller W, Myers EW, Lipman DJ (1990) Basic local alignment search tool. J Mol Biol 215:403–410

Altschul SF, Madden TL, Schäffer AA, Zhang J, Zhang Z, Miller W, Lipman DJ (1997) Gapped BLAST and PSI-BLAST: a new generation of protein database search programs. Nucleic Acids Res 25:3389–3402

Alva V, Nam S-Z, Söding J, Lupas AN (2016) The MPI bioinformatics Toolkit as an integrative platform for advanced protein sequence and structure analysis. Nucleic Acids Res 44:W410–W415

Andersson SGE, Zomorodipour A, Andersson JO, Sicheritz-Pontén T, Alsmark UCM, Podowski RM, Näslund AK, Eriksson A-S, Winkler HH, Kurland CG (1998) The genome sequence of *Rickettsia prowazekii* and the origin of mitochondria. Nature 396:133

Andersson GE, Karlberg O, Canbäck B, Kurland CG (2003) On the origin of mitochondria: a genomics perspective. Philos Trans R Soc London Ser B Biol Sci 358:165–179

Arisue N, Sánchez LB, Weiss LM, Müller M, Hashimoto T (2002) Mitochondrial-type hsp70 genes of the amitochondriate protists, *Giardia intestinalis*, *Entamoeba histolytica* and two microsporidians. Parasitol Int 51:9–16

Baker A, Schatz G (1987) Sequences from a prokaryotic genome or the mouse dihydrofolate reductase gene can restore the import of a truncated precursor protein into yeast mitochondria. Proc Natl Acad Sci U S A 84:3117–3121

Bateman A, Coin L, Durbin R, Finn RD, Hollich V, Griffiths-Jones S, Khanna A, Marshall M, Moxon S, Sonnhammer ELL, Studholme DJ, Yeats C, Eddy SR (2004) The Pfam protein families database. Nucleic Acids Res 32:D138–D141

Bausewein T, Mills DJ, Langer JD, Nitschke B, Nussberger S, Kühlbrandt W (2017) Cryo-EM Structure of the TOM Core Complex from *Neurospora crassa*. Cell 170:693–700.e7

Beasley EM, Müller S, Schatz G (1993) The signal that sorts yeast cytochrome b2 to the mitochondrial intermembrane space contains three distinct functional regions. EMBO J 12:2303–2311

Becker T, Pfannschmidt S, Guiard B, Stojanovski D, Milenkovic D, Kutik S, Pfanner N, Meisinger C, Wiedemann N (2008) Biogenesis of the mitochondrial TOM complex: Mim1 promotes insertion and assembly of signal-anchored receptors. J Biol Chem 283:120–127

Blobel G, Dobberstein B (1975a) Transfer of proteins across membranes. I. Presence of proteolytically processed and unprocessed nascent immunoglobulin light chains on membrane-bound ribosomes of murine myeloma. J Cell Biol 67:835–851

Blobel G, Dobberstein B (1975b) Transfer of proteins across membranes. II. Reconstitution of functional rough microsomes from heterologous components. J Cell Biol 67:852–862

Bohnert M, Pfanner N, van der Laan M (2007) A dynamic machinery for import of mitochondrial precursor proteins. FEBS Lett 581:2802–2810

Bolliger L, Deloche O, Glick BS, Georgopoulos C, Jeno P, Kronidou N, Horst M, Morishima N, Schatz G (1994) A mitochondrial homolog of bacterial GrpE interacts with mitochondrial hsp70 and is essential for viability. EMBO J 13:1998–2006

Bolliger L, Junne T, Schatz G, Lithgow T (1995) Acidic receptor domains on both sides of the outer membrane mediate translocation of precursor proteins into yeast mitochondria. EMBO J 14:6318–6326

Bömer U, Meijer M, Guiard B, Dietmeier K, Pfanner N, Rassow J (1997) The sorting route of cytochrome b branches from the general mitochondrial import pathway at the preprotein translocase of the inner membrane. J Biol Chem 272:30439–30446

Bonnefoy N, Remacle C, Fox TD (2007) Genetic transformation of *Saccharomyces cerevisiae* and *Chlamydomonas reinhardtii* mitochondria. Methods Cell Biol 80:525–548

Boorstein WR, Ziegelhoffer T, Craig EA (1994) Molecular evolution of the HSP70 multigene family. J Mol Evol 38:1–17

Boxma B, de Graaf RM, van der Staay GWM, van Alen TA, Ricard G, Gabald6n T, van Hoek AHAM, Moon-van der Staay SY, Koopman WJH, van Hellemond JJ, Tielens AGM, Friedrich T, Veenhuis M, Huynen MA, Hackstein JHP (2005) An anaerobic mitochondrion that produces hydrogen. Nature 434:74

Bozner P (1997) Immunological detection and subcellular localization of Hsp70 and Hsp60 homologs in *Trichomonas vaginalis*. J Parasitol 83:224–229

Bracha R, Nuchamowitz Y, Mirelman D (2003) Transcriptional silencing of an amoebapore gene in *Entamoeba histolytica*: molecular analysis and effect on pathogenicity. Eukaryot Cell 2:295–305

Bradley PJ, Lahti CJ, Plumper E, Johnson PJ (1997) Targeting and translocation of proteins into the hydrogenosome of the protist Trichomonas: similarities with mitochondrial protein import. EMBO J 16:3484–3493

Brás XP, Zimorski V, Bolte K, Maier U-G, Martin WF, Gould SB (2013) Knockout of the abundant *Trichomonas vaginalis* hydrogenosomal membrane protein *Tv* HMP23 increases hydrogenosome size but induces no compensatory up-regulation of paralogous copies. FEBS Lett 587:1333–1339

Brix J, Dietmeier K, Pfanner N (1997) Differential recognition of preproteins by the purified cytosolic domains of the mitochondrial import receptors Tom20, Tom22, and Tom70. J Biol Chem 272:20730–20735

Brix J, Ziegler GA, Dietmeier K, Schneider-Mergener J, Schulz GE, Pfanner N (2000) The mitochondrial import receptor Tom70: identification of a 25 kDa core domain with a specific binding site for preproteins. J Mol Biol 303:479–488

Brondijk TH, Durand R, van der Giezen M, Gottschal JC, Prins RA, Fevre M (1996) scsB, a cDNA encoding the hydrogenosomal beta subunit of succinyl-CoA synthetase from the anaerobic fungus *Neocallimastix frontalis*. Mol Gen Genet 253:315–323

Bui ET, Bradley PJ, Johnson PJ (1996) A common evolutionary origin for mitochondria and hydrogenosomes. Proc Natl Acad Sci U S A 93:9651–9656

Bui ET, Johnson PJ (1996) Identification and characterization of [Fe]-hydrogenases in the hydrogenosome of *Trichomonas vaginalis*. Mol Biochem Parasitol 76:305–310

Bukau B, Horwich AL (1998) The Hsp70 and Hsp60 chaperone machines. Cell 92:351–366

Burri L, Keeling PJ (2007) Protein targeting in parasites with cryptic mitochondria. Int J Parasitol 37:265–272

Burri L, Williams BAP, Bursac D, Lithgow T, Keeling PJ (2006) Microsporidian mitosomes retain elements of the general mitochondrial targeting system. Proc Natl Acad Sci U S A 103:15916–15920

Carlton JM, Hirt RP, Silva JC, Delcher AL, Schatz M, Zhao Q, Wortman JR, Bidwell SL, Alsmark UCM, Besteiro S, Sicheritz-Ponten T, Noel CJ, Dacks JB, Foster PG, Simillion C, Van de Peer Y, Miranda-Saavedra D, Barton GJ, Westrop GD, Müller S, Dessi D, Fiori PL, Ren Q, Paulsen I, Zhang H, Bastida-Corcuera FD, Simoes-Barbosa A, Brown MT, Hayes RD, Mukherjee M, Okumura CY, Schneider R, Smith AJ, Vanacova S, Villalvazo M, Haas BJ, Pertea M, Feldblyum TV, Utterback TR, Shu C-L, Osoegawa K, de Jong PJ, Hrdy I, Horvathova L, Zubacova Z, Dolezal P, Malik S-B, Logsdon JM, Henze K, Gupta A, Wang CC, Dunne RL, Upcroft JA, Upcroft P, White O, Salzberg SL, Tang P, Chiu C-H, Lee Y-S, Embley TM, Coombs GH, Mottram JC, Tachezy J, Fraser-Liggett CM, Johnson PJ (2007) Draft genome sequence of the sexually transmitted pathogen Trichomonas vaginalis. Science 315:207–212

Carpenter ML, Cande WZ (2009) Using morpholinos for gene knockdown in Giardia intestinalis. Eukaryot Cell 8:916–919

Cautain B, Hill R, de Pedro N, Link W (2015) Components and regulation of nuclear transport processes. FEBS J 282:445–462

Cavalier-Smith T (1987) The simultaneous symbiotic origin of mitochondria, chloroplasts, and microbodies. Ann N Y Acad Sci 503:55–71

Chacinska A, Lind M, Frazier AE, Dudek J, Meisinger C, Geissler A, Sickmann A, Meyer HE, Truscott KN, Guiard B, Pfanner N, Rehling P (2005) Mitochondrial presequence translocase: switching between TOM tethering and motor recruitment involves Tim21 and Tim17. Cell 120:817–829

Chan KW, Slotboom D-J, Cox S, Embley TM, Fabre O, van der Giezen M, Harding M, Horner DS, Kunji ERS, León-Avila G, Tovar J (2005) A novel ADP/ATP transporter in the mitosome of the microaerophilic human parasite Entamoeba histolytica. Curr Biol 15:737–742

Chan NC, Likić VA, Waller RF, Mulhern TD, Lithgow T (2006) The C-terminal TPR domain of Tom70 defines a family of mitochondrial protein import receptors found only in animals and fungi. J Mol Biol 358:1010–1022

Cheng MY, Hartl F-U, Martin J, Pollock RA, Kalousek F, Neuper W, Hallberg EM, Hallberg RL, Horwich AL (1989) Mitochondrial heat-shock protein hsp60 is essential for assembly of proteins imported into yeast mitochondria. Nature 337:620

Choi C, Liu Z, Adams KL (2006) Evolutionary transfers of mitochondrial genes to the nucleus in the *Populus* lineage and coexpression of nuclear and mitochondrial Sdh4 genes. New Phytol 172:429–439

Clark CG, Roger AJ (1995) Direct evidence for secondary loss of mitochondria in *Entamoeba histolytica*. Proc Natl Acad Sci U S A 92:6518–6521

Clements A, Bursac D, Gatsos X, Perry AJ, Civciristov S, Celik N, Likic VA, Poggio S, Jacobs-Wagner C, Strugnell RA, Lithgow T (2009) The reducible complexity of a mitochondrial molecular machine. Proc Natl Acad Sci U S A 106:15791–15795

Connolly T, Gilmore R (1986) Formation of a functional ribosome-membrane junction during translocation requires the participation of a GTP-binding protein. J Cell Biol 103:2253–2261

Craig EA, Kramer J, Kosic-Smithers J (1987) SSC1, a member of the 70-kDa heat shock protein multigene family of *Saccharomyces cerevisiae*, is essential for growth. Proc Natl Acad Sci U S A 84:4156–4160

Curran SP, Leuenberger D, Leverich EP, Hwang DK, Beverly KN, Koehler CM (2004) The role of Hot13p and redox chemistry in the mitochondrial TIM22 import pathway. J Biol Chem 279:43744–43751

D'Silva PD, Schilke B, Walter W, Andrew A, Craig EA (2003) J protein cochaperone of the mitochondrial inner membrane required for protein import into the mitochondrial matrix. Proc Natl Acad Sci U S A 100:13839–13844
Dagley MJ, Dolezal P, Likic VA, Smid O, Purcell AW, Buchanan SK, Tachezy J, Lithgow T (2009) The protein import channel in the outer mitosomal membrane of *Giardia intestinalis*. Mol Biol Evol 26:1941–1947
Dalbey RE, Lively MO, Bron S, Van Dijl JM (1997) The chemistry and enzymology of the type I signal peptidases. Protein Sci 6:1129–1138
Daley DO, Clifton R, Whelan J (2002) Intracellular gene transfer: reduced hydrophobicity facilitates gene transfer for subunit 2 of cytochrome c oxidase. Proc Natl Acad Sci U S A 99:10510–10515
Dan M, Wang AL, Wang CC (2000) Inhibition of pyruvate-ferredoxin oxidoreductase gene expression in *Giardia lamblia* by a virus-mediated hammerhead ribozyme. Mol Microbiol 36:447–456
Davis AJ, Ryan KR, Jensen RE (1998) Tim23p contains separate and distinct signals for targeting to mitochondria and insertion into the inner membrane. Mol Biol Cell 9:2577–2593
Dekker PJT, Keil P, Rassow J, Maarse AC, Pfanner N, Meijer M (1993) Identification of MIM23, a putative component of the protein import machinery of the mitochondrial inner membrane. FEBS Lett 330:66–70
Delgadillo MG, Liston DR, Niazi K, Johnson PJ (1997) Transient and selectable transformation of the parasitic protist *Trichomonas vaginalis*. Proc Natl Acad Sci U S A 94:4716–4720
Dietmeier K, Honlinger A, Bomer U, Dekker PJ, Eckerskorn C, Lottspeich F, Kubrich M, Pfanner N (1997) Tom5 functionally links mitochondrial preprotein receptors to the general import pore. Nature 388:195–200
Dolezal P, Smid O, Rada P, Zubacova Z, Bursac D, Sutak R, Nebesarova J, Lithgow T, Tachezy J (2005) *Giardia* mitosomes and *trichomonad* hydrogenosomes share a common mode of protein targeting. Proc Natl Acad Sci U S A 102:10924–10929
Dolezal P, Likic V, Tachezy J, Lithgow T (2006) Evolution of the molecular machines for protein import into mitochondria. Science 313:314–318
Dyall SD, Koehler CM, Delgadillo-Correa MG, Bradley PJ, Plümper E, Leuenberger D, Turck CW, Johnson PJ (2000) Presence of a member of the mitochondrial carrier family in hydrogenosomes: conservation of membrane-targeting pathways between hydrogenosomes and mitochondria. Mol Cell Biol 20:2488–2497
Dyall SD, Lester DC, Schneider RE, Delgadillo-Correa MG, Plümper E, Martinez A, Koehler CM, Johnson PJ (2003) *Trichomonas vaginalis* Hmp35, a putative pore-forming hydrogenosomal membrane protein, can form a complex in yeast mitochondria. J Biol Chem 278:30548–30561
Dyall SD, Yan W, Delgadillo-Correa MG, Lunceford A, Loo JA, Clarke CF, Johnson PJ (2004) Non-mitochondrial complex I proteins in a hydrogenosomal oxidoreductase complex. Nature 431:1103–1107
Ebneter JA, Heusser SD, Schraner EM, Hehl AB, Faso C (2016) Cyst-Wall-Protein-1 is fundamental for Golgi-like organelle neogenesis and cyst-wall biosynthesis in *Giardia lamblia*. Nat Commun 7:13859
Eddy SR (1998) Profile hidden Markov models. Bioinformatics 14:755–763
Emtage JL, Jensen RE (1993) MAS6 encodes an essential inner membrane component of the yeast mitochondrial protein import pathway. J Cell Biol 122:1003–1012
Esser K, Jan P-S, Pratje E, Michaelis G (2004) The mitochondrial IMP peptidase of yeast: functional analysis of domains and identification of Gut2 as a new natural substrate. Mol Gen Genomics 271:616–626
Falah M, Gupta RS (1994) Cloning of the hsp70 (dnaK) genes from *Rhizobium meliloti* and *Pseudomonas cepacia*: phylogenetic analyses of mitochondrial origin based on a highly conserved protein sequence. J Bacteriol 176:7748–7753
Finn RD, Clements J, Eddy SR (2011) HMMER web server: interactive sequence similarity searching. Nucleic Acids Res 39:W29–W37

Folsch H, Guiard B, Neupert W, Stuart RA (1996) Internal targeting signal of the BCS1 protein: a novel mechanism of import into mitochondria. EMBO J 15:479–487

Frazier AE, Dudek J, Guiard B, Voos W, Li Y, Lind M, Meisinger C, Geissler A, Sickmann A, Meyer HE, Bilanchone V, Cumsky MG, Truscott KN, Pfanner N, Rehling P (2004) Pam16 has an essential role in the mitochondrial protein import motor. Nat Struct Mol Biol 11:226

Freibert S-A, Goldberg AV, Hacker C, Molik S, Dean P, Williams TA, Nakjang S, Long S, Sendra K, Bill E, Heinz E, Hirt RP, Lucocq JM, Embley TM, Lill R (2017) Evolutionary conservation and in vitro reconstitution of microsporidian iron–sulfur cluster biosynthesis. Nat Commun 8:13932

Fukasawa Y, Oda T, Tomii K, Imai K (2017) Origin and evolutionary alteration of the mitochondrial import system in eukaryotic lineages. Mol Biol Evol 34:1574–1586

Gabaldón T, Huynen MA (2003) Reconstruction of the proto-mitochondrial metabolism. Science 301:609

Gakh O, Cavadini P, Isaya G (2002) Mitochondrial processing peptidases. Biochim Biophys Acta, Mol Cell Res 1592:63–77

García-Rodríguez LJ, Gay AC, Pon LA (2007) Puf3p, a Pumilio family RNA binding protein, localizes to mitochondria and regulates mitochondrial biogenesis and motility in budding yeast. J Cell Biol 176:197–207

Garg SG, Gould SB (2016) The role of charge in protein targeting evolution. Trends Cell Biol 26:894–905

Garg S, Stölting J, Zimorski V, Rada P, Tachezy J, Martin WF, Gould SB (2015) Conservation of transit peptide-independent protein import into the mitochondrial and hydrogenosomal matrix. Genome Biol Evol 7:2716–2726

Gavel Y, von Heijne G (1990) Cleavage-site motifs in mitochondrial targeting peptides. Protein Eng Des Sel 4:33–37

Geissler A, Chacinska A, Truscott KN, Wiedemann N, Brandner K, Sickmann A, Meyer HE, Meisinger C, Pfanner N, Rehling P (2002) The mitochondrial presequence translocase: an essential role of Tim50 in directing preproteins to the import channel. Cell 111:507–518

Geli V, Yang MJ, Suda K, Lustig A, Schatz G (1990) The MAS-encoded processing protease of yeast mitochondria. Overproduction and characterization of its two nonidentical subunits. J Biol Chem 265:19216–19222

Gentle I, Gabriel K, Beech P, Waller R, Lithgow T (2004) The Omp85 family of proteins is essential for outer membrane biogenesis in mitochondria and bacteria. J Cell Biol 164:19–24

Gentle IE, Perry AJ, Alcock FH, Likić VA, Dolezal P, Ng ET, Purcell AW, McConnville M, Naderer T, Chanez A-L, Charrière F, Aschinger C, Schneider A, Tokatlidis K, Lithgow T (2007) Conserved motifs reveal details of ancestry and structure in the small TIM chaperones of the mitochondrial intermembrane Space. Mol Biol Evol 24:1149–1160

Germot A, Philippe H, Le Guyader H (1996) Presence of a mitochondrial-type 70-kDa heat shock protein in *Trichomonas vaginalis* suggests a very early mitochondrial endosymbiosis in eukaryotes. Proc Natl Acad Sci U S A 93:14614–14617

Germot A, Philippe H, Le Guyader H (1997) Evidence for loss of mitochondria in *Microsporidia* from a mitochondrial-type HSP70 in *Nosema locustae*. Mol Biochem Parasitol 87:159–168

Glaser E, Sjoling S, Tanudji M, Whelan J (1998) Mitochondrial protein import in plants. Signals, sorting, targeting, processing and regulation. Plant Mol Biol 38:311–338

Glick BS, Brandt A, Cunningham K, Müller S, Hallberg RL, Schatz G (1992) Cytochromes c1 and b2 are sorted to the intermembrane space of yeast mitochondria by a stop-transfer mechanism. Cell 69:809–822

Goldberg AV, Molik S, Tsaousis AD, Neumann K, Kuhnke G, Delbac F, Vivares CP, Hirt RP, Lill R, Embley TM (2008) Localization and functionality of microsporidian iron–sulphur cluster assembly proteins. Nature 452:624–628

Gray MW (2015) Mosaic nature of the mitochondrial proteome: Implications for the origin and evolution of mitochondria. Proc Natl Acad Sci U S A 112:10133–10138

Gray MW, Burger G, Lang BF (1999) Mitochondrial evolution. Science 283:1476–1481

Gupta RS (2006) The phylogeny of proteobacteria: relationships to other eubacterial phyla and eukaryotes. FEMS Microbiol Rev 24:367–402

Gupta RS (2018) Evolution of the chaperonin families (HSP60, HSP 10 and TCP-1) of proteins and the origin of eukaryotic cells. Mol Microbiol 15:1–11

Gupta RS, Singh B (1994) Phylogenetic analysis of 70 kD heat shock protein sequences suggests a chimeric origin for the eukaryotic cell nucleus. Curr Biol 4:1104–1114

Hahne K, Haucke V, Ramage L, Schatz G (1994) Incomplete arrest in the outer membrane sorts NADH-cytochrome b5 reductase to two different submitochondrial compartments. Cell 79:829–839

Hartl F-U, Hlodan R, Langer T (1994) Molecular chaperones in protein folding: the art of avoiding sticky situations. Trends Biochem Sci 19:20–25

Hausler T, Stierhof YD, Blattner J, Clayton C (1997) Conservation of mitochondrial targeting sequence function in mitochondrial and hydrogenosomal proteins from the early-branching eukaryotes *Crithidia*, *Trypanosoma* and *Trichomonas*. Eur J Cell Biol 73:240–251

Heins L, Schmitz UK (1996) A receptor for protein import into potato mitochondria. Plant J 9:829–839

Heinz E, Lithgow T (2013) Back to basics: a revealing secondary reduction of the mitochondrial protein import pathway in diverse intracellular parasites. Biochim Biophys Acta, Mol Cell Res 1833:295–303

Henriquez FL, Richards TA, Roberts F, McLeod R, Roberts CW (2005) The unusual mitochondrial compartment of *Cryptosporidium parvum*. Trends Parasitol 21:68–74

Herrmann JM (2003) Converting bacteria to organelles: evolution of mitochondrial protein sorting. Trends Microbiol 11:74–79

Hill K, Model K, Ryan MT, Dietmeier K, Martin F, Wagner R, Pfanner N (1998) Tom40 forms the hydrophilic channel of the mitochondrial import pore for preproteins. Nature 395:516

Höhr AIC, Lindau C, Wirth C, Qiu J, Stroud DA, Kutik S, Guiard B, Hunte C, Becker T, Pfanner N, Wiedemann N (2018) Membrane protein insertion through a mitochondrial β-barrel gate. Science 359:eaah6834

Hönlinger A, Kübrich M, Moczko M, Gärtner F, Mallet L, Bussereau F, Eckerskorn C, Lottspeich F, Dietmeier K, Jacquet M (1995) The mitochondrial receptor complex: Mom22 is essential for cell viability and directly interacts with preproteins. Mol Cell Biol 15:3382–3389

Hoogenraad NJ, Ward LA, Ryan MT (2002) Import and assembly of proteins into mitochondria of mammalian cells. Biochim Biophys Acta, Mol Cell Res 1592:97–105

Hoppins SC, Nargang FE (2004) The Tim8-Tim13 complex of Neurospora crassa functions in the assembly of proteins into both mitochondrial membranes. J Biol Chem 279:12396–12405

Horie C, Suzuki H, Sakaguchi M, Mihara K (2002) Characterization of signal that directs C-tail–anchored proteins to mammalian mitochondrial outer membrane. Mol Biol Cell 13:1615–1625

Horner DS, Hirt RP, Kilvington S, Lloyd D, Embley TM (1996) Molecular data suggest an early acquisition of the mitochondrion endosymbiont. Proc Biol Sci 263:1053–1059

Horst M, Oppliger W, Rospert S, Schönfeld H, Schatz G, Azem A (1997) Sequential action of two hsp70 complexes during protein import into mitochondria. EMBO J 16:1842–1849

Hrdy I, Hirt RP, Dolezal P, Bardonová L, Foster PG, Tachezy J, Martin Embley T (2004) Trichomonas hydrogenosomes contain the NADH dehydrogenase module of mitochondrial complex I. Nature 432:618

Hrdy I, Müller M (1995a) Primary structure and eubacterial relationships of the pyruvate: ferredoxin oxidoreductase of the amitochondriate eukaryote *Trichomonas vaginalis*. J Mol Evol 41:388–396

Hrdy I, Müller M (1995b) Primary structure of the hydrogenosomal malic enzyme of *Trichomonas vaginalis* and its relationship to homologous enzymes. J Eukaryot Microbiol 42:593–603

Hulett JM, Lueder F, Chan NC, Perry AJ, Wolynec P, Likić VA, Gooley PR, Lithgow T (2008) The transmembrane segment of Tom20 is recognized by Mim1 for docking to the mitochondrial TOM complex. J Mol Biol 376:694–704

Hwang DK, Claypool SM, Leuenberger D, Tienson HL, Koehler CM (2007) Tim54p connects inner membrane assembly and proteolytic pathways in the mitochondrion. J Cell Biol 178:1161–1175

Ishikawa D, Yamamoto H, Tamura Y, Moritoh K, Endo T (2004) Two novel proteins in the mitochondrial outer membrane mediate β-barrel protein assembly. J Cell Biol 166:621–627

Janssen BD, Chen Y-P, Molgora BM, Wang SE, Simoes-Barbosa A, Johnson PJ (2018) CRISPR/Cas9-mediated gene modification and gene knock out in the human-infective parasite *Trichomonas vaginalis*. Sci Rep 8:270

Jarosch E, Rodel G, Schweyen RJ (1997) A soluble 12-kDa protein of the mitochondrial intermembrane space, Mrs11p, is essential for mitochondrial biogenesis and viability of yeast cells. Mol Gen Genet 255:157–165

Johnson PJ, Oliveira CE, Gorrell TE, Müller M (1990) Molecular analysis of the hydrogenosomal ferredoxin of the anaerobic protist *Trichomonas vaginalis*. Proc Natl Acad Sci U S A 87:6097–6101

Jores T, Klinger A, Groß LE, Kawano S, Flinner N, Duchardt-Ferner E, Wöhnert J, Kalbacher H, Endo T, Schleiff E, Rapaport D (2016) Characterization of the targeting signal in mitochondrial β-barrel proteins. Nat Commun 7:12036

Kang Y, Baker MJ, Liem M, Louber J, McKenzie M, Atukorala I, Ang C-S, Keerthikumar S, Mathivanan S, Stojanovski D (2016) Tim29 is a novel subunit of the human TIM22 translocase and is involved in complex assembly and stability. Elife 5

Katinka MD, Duprat S, Cornillot E, Méténier G, Thomarat F, Prensier G, Barbe V, Peyretaillade E, Brottier P, Wincker P, Delbac F, El Alaoui H, Peyret P, Saurin W, Gouy M, Weissenbach J, Vivarès CP (2001) Genome sequence and gene compaction of the eukaryote parasite Encephalitozoon cuniculi. Nature 414:450

Kerscher O, Holder J, Srinivasan M, Leung RS, Jensen RE (1997) The Tim54p–Tim22p Complex Mediates Insertion of Proteins into the Mitochondrial Inner Membrane. J Cell Biol 139:1663–1675

Kerscher O, Sepuri NB, Jensen RE, Fox TD (2000) Tim18p is a new component of the Tim54p-Tim22p translocon in the mitochondrial inner membrane. Mol Biol Cell 11:103–116

Kitada S, Uchiyama T, Funatsu T, Kitada Y, Ogishima T, Ito A (2007) A protein from a parasitic microorganism, *Rickettsia prowazekii*, can cleave the signal sequences of proteins targeting mitochondria. J Bacteriol 189:844–850

Koehler CM (2004) New developments in mitochondrial assembly. Annu Rev Cell Dev Biol 20:309–335

Koehler CM, Murphy MP, Bally NA, Leuenberger D, Oppliger W, Dolfini L, Junne T, Schatz G, Or E (2000) Tim18p, a new subunit of the TIM22 complex that mediates insertion of imported proteins into the yeast mitochondrial inner membrane. Mol Cell Biol 20:1187–1193

Kornmann B, Currie E, Collins SR, Schuldiner M, Nunnari J, Weissman JS, Walter P (2009) An ER-mitochondria tethering complex revealed by a synthetic biology screen. Science 325:477–481

Kovermann P, Truscott KN, Guiard B, Rehling P, Sepuri NB, Müller H, Jensen RE, Wagner R, Pfanner N (2002) Tim22, the essential core of the mitochondrial protein insertion complex, forms a voltage-activated and signal-gated channel. Mol Cell 9:363–373

Kozany C, Mokranjac D, Sichting M, Neupert W, Hell K (2004) The J domain–related cochaperone Tim16 is a constituent of the mitochondrial TIM23 preprotein translocase. Nat Struct Mol Biol 11:234

Kozjak V, Wiedemann N, Milenkovic D, Lohaus C, Meyer HE, Guiard B, Meisinger C, Pfanner N (2003) An essential role of Sam50 in the protein sorting and assembly machinery of the mitochondrial outer membrane. J Biol Chem 278:48520–48523

Kumar A, Harrison PM, Cheung K-H, Lan N, Echols N, Bertone P, Miller P, Gerstein MB, Snyder M (2002) An integrated approach for finding overlooked genes in yeast. Nat Biotechnol 20:58–63

Künkele K, Heins S, Dembowski M, Nargang FE, Benz R, Thieffry M, Walz J, Lill R, Nussberger S, Neupert W (1998) The preprotein translocation channel of the outer membrane of mitochondria. Cell 93:1009–1019

LaGier MJ, Tachezy J, Stejskal F, Kutisova K, Keithly JS (2003) Mitochondrial-type iron-sulfur cluster biosynthesis genes (IscS and IscU) in the apicomplexan *Cryptosporidium parvum*. Microbiology 149:3519–3530

Lahti CJ, Bradley PJ, Johnson PJ (1994) Molecular characterization of the alpha-subunit of *Trichomonas vaginalis* hydrogenosomal succinyl CoA synthetase. Mol Biochem Parasitol 66:309–318

Lahti CJ, d'Oliveira CE, Johnson PJ (1992) Beta-succinyl-coenzyme A synthetase from Trichomonas vaginalis is a soluble hydrogenosomal protein with an amino-terminal sequence that resembles mitochondrial presequences. J Bacteriol 174:6822–6830

Land KM, Delgadillo-Correa MG, Tachezy J, Vanacova S, Hsieh CL, Sutak R, Johnson PJ (2003) Targeted gene replacement of a ferredoxin gene in *Trichomonas vaginalis* does not lead to metronidazole resistance. Mol Microbiol 51:115–122

Lange A, Mills RE, Lange CJ, Stewart M, Devine SE, Corbett AH (2007) Classical nuclear localization signals: definition, function, and interaction with importin alpha. J Biol Chem 282:5101–5105

Länge S, Rozario C, Müller M (1994) Primary structure of the hydrogenosomal adenylate kinase of *Trichomonas vaginalis* and its phylogenetic relationships. Mol Biochem Parasitol 66:297–308

Lee CM, Sedman J, Neupert W, Stuart RA (1999) The DNA helicase, Hmi1p, is transported into mitochondria by a C-terminal cleavable targeting signal. J Biol Chem 274:20937–20942

Leger MM, Kolisko M, Kamikawa R, Stairs CW, Kume K, Čepička I, Silberman JD, Andersson JO, Xu F, Yabuki A, Eme L, Zhang Q, Takishita K, Inagaki Y, Simpson AGB, Hashimoto T, Roger AJ (2017) Organelles that illuminate the origins of *Trichomonas* hydrogenosomes and *Giardia* mitosomes. Nat Ecol Evol 1:0092

Li Y, Dudek J, Guiard B, Pfanner N, Rehling P, Voos W (2004) The presequence translocase-associated protein import motor of mitochondria. J Biol Chem 279:38047–38054

Likić VA, Perry A, Hulett J, Derby M, Traven A, Waller RF, Keeling PJ, Koehler CM, Curran SP, Gooley PR, Lithgow T (2005) Patterns that define the four domains conserved in known and novel isoforms of the protein import receptor Tom20. J Mol Biol 347:81–93

Likic VA, Dolezal P, Celik N, Dagley M, Lithgow T (2010) Using hidden Markov models to discover new protein transport machines. Methods Mol Biol 619:271–284

Lin L, Pan G, Li T, Dang X, Deng Y, Ma C, Chen J, Luo J, Zhou Z (2012) The protein import pore Tom40 in the microsporidian *Nosema bombycis*. J Eukaryot Microbiol 59:251–257

Linford AS, Moreno H, Good KR, Zhang H, Singh U, Petri WA (2009) Short hairpin RNA-mediated knockdown of protein expression in *Entamoeba histolytica*. BMC Microbiol 9:38

Lithgow T, Schneider A (2010) Evolution of macromolecular import pathways in mitochondria, hydrogenosomes and mitosomes. Philos Trans R Soc Lond Ser B Biol Sci 365:799–817

Liu S, Roellig DM, Guo Y, Li N, Frace MA, Tang K, Zhang L, Feng Y, Xiao L (2016) Evolution of mitosome metabolism and invasion-related proteins in *Cryptosporidium*. BMC Genomics 17:1006

Loftus B, Anderson I, Davies R, Alsmark UCM, Samuelson J, Amedeo P, Roncaglia P, Berriman M, Hirt RP, Mann BJ, Nozaki T, Suh B, Pop M, Duchene M, Ackers J, Tannich E, Leippe M, Hofer M, Bruchhaus I, Willhoeft U, Bhattacharya A, Chillingworth T, Churcher C, Hance Z, Harris B, Harris D, Jagels K, Moule S, Mungall K, Ormond D, Squares R, Whitehead S, Quail MA, Rabbinowitsch E, Norbertczak H, Price C, Wang Z, Guillén N, Gilchrist C, Stroup SE, Bhattacharya S, Lohia A, Foster PG, Sicheritz-Ponten T, Weber C, Singh U, Mukherjee C, El-Sayed NM, Petri WA, Clark CG, Embley TM, Barrell B, Fraser CM, Hall N (2005) The genome of the protist parasite *Entamoeba histolytica*. Nature 433:865–868

Lucattini R, Likic VA, Lithgow T (2004) Bacterial proteins predisposed for targeting to mitochondria. Mol Biol Evol 21:652–658

Luciano P, Géli V (1996) The mitochondrial processing peptidase: function and specificity. Experientia 52:1077–1082

Maarse AC, Blom J, Grivell LA, Meijer M (1992) MPI1, an essential gene encoding a mitochondrial membrane protein, is possibly involved in protein import into yeast mitochondria. EMBO J 11:3619–3628

Maarse AC, Blom J, Keil P, Pfanner N, Meijer M (1994) Identification of the essential yeast protein MIM17, an integral mitochondrial inner membrane protein involved in protein import. FEBS Lett 349:215–221

Maćasev D, Whelan J, Newbigin E, Silva-Filho MC, Mulhern TD, Lithgow T (2004) Tom22', an 8-kDa trans-site receptor in plants and protozoans, is a conserved feature of the TOM complex that appeared early in the evolution of eukaryotes. Mol Biol Evol 21:1557–1564

Maduke M, Roise D (1993) Import of a mitochondrial presequence into protein-free phospholipid vesicles. Science 260:364–367

Mai Z, Ghosh S, Frisardi M, Rosenthal B, Rogers R, Samuelson J (1999) Hsp60 is targeted to a cryptic mitochondrion-derived organelle in the microaerophilic protozoan parasite *Entamoeba histolytica*. Mol Cell Biol 19:2198–2205

Makiuchi T, Nozaki T (2014) Highly divergent mitochondrion-related organelles in anaerobic parasitic protozoa. Biochimie 100:3–17

Makiuchi T, Mi-ichi F, Nakada-Tsukui K, Nozaki T (2013) Novel TPR-containing subunit of TOM complex functions as cytosolic receptor for *Entamoeba* mitosomal transport. Sci Rep 3:1129

Makki A, Rada P, Žárský V, Kereïche S, Kováčik L, Novotný M, Jores T, Rapaport D, Tachezy J (2019) Triplet-pore structure of a highly divergent TOM complex of hydrogenosomes in *Trichomonas vaginalis*. PLoS Biol 17:32

Mani J, Desy S, Niemann M, Chanfon A, Oeljeklaus S, Pusnik M, Schmidt O, Gerbeth C, Meisinger C, Warscheid B, Schneider A (2015) Mitochondrial protein import receptors in Kinetoplastids reveal convergent evolution over large phylogenetic distances. Nat Commun 6:6646

Mani J, Meisinger C, Schneider A (2016) Peeping at TOMs—diverse entry gates to mitochondria provide insights into the evolution of eukaryotes. Mol Biol Evol 33:337–351

Manning-Krieg UC, Scherer PE, Schatz G (1991) Sequential action of mitochondrial chaperones in protein import into the matrix. EMBO J 10:3273–3280

Martijn J, Vosseberg J, Guy L, Offre P, Ettema TJG (2018) Deep mitochondrial origin outside the sampled alphaproteobacteria. Nature 557:101–105

Martin J, Langer T, Boteva R, Schramel A, Horwich AL, Hartl FU (1991) Chaperonin-mediated protein folding at the surface of groEL through a 'molten globule'-like intermediate. Nature 352:36–42

Martin WF, Garg S, Zimorski V (2015) Endosymbiotic theories for eukaryote origin. Philos Trans R Soc B Biol Sci 370:20140330

Martincová E, Voleman L, Pyrih J, Žárský V, Vondráčková P, Kolísko M, Tachezy J, Doležal P (2015) Probing the biology of *Giardia intestinalis* mitosomes using in vivo enzymatic tagging. Mol Cell Biol 35:2864–2874

Martinez-Caballero S, Grigoriev SM, Herrmann JM, Campo ML, Kinnally KW (2007) Tim17p regulates the twin pore structure and voltage gating of the mitochondrial protein import complex TIM23. J Biol Chem 282:3584–3593

Marvin-Sikkema FD, Kraak MN, Veenhuis M, Gottschal JC, Prins RA (1993) The hydrogenosomal enzyme hydrogenase from the anaerobic fungus Neocallimastix sp. L2 is recognized by antibodies, directed against the C-terminal microbody protein targeting signal SKL. Eur J Cell Biol 61:86–91

McArthur AG, Morrison HG, Nixon JE, Passamaneck NQ, Kim U, Hinkle G, Crocker MK, Holder ME, Farr R, Reich CI, Olsen GE, Aley SB, Adam RD, Gillin FD, Sogin ML (2000) The *Giardia* genome project database. FEMS Microbiol Lett 189:271–273

McFadden GI (1999) Endosymbiosis and evolution of the plant cell. Curr Opin Plant Biol 2:513–519

McInally S, Hagen K, Nosala C, Williams J, Nguyen K, Booker J, Jones K, Dawson SC (2018) Robust and stable transcriptional repression in *Giardia* using CRISPRi. Mol Biol Cell 30:119–130

Meier S, Neupert W, Herrmann JM (2005) Conserved N-terminal negative charges in the Tim17 subunit of the TIM23 translocase play a critical role in the import of preproteins into mitochondria. J Biol Chem 280:7777–7785

Meinecke M, Wagner R, Kovermann P, Guiard B, Mick DU, Hutu DP, Voos W, Truscott KN, Chacinska A, Pfanner N, Rehling P (2006) Tim50 maintains the permeability barrier of the mitochondrial inner membrane. Science 312:1523–1526

Meisinger C, Ryan MT, Hill K, Model K, Lim JH, Sickmann A, Müller H, Meyer HE, Wagner R, Pfanner N (2001) Protein import channel of the outer mitochondrial membrane: a highly stable Tom40-Tom22 core structure differentially interacts with preproteins, small tom proteins, and import receptors. Mol Cell Biol 21:2337–2348

Meisinger C, Rissler M, Chacinska A, Szklarz LKS, Milenkovic D, Kozjak V, Schönfisch B, Lohaus C, Meyer HE, Yaffe MP, Guiard B, Wiedemann N, Pfanner N (2004) The mitochondrial morphology protein Mdm10 functions in assembly of the preprotein translocase of the outer membrane. Dev Cell 7:61–71

Meisinger C, Pfannschmidt S, Rissler M, Milenkovic D, Becker T, Stojanovski D, Youngman MJ, Jensen RE, Chacinska A, Guiard B, Pfanner N, Wiedemann N (2007) The morphology proteins Mdm12/Mmm1 function in the major β-barrel assembly pathway of mitochondria. EMBO J 26:2229–2239

Ménétret J-F, Schaletzky J, Clemons WM, Osborne AR, Skånland SS, Denison C, Gygi SP, Kirkpatrick DS, Park E, Ludtke SJ, Rapoport TA, Akey CW (2007) Ribosome binding of a Single copy of the SecY complex: implications for protein translocation. Mol Cell 28:1083–1092

Mesecke N, Terziyska N, Kozany C, Baumann F, Neupert W, Hell K, Herrmann JM (2005) A disulfide relay system in the intermembrane space of mitochondria that mediates protein import. Cell 121:1059–1069

Mi-ichi F, Abu Yousuf M, Nakada-Tsukui K, Nozaki T (2009) Mitosomes in *Entamoeba histolytica* contain a sulfate activation pathway. Proc Natl Acad Sci U S A 106:21731–21736

Milenkovic D, Kozjak V, Wiedemann N, Lohaus C, Meyer HE, Guiard B, Pfanner N, Meisinger C (2004) Sam35 of the mitochondrial protein sorting and assembly machinery is a peripheral outer membrane protein essential for cell viability. J Biol Chem 279:22781–22785

Model K, Meisinger C, Prinz T, Wiedemann N, Truscott KN, Pfanner N, Ryan MT (2001) Multistep assembly of the protein import channel of the mitochondrial outer membrane. Nat Struct Biol 8:361–370

Model K, Meisinger C, Kühlbrandt W (2008) Cryo-electron microscopy structure of a yeast mitochondrial preprotein translocase. J Mol Biol 383:1049–1057

Mokranjac D, Paschen SA, Kozany C, Prokisch H, Hoppins SC, Nargang FE, Neupert W, Hell K (2003a) Tim50, a novel component of the TIM23 preprotein translocase of mitochondria. EMBO J 22:816–825

Mokranjac D, Sichting M, Neupert W, Hell K (2003b) Tim14, a novel key component of the import motor of the TIM23 protein translocase of mitochondria. EMBO J 22:4945–4956

Morrison HG, Roger AJ, Nystul TG, Gillin FD, Sogin ML (2001) *Giardia lamblia* expresses a proteobacterial-like DnaK homolog. Mol Biol Evol 18:530–541

Müller A, Rassow J, Grimm J, Machuy N, Meyer TF, Rudel T (2002) VDAC and the bacterial porin PorB of Neisseria gonorrhoeae share mitochondrial import pathways. EMBO J 21:1916–1929

Mukherjee M, Brown MT, McArthur AG, Johnson PJ (2006a) Proteins of the glycine decarboxylase complex in the hydrogenosome of *Trichomonas vaginalis*. Eukaryot Cell 5:2062–2071

Mukherjee M, Sievers SA, Brown MT, Johnson PJ (2006b) Identification and biochemical characterization of serine hydroxymethyl transferase in the hydrogenosome of *Trichomonas vaginalis*. Eukaryot Cell 5:2072–2078

Murcha MW, Millar AH, Whelan J (2005a) The N-terminal cleavable extension of plant carrier proteins is responsible for efficient insertion into the inner mitochondrial membrane. J Mol Biol 351:16–25
Murcha MW, Rudhe C, Elhafez D, Adams KL, Daley DO, Whelan J (2005b) Adaptations required for mitochondrial import following mitochondrial to nucleus gene transfer of ribosomal protein S10. Plant Physiol 138:2134–2144
Neupert W (1997) Protein import into mitochondria. Annu Rev Biochem 66:863–917
Nixon JEJ, Wang A, Morrison HG, McArthur AG, Sogin ML, Loftus BJ, Samuelson J (2002) A spliceosomal intron in *Giardia lamblia*. Proc Natl Acad Sci U S A 99:3701–3705
Nomura H, Athauda SBP, Wada H, Maruyama Y, Takahashi K, Inoue H (2006) Identification and reverse genetic analysis of mitochondrial processing peptidase and the core protein of the cytochrome bc1 complex of *Caenorhabditis elegans*, a model parasitic nematode. J Biochem 139:967–979
Nunnari J, Fox TD, Walter P (1993) A mitochondrial protease with two catalytic subunits of nonoverlapping specificities. Science 262:1997–2004
Nyathi Y, Wilkinson BM, Pool MR (2013) Co-translational targeting and translocation of proteins to the endoplasmic reticulum. Biochim Biophys Acta, Mol Cell Res 1833:2392–2402
Osborne AR, Rapoport TA, van den Berg B (2005) Protein translocation by the Sec61/SecY channel. Annu Rev Cell Dev Biol 21:529–550
Park E, Rapoport TA (2012) Mechanisms of Sec61/SecY-mediated protein translocation across membranes. Annu Rev Biophys 41:21–40
Paschen SA, Waizenegger T, Stan T, Preuss M, Cyrklaff M, Hell K, Rapaport D, Neupert W (2003) Evolutionary conservation of biogenesis of beta-barrel membrane proteins. Nature 426:862–866
Paschen SA, Neupert W, Rapaport D (2005) Biogenesis of β-barrel membrane proteins of mitochondria. Trends Biochem Sci 30:575–582
Pemberton LF, Paschal BM (2005) Mechanisms of receptor-mediated nuclear import and nuclear export. Traffic 6:187–198
Perry AJ, Hulett JM, Likić VA, Lithgow T, Gooley PR (2006) Convergent evolution of receptors for protein import into mitochondria. Curr Biol 16:221–229
Peyretaillade E, Broussolle V, Peyret P, Metenier G, Gouy M, Vivares CP (1998) Microsporidia, amitochondrial protists, possess a 70-kDa heat shock protein gene of mitochondrial evolutionary origin. Mol Biol Evol 15:683–689
Pfanner N, Geissler A (2001) Versatility of the mitochondrial protein import machinery. Nat Rev Mol Cell Biol 2:339–349
Picotti P, Clément-Ziza M, Lam H, Campbell DS, Schmidt A, Deutsch EW, Röst H, Sun Z, Rinner O, Reiter L, Shen Q, Michaelson JJ, Frei A, Alberti S, Kusebauch U, Wollscheid B, Moritz RL, Beyer A, Aebersold R (2013) A complete mass-spectrometric map of the yeast proteome applied to quantitative trait analysis. Nature 494:266–270
Plümper E, Bradley PJ, Johnson PJ (2000) Competition and protease sensitivity assays provide evidence for the existence of a hydrogenosomal protein import machinery in *Trichomonas vaginalis*. Mol Biochem Parasitol 106:11–20
Putignani L, Tait A, Smith HV, Horner D, Tovar J, Tetley L, Wastling JM (2004) Characterization of a mitochondrion-like organelle in *Cryptosporidium parvum*. Parasitology 129:1–18
Pütz S, Dolezal P, Gelius-Dietrich G, Bohacova L, Tachezy J, Henze K (2006) Fe-hydrogenase maturases in the hydrogenosomes of *Trichomonas vaginalis*. Eukaryot Cell 5:579–586
Pyrihová E, Motyčková A, Voleman L, Wandyszewska N, Fišer R, Seydlová G, Roger A, Kolísko M, Doležal P (2018) A single Tim translocase in the mitosomes of *Giardia intestinalis* illustrates convergence of protein import machines in anaerobic eukaryotes. Genome Biol Evol 10:2813–2822
Qiu J, Wenz L-S, Zerbes RM, Oeljeklaus S, Bohnert M, Stroud DA, Wirth C, Ellenrieder L, Thornton N, Kutik S, Wiese S, Schulze-Specking A, Zufall N, Chacinska A, Guiard B, Hunte C, Warscheid B, van der Laan M, Pfanner N, Wiedemann N, Becker T (2013) Coupling of

mitochondrial import and export translocases by receptor-mediated supercomplex formation. Cell 154:596–608

Rada P, Doležal P, Jedelský PL, Bursac D, Perry AJ, Šedinová M, Smíšková K, Novotný M, Beltrán NC, Hrdý I, Lithgow T, Tachezy J (2011) The core components of organelle biogenesis and membrane transport in the hydrogenosomes of *Trichomonas vaginalis*. PLoS One 6:e24428

Rada P, Makki AR, Zimorski V, Garg S, Hampl V, Hrdý I, Gould SB, Tachezy J (2015) N-terminal presequence-independent import of phosphofructokinase into hydrogenosomes of *Trichomonas vaginalis*. Eukaryot Cell 14:1264–1275

Rada P, Makki A, Žárský V, Tachezy J (2019) Targeting of tail-anchored proteins to *Trichomonas vaginalis* hydrogenosomes. Mol Microbiol

Rassow J, Dekker PJ, van Wilpe S, Meijer M, Soll J (1999) The preprotein translocase of the mitochondrial inner membrane: function and evolution. J Mol Biol 286:105–120

Rawlings ND, Barrett AJ (1995) Evolutionary families of metallopeptidases. Methods Enzymol 248:183–228

Regoes A, Egoes A, Zourmpanou D, León-Avila G, van der Giezen M, Tovar J, Hehl AB (2005) Protein import, replication, and inheritance of a vestigial mitochondrion. J Biol Chem 280:30557–30563

Rehling P, Model K, Brandner K, Kovermann P, Sickmann A, Meyer HE, Kühlbrandt W, Wagner R, Truscott KN, Pfanner N (2003) Protein insertion into the mitochondrial inner membrane by a twin-pore translocase. Science 299:1747–1751

Richards TA, van der Giezen M (2006) Evolution of the Isd11-IscS complex reveals a single alpha-proteobacterial endosymbiosis for all eukaryotes. Mol Biol Evol 23:1341–1344

Riordan CE, Ault JG, Langreth SG, Keithly JS (2003) *Cryptosporidium parvum* Cpn60 targets a relict organelle. Curr Genet 44:138–147

Roger AJ, Clark CG, Doolittle WF (1996) A possible mitochondrial gene in the early-branching amitochondriate protist *Trichomonas vaginalis*. Proc Natl Acad Sci U S A 93:14618–14622

Roger AJ, Svärd SG, Tovar J, Clark CG, Smith MW, Gillin FD, Sogin ML (1998) A mitochondrial-like chaperonin 60 gene in *Giardia lamblia*: evidence that diplomonads once harbored an endosymbiont related to the progenitor of mitochondria. Proc Natl Acad Sci U S A 95:229–234

Roger AJ, Muñoz-Gómez SA, Kamikawa R (2017) The origin and diversification of mitochondria. Curr Biol 27:R1177–R1192

Röhl T, Motzkus M, Soll J (1999) The outer envelope protein OEP24 from pea chloroplasts can functionally replace the mitochondrial VDAC in yeast. FEBS Lett 460:491–494

Roise D, Horvath SJ, Tomich JM, Richards JH, Schatz G (1986) A chemically synthesized pre-sequence of an imported mitochondrial protein can form an amphiphilic helix and perturb natural and artificial phospholipid bilayers. EMBO J 5:1327–1334

Rospert S, Junne T, Glick BS, Schatz G (1993) Cloning and disruption of the gene encoding yeast mitochondrial chaperonin 10, the homolog of *E. coli* groES. FEBS Lett 335:358–360

Rospert S, Looser R, Dubaquie Y, Matouschek A, Glick BS, Schatz G (1996) Hsp60-independent protein folding in the matrix of yeast mitochondria. EMBO J 15:764–774

Rout S, Zumthor JP, Schraner EM, Faso C, Hehl AB (2016) An interactome-centered protein discovery approach reveals novel components involved in mitosome function and homeostasis in *Giardia lamblia*. PLoS Pathog 12:e1006036

Ryan KR, Menold MM, Garrett S, Jensen RE (1994) SMS1, a high-copy suppressor of the yeast mas6 mutant, encodes an essential inner membrane protein required for mitochondrial protein import. Mol Biol Cell 5:529–538

Rye HS, Burston SG, Fenton WA, Beechem JM, Xu Z, Sigler PB, Horwich AL (1997) Distinct actions of cis and trans ATP within the double ring of the chaperonin GroEL. Nature 388:792–798

Sandoval P, León G, Gómez I, Carmona R, Figueroa P, Holuigue L, Araya A, Jordana X (2004) Transfer of RPS14 and RPL5 from the mitochondrion to the nucleus in grasses. Gene 324:139–147

Schatz G, Dobberstein B (1996) Common principles of protein translocation across membranes. Science 271:1519–1526
Schleyer M, Schmidt B, Neupert W (1982) Requirement of a membrane potential for the posttranslational transfer of proteins into mitochondria. Eur J Biochem 125:109–116
Schneider A, Behrens M, Scherer P, Pratje E, Michaelis G, Schatz G (1991) Inner membrane protease I, an enzyme mediating intramitochondrial protein sorting in yeast. EMBO J 10:247–254
Schneider RE, Brown MT, Shiflett AM, Dyall SD, Hayes RD, Xie Y, Loo JA, Johnson PJ (2011) The *Trichomonas vaginalis* hydrogenosome proteome is highly reduced relative to mitochondria, yet complex compared with mitosomes. Int J Parasitol 41:1421–1434
Shiota T, Imai K, Qiu J, Hewitt VL, Tan K, Shen H-H, Sakiyama N, Fukasawa Y, Hayat S, Kamiya M, Elofsson A, Tomii K, Horton P, Wiedemann N, Pfanner N, Lithgow T, Endo T (2015) Molecular architecture of the active mitochondrial protein gate. Science 349:1544–1548
Sirrenberg C, Bauer MF, Guiard B, Neupert W, Brunner M (1996) Import of carrier proteins into the mitochondrial inner membrane mediated by Tim22. Nature 384:582–585
Slamovits CH, Fast NM, Law JS, Keeling PJ (2004) Genome compaction and stability in microsporidian intracellular parasites. Curr Biol 14:891–896
Slapeta J, Keithly JS (2004) *Cryptosporidium parvum* mitochondrial-type HSP70 targets homologous and heterologous mitochondria. Eukaryot Cell 3:483–494
Šmíd O, Matušková A, Harris SR, Kučera T, Novotný M, Horváthová L, Hrdý I, Kutějová E, Hirt RP, Embley TM, Janata J, Tachezy J (2008) Reductive evolution of the mitochondrial processing peptidases of the unicellular parasites *Trichomonas vaginalis* and *Giardia intestinalis*. PLoS Pathog 4:e1000243
Söding J, Biegert A, Lupas AN (2005) The HHpred interactive server for protein homology detection and structure prediction. Nucleic Acids Res 33:W244–W248
Söllner T, Griffiths G, Pfaller R, Pfanner N, Neupert W (1989) MOM19, an import receptor for mitochondrial precursor proteins. Cell 59:1061–1070
Stewart M (2007) Molecular mechanism of the nuclear protein import cycle. Nat Rev Mol Cell Biol 8:195–208
Stojanovski D, Pfanner N, Wiedemann N (2007) Import of proteins into mitochondria. Methods Cell Biol 80:783–806
Stojanovski D, Bragoszewski P, Chacinska A (2012) The MIA pathway: a tight bond between protein transport and oxidative folding in mitochondria. Biochim Biophys Acta, Mol Cell Res 1823:1142–1150
Stuart R (2002) Insertion of proteins into the inner membrane of mitochondria: the role of the Oxa1 complex. Biochim Biophys Acta 1592:79–87
Stuart RA, Cyr DM, Craig EA, Neupert W (1994) Mitochondrial molecular chaperones: their role in protein translocation. Trends Biochem Sci 19:87–92
Sutak R, Dolezal P, Fiumera HL, Hrdy I, Dancis A, Delgadillo-Correa M, Johnson PJ, Müller M, Tachezy J (2004) Mitochondrial-type assembly of FeS centers in the hydrogenosomes of the amitochondriate eukaryote *Trichomonas vaginalis*. Proc Natl Acad Sci U S A 101:10368–10373
Szabo A, Langer T, Schröder H, Flanagan J, Bukau B, Hartl FU (1994) The ATP hydrolysis-dependent reaction cycle of the Escherichia coli Hsp70 system DnaK, DnaJ, and GrpE. Proc Natl Acad Sci U S A 91:10345–10349
Tachezy J, Sánchez LB, Müller M (2001) Mitochondrial type iron-sulfur cluster assembly in the amitochondriate eukaryotes *Trichomonas vaginalis* and *Giardia intestinalis*, as indicated by the phylogeny of IscS. Mol Biol Evol 18:1919–1928
Tovar J, Fischer A, Clark CG (1999) The mitosome, a novel organelle related to mitochondria in the amitochondrial parasite *Entamoeba histolytica*. Mol Microbiol 32:1013–1021
Tovar J, León-Avila G, Sánchez LB, Sutak R, Tachezy J, van der Giezen M, Hernández M, Müller M, Lucocq JM (2003) Mitochondrial remnant organelles of *Giardia* function in iron-sulphur protein maturation. Nature 426:172–176

Truscott KN, Kovermann P, Geissler A, Merlin A, Meijer M, Driessen AJ, Rassow J, Pfanner N, Wagner R (2001) A presequence- and voltage-sensitive channel of the mitochondrial preprotein translocase formed by Tim23. Nat Struct Biol 8:1074–1082

Truscott KN, Voos W, Frazier AE, Lind M, Li Y, Geissler A, Dudek J, Müller H, Sickmann A, Meyer HE, Meisinger C, Guiard B, Rehling P, Pfanner N (2003) A J-protein is an essential subunit of the presequence translocase–associated protein import motor of mitochondria. J Cell Biol 163:707–713

Tsaousis AD, Gaston D, Stechmann A, Walker PB, Lithgow T, Roger AJ (2011) A functional Tom70 in the human parasite *Blastocystis* sp.: implications for the evolution of the mitochondrial import apparatus. Mol Biol Evol 28:781–791

van der Giezen M, Rechinger KB, Svendsen I, Durand R, Hirt RP, Fèvre M, Embley TM, Prins RA (1997) A mitochondrial-like targeting signal on the hydrogenosomal malic enzyme from the anaerobic fungus *Neocallimastix frontalis*: support for the hypothesis that hydrogenosomes are modified mitochondria. Mol Microbiol 23:11–21

van der Giezen M, Kiel JA, Sjollema KA, Prins RA (1998) The hydrogenosomal malic enzyme from the anaerobic fungus *Neocallimastix frontalis* is targeted to mitochondria of the methylotrophic yeast *Hansenula polymorpha*. Curr Genet 33:131–135

van der Giezen M, Slotboom DJ, Horner DS, Dyal PL, Harding M, Xue G-P, Embley TM, Kunji ERS (2002) Conserved properties of hydrogenosomal and mitochondrial ADP/ATP carriers: a common origin for both organelles. EMBO J 21:572–579

van der Giezen M, Birdsey GM, Horner DS, Lucocq J, Dyal PL, Benchimol M, Danpure CJ, Embley TM (2003) Fungal hydrogenosomes contain mitochondrial heat-shock proteins. Mol Biol Evol 20:1051–1061

van der Giezen M, León-Avila G, Tovar J (2005) Characterization of chaperonin 10 (Cpn10) from the intestinal human pathogen *Entamoeba histolytica*. Microbiology 151:3107–3115

van der Laan M, Chacinska A, Lind M, Perschil I, Sickmann A, Meyer HE, Guiard B, Meisinger C, Pfanner N, Rehling P (2005) Pam17 is required for architecture and translocation activity of the mitochondrial protein import motor. Mol Cell Biol 25:7449–7458

van der Laan M, Rissler M, Rehling P (2006a) Mitochondrial preprotein translocases as dynamic molecular machines. FEMS Yeast Res 6:849–861

van der Laan M, Wiedemann N, Mick DU, Guiard B, Rehling P, Pfanner N (2006b) A role for Tim21 in membrane-potential-dependent preprotein sorting in mitochondria. Curr Biol 16:2271–2276

van Wilpe S, Ryan MT, Hill K, Maarse AC, Meisinger C, Brix J, Dekker PJ, Moczko M, Wagner R, Meijer M, Guiard B, Hönlinger A, Pfanner N (1999) Tom22 is a multifunctional organizer of the mitochondrial preprotein translocase. Nature 401:485–489

Vinayak S, Pawlowic MC, Sateriale A, Brooks CF, Studstill CJ, Bar Peled Y, Cipriano MJ, Striepen B (2015) Genetic modification of the diarrhoeal pathogen *Cryptosporidium parvum*. Nature 523:477

von Heijne G (1990) The signal peptide. J Membr Biol 115:195–201

von Heijne G, Steppuhn J, Herrmann RG (1989) Domain structure of mitochondrial and chloroplast targeting peptides. Eur J Biochem 180:535–545

Voncken FGJ, Boxma B, van Hoek AHAM, Akhmanova AS, Vogels GD, Huynen M, Veenhuis M, Hackstein JHP (2002) A hydrogenosomal [Fe]-hydrogenase from the anaerobic chytrid *Neocallimastix* sp. L2. Gene 284:103–112

Voos W, Röttgers K (2002) Molecular chaperones as essential mediators of mitochondrial biogenesis. Biochim Biophys Acta 1592:51–62

Voulhoux R, Bos MP, Geurtsen J, Mols M, Tommassen J (2003) Role of a highly conserved bacterial protein in outer membrane protein assembly. Science 299:262–265

Waizenegger T, Habib SJ, Lech M, Mokranjac D, Paschen SA, Hell K, Neupert W, Rapaport D (2004) Tob38, a novel essential component in the biogenesis of β-barrel proteins of mitochondria. EMBO Rep 5:704–709

Waizenegger T, Schmitt S, Zivkovic J, Neupert W, Rapaport D (2005) Mim1, a protein required for the assembly of the TOM complex of mitochondria. EMBO Rep 6:57–62
Waller RF, Jabbour C, Chan NC, Celik N, Likić VA, Mulhern TD, Lithgow T (2009) Evidence of a reduced and modified mitochondrial protein import apparatus in microsporidian mitosomes. Eukaryot Cell 8:19–26
Wenz L-S, Ellenrieder L, Qiu J, Bohnert M, Zufall N, van der Laan M, Pfanner N, Wiedemann N, Becker T (2015) Sam37 is crucial for formation of the mitochondrial TOM–SAM supercomplex, thereby promoting β-barrel biogenesis. J Cell Biol 210:1047–1054
Werhahn W, Niemeyer A, Jänsch L, Kruft V, Schmitz UK, Braun H (2001) Purification and characterization of the preprotein translocase of the outer mitochondrial membrane from *Arabidopsis*. Identification of multiple forms of TOM20. Plant Physiol 125:943–954
Wickner W, Schekman R (2005) Protein translocation across biological membranes. Science 310:1452–1456
Wiedemann N, Pfanner N (2017) Mitochondrial machineries for protein import and assembly. Annu Rev Biochem 86:685–714
Wiedemann N, Pfanner N, Ryan MT (2001) The three modules of ADP/ATP carrier cooperate in receptor recruitment and translocation into mitochondria. EMBO J 20:951–960
Wiedemann N, Kozjak V, Chacinska A, Schönfisch B, Rospert S, Ryan MT, Pfanner N, Meisinger C (2003) Machinery for protein sorting and assembly in the mitochondrial outer membrane. Nature 424:565–571
Wiedemann N, Truscott KN, Pfannschmidt S, Guiard B, Meisinger C, Pfanner N (2004) Biogenesis of the protein import channel Tom40 of the mitochondrial outer membrane. J Biol Chem 279:18188–18194
Williams BAP, Hirt RP, Lucocq JM, Embley TM (2002) A mitochondrial remnant in the microsporidian *Trachipleistophora hominis*. Nature 418:865–869
Wimley WC (2003) The versatile beta-barrel membrane protein. Curr Opin Struct Biol 13:404–411
Xu Z, Horwich AL, Sigler PB (1997) The crystal structure of the asymmetric GroEL–GroES–(ADP)7 chaperonin complex. Nature 388:741–750
Xu P, Widmer G, Wang Y, Ozaki LS, Alves JM, Serrano MG, Puiu D, Manque P, Akiyoshi D, Mackey AJ, Pearson WR, Dear PH, Bankier AT, Peterson DL, Abrahamsen MS, Kapur V, Tzipori S, Buck GA (2004) The genome of *Cryptosporidium hominis*. Nature 431:1107–1112
Yaffe MP, Ohta S, Schatz G (1985) A yeast mutant temperature-sensitive for mitochondrial assembly is deficient in a mitochondrial protease activity that cleaves imported precursor polypeptides. EMBO J 4:2069–2074
Yamamoto H, Esaki M, Kanamori T, Tamura Y, Nishikawa S, Endo T (2002) Tim50 is a subunit of the TIM23 complex that links protein translocation across the outer and inner mitochondrial membranes. Cell 111:519–528
Zara V, Dolce V, Capobianco L, Ferramosca A, Papatheodorou P, Rassow J, Palmieri F (2007) Biogenesis of eel liver citrate carrier (CIC): negative charges can substitute for positive charges in the presequence. J Mol Biol 365:958–967
Žárský V, Doležal P (2016) Evolution of the Tim17 protein family. Biol Direct 11:54
Zimorski V, Major P, Hoffmann K, Brás XP, Martin WF, Gould SB (2013) The N-terminal sequences of four major hydrogenosomal proteins are not essential for import into hydrogenosomes of *Trichomonas vaginalis*. J Eukaryot Microbiol 60:89–97

Structure of the Hydrogenosome

Marlene Benchimol and Wanderley de Souza

Contents

1 Introduction 86
2 Where Are Hydrogenosomes Found? 86
3 The Hydrogenosome Shape 90
4 Hydrogenosome Size 90
5 Hydrogenosome Components 90
 5.1 The Hydrogenosomes Envelope 90
 5.2 The Peripheral Vesicle 93
 5.3 The Matrix of the Hydrogenosome 99
6 Fungal Hydrogenosomes 102
7 Proximity with Other Cellular Structures 102
 7.1 Hydrogenosomes and Endoplasmic Reticulum 103
8 Hydrogenosome Autophagy 104
9 Hydrogenosome Division 104
10 Hydrogenosome Behavior in the Cell Cycle 106
11 Hydrogenosomes Connection to Microtubules 106
12 Immunolabeling 107
References 107

Abstract Hydrogenosomes are very interesting organelles found in non-mitochondrial organisms. They display similarities and differences with mitochondria. Hydrogenosomes are spherical or slightly elongated organelles, although very elongated hydrogenosomes are also found. They measure between 200 and 1000 nm but under stress conditions can reach 2 μm. Hydrogenosomes divide in three different ways, like mitochondria: segmentation, partition, and the heart form. They may divide at any phase of the cell cycle. Nucleoid or electron-dense deposits

M. Benchimol (✉)
Universidade do Grande Rio, UNIGRANRIO, Duque de Caxias, Rio de Janeiro, Brazil

W. de Souza
Laboratório de Ultraestrutura Celular Hertha Meyer, Instituto de Biofísica Carlos Chagas Filho, Núcleo de Biologia Estrutural e Bioimagens-CENABIO, Universidade Federal do Rio de Janeiro, Rio de Janeiro, Brazil

J. Tachezy (ed.), *Hydrogenosomes and Mitosomes: Mitochondria of Anaerobic Eukaryotes*, Microbiology Monographs 9,
https://doi.org/10.1007/978-3-030-17941-0_4

are not considered part of the normal structure of the hydrogenosomes. Hydrogenosomes are surrounded by two closely apposed membranes and present a granular matrix. Hydrogenosomes have one or multiple peripheral vesicles, which incorporate calcium. The peripheral vesicle can be isolated from the hydrogenosomal matrix and is considered a distinct hydrogenosomal compartment. Dysfunctional hydrogenosomes are removed by an autophagic process and further digested in lysosomes. Similarities and differences with mitochondria are presented.

1 Introduction

During the evolutionary process, new structures and organelles appeared in eukaryotic microorganisms (Fig. 1a–d). One example is the hydrogenosome, initially found in protozoa of the Trichomonadida order and which contains enzymes that participate in the metabolism of pyruvate formed during glycolysis and was the site of molecular hydrogen and ATP formation (reviewed in Müller 1993). In trichomonads, hydrogenosomes have been recognized by light microscopists for a long time, as paraxostylar and paracostal granules, due to their proximity to the axostyle (bundle of microtubules) and the costa, a periodic proteinaceous structure (Fig. 1c–d). Only biochemistry revealed their functional significance, which showed molecular hydrogen production as a metabolic end product. Consequently, they were named hydrogenosomes by Lindmark and Müller in 1973. Hydrogenosomes present an unusual function: under anaerobic conditions they produce molecular hydrogen by oxidizing pyruvate to malate (Müller 1993). Indeed, presently it has been established that hydrogenosomes are members of a large group of organelles collectively classified according to their role in energy metabolism designed as mitochondria-related organelles (MROs) (reviewed in Makiuchi and Nozaki 2014; Stairs et al. 2015). They are classified into five different types: Class 1, typical mitochondria only capable of generating ATP through oxidative phosphorylation and morphologically characterized by the presence of two membranes and the formation of internal cristae; Class 2, those which are, additionally, capable of functioning anaerobically using alternative electron acceptors such as fumarate; Class 3, possesses both an electron transport chain and a bacterial-like anaerobic ATP-generation pathway; Class 4, the hydrogenosomes that possess the anaerobic ATP-generation pathway but have most components of the electron transport chain such that they are no longer able to generate ATP through aerobic phosphorylation; and Class 5, lack any role in ATP generation and is known as mitosomes.

2 Where Are Hydrogenosomes Found?

The Class 4 MROs, known as hydrogenosome, is presently identified based on several criteria such as:

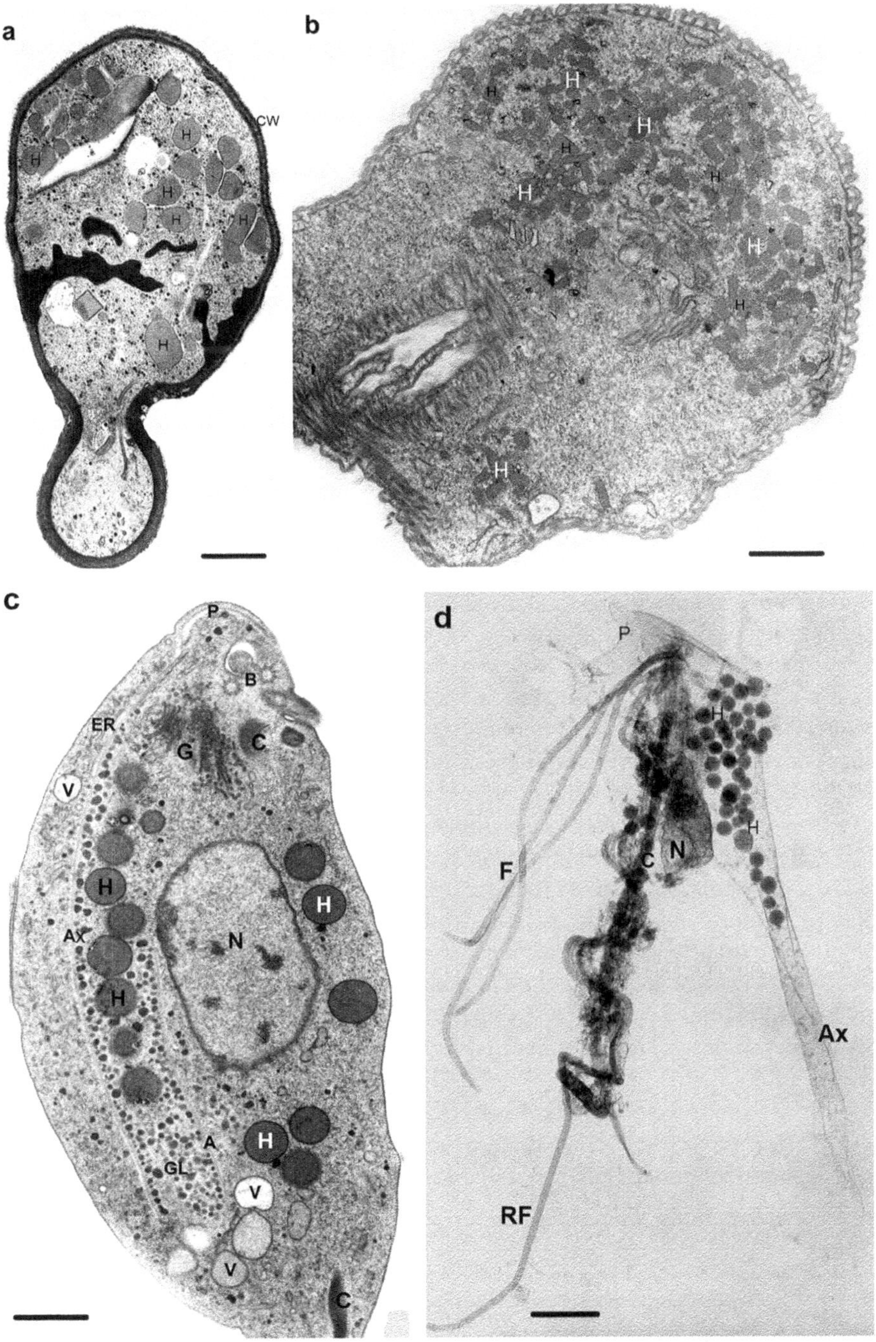

Fig. 1 General view of different cells that possess hydrogenosomes (H): (**a**) the anaerobic fungus *Neocallimastix*, (**b**) a rumen ciliate, and (**c–d**) *Tritrichomonas foetus* as seen by transmission electron microscopy (**a–c**) and high-voltage electron microscope (**d**). In (**d**), the plasma membrane of *T. foetus* was removed, and the preparation was critical point dried and observed in a high-

(a) Its morphology as seen by transmission electron microscopy of thin sections, where two membranes surrounding the organelle can be clearly identified, especially if the cells are fixed in calcium-containing glutaraldehyde solution and post-fixed in osmium tetroxide solution containing potassium ferrocyanide (Benchimol and De Souza 1983).
(b) The presence of a set of proteins displaying enzymatic activity, such as pyruvate-ferredoxin oxidoreductase, (Fe-Fe)-hydrogenase and associated hydrogenase maturases, acetate:succinate CoA-transferase, acetyl-CoA synthetase, succinyl-CoA synthase, and arginine dihydrolase pathway, that has an analogous function to the urea cycle in mitochondria-containing cell, while other enzymes of this pathway (ornithine carbamyl transferase and carbamate kinase) are localized in the cytosol (Morada et al. 2011). Three gene sequences of the hydrogenosomal enzyme were identified in the *T. vaginalis* genome all having putative mitochondrial targeting sequence. The enzyme was found enriched in a subcellular fraction of hydrogenosomes that could be identified based on their enzymatic activity both in whole cells and in isolated subcellular fractions.
(c) Identification of proteins based on genomic, RNAseq, and proteomic analysis.

Investigation the targeting behavior of *T. vaginalis* hydrogenosome protein in *Saccharomyces* and vice versa showed that hydrogenosomes import yeast mitochondrial proteins even in the absence of their native NTSs (N-terminal targeting sequences that are recognized by the translocon of the mitochondrial outer membrane—TOM40) but do not import yeast cytosolic proteins (Garg et al. 2015). Conversely, yeast mitochondria import hydrogenosomal proteins with and without their short NTSs. Thus, it was demonstrated that the two membranes that surround hydrogenosomes harbor many homologs of the TOM/TIM machinery.

Rada et al. (2015) studied the cellular localization of two glycolytic enzymes in *T. vaginalis*: PPi-dependent phosphofructokinase, which is the main glycolytic PFK activity of the protist, and ATP-dependent PFK. TvPPi-PFK was detected predominantly in the cytosol, as expected, while all four TvATP-PFK paralogues were imported into *T. vaginalis* hydrogenosomes although none of them possesses an NTS (see Fig. 2 in the paper of this group).

One important evidence of symbiotic origin is the presence of cardiolipin in hydrogenosomes, which was detected by Rosa et al. (2006).

In addition to protozoa of the Trichomonadida order (Fig. 1c) where hydrogenosomes were initially described (Lindmark and Müller 1973; Kulda et al. 1987), they have been found in anaerobic protists including excavated *Sawyeria marylandensis* (Barbera et al. 2010) and *Psalteriomonas lanterna* (de Graaf et al.

Fig. 1 (continued) voltage electron microscopy with 1000 KV of electron acceleration. Note that hydrogenosomes are preferentially located along the axostyle (Ax) and costa (C). *ER* endoplasmic reticulum, *GL* glycogen granules, *F* anterior flagella, *RF* recurrent flagellum, *N* nucleus, *P* pelta, *V* vacuoles. Bars = (**a**) 2 μm; (**b**) 300 nm; (**c**) 500 nm; (**d**) 1 μm. (**a**) From Benchimol et al. (1997), (**b–d**) Benchimol, unpublished

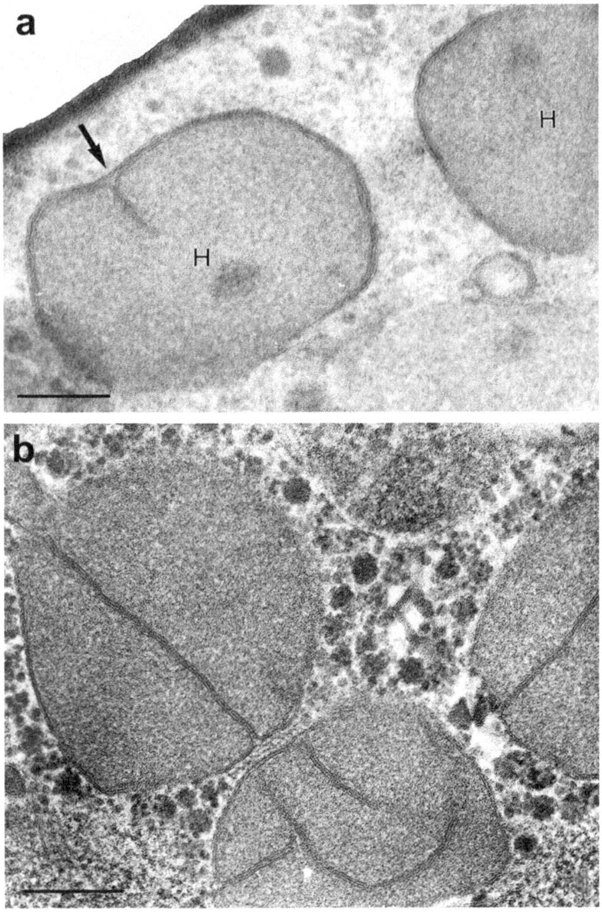

Fig. 2 Thin sections of the fungal hydrogenosome (H) from *Neocallimastix frontalis*. In (**a**), there is one hydrogenosome in the process of division, and the inner hydrogenosomal membrane is in process of septum formation. Note that there is another hydrogenosome with a double membrane, which is not in process of division. In (**b**), all the hydrogenosomes are dividing, presenting internal septa. Bars = 100 nm. From Benchimol et al. (1997)

2009) and in the salmon parasite *Spironucleus salmonicida* (Jerstrom-Hultqvist et al. 2013) and *S. vortens* (Horner et al. 2000). In addition, hydrogenosomes were also detected in some anaerobic fungi (Fig. 1a) such as *Neocallimastix* spp. (Bowman et al. 1992; Yarlett et al. 1986; Marvin-Sikkema et al. 1992; Benchimol et al. 1997) and some anaerobic ciliates (Yarlett et al. 1981), in *Acanthamoeba castellanii*, in the free-living anaerobic amoebozoan *Mastigamoeba balamuthi*, in free-living anaerobic jakobid excavate *Stygiella incarcerata*, in the anaerobic heterolobosid amoeba *Sawyeria marylandensis* (from anaerobic sediments), in the free-living anaerobic protist *Cantina marsupialis* in rumen ciliates (Fig. 1b) (Yarlett et al. 1981, 1984; Snyers et al. 1982), and also in some free-living ciliates (van Bruggen et al. 1984). The hydrogenosomes-containing microorganisms do not present typical mitochondria (Embley et al. 2003a, b). Until today, hydrogenosomes have not been found in multicellular animals or plants nor in other anaerobic protists, such as amoebas and giardias. The most extensive studies of this organelle have been carried out in the trichomonad species. The distribution of hydrogenosomes in different phyla raised questions concerning their evolutionary origin.

3 The Hydrogenosome Shape

In some fungi such as *Neocallimastix frontalis* (Figs. 1a, 2, and 3b) (Yarlett et al. 1986; Benchimol et al. 1997; van der Giezen et al. 1997), in rumen ciliates (Fig. 1b) (Yarlett et al. 1981, 1984), in trichomonads (Figs. 1c–d, 3, 4, 5, 6a, and 7) (Benchimol et al. 1996a), and in free-living ciliates (Finlay and Fenchel 1989) hydrogenosomes are spherical or slightly elongated granules. Using freeze-etching, filaments are seen connecting hydrogenosomes to cytoskeletal structures such as the axostyle and costa structures (Benchimol et al. 2000). In some cells, hydrogenosomes are not spherical but are very elongated structures, such as those found in *Monocercomonas* sp. (Diniz and Benchimol 1998), which can reach 2.0 μm in length (Fig. 4c).

4 Hydrogenosome Size

Hydrogenosome sizes vary according to the species or if the cell is under stress or drug treatment (Figs. 8 and 9) (Benchimol 1999, 2001; Madeiro and Benchimol 2004). In trichomonads without drug treatment (Figs. 1, 3a, 4, 5, 6a, 7 and 10), hydrogenosomes present an average diameter of 300 nm (Figs. 3a and 4a, b), but may reach 2 μm in *Monocercomonas* sp. (Diniz and Benchimol 1998) (Fig. 4c) or when trichomonads are subjected to drug treatment. In this last situation when drugs such as metronidazol, hydroxyurea, cytochalasin and others were used, abnormal or bizarre shapes and sizes were found (Figs. 8 and 9) (Benchimol 1999, 2001; Ribeiro et al. 2002; Mariante et al. 2003). It was reported that cells expressing the dynamin dominant negative protein (Wexler-Cohen et al. 2014) display hydrogenosomes increased in size and reduced in number thus indicating a role of DRP in organelle division.

5 Hydrogenosome Components

5.1 The Hydrogenosomes Envelope

It has been shown that trichomonads (Figs. 3a, 4a–b, 11c, and 12b) and fungal hydrogenosomes are enveloped by two unit membranes (Figs. 2 and 3b) (Benchimol and De Souza 1983; van der Giezen et al. 1997; Díaz and De Souza 1997). These membranes are very thin and very closely opposed to each other (Figs. 2, 3, 4, and

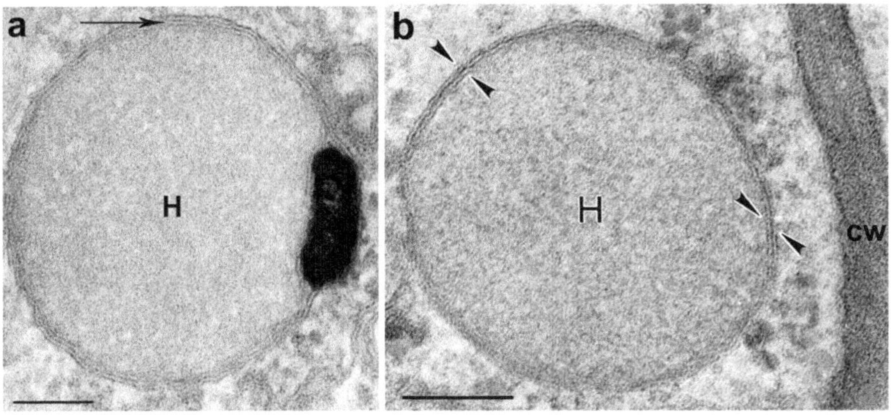

Fig. 3 Thin sections of a *T. foetus* (**a**) and a fungal (**b**) hydrogenosome (H) from *Neocallimastix frontalis*. Both hydrogenosomes are clearly enveloped by a double membrane [arrow in (**a**), arrowheads in (**b**)]. Invaginations of the hydrogenosome membrane were occasionally observed (arrow in **a**). The black spot in (**a**) is a calcium deposit in the peripheral vesicle (asterisks). *CW* cell wall. Bars = 100 nm. (**a**) From Benchimol, unpublished; (**b**) from Benchimol et al. (1997)

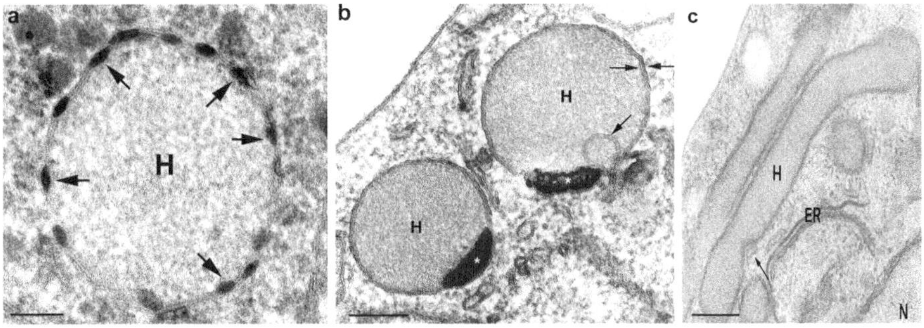

Fig. 4 Routine preparation of a hydrogenosome (H) in *Trichomonas vaginalis* (**a**), *Tritrichomonas foetus* (**b**), and *Monocercomonas* sp. (**c**). Note that in *T. foetus* hydrogenosome is spherical, enveloped by a double membrane (arrows), and presents a single peripheral vesicle, whereas in *T. vaginalis* several peripheral vesicles are seen surrounding the organelle (arrows), and in *Monocercomonas* the hydrogenosome is very elongated. *ER* endoplasmic reticulum, *N* nucleus. Bars = 100 nm. From Benchimol (2001), from Diniz and Benchimol (1998)

12b). As a general rule, no space is observed between the two membranes. Each membrane has a thickness of 6 nm and presents a certain undulation (Benchimol and De Souza 1983; Benchimol et al. 1996a). The presence of $CaCl_2$ in the fixation solution, as well as the use of reduced osmium, is important for good visualization of the two membranes (Benchimol and De Souza 1983; Benchimol et al. 1996a) (Figs. 2, 3, and 4). When the sections are thicker, the two membranes are hardly visualized.

Conventional freeze-fracture replicas revealed the presence of the two membranes enveloping the hydrogenosomes presenting a different number and

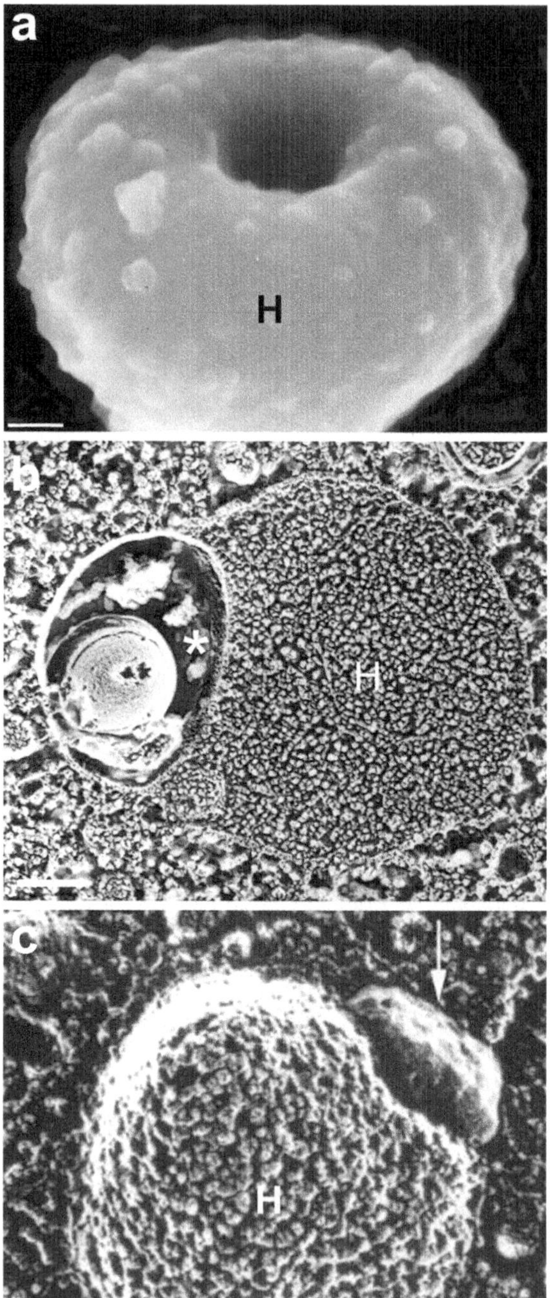

Fig. 5 Different views of *T. foetus* hydrogenosomes (H) after FESEM (**a**) and freeze-etching (**b**–**c**). An isolated hydrogenosome obtained from *T. foetus* observed by field emission scanning electron microscopy (FESEM), where details of its surface can be seen. (**b**) shows a calcium deposit in the peripheral vesicle (asterisk), and (**c**) shows that the peripheral vesicle (arrow) presents a smooth surface, distinct from the organelle body. Bars = 50 nm. From Benchimol (2000)

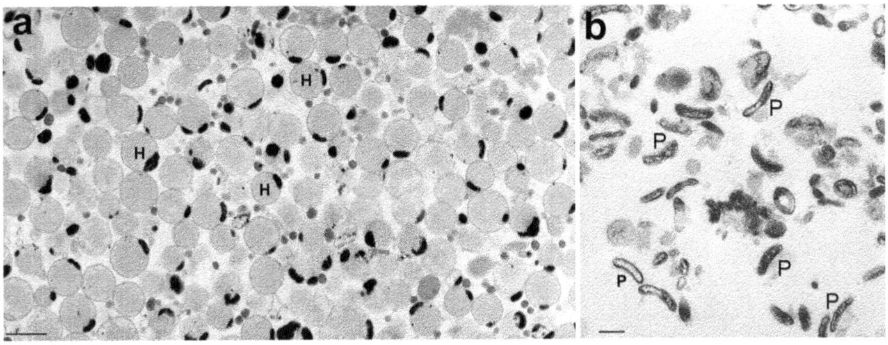

Fig. 6 (**a**) Hydrogenosomes isolated by Percoll-sucrose density centrifugation. The hydrogenosomes (H) are seen as spherical organelles presenting a peripheral vesicle. Bar = 300 nm. (**b**) Hydrogenosomes peripheral vesicles (P) isolated from a pure hydrogenosomes fraction (**a**) show that they form a distinct sub-compartment. Bar = 100 nm. Courtesy of Dr. José Andrés Morgado Díaz

distribution of intramembranous particles (Fig. 11a–c) (Benchimol et al. 1996a; Benchimol 2001). Four fracture faces were identified: two concave faces representing the P faces of the outer and the inner membranes and two convex faces representing the E faces of the two membranes (Fig. 11a, c). The P and E faces of the outer membrane were frequently found. Although no quantitative analyses were carried out, the P face seems to have a higher particle density than the E face. The E face of the inner membrane was observed only in a few cases (Benchimol et al. 1996a). Special arrangement of intramembranous particles (rosettes) was found only when isolated hydrogenosomes were freeze-fractured, after cell fractionation (Fig. 11b) (Benchimol 2000).

Invaginations of the hydrogenosome membrane delimitating inner compartments were observed when trichomonads where drug treated (Fig. 8). Some of the compartments had the same morphology and electron density as the hydrogenosomal matrix, while others had a lower density and presented tubular structures (Benchimol et al. 1996a).

Ciliate hydrogenosomes in *Metopus contortus* and *Cyclidium porcatum* present internal membranes and look like mitochondria (Fenchel and Finlay 1995). In addition, these hydrogenosomes are also calcium stores and displayed a membrane potential, which are similar to features found in mitochondria (Biagini et al. 1997).

5.2 *The Peripheral Vesicle*

A special compartment is found at the periphery of most hydrogenosomes (Figs. 3a, 4a–b, 5b–c, 6, 11b–c, 12, 13b, and 14a, c). *T. foetus* hydrogenosomes present one or two peripheral vesicles (Figs. 3a and 4b), whereas *T. vaginalis* hydrogenosomes

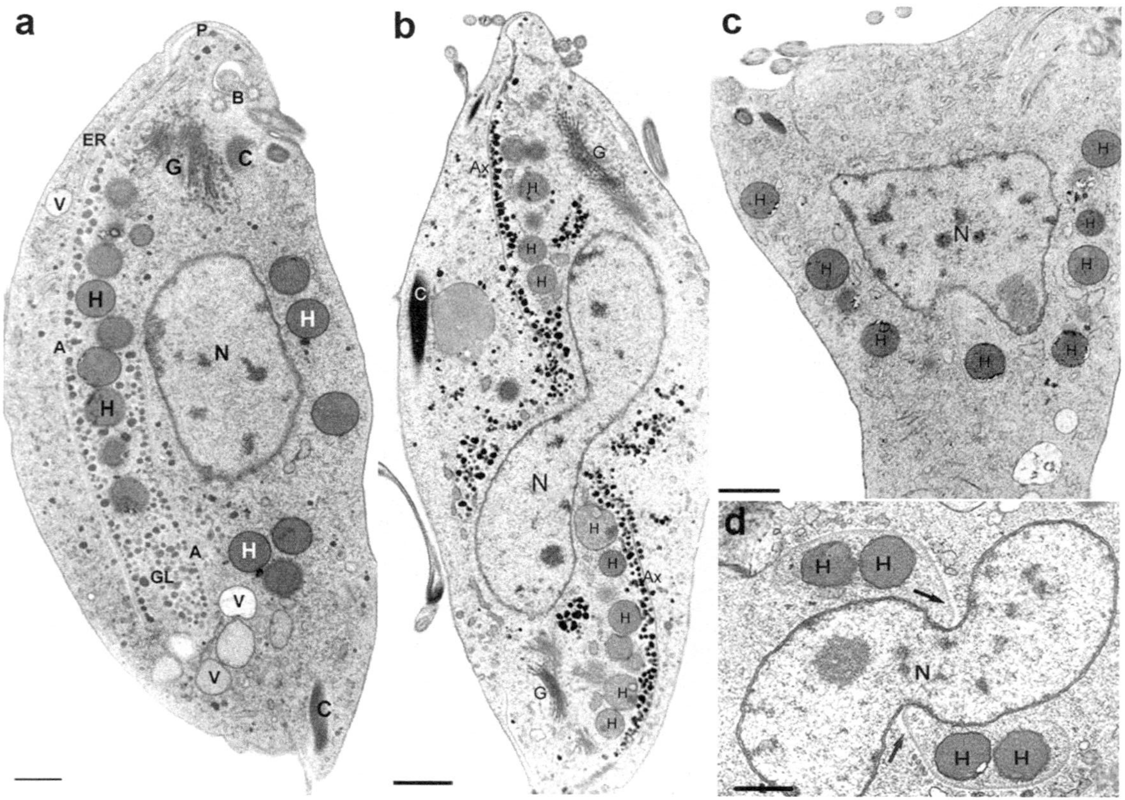

Fig. 7 General aspect of *T. foetus* in interphase (**a**) and under division (**b–d**). Notice that in interphase, the hydrogenosomes are aligned on the costa and axostyle (**a**), whereas during the division, they are close to the nucleus (**b–d**). Bars = 300 nm. (**a**) From Benchimol, unpublished; (**b–c**) from Benchimol and Engelke (2003); (**d**) Ribeiro et al. (2000)

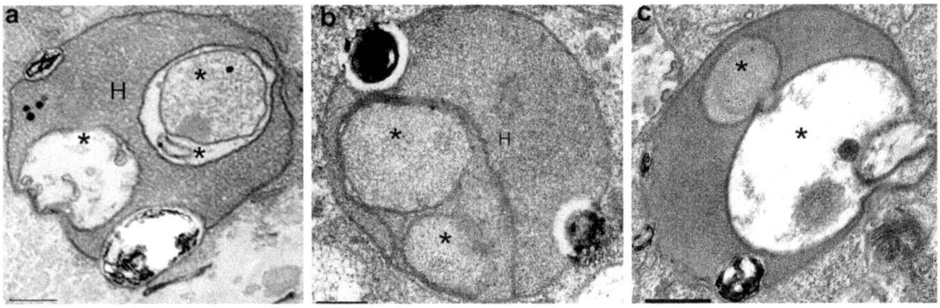

Fig. 8 Abnormal hydrogenosomes. Note that the hydrogenosomes (H) are not spherical as in the routine preparations. They are giant and present internal compartments and abnormal peripheral vesicles. Dense spots are seen in the hydrogenosomal matrix, which represent calcium deposits. When cells are submitted to stress conditions, such as incubation with fibronectin (Benchimol 2001) or other drugs (Madeiro and Benchimol 2004), the hydrogenosomes present such abnormal shapes and sizes, reaching 2 μm. Enlarged vesicles and internal membranes are also seen (asterisks). Bars = 150 nm. From Benchimol et al. (1996a)

exhibit several vesicles at the organelle periphery (Fig. 4a). Thus, the number of these compartments can be useful in taxonomic studies (Benchimol, unpublished). The hydrogenosome peripheral vesicle varies in size and electron density. Morphometric analysis showed that it represents 8.6% of the whole organelle in *T. foetus*. The peripheral vesicle is surrounded along its full extension by two closely apposed unit membranes. The inner portion of this compartment is completely distinct from the hydrogenosomal matrix (Figs. 5c and 6b). In some preparations it appears empty (Fig. 13b), whereas in others it presents a certain content, such as when the cells are fixed in a solution containing calcium (Figs. 3a, 4a–b, 6a, 12a–b, and 14a, c) or when the cells are submitted to freeze-fracture and deep-etching (Fig. 5b). This compartment is occupied by electron-dense reaction products after various cytochemical detections for calcium (Figs. 3a and 4a–b), phosphatases such as acid phosphatase (Fig. 15b), Mg^{++}-ATPase, or 5′-nucleotidase (Queiroz et al. 1991).

Carbohydrates were cytochemically detected in hydrogenosome membranes using the periodic acid-thiosemicarbazide-silver proteinate technique (Fig. 13a) and gold-labeled lectins, such as WGA (Fig. 13b) (Benchimol et al. 1996a; Benchimol and Bernardino 2002). The membrane of the peripheral vesicle compartment is intensely labeled by WGA-gold, indicating the presence of *N*-acetyl-glucosamine (Fig. 13b). Interestingly, the membrane lining the outer portion of the hydrogenosome is not labeled (Benchimol et al. 1996a). The membrane surrounding the peripheral vesicle is smooth when analyzed in freeze-fracture and deep-etched cells (Figs. 5c and 11c) and thus rather distinct from the whole organelle, which is rough (Fig. 5a, c). With field emission scanning electron microscopy (FESEM), the hydrogenosomal surface exhibits a distinct morphology with a rough surface (Fig. 5a) (Benchimol 2000).

In a study by Díaz and De Souza (1997), the authors were able to purify a hydrogenosomal fraction (Fig. 6a) and then another hydrogenosomal subfraction

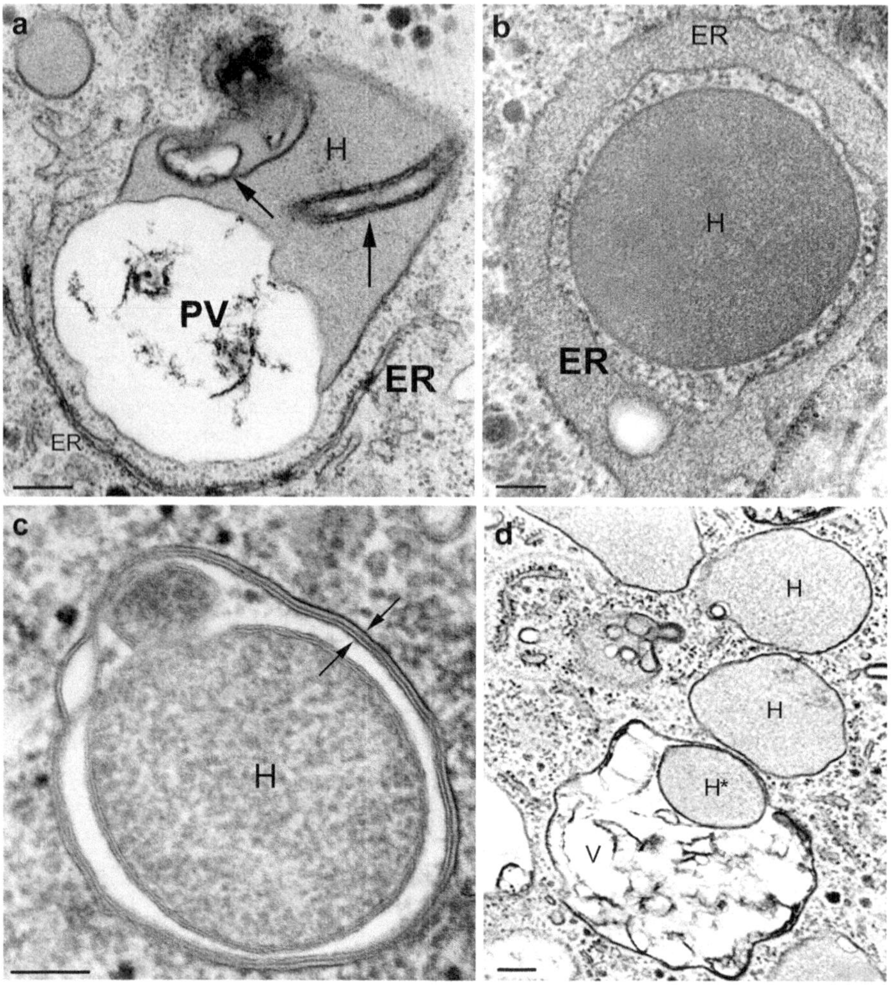

Fig. 9 Hydrogenosomes in process of autophagy. (**a**) View of a hydrogenosome from a *T. foetus* cell grown in a culture medium containing 4 mM hydroxyurea for 15 h. One giant hydrogenosome presenting an enlarged peripheral vesicle (PV) and internal membranes (arrows) is seen. Profiles of the endoplasmic reticulum (ER) are seen close and surrounding the organelle. Bar = 80 nm. (**b**) Thin section of a hydrogenosome (H) enclosed by the rough endoplasmic reticulum (ER). Bar = 50 nm. (**c**) Observation of a double-layered autophagic vacuole (arrows), containing an intact hydrogenosome (H) after high-pressure freezing and freeze substitution. *T. foetus* cell was grown under normal conditions. Arrows point to the double membrane of the enclosing vacuole. Bar = 200 nm. (**d**) Routine preparation of *T. foetus* without any drug treatment. Several hydrogenosomes (H) are seen all around, while one of them (asterisk) is seen inside a vacuolar lysosome-like structure (V), suggesting an autophagic process. Bar = 200 nm. From Benchimol (1999)

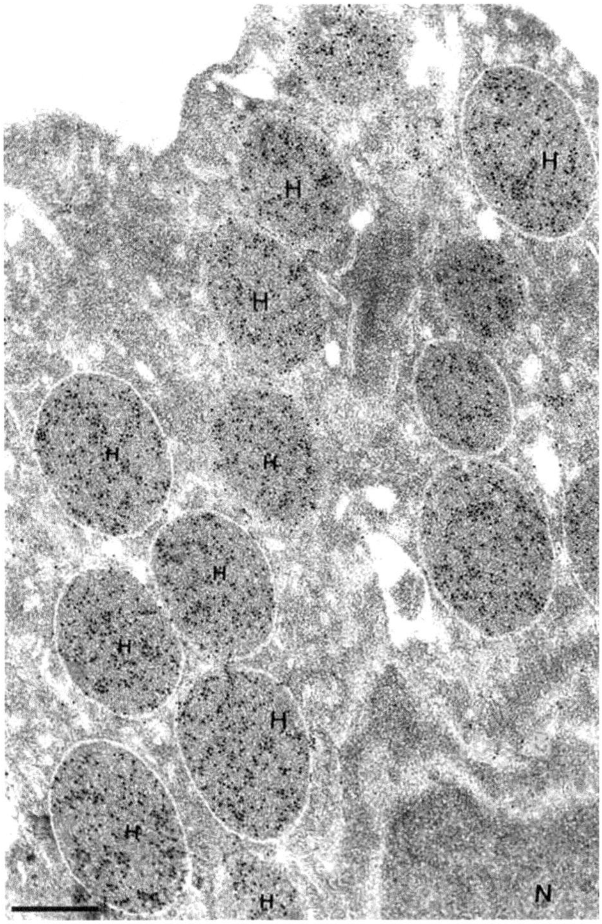

Fig. 10 Immuno-electron microscopy showing detection of malic enzyme in the hydrogenosomes (H) of *T. vaginalis*. *N* nucleus. Bar = 100 nm (Benchimol, unpublished)

containing only the peripheral vesicles (Fig. 6b), showing that it is a distinct compartment. The authors used *Tritrichomonas foetus* and showed that a further treatment with proteinase K solubilized the matrix components, leaving a pure peripheral vesicle fraction (Fig. 6b). The isolated peripheral vesicles maintained their flattened morphology, suggesting that each individual vesicle has its own inherent structural framework (Fig. 6b). After hydrogenosomal membrane solubilization, the hydrogenosome matrix appeared attached to the peripheral vesicle. SDS-PAGE showed that proteins of 66, 45, and 32 kDa were localized in the peripheral vesicle (Díaz and De Souza 1997). Western blot analysis revealed the presence of glycoproteins, with a major one of 45 kDa in the peripheral vesicle of the

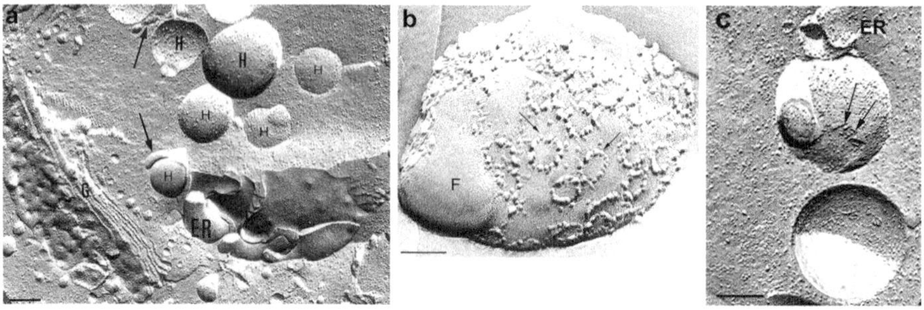

Fig. 11 Freeze-fracture images of hydrogenosomes from *T. foetus*. (**a**) shows a fractured cell showing a prominent Golgi (G) with several lamellae and fenestrae, as well as profiles of endoplasmic reticulum (ER) in close proximity (arrows) with hydrogenosomes (H). Bar = 100 nm. (**b**) Hydrogenosomes from an isolated fraction observed by freeze-fracture. Note the clusters of intramembranous particles forming rosettes (arrow). The peripheral vesicle is smooth and does not present clusters of particles or rosettes. Bar = 50 nm. (**c**) Shows two freeze-fractured hydrogenosomes exhibiting different fracture planes (arrows). Bar = 100 nm. From Benchimol et al. (1996a)

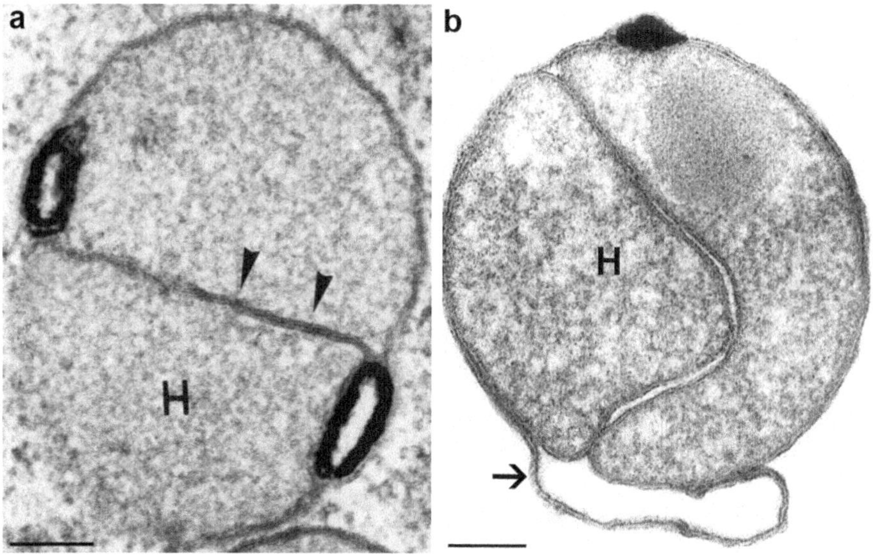

Fig. 12 Thin section of two hydrogenosomes (H) dividing via the partition process. The hydrogenosome (H) becomes larger and an invagination of the inner hydrogenosomal membrane is observed (arrow), gradually dividing the hydrogenosomal matrix in two compartments. (**b**) Shows that initially the inner membrane separates the hydrogenosome (H) in two compartments, but they are still joined by the outer hydrogenosomal membrane (arrows). Bars = 100 nm. From Benchimol et al. (1996b)

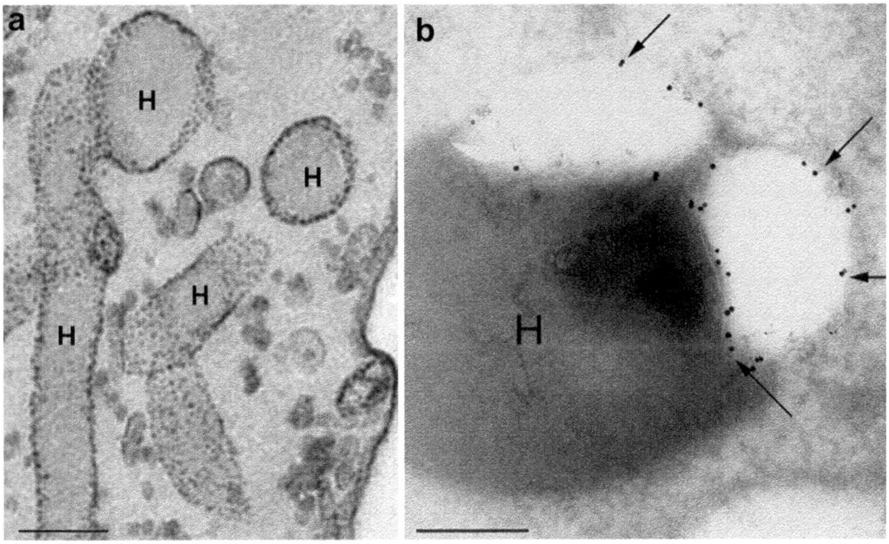

Fig. 13 Carbohydrates in hydrogenosomes. *Monocercomonas* sp. after the Thiéry technique (**a**). The hydrogenosomal membranes are positive for carbohydrates. (**b**) Shows *T. foetus* cryosection labeled with gold-conjugated WGA. Hydrogenosome (H) shows that the membrane lining the peripheral vesicle, but not other portions of the organelle, is labeled. Bars = 100 nm. (**a**) from Diniz and Benchimol (1998); (**b**) from Benchimol et al. (1996a)

hydrogenosome. The authors concluded that the peripheral vesicle is a distinct hydrogenosomal compartment (Díaz and De Souza 1997).

It was shown that cells transfected with the mutant protein dynamin TVAG_350040 K.38A has a larger number of peripheral vesicles on the surface of the hydrogenosomes (Wexler-Cohen et al. 2014). The average number of PV increased from 0.48 in cells transfected the EV to 1.1 in cells transfected with above-indicated protein.

5.3 *The Matrix of the Hydrogenosome*

The hydrogenosome matrix is homogeneous, presenting a granular appearance, which is different from the cytoplasmic matrix (Figs. 1, 3, 4, and 5b). The hydrogenosome matrix was described in previous studies as homogeneously granular, occasionally presenting a dense amorphous or crystalline core, also known as nucleoid (Fig. 16a) (Honigberg and Brugerolle 1990). Observations indicated that this core is not a usual structure, appearing however either when the protozoa were incubated in the presence of drugs or when good fixation is not achieved (Benchimol et al. 1996a). The electron-dense core is frequently seen in not-well-preserved cells,

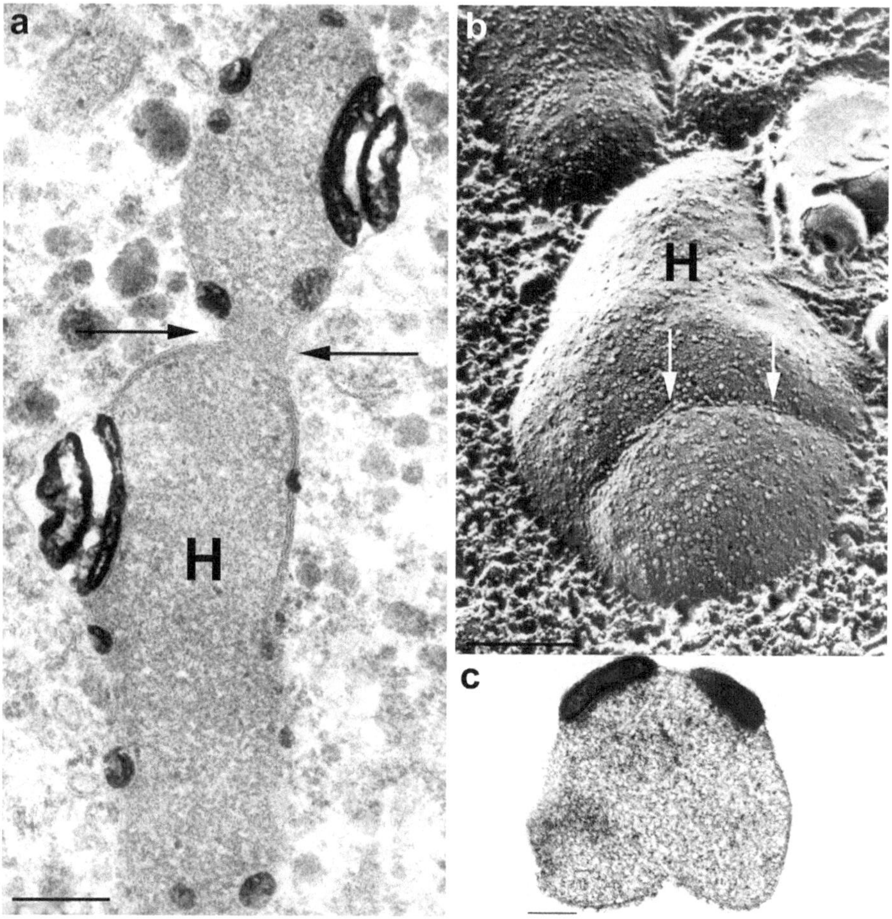

Fig. 14 Views of the segmentation process (**a**–**b**) of a dividing hydrogenosome (H). (**a**) The organelle is elongated showing a constriction in the central region (arrows). Bar = 100 nm. From Benchimol et al. (1996b). (**b**) Freeze-etching of a *T. foetus* hydrogenosome after quick freezing of living cells. Arrows point to a linear array of particles at the septum constriction. Bar = 200 nm. (**c**) *T. foetus* showing a hydrogenosome in the process of heart division. Bar = 100 nm. From Benchimol and Engelke (2003)

a situation in which the hydrogenosome proteins could coagulate and precipitate, leading to the formation of the core (Fig. 16a). In healthy or well-preserved cells (Figs. 1, 3, and 4), as well as in nondrug-treated cells, it is very unusual to find this electron-dense amorphous core, and thus it is not considered as a hydrogenosomal structure any more (Benchimol et al. 1996a). In the case of *Psalteriomonas lantern* (Page and Blanton 1985) and *Sawyeria marylandensis*, the hydrogenosomes appear as densely stained 1–1.5 μm organelles (Barbera et al. 2010). Three-dimensional reconstruction showed two different shapes, and both harbor an invagination which forms a cavity, often showing a cup shape (see Fig. 8 in the paper of this group). In

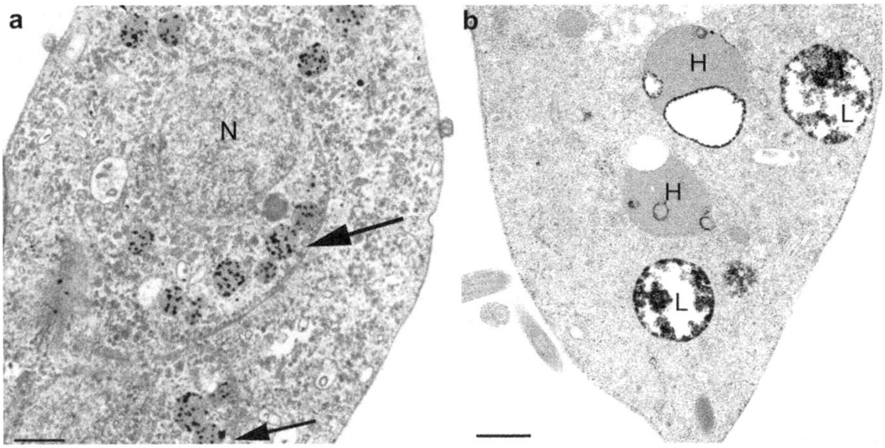

Fig. 15 (**a**) *T. foetus* after cytochemistry for calcium using potassium pyroantimonate showing an intense labeling in the hydrogenosomes matrix (arrows). *N* nucleus. Bar = 300 nm. From Benchimol et al. (1982b). (**b**) Cytochemical detection of acid phosphatase presented positive reactions not only in lysosomes (L) but also in the peripheral vesicles of the hydrogenosomes (H). Bar = 100 nm. Bar = 100 nm

Mastigamoeba balamuthi isolated hydrogenosomes showed a nonhomogeneous matrix (Nyvltova et al. 2015).

The hydrogenosome matrix exhibited basic proteins when cytochemistry for ammoniacal silver and PTA (phosphotungstic acid) (Fig. 16b) was used (Benchimol et al. 1982a). The granular structure of the hydrogenosome matrix may be clearly visualized in replicas of quick-frozen, freeze-fractured, deep-etched, and rotary-replicated cells (Fig. 5b). When the fracture plane exposed the internal portion of the hydrogenosome matrix, a large number of particles were seen. Most of them had a diameter of 6 nm. Some, however, were larger, with a diameter of 20 nm. These particles were not randomly distributed. A certain orientation in their array was noted.

Calcium deposits are seen as electron-dense spots in the matrix of some hydrogenosomes when cells are incubated in the presence of calcium ions in the fixative (Fig. 8a) or processed cytochemically for the localization of Ca^{++} (Fig. 15a), such as with the pyroantimonate technique (Benchimol et al. 1982b).

When cells are stressed with drugs, the hydrogenosomal matrix, which normally appears finely granular and homogeneous, frequently contains internal elements, such as concentric membranes, internal sub-compartments, vesicles, grains, and fibrils (Fig. 8). A large dense precipitate or nucleoid (Fig. 16a) is sometimes observed in the hydrogenosome matrix, which means that the organelle is dysfunctional (Benchimol et al. 1996a).

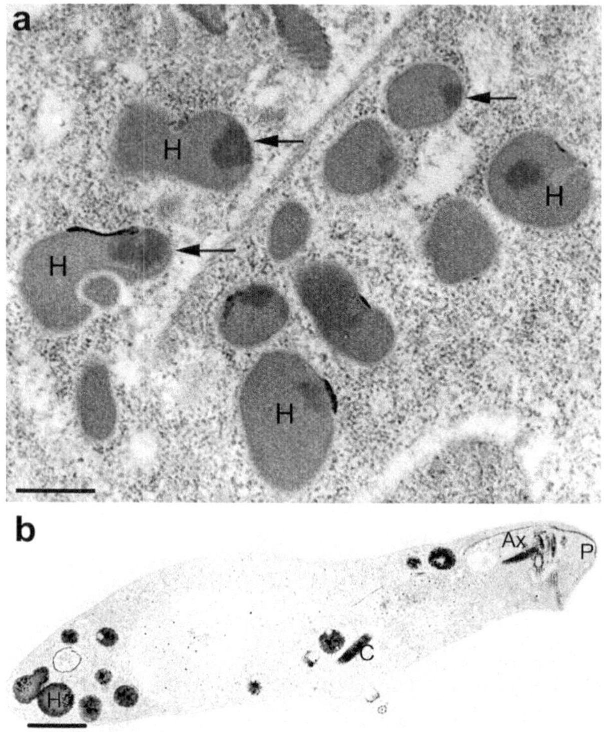

Fig. 16 *T. foetus* showing a dense material in the hydrogenosomes (H) matrix, named core or nucleoid (arrow) in early literature. The nucleoid is considered a fixation artifact or is a result of a disturbance in hydrogenosome metabolism, being a sign of organelle death (Benchimol, unpublished). Bar = 300 nm. (**b**) Cytochemistry for basic proteins (PTA), showing positive reaction in the hydrogenosomes matrix (H), in the axostyle (Ax), costa (C), and pelta (P). Bar = 300 nm. From Benchimol et al. (1982a)

6 Fungal Hydrogenosomes

The hydrogenosomes of the anaerobic fungus *Neocallimastix frontalis* are round or elongated structures, always enveloped by two distinct, but tightly apposed membranes (Figs. 2a–b and 3b) (van der Giezen et al. 1997; Benchimol et al. 1997). *N. frontalis* hydrogenosomes were studied in routine preparation for transmission electron microscopy, freeze-fracture, and immunocytochemistry. In addition, images of organelle division (Fig. 2b) were very similar to those observed in trichomonad protozoa (Fig. 12a). These observations suggested that hydrogenosomes are homologous organelles in unrelated species weakening a previous hypothesis of a polyphyletic origin (Marvin-Sikkema et al. 1992) and reinforcing the hypothesis that fungal and trichomonad hydrogenosomes are derived from an ancestral endosymbiont.

7 Proximity with Other Cellular Structures

Glycogen particles although distributed throughout the protozoan are concentrated in the region where hydrogenosomes are located (Fig. 1c).

7.1 Hydrogenosomes and Endoplasmic Reticulum

Not only continuity but also chemical similarities between hydrogenosomes and the ER have been demonstrated. ER is closely associated, though not necessarily continuous, with mitochondria, peroxisomes (Frederick et al. 1968), and plastids. Data indicated that certain mitochondrial phospholipids were formed in ER and then transferred to the mitochondrion (Jungalwala and Dawson 1970). Since hydrogenosome has been considered a modified mitochondrion (Embley et al. 1997) and presents several similarities to this organelle, it has been suggested that at least the peripheral vesicle of the hydrogenosomes originated from the ER.

The continuities or close associations between hydrogenosomes and the rough or smooth endoplasmic reticulum are often observable (Figs. 9a–b, 11a, c, 17a–b, and 18a–b). Some hydrogenosomes present membranous cisternae projecting to the cytoplasm, conferring bizarre images of the organelle (Fig. 17a). In some cases, the hydrogenosomal outer membrane displayed attached granules, similar to ribosomes, when cells were treated with cytochalasin B.

An intimate association of the hydrogenosome with the endoplasmic reticulum (ER) has been described (Benchimol et al. 1996a; Benchimol 1999). It has been proposed that the ER could provide new membranes for hydrogenosome growth, since this organelle enlarges before its division. The ER could participate by providing the membrane lipids. Three-dimensional reconstruction of a *T. foetus* in division (not shown) and in interphase demonstrated a close association of the membranous profiles of the ER and hydrogenosomes, in a similar way to mitochondria and peroxisomes (Franke and Kartenbeck 1971).

Also it has been suggested that membranous structures seen in close contact with or in the vicinity of the hydrogenosomes might be a source of membrane lipids for hydrogenosome growth (Benchimol et al. 1996a).

In fungi a protein assemblage termed the ER-mitochondria encounter structures (ERMES) tethers the ER to mitochondria (Kornmann et al. 2009). It is composed of four proteins: the mitochondrial outer membrane beta-barrel Mdm 10 and three synaptotagmin-like mitochondrial lipid-binding proteins. In addition, there is the ubiquitous dynamin-related protein-anchored GTPase. Using iterative basic local

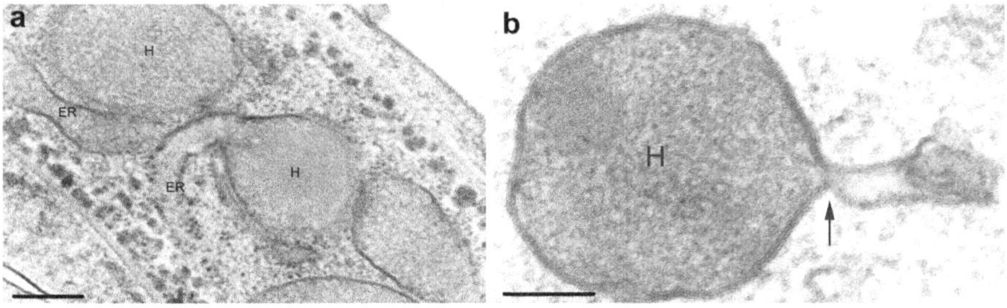

Fig. 17 Thin sections of *T. foetus* showing that profiles of endoplasmic reticulum (ER) are seen in continuity (arrow) to the hydrogenosomes (H). Bars = 100 nm

alignment search tool, some ERMES proteins were found in *T. vaginalis* as well as in the hydrogenosome-containing fungus *Piromyces* sp. (Widermann et al. 2013). The search of the genomes of organisms that contain mitosomes (*Giardia lamblia*, *Entamoeba histolytica*, *Nosema ceranae*, and *Encephalitozoon cuniculi*) did not identify ERMES components.

8 Hydrogenosome Autophagy

T. foetus under serum deprivation and drug treatment (hydroxyurea, zinc sulfate) and also in normal conditions presents autophagy (Benchimol 1999). Apparently normal hydrogenosomes and also giant, abnormal, and those presenting internal membranes were observed in the autophagic process (Fig. 9). The first event observed was the presence of cisternae of the rough endoplasmic reticulum surrounding and enclosing the altered hydrogenosome, forming an isolation membrane (Fig. 9a–c). The hydrogenosome is first sequestered from the remaining cytoplasm (Fig. 9b–c) and then degraded within lysosomes (Fig. 9d). The autophagic vacuole is limited by double (Fig. 9c) or multiple concentric membranes and may contain a recognizable hydrogenosome (Fig. 9b–d), probably in the preliminary steps of degradation. Lysosomes fuse with the autophagic vacuole forming a degradative structure bounded by a single membrane (Fig. 9d) and containing the hydrogenosome in various stages of degeneration. Hydrogenosomes have been found within lysosomes, partially degraded, forming hydrogenosomal remnants (Benchimol 1999).

9 Hydrogenosome Division

Hydrogenosomes, as almost all other organelles, always grow by proliferation of preexisting organelles. Thus, each daughter cell will receive a complete set of organelles during cell division. Morphological evidence was presented showing that trichomonad hydrogenosomes, like mitochondria, may divide by three distinct processes: (1) segmentation (Fig. 14a–b), (2) heart form (Fig. 14c), and partition (Fig. 12a–b). In the segmentation process, the hydrogenosome grows, becoming elongated with the appearance of a constriction in the central portion (Fig. 14a–b) (Benchimol et al. 1996b). Microfibrillar structures appear to help the furrowing process, ending with a total fission of the organelle. In the partition process, rounded hydrogenosomes, in a bulky form, are further separated by a membranous internal septum (Fig. 12a–b). The division begins by an invagination of the inner hydrogenosome membrane, forming a transversal septum, separating the organelle matrix into two compartments (Benchimol et al. 1996b). A necklace of intramembranous particles delimiting the outer hydrogenosomal membrane in the region of organelle division was observed by freeze-etching (Fig. 14b).

In the hydrogenosome heart-shaped process (Benchimol and Engelke 2003), the organelle gradually presents a membrane invagination on one side, leading to the organelle division (Fig. 14c). In this case, the hydrogenosome grows anteriorly, and the process of division starts at one of the organelle poles, which becomes larger than the remaining organelle. Gradually, the membrane at this pole is seen inwarding. Although the organelle division begins with an inward furrowing, no septum is formed. The heart-shaped process is also distinct from segmentation, since in this process an elongation of the organelle occurs first, giving a sausage-shape and a progressive attenuation of its mid-region. In the heart-shaped process, the organelle neither forms a septum nor elongates and in this way is considered a new organellar division process.

The most common form of division observed in *T. foetus* was the segmentation process, whereas the partition is the most unusual division process observed in this protist. However, in the hydrogenosomes of the fungus *Neocallimastix frontalis*, the most common division process was the partition (Benchimol, unpublished observations). On the other hand, in *Trichomonas augusta* the heart shape was the most frequently observed division process (Benchimol, unpublished observations). It is important to point out that all three forms of division could be found during any phase of the cell cycle and that the segmentation and heart form could be observed in the same cell. In addition, the organellar division is not synchronized (Benchimol and Engelke 2003). Interestingly, these three modes of division and timing were also described previously in mitochondria (Tandler and Hoppel 1973). It is not known why and how the cell chooses which form of hydrogenosomal division to perform.

The mitochondrial division process has been compared with the division in bacteria, since the mitochondrial inner membrane is likely to be of bacterial descent. In bacteria a FtsZ ring adheres to the inside of the bacterial membrane and constricts it to mediate division. The use of FtsZ homologs (van der Bliek 2000) and dynamin has been shown to participate in mitochondrial division (Koch et al. 2005). However, these proteins have not yet been studied in dividing hydrogenosomes.

In hydrogenosomes, the process of division of the inner membrane is unknown, whereas the division of the outer membrane appears to be mediated by membranous profiles, probably of endoplasmic reticulum origin. These structures could aid the hydrogenosome division or participate in another function, such as providing membranes for hydrogenosome growth. Dynamin is a GTPase protein that has been involved in several processes involving membrane invagination and fission coupled to division processes and invaginations. Eight dynamin-related proteins (DRPs), grouped into three classes, were found in *T. vaginalis*. Five TvDRPs were examined by introducing dominant negative mutations analogous to those that interfere with mitochondrial division in yeasts (Wexler-Cohen et al. 2014). Microscopic and cell fractionation analysis showed that this protein localizes in hydrogenosomes. Cells expressing the dominant negative protein display hydrogenosomes increased in size and reduced in number, thus indicating a role of DRP in organelle division (Wexler-Cohen et al. 2014).

10 Hydrogenosome Behavior in the Cell Cycle

Dividing hydrogenosomes can be found in all phases of the trichomonad cell cycle, similar to mitochondria (Suzuki et al. 1994). They were even observed during the mitotic process (Benchimol and Engelke 2003). During the interphase the hydrogenosomes in trichomonads are distributed mainly along the axostyle and costa (Figs. 1c–d and 7a), and at mitosis onset the hydrogenosomes migrate to and around the nucleus (Fig. 7b–c). In nondividing cells, small hydrogenosomes (100–200 nm in diameter) were found together with normal-sized organelles (about 300 nm). They were located mainly along the duplicated axostyle (Fig. 7b). During karyokinesis, hydrogenosomes follow the axostyle (Fig. 7d) (Benchimol and Engelke 2003; Ribeiro et al. 2000).

11 Hydrogenosomes Connection to Microtubules

Delicate bridges were observed connecting the outer hydrogenosomal membrane to the microtubules of the axostyle. These bridges were also seen when fast-freezing and freeze-etching were used and could explain the hydrogenosome alignment along the axostyle. In addition, close proximity to free microtubules was also seen (Fig. 18), both in thin sections (Fig. 18a) and after quick-freezing followed by freeze-etching (Fig. 18b) (Benchimol, unpublished).

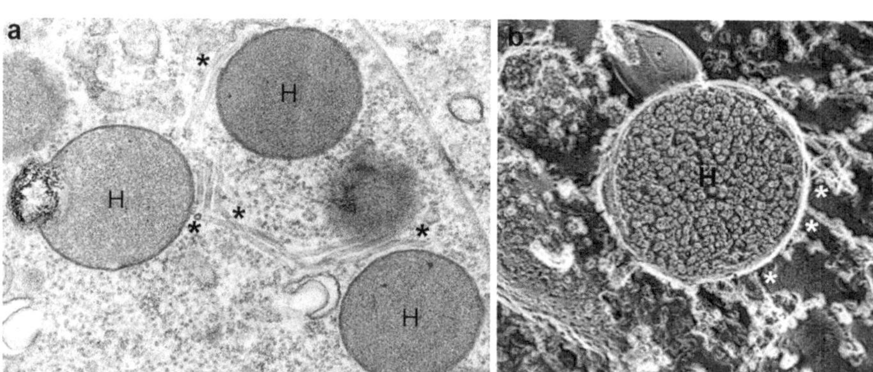

Fig. 18 Association of hydrogenosomes and microtubules in *T. foetus*. (**a**) Thin section of hydrogenosomes in close proximity with microtubules (asterisks). (**b**) Freeze-etching after quick-freezing and rotatory shadowing showing a hydrogenosome (H) in close association with cytoskeletal structures, probably microtubules. Bars = 200 nm. Unpublished material from the author

12 Immunolabeling

Hydrogenosomes can be immunolabeled using two different but complementary approaches. First, the classical immunocytochemistry where specific anti-hydrogenosomal protein antibodies are used, followed by incubation in the presence of secondary antibodies labeled with fluorochromes or colloidal gold particles and observed in fluorescence or electron microscopy microscopes, respectively. Proteins such as malic enzyme (Fig. 10), IscS, and (Fe-Fe)-hydrogenase (Leger et al. 2016) and other proteins are found in the hydrogenosomal matrix. Interestingly, similar labeling is also observed when antibodies anti-AP65, an anti-adhesin protein, are used. Alderete et al. (2001) explain this result as an example of molecular mimicry and functional diversity. A second approach starts with the transfection of the cells with hemagglutinin tag of fusion constructs of proteins which are subsequently revealed using immunofluorescence microscopy with fluorochrome-labeled anti-hemagglutinin antibodies. Using this approach, proteins such as Cpn60, PFOR5, HydF, (Fe-Fe)-hydrogenase, acetyl-CoA synthetase, PPi-dependent-phosphofructo-kinase, ATP-dependent PFK, arginine deaminase, Tom 40-3, and Pam 16 have been localized in the hydrogenosomes (Jerstrom-Hultqvist et al. 2013; Rada et al. 2011, 2015; Morada et al. 2011; Zimorski et al. 2014).

Acknowledgments This work was supported by CNPq (Conselho Nacional de Desenvolvimento Científico e Tecnológico), PRONEX (Programa de Núcleo de Excelência), and FAPERJ (Fundação Carlos Chagas Filho de Amparo à Pesquisa do Estado do Rio de Janeiro).

References

Alderete JF, Millsap KW, Lehker MW, Benchimol M (2001) Enzymes on microbial pathogens and *Trichomonas vaginalis*: molecular mimicry and functional diversity. Cell Microbiol 3:359–370

Barbera MJ, Ruiz-Trillo I, Tufts JYA, Bery A, Slberman JD, Roger AJ (2010) *Sawyeria marylandensis* (Heterolobosea) has a hydrogenosome with novel metabolic properties. Eukaryot Cell 9:1913–1924

Benchimol M (1999) Hydrogenosome autophagy in *Tritrichomonas foetus*: an ultrastructural and cytochemical study. Biol Cell 91:165–174

Benchimol M (2000) Ultrastructural Characterization of the isolated hydrogenosome in *Tritrichomonas foetus*. Tissue Cell 32:1–9

Benchimol M (2001) Hydrogenosome morphological variation induced by fibronectin and other drugs in *Tritrichomonas foetus* and *Trichomonas vaginalis*. Parasitol Res 87:215–222

Benchimol M, Bernardino MV (2002) Ultrastructural localization of glycoconjugates in *Tritrichomonas foetus*. Parasitol Res 88:134–143

Benchimol M, De Souza W (1983) Fine structure and cytochemistry of the hydrogenosome of *Tritrichomonas foetus*. J Protozool 30:422–425

Benchimol M, Engelke F (2003) Hydrogenosome behavior during the cell cycle in *Tritrichomonas foetus*. Biol Cell 95:283–293

Benchimol M, Elias CA, De Souza W (1982a) *Tritrichomonas foetus*: ultrastructural localization of basic proteins and carbohydrates. Exp Parasitol 54:135–144

Benchimol M, Elias CA, De Souza W (1982b) Ultrastructural localization of calcium in the plasma membrane and in the hydrogenosome of *Tritrichomonas foetus*. Exp Parasitol 54:277–284

Benchimol M, Almeida JCA, De Souza W (1996a) Further studies on the organization of the hydrogenosome in *Tritrichomonas foetus*. Tissue Cell 28:287–299

Benchimol M, Johnson PJ, De Souza W (1996b) Morphogenesis of the hydrogenosome: an ultrastructural study. Biol Cell 87:197–205

Benchimol M, Durand R, Almeida J (1997) A double membrane surrounds the hydrogenosomes of the anaerobic fungus *Neocallimastix frontalis*. FEMS Microbiol 154:277–282

Benchimol M, Diniz JAP, Ribeiro K (2000) The fine structure of the axostyle and its associations with organelles in trichomonads. Tissue Cell 32:178–187

Biagini GA, Hayes AJ, Suller MTE, Winters C, Finlay BJ, Lloyd D (1997) Hydrogenosomes of *Metopus contortus* physiologically resemble mitochondria. Microbiology 143:1623–1629

Bowman BH, Taylor JW, Brownlec AG, Lee J, Lu SD, White TJ (1992) Molecular evolution of the fungi relationships of the Basidiomycetes, Ascomycetes and Chytridiomycetes. Mol Biol Evol 9:285–296

De Graaf RM, Duarte I, van Alen TA, Kuiper JW, Schotanus K, Rosenberg J, Huynen MA, Hackstein JH (2009) The hydrogenosome of *Psalteriomonas lantern*. BMC Evol Biol 9:287. https://doi.org/10.1186/1471-2148-9-287

Díaz JAM, De Souza W (1997) Purification and biochemical characterization of the hydrogenosomes of the flagellate protist *Tritrichomonas foetus*. Eur J Cell Biol 74:85–91

Diniz JA, Benchimol M (1998) Monocercomonas sp. cytochemistry and fine structure of freeze-fractured membranes. J Eukaryot Microbiol 45:314–322

Embley TM, Horner DA, Hirt RP (1997) Anaerobic eukaryote evolution: hydrogenosomes as biochemically modified mitochondria? Trends Ecol Evol 12:437–441

Embley TM, van der Gienzen M, Horner DA, Hirt RP, Dyal PL, Bell S, Foster PG (2003a) Hydrogenosomes, mitochondria and early eukaryotic evolution. IUBMB Life 55:387–395

Embley TM, van der Gienzen M, Horner DA, Hirt RP, Dyal PL, Bell S, Foster PG (2003b) Mitochondria and hydrogenosomes are two forms of the same fundamental organelle. Philos Trans R Soc Lond B Biol Sci 358:191–201

Fenchel T, Finlay BJ (1995) Ecology and evolution in anoxic worlds. Oxford University Press, Oxford

Finlay BJ, Fenchel T (1989) Hydrogenosomes in some anaerobic protozoa resemble mitochondria. FEMS Microb Lett 65:311–314

Franke WW, Kartenbeck J (1971) Outer mitochondrial membrane continuous with endoplasmic reticulum. Protoplasma 73:35–41

Frederick SE, Newcomb EH, Vigil EL, Wergin WP (1968) Fine-structural characterization of plant microbodies. Planta (Berl.) 81:229–252

Garg S, Stölting J, Zimorski V, Rada P, Tachezy J, Martin WF, Gould SB (2015) Conservation of transit peptide-independent protein import into the mitochondrial and hydrogenosomal matrix. Genome Biol Evol 7:2716–2726

Honigberg MB, Brugerolle G (1990) Structure. In: Honigberg BM (ed) Trichomonads parasitic in humans. Springer, New York, pp 5–35

Horner DS, Foster PG, Embley TM (2000) Iron hydrogenase and the evolution of anaerobic eukaryotes. Mol Biol Evol 17:1695–1709

Jerstrom-Hultqvist J, Einarsson E, Xu F, Hjort K, Ek B, Steinhauf D, Hultenby K, Bergquist J, Anderson JO, Svard SG (2013) Hydrogenases in the diplomonad *Spironucleus salmonicida*. Nat Commun 4:2493

Jungalwala FB, Dawson RMC (1970) Phospholipid synthesis and exchange in isolated liver cells. Biochem J 117:481–490

Koch A, Yoon Y, Bonekamp NA, McNiven MA, Schrader M (2005) A role for Fis1 in both mitochondrial and peroxisomal fission in mammalian cells. Mol Biol Cell 16:5077–5086

Kornmann B, Currie E, Collins SR, Schuldiner M, Nunnari J, Weissman JR, Walter P (2009) An ER-mitochondria tethering complex revealed by a synthetic biology screen. Science 325:477–481

Kulda J, Nohýnková E, Ludvik J (1987) Basic structure and function of the trichomonad cell. Acta Univ Carol 30:181–198

Leger MM, Eme L, Hug LA, Roger AJ (2016) Novel hydrogenosomes in the microaerophilic jakobid *Stygiella incarcerata*. Mol Biol Evol 33:2318–2336

Lindmark DG, Müller M (1973) Hydrogenosome, a cytoplasmic organelle of the anaerobic flagellate, *Tritrichomonas foetus*, and its role in pyruvate metabolism. J Biol Chem 248:7724–7728

Madeiro RF, Benchimol M (2004) The effect of drugs in *Tritrichomonas foetus*. Parasitol Res 92:159–170

Makiuchi T, Nozaki T (2014) Highly divergent mitochondrion-related organelles in anaerobic protozoa. Biochimie 100:3–17

Mariante RM, Guimarães CA, Linden R, Benchimol M (2003) Hydrogen peroxide induces caspase activation and programmed cell death in the amitochondrial *Tritrichomonas foetus*. Histochem Cell Biol 120:129–141

Marvin-Sikkema FD, Lahpor GA, Kraak MN, Gottschal JC, Prins R (1992) Characterization of an anaerobic fungus from llama faeces. J Gen Microbiol 138:2235–2241

Morada M, Smid O, Hampl V, Sutak R, Lam B, Rappelli P, Dessi D, Fiori PL, Tachezy J, Yarlett N (2011) Hydrogenosome-localization of arginine deiminase in *Trichomonas vaginalis*. Mol Biochem Parasitol 176:51–54

Müller M (1993) The hydrogenosome. J Gen Microbiol 139:2879–2889

Nyvltova E, Stairs CW, Hrdy I, Ridl J, Mach J, Paces J, Roger AJ, Tachezy J (2015) Lateral gene transfer and gene duplication played a key role in the evolution of *Mastigamoeba balamuthi* hydrogenosomes. Mol Biol Evol 32:1039–1055

Page FC, Blanton RL (1985) The Heterolobosea (Sarcodina: Rhizopoda), a new class uniting the Schyzopyrenida and the Acrasidae (Acraosida). Protistologica 21:121–132

Queiroz RC, Santos LM, Benchimol M (1991) Cytochemical localization of enzyme markers in *Tritrichomonas foetus*. Parasitol Res 77:561–566

Rada P, Doležal P, Jedelský PL, Bursac D, Perry AJ, Šedinová M, Smíšková K, Novotný M, Beltrán NC, Hrdý I, Lithgow T, Tachezy J (2011) The core components of organelle biogenesis and membrane transport in the hydrogenosomes of *Trichomonas vaginalis*. PLoS One 6(9)

Rada P, Makki AR, Zimorski V, Garg S, Hampl V, Herdy I, Gould SB, Tachezi J (2015) N-terminal pre-sequence-independent import of phosphofructokinase into hydrogenosomes of *Trichomnas vaginalis*. Eukaryot Cell 14:1264–1275

Ribeiro KC, Monteiro-Leal LH, Benchimol M (2000) Contributions of the axostyle and flagella on the division process of *Tritrichomonas foetus*. J Eukaryot Microbiol 47:481–492

Ribeiro KC, Vetö Arnholdt AC, Benchimol M (2002) *Tritrichomonas foetus*: induced synchrony by hydroxyurea. Parasitol Res 88:627–631

Rosa IA, Einicker-Lamas M, Bernardo RR, Previatto LM, Mohana-Borges R, Díaz JAM, Benchimol M (2006) Cardiolipin in hydrogenosomes: evidence of symbiotic origin. Eukaryot Cell 5:784–787

Snyers S, Hellings P, Bovy-Kesler C, Thines-Sempoux D (1982) Occurrence of hydrogenosomes in the rumen ciliates Ophryoscolecidae. FEBS Lett 137:35–39

Stairs CW, Leger MM, Roger A (2015) Diversity and origins of anaerobic metabolism in the mitochondria and related organelles. Philos Trans R Soc Lond B Biol Sci 370(1678):201400326

Suzuki K, Ehara T, Osafune T, Kuroiwa H, Kawano S, Kuroiwa T (1994) Behavior of mitochondria, chloroplasts and their nuclei during the mitotic cycle in the ultramicroalga *Cyanidioschyzon merolae*. Eur J Cell Biol 63:280–288

Tandler B, Hoppel L (1973) Division of giant mitochondria during recovery from cuprizone intoxication. J Cell Biol 56:266–272 van Bruggen JJA, Zwart KD, van Assema RM, Stumm CK, Vogels GD (1984) *Methanobacterium formicium*, an endosymbiont of the anaerobic ciliate *Metopus striatus* Mc. Murrich Arch Microbiol 139:1–7

van der Bliek AM (2000) A mitochondrial division apparatus take place. J Cell Biol 151(2):F1–F4

van der Giezen M, Sjollema KA, Artz RR, Alkema W, Prins RA (1997) Hydrogenosomes in the anaerobic fungus *Neocallimastix frontalis* have a double membrane but lack an associated organelle genome. FEBS Lett. 408:147–150

Wexler-Cohen Y, Stevens GC, Barnoy E, van der Bliek AM, Johnson P (2014) A dynamin-related protein contributes to *Trichomonas vaginalis* hydrogenosomal fission. FASEB J 28:1113–1121

Widermann JG, Gawryluk RMR, Gray MW, Dacks JB (2013) The ancient and widespread nature of the ER-mitochondria encounter structure. Mol Biol Evol 30:2044–2049

Yarlett N, Hann AC, Lloyd D, Williams AG (1981) Hydrogenosomes in the rumen protozoan *Dasytricha ruminantium* Schuberg. Biochem J 200:365–372

Yarlett N, Coleman GS, Williams AG, Lloyd D (1984) Hydrogenosomes in known species of rumen entodiniomorphid protozoa. FEMS Microbiol Lett 21:15–19

Yarlett N, Orpin CG, Munn EA, Yarlett NC, Greenwood CA (1986) Hydrogenosomes in the rumen fungus *Neocallimastix patriciarum*. Biochem J 236:729–739

Zimorski V, Ku C, Martin WF, Gould SB (2014) Endosymbiotic theory for organelle origins. Curr Opin Microbiol 22:38–48

Hydrogenosomes of Anaerobic Ciliates

Johannes H. P. Hackstein, Rob M. de Graaf, Jaap J. van Hellemond, and Aloysius G. M. Tielens

Contents

1 Introduction 112
2 *Nyctotherus ovalis* 113
2.1 The Energy Metabolism of *N. ovalis* 116
2.2 In Silico Reconstruction of the Basal Hydrogenosomal Metabolism of *N. ovalis* ... 117
3 The Hydrogenosomes of Other Ciliates 118
4 Can the Methanogenic Symbionts Tell Us More About the Origin and Function of Ciliate Hydrogenosomes 121
5 Evolutionary Aspects 122
References 123

Abstract Ciliates are highly complex unicellular eukaryotes. Most of them live in aerobic environments and possess mitochondria that use oxygen as final electron acceptor. However, in several orders of ciliates, anaerobic species evolved that contain hydrogenosomes, ATP-producing organelles that use protons as final electron acceptor, producing hydrogen in this process. These hydrogenosomes of ciliates have not been studied in the same detail as those of trichomonads and anaerobic fungi.

J. H. P. Hackstein · R. M. de Graaf
Faculty of Science, Institute of Water and Wetland Research (IWWR), Radboud University Nijmegen, Nijmegen, The Netherlands
e-mail: r.degraaf@science.ru.nl

J. J. van Hellemond
Department of Medical Microbiology and Infectious Diseases, Erasmus MC, University Medical Center Rotterdam, Rotterdam, The Netherlands
e-mail: j.vanhellemond@erasmusmc.nl

A. G. M. Tielens (✉)
Department of Medical Microbiology and Infectious Diseases, Erasmus MC, University Medical Center Rotterdam, Rotterdam, The Netherlands

Faculty of Veterinary Medicine, Department of Biochemistry and Cell Biology, Utrecht University, Utrecht, The Netherlands
e-mail: a.tielens@erasmusmc.nl

J. Tachezy (ed.), *Hydrogenosomes and Mitosomes: Mitochondria of Anaerobic Eukaryotes*, Microbiology Monographs 9,
https://doi.org/10.1007/978-3-030-17941-0_5

Therefore, generalizations on the characteristics of hydrogenosomes of ciliates are somewhat premature, especially since phylogenetic studies suggest that hydrogenosomes have evolved independently several times in ciliates. In this chapter, the hydrogenosomes of the anaerobic, heterotrichous ciliate *Nyctotherus ovalis* from the hindgut of cockroaches will mainly be described as these are the ones that are, at the moment, the most thoroughly studied. This was the first hydrogenosome shown to possess a genome, and this genome is clearly of mitochondrial origin. In fact, the hydrogenosome of *N. ovalis* unites typical mitochondrial features such as a genome and an electron-transport chain with the characteristic hydrogenosomal property, the production of hydrogen. The hydrogenosomal metabolism of *N. ovalis* will be compared with that of two other ciliates that have been studied in less detail, i.e. the holotrichous rumen ciliate *Dasytricha* and the free-living plagiopylid ciliate *Trimyema*. All studies combined show that the hydrogenosomes of anaerobic ciliates are different from those of anaerobic fungi and from the well-studied ones in trichomonads.

1 Introduction

Ciliates represent an extremely species-rich, monophyletic group of highly complex unicellular eukaryotes. They are characterized by a nuclear dimorphism and rather complex patterns of morphologically distinct, cortical cilia. Most ciliates thrive in aerobic environments and possess canonical aerobically functioning mitochondria (class 1 in Table 1). In at least 8 of the 22 orders of ciliates as classified by Corliss

Table 1 Classification of the various types of MROs is based on whether or not they produce ATP, possess a proton-translocating electron-transport chain and use oxygen, protons or another compound as final electron acceptor (Müller et al. 2012)

		ATP production	Electron transport chain	Final electron acceptor		
				O_2	H^+	'Fumarate'
Class 1	Aerobic mitochondrion	+	+	+	–	–
Class 2	Anaerobic mitochondrion	+	+	–	–	+
Class 3	H_2-producing anaerobic mitochondrion	+	+	–	+	+
Class 4	Hydrogenosome	+	–	–	+	–
Class 5	Mitosome	–	–	–	–	–

Red colour indicates an aerobic energy metabolism, blue an anaerobic one and yellow indicates that these MROs are not involved in ATP production

(1979), Fenchel and Finlay (1995) evolved anaerobic species. Certain ciliates in seven of these eight orders possess mitochondrion-related organelles (MROs) called hydrogenosomes, organelles that produce hydrogen and ATP (class 4 MRO in Table 1) (Müller 1993; Hackstein et al. 1999, 2001; Embley et al. 2003; Dyall et al. 2004; Embley and Martin 2006; Embley 2006; Müller et al. 2012). However, the identification of these hydrogenosomes was frequently based solely on the presence of symbiotic methanogenic archaea. Such an association is indicative of an interspecies hydrogen transfer and could reveal the presence of intracellular hydrogen sources, i.e. hydrogenosomes (Hackstein et al. 2002).

The development of fluorescence microscopy, electron microscopy, cytobiochemistry and techniques for cellular fractionation soon allowed the discovery of hydrogenosomes in free-living anaerobic ciliates such as *Plagiopyla* and *Trimyema* (Plagiopylea) and *Metopus* (Armophorea) (van Bruggen et al. 1983, 1984, 1986; Goosen et al. 1988, 1990; Zwart et al. 1988; Finlay and Fenchel 1989; Fenchel and Finlay 1995; Biagini et al. 1997). Ciliates with hydrogenosomes were also identified in the gastrointestinal tract of ruminants and marsupials (e.g. *Isotricha*, *Dasytricha*, *Epidinium*, *Eudiplodinium*, *Polyplastron*, *Amylovorax*; Trichostomatia) (Vogels et al. 1980; Snyers et al. 1982; Yarlett et al. 1981, 1982, 1983, 1984, 1985; Lloyd et al. 1989; Paul et al. 1990; Ellis et al. 1991a, b, c; Cameron and O'Donoghue 2002a) and in the hindgut of cockroaches (*Nyctotherus*; Armophorea) (Gijzen et al. 1991). See Fig. 1 for the distribution of hydrogenosomes and mitochondria in various orders of ciliates. All these hydrogenosomes are surrounded by a double membrane, and under optimal fixation conditions, cristae-like protrusions can be seen in these organelles of a number of ciliate species. In this respect, they clearly resemble mitochondria (Fig. 2). The hydrogenosomes of ciliates have not been studied in the same detail as those of trichomonads and anaerobic fungi because culturing of these ciliates is difficult (Müller 1993; Hackstein et al. 2001).

2 *Nyctotherus ovalis*

Nyctotherus species (Armophorea) are anaerobic, heterotrichous ciliates with hydrogenosomes that thrive in the intestinal tract of cockroaches, millipedes, frogs and reptiles. *N. ovalis* from the hindgut of cockroaches is the only species that has been studied in more detail (van Hoek et al. 1998, 1999, 2000b). Notably, the presence of a mitochondrial genome has been demonstrated in the hydrogenosomes of *N. ovalis* (Akhmanova et al. 1998; van Hoek et al. 2000a; Boxma et al. 2005; De Graaf et al. 2011). This genome, which has been identified by immunocytochemistry, fluorescence in situ hybridization, Southern blotting, cDNA analysis and DNA sequencing, is a typical mitochondrial genome (Boxma et al. 2005; De Graaf et al. 2011). Since the organelle unites the hallmark features of mitochondria (a DNA genome and a proton-translocating electron-transport chain) and those of hydrogenosomes (hydrogen production), it is a class 3 MRO, an anaerobic

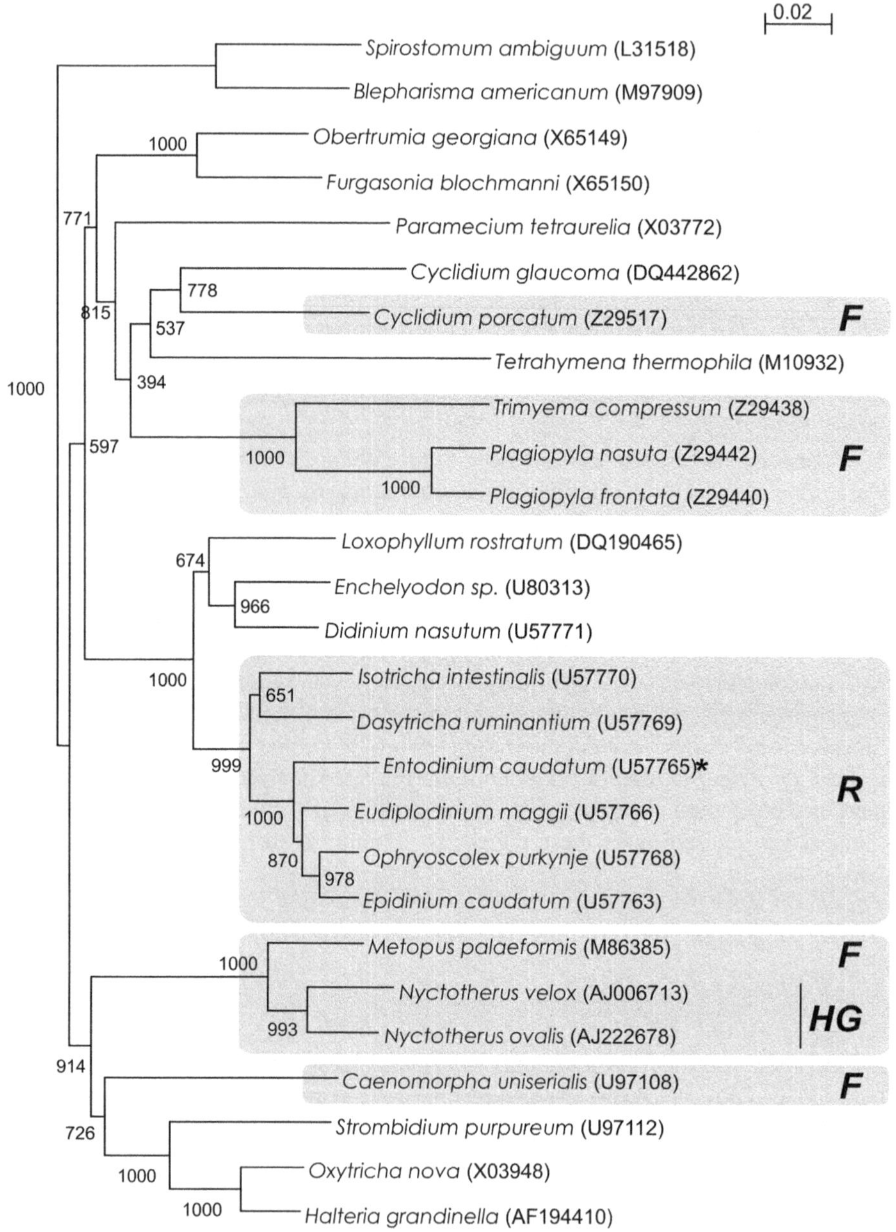

Fig. 1 Neighbour-joining phylogenetic tree of 18S ribosomal RNA of ciliates. Ribosomal RNA sequences were aligned using Clustal X (Jeanmougin et al. 1998), and phylogenetic trees were prepared by neighbour-joining (Saitou and Nei 1987). The accession numbers of used sequences and the bootstrap values for 1000 independent analyses are shown. Shaded boxes indicate anaerobic ciliates with hydrogenosomes, MRO class 3 or 4, whereas all other ciliates contain mitochondria (MRO class 1) and function aerobically. The natural habitat of the hydrogenosome-containing ciliates is indicated by the following abbreviations: *F* free-living, *HG* hindgut, *R* rumen. Ciliate species that might possess mitosomes instead of hydrogenosomes are indicated by an asterisk. The length of the branches represents the relative divergence in nucleotide sequence. The bar represents 2% diversity

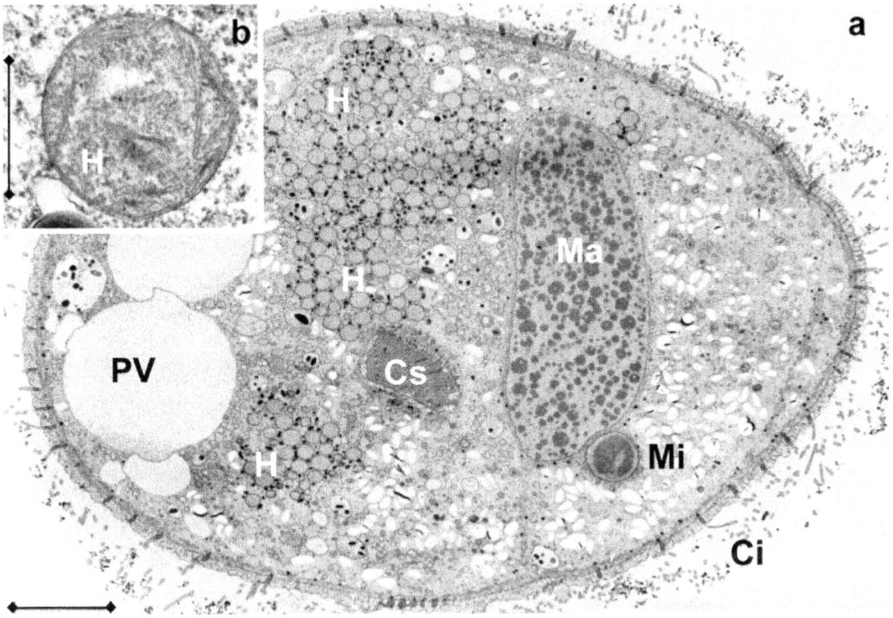

Fig. 2 Electron micrograph of *N. ovalis* (**a**) with a hydrogenosome (**b**). The *scale bar* in (**a**): represents 10 µm, the *scale bar* in (**b**): represents 0.5 µm. *H* hydrogenosomes, *Ma* macronucleus, *Mi* micronucleus, *Cs* cytostome, *PV* pulsating vacuole

mitochondrion that produces hydrogen (Table 1). This organelle represents a true 'missing link' connecting the aerobic and anaerobic forms of organelles of mitochondrial origin which are now recognized to form an array of organelles with a wide spectrum of metabolic properties (Tielens et al. 2002; Martin 2005; Embley and Martin 2006; Hackstein et al. 2006; Müller et al. 2012). A comparable MRO is found in the phylogenetically only distantly related Stramenopile *Blastocystis hominis*, a common inhabitant of the human gastrointestinal tract. *Blastocystis* also retained a mitochondrial genome and parts of an electron-transport chain (Perez-Brocal and Clark 2008; Stechmann et al. 2008; Wawrzyniak et al. 2008). As the MROs of this anaerobe also contain a hydrogenase, it can be regarded as having an MRO similar to the one of *N. ovalis*, an anaerobic mitochondrion that produces hydrogen, although in the case of *Blastocystis* thus far no hydrogenase activity itself has been demonstrated (Lantsman et al. 2008; Stechmann et al. 2008).

N. ovalis has numerous MROs (class 3) with typical well-defined mitochondrial cristae, and these MROs are closely associated with endosymbiotic methane-producing archaea that depend for their growth and survival upon the hydrogen produced by the organelles (Boxma et al. 2005). Since the phylogenetic analysis of the hydrogenosomal genome of *N. ovalis* unequivocally revealed a ciliate origin, its relation to an ancestral ciliate mitochondrion can be taken for sure. This ciliate origin was reinforced by the analysis of some 80 nuclear-encoded genes that encode mitochondrial proteins (Boxma et al. 2005; De Graaf et al. 2011).

2.1 The Energy Metabolism of N. ovalis

Metabolic studies using tracer amounts of radioactive labelled glucose (labelled in all (C_1 to C_6) positions or at position C_6 only) were performed with isolated *N. ovalis* cells to investigate their energy metabolism. These studies revealed that part of the glucose was degraded to typical end products of a glycolytic fermentation, as approximately 24% of the degraded glucose was excreted as lactate and 5% as ethanol (Boxma et al. 2005). The major part of the glucose was degraded to acetate (60%) and succinate (12%), most likely within the hydrogen-producing mitochondrion. Labelled carbon dioxide was only produced in the incubations with uniformly labelled glucose and not in the incubations with [6-^{14}C]glucose.

These results prompt several important conclusions (Fig. 3). First, the absence of $^{14}CO_2$ production in incubations with [6-^{14}C]glucose implies that *N. ovalis* does not use a complete Krebs cycle for the degradation of glucose, as ^{14}C-labelled CO_2 is released from [6-^{14}C]glucose exclusively by successive decarboxylations in subsequent rounds in the Krebs cycle. Secondly, because formate is not one of the end

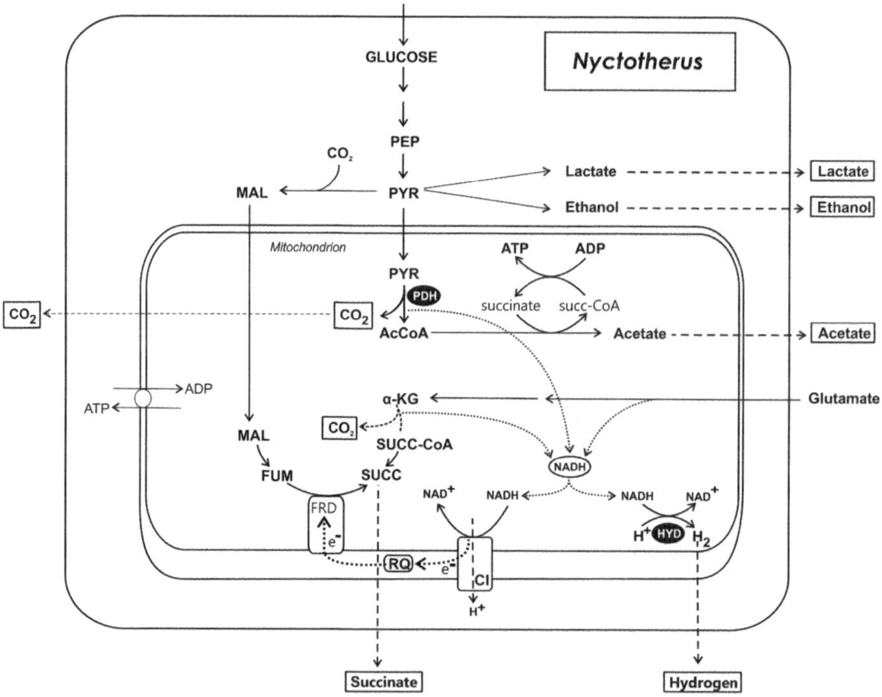

Fig. 3 Energy metabolism of *Nyctotherus ovalis*. Shown is a scheme of the metabolic pathways involved in the production of the major end products by *N. ovalis*. End products are in boxes. *AcCoA* acetyl-CoA, *CI* complex I, *FRD* fumarate reductase, *FUM* fumarate, *HYD* hydrogenase, α-KG α-ketoglutarate, *MAL* malate, *PDH* pyruvate dehydrogenase, *PEP* phosphoenolpyruvate, *PYR* pyruvate, *RQ* rhodoquinone, *SUCC* succinate, *SUCC-CoA* succinyl-CoA

products, *N. ovalis* does not use pyruvate formate lyase (PFL) activity in its pyruvate metabolism, as is the case in hydrogenosomes of anaerobic fungi (see Hackstein et al. 2019). Thirdly, the product of glycolysis in the cytosol, pyruvate, is apparently either converted into lactate or ethanol or is transported into the hydrogenosome to be converted into acetate or succinate. Fourthly, for the production of acetate, this pyruvate is decarboxylated, which is confirmed by the release of ^{14}C-labelled carbon dioxide from incubations with uniformly labelled glucose. In principle, this decarboxylation could be performed either by a pyruvate dehydrogenase complex (PDH), as in 'normal' mitochondria (see below), or by a pyruvate/ferredoxin oxidoreductase (PFO) that occurs in certain other types of hydrogenosomes. Fifthly, the excretion of significant amounts of succinate indicates that endogenously produced fumarate is used as a terminal electron acceptor. Protons act as another hydrogenosomal electron acceptor, which results in the formation of hydrogen. Fumarate reduction is most likely catalysed by a membrane-bound fumarate reductase (an anaerobically functioning variant of complex II), coupled to complex I of the electron-transport chain via quinones (Fig. 3). The reduction of fumarate requires the presence of rhodoquinone, which has a lower redox potential than ubiquinone, the quinone used by aerobic mitochondria for the reverse reaction, the acceptance of electrons from succinate (Tielens et al. 2002). Consistent with these biochemical/biophysical requirements, small amounts of rhodoquinone 9 and menaquinone 8 were detected, whereas ubiquinone 7 and 8 (which are found in large amounts in the aerobic ciliates *Euplotes* and *Tetrahymena*, respectively) were not detected in *N. ovalis* (Boxma et al. 2005).

2.2 *In Silico Reconstruction of the Basal Hydrogenosomal Metabolism of* N. ovalis

These experimental (metabolic) data are supported by molecular (DNA sequence) data on mitochondrial as well as nuclear-encoded proteins (Boxma et al. 2005; De Graaf et al. 2011). As in other classes of mitochondria, most of the organellar proteins in *N. ovalis* are encoded by the nuclear genome, synthesized in the cytoplasm and then imported into the organelle.

Notably, for the degradation of pyruvate, no gene coding for a protein similar to PFO or PFL could be detected. However, genes for the three different enzymes in PDH complexes are present and are expressed. The use of PDH instead of PFO or PFL for the oxidation of pyruvate is a typical property of class 1 aerobic mitochondria, and this enzyme is not present in class 4 MROs, hydrogenosomes such as the ones in anaerobic fungi and trichomonads. In *N. ovalis* the acetyl-CoA is then further metabolized to acetate by an acetate/succinate CoA-transferase (ASCT) belonging to subfamily 1A, with sequence similarity to the ASCT from *Trypanosoma brucei* (Tielens et al. 2010). In this process the CoA moiety is transferred to succinate and the resulting succinyl-CoA is converted back to succinate by succinyl-CoA synthetase (SCS) with the concomitant production of ATP (Fig. 3). SCS is an enzyme that is also part of the Krebs cycle, but not in *N. ovalis* which does not possess a functional full cycle, but only a very small part of it (see below).

The above described metabolic studies indicated that in *N. ovalis* the Krebs cycle is not complete, using only the malate—fumarate—succinate part in the reductive direction. This is reflected in the presence of genes for enzymes discovered in its genome. Thus far only the α and β subunits of SCS, as well as the α and β subunits of succinate dehydrogenase/fumarate reductase have been identified in the nuclear genome (De Graaf et al. 2011). A gene encoding a malate dehydrogenase was encountered, but this enzyme is orthologous to cytosolic malate dehydrogenases from other species and was therefore considered to be also cytoplasmic (De Graaf et al. 2011).

A true hallmark of mitochondria is the presence of a proton-translocating electron-transport chain. Twelve subunits of the mitochondrial complex I were found in *N. ovalis*, among them three that are encoded by the nuclear genome, and the others are encoded in the organellar genome (De Graaf et al. 2011). All 12 are part of the 14 proteins that compose the core bacterial Complexes I. Analysis of the *N. ovalis* genome did not provide any evidence for the presence of genes encoding components of Complexes III and IV in *N. ovalis*, but these complexes are also absent in the electron-transport chains of anaerobic mitochondria class 2 and in the hydrogenosomes of *Blastocystis* (Tielens et al. 2002; Perez-Brocal and Clark 2008; Stechmann et al. 2008; Wawrzyniak et al. 2008; Müller et al. 2012; Gentekaki et al. 2017). In contrast, the mitochondrial genomes of the aerobic ciliates *P. aurelia*, *Tetrahymena* spp. and *Euplotes minuta* do code for components of the complexes III and IV (De Graaf et al. 2009, 2011).

Apparently the NADH formed during the oxidation of pyruvate to acetyl-CoA by PDH is reoxidized by a truncated electron-transport chain in which complex I passes the electrons from NADH through rhodoquinone to fumarate reductase, which then uses the electrons to reduce fumarate to succinate. Another part of NADH is reoxidized by the hydrogenase, thereby producing hydrogen (Fig. 3). The proton motive force (PMF) formed by complex I is, however, most likely used for other purposes than the production of ATP, as no evidence for components of the F_oF_1-ATP synthase were found in the *N. ovalis* genome (Boxma et al. 2005; De Graaf et al. 2011).

Analysis of the organellar genome of *N. ovalis* revealed the presence of components of an extensive protein network for the metabolism of amino acids, defence against oxidative stress, mitochondrial protein synthesis, mitochondrial protein import and processing and transport of metabolites across the mitochondrial membrane (De Graaf et al. 2011). Thus, the hydrogenosome of *N. ovalis* is not simply a rudimentary mitochondrion. It is a highly specialized organelle of considerable complexity.

3 The Hydrogenosomes of Other Ciliates

Metabolic studies have also been carried out on hydrogenosomes of rumen ciliates such as *Dasytricha*, *Isotricha*, *Epidinium* and *Eudiplodinium*. All rumen ciliates form a monophyletic group (Fig. 1; Struder-Kypke et al. 2006), but not all of them possess hydrogenosomes, e.g. *Entodinium simplex*, *Diploplastron affine*, *Ophryoscolex caudatus*, *Eremoplastron bovis* and *Ostracodinium obtusum bilobum* possess MROs type 1, aerobic mitochondria (Yarlett et al. 1984, 1985). Many rumen ciliates utilize cellulose and starch besides being predators of bacteria and smaller protozoa. Glucose is the major monosaccharide liberated by degradation of these plant polymers and can be used by these rumen protozoa as fermentation substrate. The main end products of the metabolism of exogenously added glucose as well as of intracellular amylopectin of rumen ciliates with hydrogenosomes are hydrogen, acetate, lactate, butyrate and CO_2 (Yarlett et al. 1985; Ellis et al. 1991a, b, c). The ratio in which these end products are formed is influenced by O_2 and CO_2 at those concentrations present in the rumen. The investigated rumen ciliates are able to use oxygen as terminal electron acceptor. The nature of this terminal oxidase is still unknown, but cytochromes do not appear to be involved. *Dasytricha ruminantium* is the most thoroughly studied rumen ciliate; however, even knowledge regarding the metabolism of this rumen ciliate is still far from complete. The enzyme used for the degradation of pyruvate to acetyl-CoA in this protozoon is suggested to be PFO, which has been identified tentatively in the hydrogenosomal fraction (Yarlett et al. 1981, 1982, 1985). This acetyl-CoA is the substrate for the hydrogenosomal formation of acetate, but seems also to be exported from the hydrogenosomes for the formation of butyrate (Yarlett et al. 1985; Ellis et al. 1991b). A hypothetical scheme for the hydrogenosomal metabolism of the holotrich (rumen) ciliate *Dasytricha* that is based on various studies is shown in Fig. 4. The

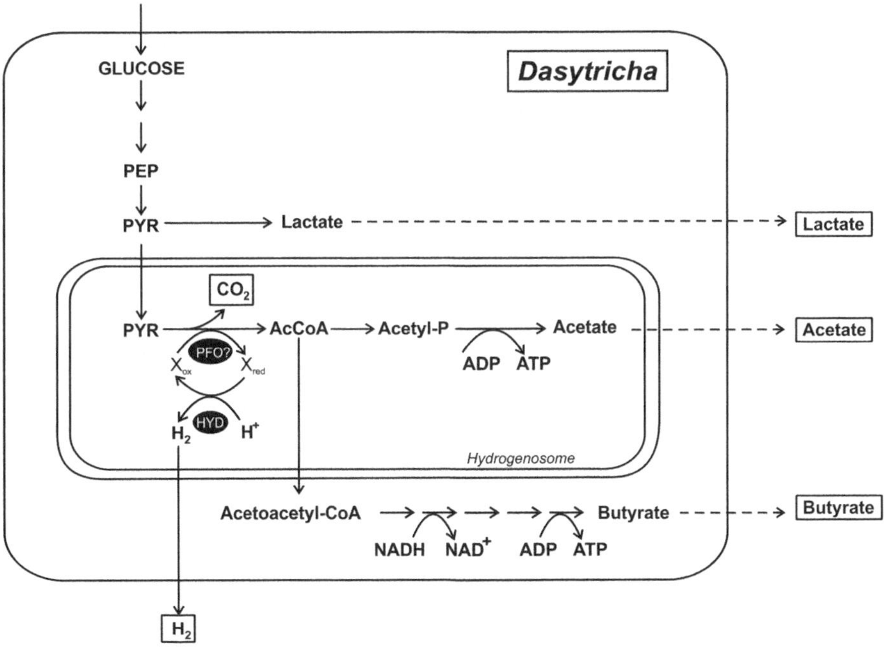

Fig. 4 Speculative metabolic scheme of the main pathways in carbohydrate metabolism in *Dasytricha* sp. (after Ellis et al. 1991b). End products are in boxes. *AcCoA* acetyl-CoA, *HYD* hydrogenase, *PEP* phosphoenolpyruvate, *PFO* pyruvate/ferredoxin oxidoreductase, *PYR* pyruvate; Xox, red, unknown electron carrier

scheme is remarkable as it requires the export of acetyl-CoA from the hydrogenosome for the formation of butyryl-CoA. This butyryl-CoA is then used for the production of butyrate, which is accompanied by the production of ATP (Yarlett et al. 1985; Ellis et al. 1991a, b, c). These aspects make this hydrogenosome of rumen ciliates very different from that of *Nyctotherus* (Fig. 3) and also different from the hydrogenosomes of *Trichomonas* (see Hrdý et al. 2019).

The only other published metabolic study on hydrogenosomes of ciliates deals with the free-living Plagiopylid ciliate *Trimyema* (Goosen et al. 1988, 1990). This free-living ciliate produces neither butyrate nor significant amounts of succinate, as rumen ciliates and *N. ovalis*, respectively, do. *Trimyema* consumes oxygen under microaerobic conditions and is reported to produce formate under such conditions as the major end product with minor amounts of acetate and lactate (Goosen et al. 1990). Under those microaerobic conditions, hydrogen and ethanol are not produced. Under strictly anaerobic conditions, however, ethanol is the main end product, while acetate, lactate, formate and hydrogen are then formed in minor amounts (Goosen et al. 1990). For easy comparison with the metabolism of the other hydrogenosome-containing ciliates (Figs. 3 and 4), a very speculative scheme of the carbohydrate metabolism of *Trimyema* is shown in Fig. 5. This pattern of anaerobic fermentation products resembles the one found in anaerobic fungi from

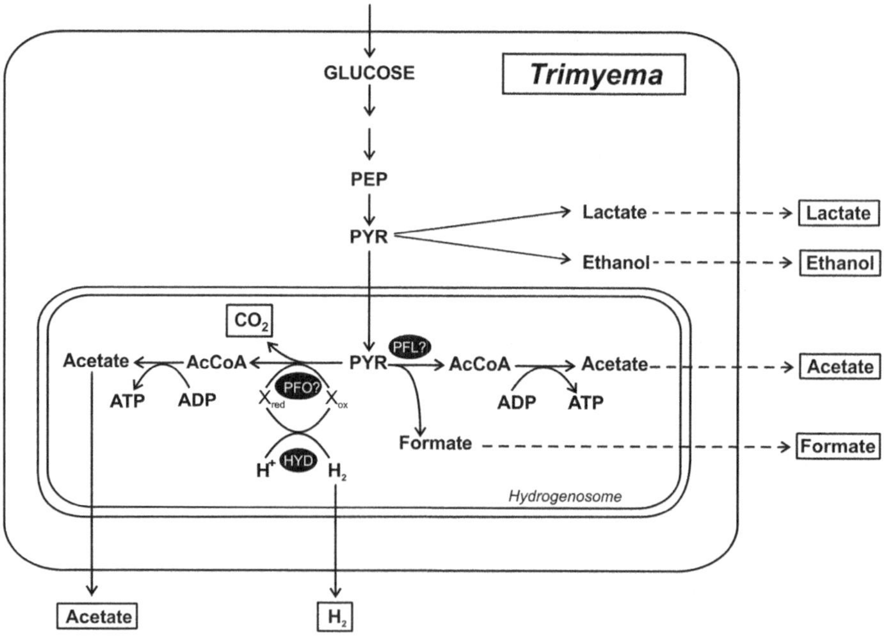

Fig. 5 Speculative metabolic scheme of the main pathways in carbohydrate metabolism in *Trimyema compressum* (after Goosen et al. 1990). End products are in boxes. *AcCoA* acetyl-CoA, *HYD* hydrogenase, *PEP* phosphoenolpyruvate, *PFL* pyruvate formate lyase, *PFO* pyruvate/ferredoxin oxidoreductase, *PYR* pyruvate; Xox, red unknown electron carrier

the gastrointestinal tract of large herbivores (Boxma et al. 2004). These fungi perform a bacterial type mixed acids fermentation using PFL for the degradation of pyruvate instead of PDH or PFO, which is used by *N. ovalis* and *Trichomonads*, respectively. Even though no additional biochemical data are available and no cell fractionation studies have been performed, it is likely that the plagiopylids evolved a type of hydrogenosome that is clearly different from those of *Nyctotherus* and *Dasytricha*.

4 Can the Methanogenic Symbionts Tell Us More About the Origin and Function of Ciliate Hydrogenosomes

Additionally, the nature of the methanogenic (endo)symbionts supports the conclusion that ciliates host different types of hydrogenosomes. While *Nyctotherus* and *Metopus*, but also *Plagiopyla* and *Trimyema*, host endosymbiotic methanogens, certain rumen ciliates seem to host episymbiotic methanogens. Whether this episymbiotic association is specific, and whether there is any rumen ciliate (except *Dasytricha* and *Isotricha*) with symbiotic methanogens is still a matter of debate (Fenchel and Finlay 1995; Tokura et al. 1999; Regensbogenova et al. 2004).

Because the methanogens (regardless of being endo- or episymbiotic) rely on substrates provided by the host, the properties of the endosymbiont might provide some information about the metabolic characteristics of the host. The Vogels and Stumm group succeeded in cultivating a number of putative methanogenic endosymbionts from the anaerobic ciliates *Metopus striatus*, *Metopus contortus* and *Plagiopyla nasuta*, from the amoeboflagellate *Psalteriomonas vulgaris* and from the giant amoeba *Pelomyxa palustris* (van Bruggen et al. 1984, 1986, 1988; Goosen et al. 1988; see Fenchel and Finlay 1995 for more references and discussion). The conclusion from these studies was that many endosymbionts were similar if not identical to well-known free-living methanogens, e.g. *Methanobacterium formicicum*. Only the putative endosymbiont from *Metopus contortus* seemed to represent a new type of methanogen, i.e. *Methanoplanus endosymbiosus*. The latter host, *Metopus*, belongs to the same taxon as *N. ovalis*, which makes it likely that this ciliate (*Metopus*) possesses a similar mode of pyruvate metabolism. The metabolic properties of the methanogenic endosymbiont *M. formicium* suggested that these might be capable of using other substrates besides hydrogen and CO_2, e.g. formate (Dong et al. 1994). This argues again for metabolic diversity among ciliate hydrogenosomes. This metabolic diversity could provide hard arguments for multiple origins of the hydrogenosomes, but unfortunately, metabolic data of both hosts and symbionts are scarce.

5 Evolutionary Aspects

Phylogenetic studies indicate that hydrogenosomes have evolved independently at least three to four times in ciliates (Fig. 1; Clarke et al. 1993; Embley and Finlay 1994; Embley et al. 1995, 2003; Fenchel and Finlay 1995; Hirt et al. 1998; Hackstein et al. 2001, 2002). The existence in *N. ovalis* of a 'missing link', an organel with characteristics of anaerobic mitochondria as well as of hydrogenosomes, demonstrates that hydrogenosome-bearing ciliates can evolve from mitochondriate ciliates (Martin 2005). Albeit that the patchy distribution of hydrogenosomes alone is not sufficient to prove multiple, independent origins of ciliate hydrogenosomes, the existence of such a missing link like the hydrogenosome of *N. ovalis* provides a possible scenario for the evolution of hydrogenosomes from mitochondria, but could also be used to promote the possibility that 'anaerobic enzymes' were already present in the eukaryote common ancestor (Müller et al. 2012). Apparently, hydrogenosomes in ciliates can evolve 'easily' by evolutionary tinkering from mitochondria in the course of the adaptation of their hosts to anaerobic/microaerobic environments. This happened several times independently in the evolution of ciliates—at least three independent origins are supported by the existence of three different types of hydrogenosomes in the ciliates that have been studied so far.

It has remained unclear until now whether or not all anaerobic ciliates possess hydrogenosomes, in particular those anaerobes that do not possess endosymbiotic methanogens. Theoretically, anaerobic ciliates could possess anaerobic mitochondria, or hydrogenosomes, or they could even have lost ATP-generating organelles completely. In that case, they most likely host MROs class 5, mitosomes, just like *Giardia* and *Entamoeba* spp., which are completely dependent on cytosolic reactions for the production of ATP (Table 1; see, e.g. Hackstein et al. 2006 for discussion). However, the presence of these elusive organelles has not been studied systematically and in more detail in ciliates thus far—with a few remarkable exceptions to be discussed below.

It has already been addressed that (at least) among the rumen ciliates, species with mitosomes might exist since there is evidence that certain rumen ciliates, such as *Entodinium simplex*, *Entodinium caudatum*, *Diploplastron affine*, *Ophryoscolex caudatus*, *Eremoplastron bovis* and *Ostracodinium obtusum bilobum*, did not exhibit detectable hydrogenase activity in the particulate cell fraction and, therefore, by definition do not contain hydrogenosomes (Yarlett et al. 1984). Also, electron microscopy did not reveal the presence of mitochondrial-shaped organelles or typical hydrogenosomes in certain species of rumen and marsupial gut ciliates; a systematic search for mitosomes, however, has not been performed (Williams and Coleman 1992; Cameron and O'Donoghue 2002b). The observation that also PFO and malate dehydrogenase (decarboxylating) activities (Yarlett et al. 1984) are not enhanced in the particulate cell fraction, together with a low cytoplasmic hydrogenase activity, might argue for the absence of hydrogenosomes, but potentially for the presence of mitosomes. However, until now, there are no additional data that could support this speculation.

In recent years it has become clear that there are not only mitochondria that function as described in biochemical textbooks, but that many different types of mitochondria exist (Table 1), which exhibit a large variety in metabolic properties (Tielens et al. 2002; Müller et al. 2012). It is now also clear that the same holds true for hydrogenosomes. Firstly, there are obvious metabolic differences between the hydrogenosomes of *Trichomonads* and those of anaerobic fungi and of anaerobic ciliates. Secondly, substantial differences in metabolism also exist within the large community of anaerobic ciliates (Figs. 3, 4 and 5).

References

Akhmanova A, Voncken F, van Alen T, van Hoek A, Boxma B, Vogels G, Veenhuis M, Hackstein JHP (1998) A hydrogenosome with a genome. Nature 396:527–528

Biagini GA, Hayes AJ, Suller MTE, Winters C, Finlay BJ, Lloyd D (1997) Hydrogenosomes of *Metopus contortus* physiologically resemble mitochondria. Microbiology 143:1623–1629

Boxma B, Voncken F, Jannink S, van Alen T, Akhmanova A, van Weelden SWH, van Hellemond JJ, Ricard G, Huynen M, Tielens AGM, Hackstein JHP (2004) The anaerobic chytridiomycete fungus *Piromyces* sp. E2 produces ethanol via pyruvate: formate lyase (PFL) and an alcohol dehydrogenase E (ADHE). Mol Microbiol 51:1389–1399

Boxma B, deGraaf RM, van der Staay GWM, van Alen TA, Ricard G, Gabaldon T, van Hoek AHAM, Moon-van der Staay SY, Koopman WJH, van Hellemond JJ, Tielens AGM, Friedrich T, Veenhuis M, Huynen MA, Hackstein JHP (2005) An anaerobic mitochondrion that produces hydrogen. Nature 434:74–79

Cameron SL, O'Donoghue PJ (2002a) The ultrastructure of *Amylovorax dehori tyi* comb. Nov. and erection of the Amylovoracidae fam. Nov. (Ciliophora: Trichostomatia). Eur J Protistol 38:29–44

Cameron SL, O'Donoghue PJ (2002b) The ultrastructure of *Macropodinium moiri* and revised diagnosis of the Macropodiniidae (Litostomatea: Trichostomatia). Eur J Protistol 38:79–194

Clarke KJ, Finlay BJ, Esteban G, Guhl BE, Embley TM (1993) *Cyclidium porcatum* N. sp. – a free-living anaerobic scuticociliate containing a stable complex of hydrogenosomes, eubacteria and archaeobacteria. Eur J Protistol 29:262–270

Corliss JO (1979) The ciliated protozoa: characterization, classification, and guide to the literature. Pergamon Press, London

De Graaf RM, van Alen TA, Dutilh BE, Kuiper JW, van Zoggel HJ, Huynh MB, Görtz HD, Huynen MA, Hackstein JH (2009) The mitochondrial genomes of the ciliates *Euplotes minuta* and *Euplotes crassus*. BMC Genomics 10:514

De Graaf RM, Ricard G, van Alen TA, Duarte I, Dutilh BE, Burgtorf C, Kuiper JWP, van der Staay GWM, Tielens AGM, Huynen MA, Hackstein JHP (2011) The organellar genome and metabolic potential of the hydrogen-producing mitochondrion of *Nyctotherus ovalis*. Mol Biol Evol 28:2379–2391

Dong XZ, Plugge CM, Stams AJM (1994) Anaerobic degradation of propionate by a mesophilic acetogenic bacterium in coculture and triculture with different methanogens. Appl Environ Microbiol 60:2834–2838

Dyall SD, Brown MT, Johnson PJ (2004) Ancient invasions: from endosymbionts to organelles. Science 304:253–257

Ellis JE, McIntyre PS, Saleh M, Williams AG, Lloyd D (1991a) Influence of CO2 and low concentrations of O_2 on fermentative metabolism of the ruminal ciliate *Polyplastron multivesiculatum*. Appl Environ Microbiol 57:1400–1407

Ellis JE, McIntyre PS, Saleh M, Williams AG, Lloyd D (1991b) Influence of CO2 and low concentrations of O_2 on fermentative metabolism of the rumen ciliate *Dasytricha ruminantium*. J Gen Microbiol 137:1409–1417
Ellis JE, McIntyre PS, Saleh M, Williams AG, Lloyd D (1991c) The influence of ruminal concentrations of O2 and CO_2 on fermentative metabolism of the rumen entodiniomorphid ciliate *Eudiplodinium maggii*. Curr Microbiol 23:245–251
Embley TM (2006) Multiple secondary origins of the anaerobic lifestyle in eukaryotes. Philos Trans R Soc B Biol Sci 361:1055–1067
Embley TM, Finlay BJ (1994) The use of small-subunit ribosomal-RNA sequences to unravel the relationships between anaerobic ciliates and their methanogen endosymbionts. Microbiology 140:225–235
Embley TM, Martin W (2006) Eukaryotic evolution, changes and challenges. Nature 440:623–630
Embley TM, Finlay BJ, Dyal PL, Hirt RP, Wilkinson M, Williams AG (1995) Multiple origins of anaerobic ciliates with hydrogenosomes within the radiation of aerobic ciliates. Proc R Soc Lond Ser B Biol Sci 262:87–93
Embley TM, van der Giezen M, Horner DS, Dyal PL, Bell S, Foster PG (2003) Hydrogenosomes, mitochondria and early eukaryotic evolution. IUBMB Life 55:387–395
Fenchel T, Finlay BJ (1995) Ecology and evolution in anoxic worlds. Oxford University Press, New York
Finlay BJ, Fenchel T (1989) Hydrogenosomes in some anaerobic protozoa resemble mitochondria. FEMS Microbiol Lett 65:311–314
Gentekaki E, Curtis BA, Stairs CW, Klimeš V, Eliáš M, Salas-Leiva DE, Herman EK, Eme L, Arias MC, Henrissat B, Hilliou F, Klute MJ, Suga H, Malik SB, Pightling AW, Kolisko M, Rachubinski RA, Schlacht A, Soanes DM, Tsaousis AD, Archibald JM, Ball SG, Dacks JB, Clark CG, van der Giezen M, Roger AJ (2017) Extreme genome diversity in the hyper-prevalent parasitic eukaryote *Blastocystis*. PLoS Biol 15(9):e2003769
Gijzen HJ, Broers CAM, Barughare M, Stumm CK (1991) Methanogenic bacteria as endosymbionts of the ciliate *Nyctotherus ovalis* in the cockroach hindgut. Appl Environ Microbiol 57:1630–1634
Goosen NK, Horemans AMC, Hillebrand SJW, Stumm CK, Vogels GD (1988) Cultivation of the sapropelic ciliate *Plagiopyla nasuta* Stein and isolation of the endosymbiont *Methanobacterium formickum*. Arch Microbiol 150:165–170
Goosen NK, Van der Drift C, Stumm CK, Vogels GD (1990) End products of metabolism in the anaerobic ciliate *Trimyema compressutn*. FEMS Microbiol Lett 69:171–175
Hackstein JHP, Akhmanova A, Boxma B, Harhangi HR, Voncken FGJ (1999) Hydrogenosomes: eukaryotic adaptations to anaerobic environments. Trends Microbiol 7:441–447
Hackstein JHP, Akhmanova A, Voncken F, van Hoek A, van Alen T, Boxma B, Moon, van der Staay SY, van der Staay G, Leunissen J, Huynen M, Rosenberg J, Veenhuis M (2001) Hydrogenosomes: convergent adaptations of mitochondria to anaerobic environments. Zool Anal Complex Syst 104:290–302
Hackstein JHP, van Hoek AHAM, Leunissen JAM, Huynen M (2002) Anaerobic ciliates and their methanogenic endosymbionts. In: Seckbach J (ed) Symbiosis: mechanisms and model systems. Kluwer Academic, Dordrecht, pp 451–464
Hackstein JHP, Tjaden J, Huynen M (2006) Mitochondria, hydrogenosomes and mitosomes: products of evolutionary tinkering! Curr Genet 50:225–245
Hackstein JHP, Baker SE, van Hellemond JJ, Tielens AGM (2019) Hydrogenosomes of anaerobic fungi: an alternative way to adapt to anaerobic environments. In: Tachezy J (ed) Hydrogenosomes and mitosomes: mitochondria of anaerobic eukaryotes. Springer, Heidelberg
Hirt RP, Wilkinson AG, Embley TM (1998) Molecular and cellular evolution of ciliates: a phylogenetic perspective. In: Coombs GH, Vickerman K, Sleigh MA, Warren A (eds) Evolutionary relationships among protozoa. Chapman and Hall, London, pp 327–340

Hrdý I, Müller M, Tachezy J (2019) Metabolism of trichomonad hydrogenosomes. In: Tachezy J (ed) Hydrogenosomes and mitosomes: mitochondria of anaerobic eukaryotes. Springer, Heidelberg

Jeanmougin F, Thompson JD, Gouy M, Higgins DG, Gibson TJ (1998) Multiple sequence alignment with Clustal X. Trends Biochem Sci 23:403–405

Lantsman Y, Tan KS, Morada YN (2008) Biochemical characterization of a mitochondrial-like organelle from *Blastocystis* sp. subtype 7. Microbiology 154:2757–2766

Lloyd D, Hillman K, Yarlett N, Williams AG (1989) Hydrogen-production by rumen holotrich protozoa – effects of oxygen and implications for metabolic control by in situ conditions. J Protozool 36:205–213

Martin W (2005) The missing link between hydrogenosomes and mitochondria. Trends Microbiol 13:457–459

Müller M (1993) The hydrogenosome. J Gen Microbiol 139:2879–2889

Müller M, Mentel M, van Hellemond JJ, Henze K, Woehle C, Gould SB, Yu RY, van der Giezen M, Tielens AGM, Martin WF (2012) Biochemistry and evolution of anaerobic energy metabolism in eukaryotes. Microbiol Mol Biol Rev 76:444–495

Paul RG, Williams AG, Butler RD (1990) Hydrogenosomes in the rumen entodiniomorphid ciliate *Polyplastron multivesiculatum*. J Gen Microbiol 136:1981–1989

Perez-Brocal V, Clark CG (2008) Analysis of two genomes from the mitochondrion-like organelle of the intestinal parasite *Blastocystis*: complete sequences, gene content, and genome organization. Mol Biol Evol 25:2475–2482

Regensbogenova M, McEwan NR, Javorsky P, Kisidayova S, Michalowski T, Newbold CJ, Hackstein JHP, Pristas P (2004) A re-appraisal of the diversity of the methanogens associated with the rumen ciliates. FEMS Microbiol Lett 238:307–313

Saitou N, Nei M (1987) The neighbour-joining method: a new method for reconstructing phylogenetic trees. Mol Biol Evol 4:406–425

Snyers L, Hellings P, Bovy-Kesler C, Thines-Sempoux D (1982) Occurrence of hydrogenosomes in the rumen ciliates *Ophryoscolecidae*. FEBS Lett 137:35–39

Stechmann A, Hamblin K, Perez-Brocal V, Gaston D, Richmond GS, van der Giezen M, Clark CG, Roger AJ (2008) Organelles in *Blastocystis* that blur the distinction between mitochondria and hydrogenosomes. Curr Biol 18(8):580–585

Struder-Kypke MC, Wright ADG, Foissner W, Chatzinotas A, Lynn DH (2006) Molecular phylogeny of litostome ciliates (Ciliophora, Litostomatea) with emphasis on free-living Haptorian genera. Protist 157:261–278

Tielens AGM, Rotte C, van Hellemond JJ, Martin W (2002) Mitochondria as we don't know them. Trends Biochem Sci 27:564–572

Tielens AGM, van Grinsven K, Henze K, van Hellemond JJ, Martin W (2010) Acetate formation in the energy metabolism of parasitic helminths and protists. Int J Parasitol 40:387–397

Tokura M, Chagan I, Ushida K, Kojima Y (1999) Phylogenetic study of methanogens associated with rumen ciliates. Curr Microbiol 39:123–128

van Bruggen JJA, Stumm CK, Vogels GD (1983) Symbiosis of methanogenic bacteria and sapropelic protozoa. Arch Microbiol 136:89–95

van Bruggen JJA, Zwart KB, van Assema RM, Stumm CK, Vogels GD (1984) *Methanobacterium formicicum*, an endosymbiont of the anaerobic ciliate *Metopus striatus* McMurrich. Arch Microbiol 139:1–7

van Bruggen JJA, Zwart KB, Hermans JGF, van Hove EM, Stumm CK, Vogels GD (1986) Isolation and characterization of *Methanoplanus endosymbiosus* sp. nov, an endosymbiont of the marine sapropelic ciliate *Metopus contortus* Quennerstedt. Arch Microbiol 144:367–374

van Bruggen JJA, van Rens GLM, Geertman EJM, Zwart KB, Stumm CK, Vogels GD (1988) Isolation of a methanogenic endosymbiont of the sapropelic ameba *Pelomyxa palustris* Greeff. J Protozool 35:20–23 van Hoek AHAM, van Alen TA, Sprakel VSI, Hackstein JHP, Vogels GD (1998) Evolution of anaerobic ciliates from the gastrointestinal tract: phylogenetic analysis of the ribosomal repeat from *Nyctotherus ovalis* and its relatives. Mol Biol Evol 15:1195–1206

van Hoek AHAM, Sprakel VSI, Van Alen TA, Theuvenet APR, Vogels GD, Hackstein JHP (1999) Voltage-dependent reversal of anodic galvanotaxis in *Nyctotherus ovalis*. J Eukaryot Microbiol 46:427–433

van Hoek AHAM, Akhmanova AS, Huynen MA, Hackstein JHP (2000a) A mitochondrial ancestry of the hydrogenosomes *of Nyctotherus ovalis*. Mol Biol Evol 17:202–206

van Hoek AHAM, van Alen TA, Sprakel VSI, Leunissen JAM, Brigge T, Vogels GD, Hackstein JHP (2000b) Multiple acquisition of methanogenic archaeal symbionts by anaerobic ciliates. Mol Biol Evol 17:251–258

Vogels GD, Hoppe WF, Stumm CK (1980) Association of methanogenic bacteria with rumen ciliates. Appl Environ Microbiol 40:608–612

Wawrzyniak I, Roussel M, Diogon M, Couloux A, Texier C, Tan KSW, Vivares CP, Delbac F, Wincker P, El Alaoui H (2008) Complete circular DNA in the mitochondria-like organelles of *Blastocystis hominis*. Int J Parasitol 38:1377–1382

Williams AG, Coleman GS (1992) The rumen protozoa. Springer, New York

Yarlett N, Hann AC, Lloyd D, Williams A (1981) Hydrogenosomes in the rumen protozoan *Dasytricha ruminantium* Schuberg. Biochem J 200:365–372

Yarlett N, Lloyd D, Williams AG (1982) Respiration of the rumen ciliate *Dasytricha ruminantium* Schuberg. Biochem 206:259–266

Yarlett N, Hann AC, Lloyd D, Williams AG (1983) Hydrogenosomes in a mixed isolate of *Isotricha prostoma* and *Isotricha intestinalis* from ovine rumen contents. Comp Biochem Physiol B Biochem Mol Biol 74:357–364

Yarlett N, Coleman GS, Williams AG, Lloyd D (1984) Hydrogenosomes in known species of rumen entodiniomorphid protozoa. FEMS Microbiol Lett 21:15–19

Yarlett N, Lloyd D, Williams AG (1985) Butyrate formation from glucose by the rumen protozoan *Dasytricha ruminantium*. Biochem J 228:187–192

Zwart KB, Goosen NK, van Schijndel MW, Broers CAM, Stumm CK, Vogels GD (1988) Cytochemical-localization of hydrogenase activity in the anaerobic protozoa *Trichomonas vaginalis, Plagiopyla nasuta* and *Trimyema compressum*. J Gen Microbiol 134:2165–2170

Metabolism of Trichomonad Hydrogenosomes

Ivan Hrdý, Jan Tachezy, and Miklós Müller

Contents

1 Introduction 128
2 The Hydrogenosomal Membrane 128
3 Energy Metabolism 129
4 Proteins of the Core Catabolic Pathway 135
  4.1 Pyruvate:Ferredoxin Oxidoreductase 135
  4.2 Ferredoxin 136
  4.3 Hydrogenase 138
  4.4 Malic Enzyme 139
  4.5 NADH Dehydrogenase 140
  4.6 Succinyl-CoA:Acetate CoA Transferase 141
  4.7 Succinyl-CoA Synthetase 141
  4.8 Adenylate Kinase 141
5 Interaction with Oxygen and Reactive Oxygen Species 142
6 Iron-Sulfur Cluster Assembly Machinery 146
7 Amino Acid and Polyamine Metabolism 148
  7.1 Serine Hydroxymethyltransferase and the Case of Glycine Decarboxylase Complex 148
  7.2 Polyamine Metabolism 150
8 Concluding Remarks 151
References 151

Abstract Trichomonad hydrogenosomes are the best studied organelles of their kind to date. Their role in energy metabolism, antioxidant defense, and the assembly of iron-sulfur centers, the vital cofactors of a number of essential proteins, has been reliably established. In this chapter, we summarize our knowledge of trichomonad

I. Hrdý (✉) · J. Tachezy
Faculty of Science, Department of Parasitology, BIOCEV, Charles University, Vestec, Czech Republic
e-mail: hrdy@natur.cuni.cz

M. Müller
The Rockefeller University, New York, NY, USA

J. Tachezy (ed.), *Hydrogenosomes and Mitosomes: Mitochondria of Anaerobic Eukaryotes*, Microbiology Monographs 9,
https://doi.org/10.1007/978-3-030-17941-0_6

hydrogenosome biochemistry, particularly the energy-linked pathway, the oxygen- and reactive oxygen species-related biochemistry, and the iron-sulfur cluster assembly machinery. The structure of the proteins constituting the core pathway is dealt with in some detail. We also attempt to incorporate the results of the *T. vaginalis* genome annotation, proteomic, and biochemical studies into the metabolic scheme of the hydrogenosome.

1 Introduction

Hydrogenosomes are defined as eukaryotic organelles that produce molecular hydrogen, a definition that reflects a single biochemical property and does not imply any particular morphology or other common biochemical traits for such organelles from different organisms. Hydrogenosomes have been detected in various protists that live in anoxic and hypoxic habitats. The history of evolutionary appearance of hydrogenosomes in different organisms is still a much debated topic (see chapter "The Evolution of Oxygen Independent Energy Metabolism in Eukaryotes with Hydrogenosomes and Mitosomes").

Hydrogenosomes were first detected in anaerobic or microaerophilic parabasalid flagellates, the trichomonads. The first species to be explored was the cattle parasite, *Tritrichomonas foetus* (Lindmark and Müller 1973). Soon thereafter, these organelles were detected in the human parasite, *Trichomonas vaginalis* (Lindmark et al. 1975), and in a reptile parasite, *Monocercomonas* sp. (Lindmark and Müller 1974). Hydrogenosomes of other protists and fungi share the ability to form hydrogen as a metabolic end product but differ from the trichomonad organelles in many aspects of their metabolism (see chapters "Hydrogenosomes of Anaerobic Ciliates" and "Hydrogenosomes of Anaerobic Fungi: An Alternative Way to Adapt to Anaerobic Environments"). This circumstance warrants a separate discussion of metabolism and enzymology of trichomonad hydrogenosomes in this chapter. Although earlier studies concentrated more on *T. foetus*, later the main interest shifted to the medically important *T. vaginalis*. In the present chapter, we also keep the focus on *T. vaginalis* and refer to *T. foetus* only occasionally. This bias is further justified by the fact that the genome annotation of *T. vaginalis* has been completed and published in 2007 (Carlton et al. 2007), providing support to existing knowledge and additional information on putative metabolic properties of its hydrogenosome.

2 The Hydrogenosomal Membrane

Trichomonad hydrogenosomes are bounded by two unit membranes that are more closely apposed than the two membranes surrounding typical mitochondria (Benchimol and De Souza 1983). The properties of these membranes have not been studied in detail. An early work reported the presence of phosphatidylethanolamine, phosphatidylserine, phosphatidylcholine, and cardiolipin in the hydrogenosomes of *Tritrichomonas foetus* (Čerkasovová et al. 1976). The discovery of cardiolipin in the hydrogenosomal membrane was regarded as having particular significance from the evolutionary standpoint, since this phospholipid is present almost exclusively in bacterial and inner mitochondrial membranes, where it plays important functional and structural roles. However, the presence of cardiolipin in the hydrogenosomes was later refuted (Paltauf and Meingassner 1982), and this conclusion was confirmed by the work of Guschina et al. (2009). No further information exists on the lipid composition of the hydrogenosomal membrane. Ubiquinone, a lipid-soluble coenzyme mainly participating in mitochondrial electron transport but having also additional functions, has been identified at low levels (up to two orders of magnitude lower than in mitochondria-possessing cells) in extracts of *T. foetus*, but neither the subcellular localization nor the possible role of this coenzyme in *T. foetus* has been explored (Ellis et al. 1994a).

The hydrogenosomal membrane displays selective permeability, thereby presenting an effective barrier to pyridine nucleotides and coenzyme A (Lindmark and Müller 1973; Čerkasov et al. 1978; Steinbüchel and Müller 1986). Transport of metabolic substrates and products across the hydrogenosomal membrane has not been studied, but isolated *T. foetus* hydrogenosomes rapidly accumulate both radiolabeled pyruvate and malate (our unpublished data).

That hydrogenosomes maintain membrane potential was inferred from the experiments where the cell viability was assessed using the JC-1 fluorescent probe (Vilela et al. 2010; Mallo et al. 2013; Huang et al. 2015); however, the mechanism of formation or the values of hydrogenosomal membrane potential were never examined.

Hydrogenosomal membrane proteins and protein import machinery are discussed in chapter "Protein Import into Hydrogenosomes and Mitosomes."

3 Energy Metabolism

The only metabolic activity that has been for decades ascribed to the trichomonad hydrogenosome was related to energy metabolism. It has been unequivocally shown that, under anaerobic conditions, isolated intact hydrogenosomes produced roughly equimolar amounts of acetate, CO_2, and hydrogen from pyruvate in a process accompanied by substrate-level phosphorylation of ADP to ATP (Fig. 1). These findings confirmed the metabolic role of hydrogenosome as initially deduced from the enzymatic activities found in the organelle (Lindmark and Müller 1973; Steinbüchel and Müller 1986; Müller 1993) and are now fully supported by the results of the *T. vaginalis* genome sequencing project and proteomic studies (Rada et al. 2011; Schneider et al. 2011; Beltrán et al. 2013). All genes coding for enzymes of the core pathway were identified, mostly in multiple copies and with clearly recognizable hydrogenosomal targeting signals at the amino-termini of the translated

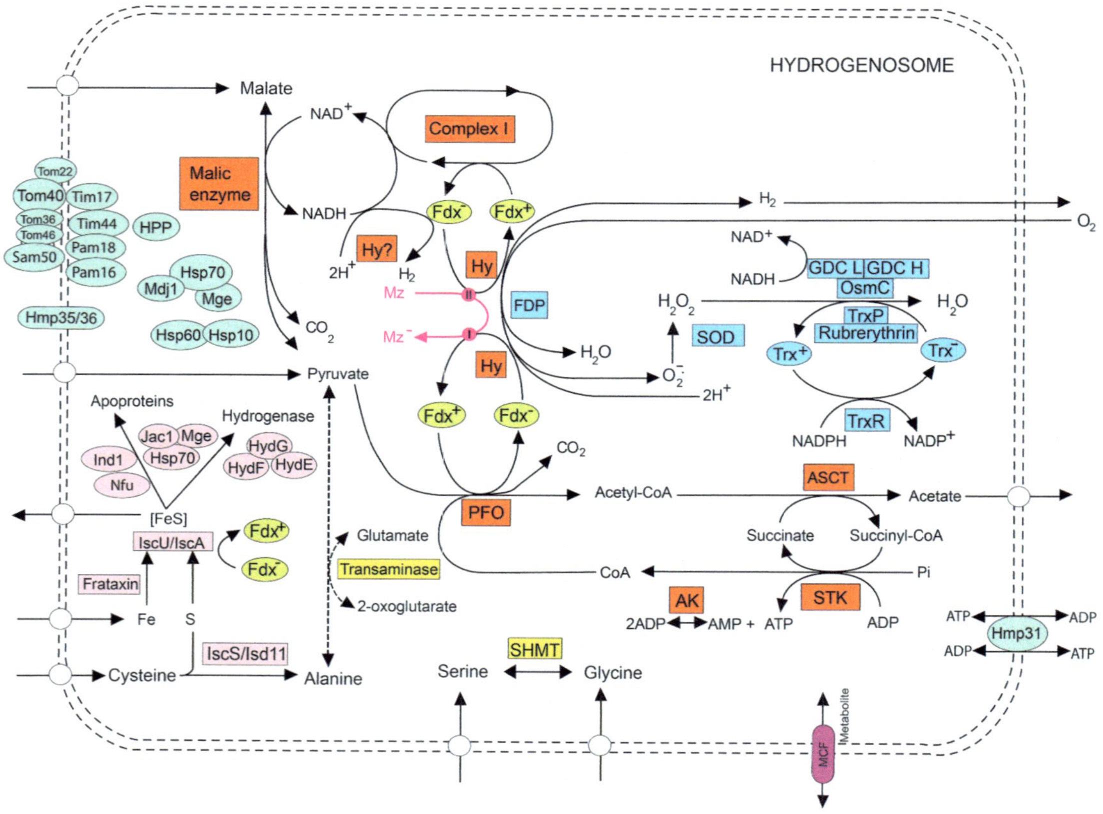

Fig. 1 Proposed metabolic map of the *T. vaginalis* hydrogenosome. Orange rectangles represent enzymes known to be involved in energy metabolism. Ferredoxins (Fdx, green ovals) donate electrons to hydrogenases which subsequently synthesize molecular hydrogen. In the presence of metronidazole (pink,

Mz), Fdx transports electrons derived via oxidative decarboxylation of malate or pyruvate to the drug, resulting in its activation. Predicted protein activities, mostly uncovered by analysis of the genome sequence, are color-coded as follows: pink, iron-sulfur cluster assembly machinery and hydrogenase maturation; blue, oxygen scavenging system; yellow, amino acid metabolism; and cyan, protein translocation and maturation. Open circles located in the membrane indicate unidentified transporters that likely facilitate substrate and metabolite transport. *AK* adenylate kinase, *ASCT* acetate:succinate CoA-transferase, *FDP* A-type flavodiiron protein, *Fdx* ferredoxin, *GDC L* glycine decarboxylase complex L protein, *GDC H* glycine decarboxylase complex H protein, *Hy* hydrogenase, *Hy?* putative bifurcating hydrogenase, *PFOR* pyruvate:ferredoxin oxidoreductase, *SOD* superoxide dismutase, *STK* succinate thiokinase (succinate-CoA ligase, succinyl-CoA synthetase), *SHMT* serine hydroxymethyltransferase, *Trx* thioredoxin, *TrxP* thioredoxin peroxidase, *TrxR* thioredoxin reductase. Dashed lines represent presumed reactions that await experimental verification

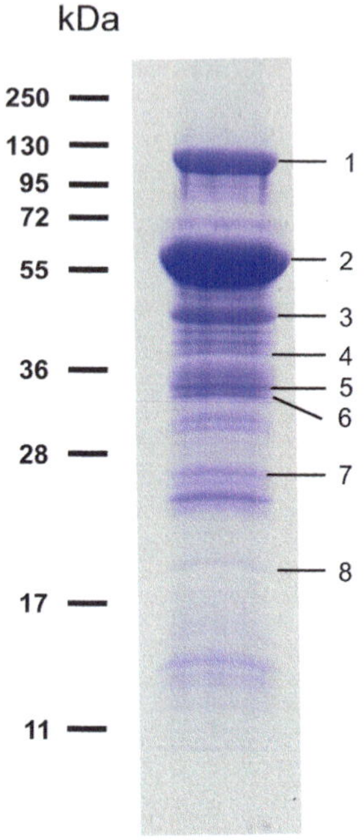

Fig. 2 SDS-polyacrylamide gel electrophoresis of *Trichomonas vaginalis* hydrogenosomes purified by isopycnic centrifugation on a Percoll gradient. (1) PFOR, (2) malic enzyme, 64 kDa hydrogenase and Cpn 60, (3) succinyl-CoA synthetase β subunit, (4) HMP35, (5) succinyl-CoA synthetase α subunit, (6) HMP31 (ATP/ADP carrier), (7) adenylate kinase, (8) thiol peroxidase. Well resolved but unmarked bands mostly belong to malic enzyme fragments or unknown proteins. Proteins were identified by mass spectroscopy. The 12% gel is stained with Coomassie Brilliant Blue R 250. (Authors' original)

gene products (Carlton et al. 2007; van Grinsven et al. 2008). Moreover, most enzymes of the core pathway could be identified on SDS-polyacrylamide gels by immunological methods and/or mass spectroscopy, showing that these proteins are the most prominent among all hydrogenosomal proteins (Fig. 2).

The substrates for the core catabolic pathway of hydrogenosomes originate from glycolysis in the cytosol, with pyruvate being the main fuel. Malate is also utilized after it is oxidatively decarboxylated to pyruvate by malic enzyme inside hydrogenosome. While pyruvate could be produced in *T. vaginalis* from phosphoenolpyruvate (PEP) by conventional pyruvate kinase (Mertens et al. 1992), pyruvate, orthophosphate dikinase (Slamovits and Keeling 2006), and cytosolic NADP-specific malic enzyme (Hrdý et al. 1993; Doležal et al. 2004) (apart from the activity of aminotransferases), malate is formed by cytosolic malate dehydrogenase reducing oxaloacetate arising from PEP by the activity of PEP carboxykinase. The relative carbon flux through the reactions starting from PEP is unknown (and may vary depending on conditions), as is the contribution of the two substrates to hydrogenosomal catabolism. It has been shown that malate can be excreted, rather than consumed, by the organelle. This observation was made during a study of isolated hydrogenosomes under an elevated concentration of CO_2, accounting for the lowered production of hydrogen (Steinbüchel and Müller 1986). Similarly, a high CO_2 concentration was observed to lower the hydrogenosomal acetate and hydrogen production in *T. vaginalis* cultures, and it was suggested that reductive carboxylation of pyruvate to malate occurs in the hydrogenosome under these conditions (Paget and Lloyd 1990). On the other hand, the thermodynamically favorable oxidative decarboxylation of malate by malic enzyme in the hydrogenosomes occurs in cultures of *T. vaginalis* strains that were forced to develop high resistance to the antibiotic metronidazole, a drug that is effective against various anaerobic bacteria and protists (Leitsch 2017). These cells lack the enzymatic activities necessary to reductively activate the drug, primarily the activity of pyruvate:ferredoxin oxidoreductase (PFOR). However, this activity is not the only one in the hydrogenosomes that is able to reduce metronidazole. It has been shown that another pathway of metronidazole activation exists in *T. vaginalis* hydrogenosomes that is dependent on the activity of NADH dehydrogenase (see below) that reduces ferredoxin (the electron transporter directly involved in organellar metronidazole reduction) and that uses NADH resulting from the activity of malic enzyme as a reductant. When high-level metronidazole resistance is developed in vitro, *T. vaginalis* strains not only lose PFOR but also malic enzyme, showing that the enzyme functions in the direction of malate decarboxylation and toward the production of reducing power in the form of NADH (Rasoloson et al. 2002; Hrdý et al. 2005). Thus, it seems that, in vivo, the reaction of malic enzyme proceeds in both directions and is affected by the concentration of reactants, possibly also by the actual concentration of CO_2. Indeed, the excretion of malate by *T. vaginalis* cultures was confirmed in a recent study (Westrop et al. 2017).

Within the hydrogenosome, pyruvate is oxidatively decarboxylated to acetyl coenzyme A and CO_2 (Fig. 1). This reaction is catalyzed by an iron-sulfur protein, pyruvate:ferredoxin oxidoreductase, an enzyme characteristic of certain anaerobic and nitrogen-fixing bacteria that otherwise occurs only in a few anaerobic or facultatively anaerobic (such as *Chlamydomonas*) eukaryotes. Electrons released from pyruvate are transferred to ferredoxin, a low molecular weight electron carrier protein. This is also an iron-sulfur protein, with a [2Fe2S] cluster. Reduced ferredoxin is reoxidized by [FeFe] hydrogenase in a reaction that reduces protons to molecular hydrogen. Energy of the thioester bond of acetyl CoA resulting from the activity of PFOR is conserved in one molecule of ATP that is formed in two consecutive steps. First, acetate:succinate CoA transferase releases acetate and transfers the CoA moiety to succinate. The succinyl-CoA thus formed serves as a substrate for energy-conserving succinyl-CoA synthetase [ADP/GDP-forming, also known as succinate thiokinase (STK)] that catalyzes the phosphorylation of ADP (GDP) to ATP (GDP) while releasing CoA-SH and succinate, which reenter the catalytic cycle. Succinyl-CoA synthetase is the only enzyme of the mitochondrial Krebs cycle that is shared by the two organelles. The ubiquitous enzyme of energy metabolism, ATP:AMP phosphotransferase (adenylate kinase), is also present in *T. vaginalis* hydrogenosomes (Declerck and Müller 1987).

When malate is used as a hydrogenosomal substrate, it is oxidatively decarboxylated to pyruvate and CO_2 by malate dehydrogenase (decarboxylating) (malic enzyme) (Fig. 1). This preferentially NAD-specific enzyme (Drmota et al. 1996) is by far the most abundant hydrogenosomal protein (Fig. 2). Utilization of malate results in the production of reduced pyridine nucleotide coenzyme NADH. In order to maintain the redox balance, the NADH needs to be reoxidized; however, the enzyme that would enable this has remained elusive for a long time. The activity that reduces the artificial electron acceptor methyl viologen at the expense of NADH, and presumably the same activity that transfers electrons between reduced ferredoxin and NAD^+, was identified in the mid-1980s (Steinbüchel and Müller 1986; Thong and Coombs 1987), but the enzyme itself was isolated only in 2004 with a surprising result. The NADH:ferredoxin (and methyl viologen, ferricyanide, and other electron acceptors) oxidoreductase activity was found to belong to a heterodimeric protein whose subunits turned out to be homologous to the 51 and 24 kDa subunits of respiratory complex 1 (also called NADH dehydrogenase or NADH:ubiquinone oxidoreductase) of mitochondria and aerobic bacteria. These two subunits are part of the hydrophilic peripheral arm of this most complicated respiratory complex and function as an electron-input site to the respiratory chain (Hrdý et al. 2004). The finding of complex 1 core subunits in trichomonad hydrogenosomes disclosed an unexpected trait shared by mitochondria and trichomonad hydrogenosomes (in addition to the similar mode of protein import and processing, see chapter "Protein Import into Hydrogenosomes and Mitosomes," and iron cluster assembly machinery, see below) showing that NADH is recycled by homologous proteins in a similar way in these two organelles. The fate of electrons derived from NADH is markedly different, however. In mitochondria, the transfer of electrons to ubiquinone by respiratory complex 1 (and eventually to oxygen in complex 4) is linked to the formation of a proton gradient across the inner mitochondrial membrane that serves as a driving force for ATP synthesis by membrane-anchored F_1F_O ATP synthase. Hydrogenosomes lack the electron transport chain-driven extrusion of protons and ATP synthase. Instead, in addition to its role in the oxidation of pyruvate, they apparently recruit ferredoxin also as an acceptor of electrons derived from NADH and may thus link the NADH oxidation to production of hydrogen by hydrogenase, for which reduced ferredoxin is a physiological electron donor. The exact mechanism by which the complex 1 remnant may catalyze thermodynamically unfavorable transfer of electrons between NADH, ferredoxin, and protons remains to be clarified, but the analogy to trimeric electron-bifurcating hydrogenase from *Thermotoga maritima* that utilizes NADH and reduced ferredoxin simultaneously to produce H_2 (Schut and Adams 2009) could be considered.

Besides the above described involvement in carbohydrate metabolism, hydrogenosomes have other, and indispensable, functions. In fact, their role in energy metabolism may not be the one which cannot be missed. Hydrogenosomes lacking their typical activities involved in ATP generation persist in trichomonad cells that exhibit extreme levels of resistance to metronidazole [induced in laboratory conditions by long cultivation of trichomonads under increasing drug pressure (Rasoloson et al. 2002)]. The true indispensable role of hydrogenosomes is likely connected to their iron-sulfur (FeS) cluster assembly system (see below) that is located in them and that is shared between hydrogenosomes, mitochondria, and certain mitosomes.

4 Proteins of the Core Catabolic Pathway

The few proteins involved in the core catabolic pathway of hydrogenosomes represent a curious patchwork of enzymes that can be found only in a very limited number of eukaryotes and of enzymes with an almost ubiquitous distribution in eukaryotes. Of course, it is the presence of enzymes with limited distribution that makes the trichomonad hydrogenosome a hydrogenosome. The proteins in question have all been studied with enzymological and phylogenetic methods. It has to be mentioned that earlier studies already have revealed the presence of multiple genes coding for most of these proteins, a point that was further supported by the analysis of the complete genome of *T. vaginalis* (Carlton et al. 2007). For some proteins, these multiple genes code for very similar isoforms, but in the case of others, the putative products are quite dissimilar and may have different functions.

4.1 *Pyruvate:Ferredoxin Oxidoreductase*

Pyruvate:ferredoxin oxidoreductase (PFOR, EC 1.2.7.1) is a hallmark enzyme of hydrogenosomes, together with hydrogenase. The protein catalyzes the reversible, CoA-dependent oxidative decarboxylation of pyruvate, releasing acetyl-CoA, CO_2, and two electrons. In *T. vaginalis*, the electron acceptor is [2Fe2S] ferredoxin (Williams et al. 1987). The reaction catalyzed by PFOR is formally identical to that of the pyruvate dehydrogenase complex (PDH) of mitochondria and aerobic bacteria (however, the reaction catalyzed by PDH is irreversible), but structurally the trichomonad enzyme has little in common with the multienzyme dehydrogenase complex. It is an oxygen-sensitive, 120 kDa-subunit dimeric protein that is relatively tightly associated with the hydrogenosomal membrane. It contains non-covalently bound thiamin pyrophosphate cofactor and [4Fe4S] clusters (Williams et al. 1987). The number of FeS clusters has not been determined for the trichomonad enzyme, but the homologous protein with identically spaced critical cysteine residues from *Desulfovibrio africanus* has six clusters per dimer (Chabrière et al. 1999). In addition to the homologues of similar structure that are present in other anaerobic eukaryotes, another homologue can be found in other eukaryotic groups. This is the pyruvate: NADP oxidoreductase (PNO) of *Euglena gracilis* (Inui et al. 1987), the parasitic alveolates *Cryptosporidium* spp. and *Perkinsus* spp. (Mogi and Kita 2010), and of stramenopile *Blastocystis* sp. (Lantsman et al. 2008). PNO possesses an additional carboxy-terminal NADPH-cytochrome P450 reductase domain that enables transfer of electrons from pyruvate to $NADP^+$. Annotation of the *T. vaginalis* genome identified seven different PFOR genes (Carlton et al. 2007), and six of them have been found in hydrogenosome proteome (Schneider et al. 2011). The hydrogenosomal targeting sequences of some PFOR paralogues are very short, consisting of only five amino acid residues, with a conserved arginine at the –3 position relative to the processing site (typically, the conserved arginine is in the –2 position) (Hrdý and Müller 1995). Phylogenetic analysis suggests that all known eukaryotic PFORs are monophyletic (Leger et al. 2013, 2016; Fig. 3). However, in which eubacterial group eukaryotic PFORs originated is still a debated topic.

A medical implication of the presence of PFOR in *T. vaginalis* should be briefly mentioned. PFOR is the key enzyme in the activation of antimicrobial 5-nitroimidazole drugs, such as metronidazole, in the trichomonad hydrogenosome. These drugs are reduced to their cytotoxic form by enzymatic systems that utilize low redox potential electron transporters, as are FeS proteins like PFOR and ferredoxins. Such pathways are absent from aerobic organisms; therefore these drugs are highly selective. Short-lived radicals resulting from the reduction of the nitro group of 5-nitroimidazoles are very reactive, inflicting cellular damage and subsequent cell death (Kulda 1999). In addition, more recent research showed that also cytosolic factors such as thioredoxin reductase are involved in metronidazole activation and toxicity (Leitsch et al. 2009, 2018).

4.2 Ferredoxin

The low molecular weight iron-sulfur protein, ferredoxin, is the key electron transport component of hydrogenosomes. Trichomonad ferredoxin is a small, approximately 10 kDa, soluble protein. It is the univalent electron transporter that links pyruvate:ferredoxin oxidoreductase and NADH dehydrogenase with hydrogenase. This protein was the first hydrogenosomal protein to be purified and characterized (Marczak et al. 1983), the first whose gene was cloned and sequenced (Johnson et al. 1990), and the only one whose crystal structure has been determined (Crossnoe et al. 2002). It was also the first hydrogenosomal protein on which the presence of a short amino-terminal targeting signal has been recognized and its function suggested (Johnson et al. 1990).

When the primary structure of *T. vaginalis* ferredoxin was established, it came as a surprise that it belonged to [2Fe2S]-type ferredoxins, which are usually found in aerobic bacteria and mitochondria. It was expected that the hydrogenosomal ferredoxin would be of the 2[4Fe4S] type, the one that is common in hydrogen-producing anaerobic bacteria, and is also present in *Entamoeba histolytica* (Reeves et al. 1980; Huber et al. 1988) and *Giardia intestinalis* (Townson et al. 1994). The latter species, however, possesses multiple ferredoxins of both types (http://giardiadb.org/giardiadb/).

For a long time it was thought that *T. vaginalis* possessed only a single ferredoxin gene. This view was challenged when it was found that the ferredoxin knockout did not result in decreased susceptibility of transformed trichomonads to metronidazole, since ferredoxin is believed to play a critical role in drug activation (Land et al. 2004). Shortly thereafter, seven ferredoxin genes were recognized in the *T. vaginalis* genome, all with predicted hydrogenosomal target peptides (Carlton et al. 2007). Dominantly expressed is ferredoxin 1, the one that was purified from *T. vaginalis* hydrogenosomes (Gorrell et al. 1984) and the one first sequenced (Johnson et al.

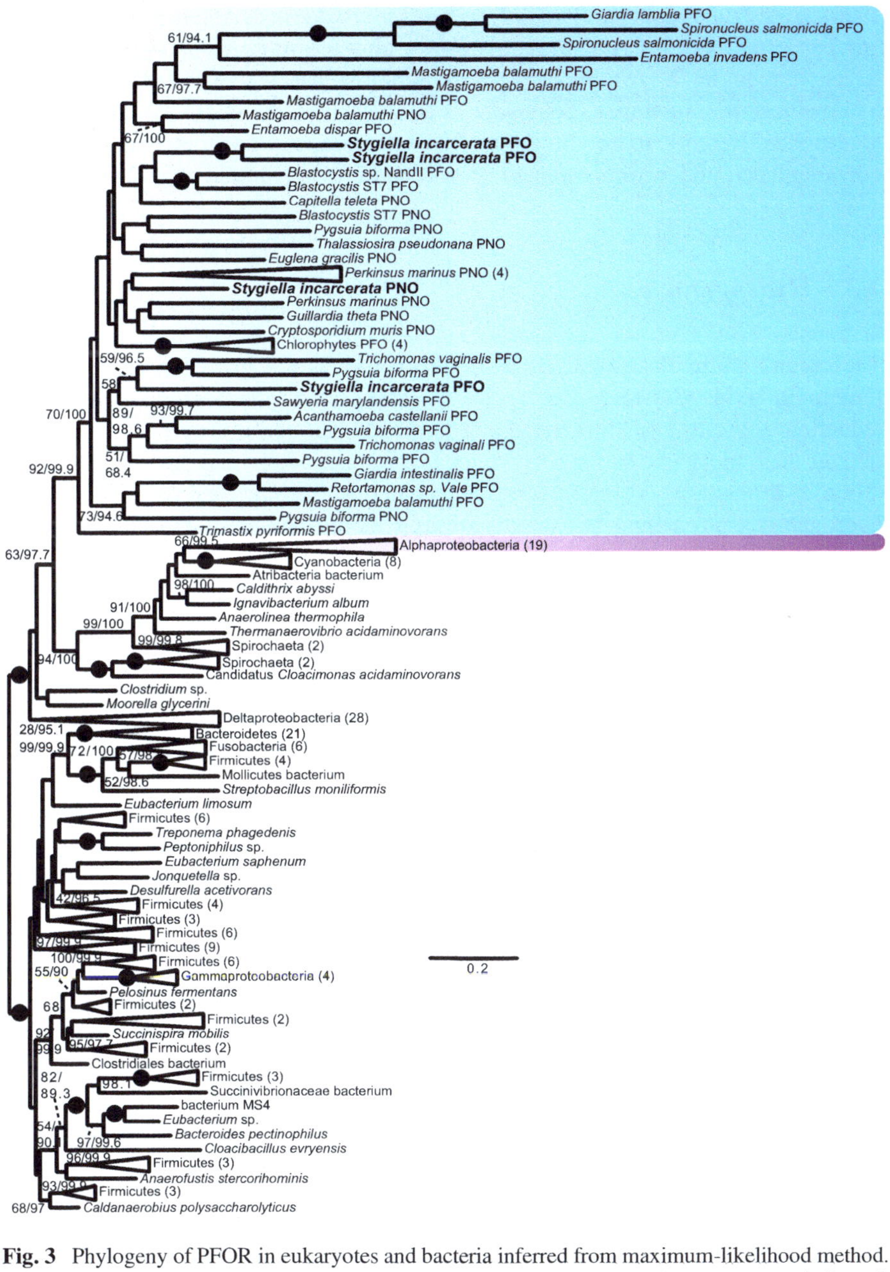

Fig. 3 Phylogeny of PFOR in eukaryotes and bacteria inferred from maximum-likelihood method. Eukaryotes are shaded blue and α-proteobacteria magenta. PFOR and PNO sequences are indicated. Bootstrap support values >50%, and ultrafast bootstrap values >95% are shown. Black circles indicate branches with 100% bootstrap and ultrabootstrap support. (Reproduced from Leger et al. 2016; more details on the analysis are given in the reference)

1990). Based on overall similarity, *T. vaginalis* ferredoxins could be assigned to two groups, ferredoxin 1–3 and ferredoxin 4–6, plus the more distant and somewhat larger ferredoxin 7. All but one (ferredoxin 5) paralogues were identified in hydrogenosome proteomes (Schneider et al. 2011; Beltrán et al. 2013) and shown to respond differently to iron limitation [three paralogues were upregulated and three downregulated under iron-restricted growth conditions (Beltrán et al. 2013)].

4.3 Hydrogenase

The second hallmark enzyme of hydrogenosomes is the hydrogenase (hydrogen: ferredoxin oxidoreductase, EC 1.12.7.2). This enzyme catalyzes the reversible reduction of protons yielding molecular hydrogen. Based on the metal composition of their diatomic active site, the hydrogenases can be divided into two broad classes: [NiFe] hydrogenases possessing one nickel and one iron atom and [FeFe] or Fe-only hydrogenases possessing two atoms of iron in the active site. A third class of enzymes has a different active site composition and reaction mechanism and has a very limited distribution among a few methanogenic Archaea. Representatives of the first two classes occur in a wide variety of microorganisms, the more prevalent [NiFe] enzymes being found in both *Bacteria* and Archaea, while the fewer representatives of [FeFe] hydrogenases are limited to *Bacteria*, a few anaerobic eukaryotes, and green algae (Meyer 2007; Peters et al. 2015).

A CO- and oxygen-sensitive 64 kDa hydrogenase was partially purified from *Trichomonas vaginalis* with the conclusion that the protein belongs to the family of [FeFe] hydrogenases (Payne et al. 1993). Subsequently, three hydrogenase genes have been identified and two of them sequenced, describing the putative proteins as being 63% identical, approx. 51.5 and 50 kDa [FeFe] hydrogenases. Western blotting demonstrated the localization of at least one (but likely both) in hydrogenosomes (Bui and Johnson 1996). While the quaternary structure has not been determined for any of these proteins, it seems reasonable to expect that they are monomers, as are the majority of [FeFe] hydrogenases characterized so far (Meyer 2007). Analysis of the *T. vaginalis* genome revealed an extraordinarily large set of up to ten [FeFe] hydrogenase homologues, seven of them with conserved all cysteine residues that are known to constitute the hydrogen-activating center consisting of the [4Fe4S] cluster and binuclear [2Fe] center (H-cluster) (Vignais et al. 2001; Meyer 2007). All genes possess signature motifs implicating the presence of additional FeS clusters (2–4, depending on particular gene) upstream of H-cluster, but not all of them have clearly recognizable hydrogenosomal targeting sequence. Of the hydrogenases identified in early studies, the 64 kDa protein is encoded by two similar genes and possesses binding motifs for one [2Fe2S] and three [4Fe4S] clusters in the amino-terminal part of the sequence. The approximately 50 kDa hydrogenase is also encoded by two genes and likely binds two auxiliary [4Fe4S] clusters. These four genes possess predicted hydrogenosomal targeting presequences. In three sequences out of ten, cysteine 1 of the H-cluster, believed to act as an acid/base near the active site (Meyer 2007), is missing, raising the question of the possible function of these putative proteins. Interestingly, two of the sequences lacking cysteine 1 possess an additional, carboxy-terminally located domain with FAD and NAD(P) binding motifs (Carlton et al. 2007) that is similar to NADPH-cytochrome P450 reductase and to the carboxy-terminal domain of pyruvate:NADP oxidoreductases of *Euglena gracilis* and *Cryptosporidium parvum* (Rotte et al. 2001; Čtrnáctá et al. 2006). The number and structural diversity of *T. vaginalis* hydrogenases suggest that the hydrogen-related metabolism of this parasite could be a more complex process than originally thought. In fact, it has been already proposed that hydrogen gas formed in the hydrogenosomes may not be a mere metabolic end product that is released into the environment. During the study of the cytosolic secondary alcohol dehydrogenase, it has been observed that some hydrogenosome-derived hydrogen is possibly utilized as a reductant in the redox processes localized in the cytoplasm (Sutak et al. 2012). This observation prompted the search for the so far unknown cytosolic hydrogenase that would function in hydrogen oxidation, and, indeed, such hydrogenase was identified and is currently under study (our unpublished data).

While the formation of accessory FeS clusters (ISC) of hydrogenases likely requires the activity of a more universal ISC assembly system (Meyer 2007; and see below), assembly and maturation of the H-cluster of [FeFe] hydrogenases depend on the activity of additional specialized proteins. Hydrogenase maturases, or Hyd proteins, catalyze the maturation of [FeFe] hydrogenases in all prokaryotes possessing this enzyme (Peters et al. 2015). All of these proteins, HydE, HydF, and HydG, have been identified in *T. vaginalis* hydrogenosomes (Pütz et al. 2006; Schneider et al. 2011; Beltrán et al. 2013). HydE and HydG are radical-SAM proteins, and HydF is a GTPase; they all contain FeS clusters (Rubach et al. 2005; Peters et al. 2015).

Comparative evaluation of the domain structure and phylogenetic reconstruction of the sequences of [FeFe] hydrogenases still leave the challenging question of the origin and history of this enzyme in eukaryotic organisms wide open (Meyer 2007; Hug et al. 2010).

4.4 Malic Enzyme

Malic enzyme [malate dehydrogenase (decarboxylating), EC 1.1.1.39] catalyzes reversible oxidative decarboxylation of malate to pyruvate. The enzyme uses NAD^+ as an electron acceptor, but it is also able to utilize $NADP^+$ with lower affinity (Drmota et al. 1996). With a subunit size of approx. 63 kDa, the *Trichomonas* hydrogenosomal malic enzyme belongs to the family of large, eukaryotic type of malic enzymes. In contrast, the approx. 40 kDa subunit malic enzyme, located in the cytosol, belongs to the prokaryotic-type malic enzymes and was apparently acquired by lateral gene transfer (Doležal et al. 2004). Hydrogenosomal malic enzyme is a membrane-associated tetramer (Drmota et al. 1996). *T. vaginalis* encodes seven similar hydrogenosomal malic enzyme genes (Carlton et al. 2007), and all protein products are present in the hydrogenosome proteome (Rada et al. 2011; Schneider et al. 2011; Beltrán et al. 2013).

Malic enzyme is the most abundant hydrogenosomal protein (Fig. 2), and since its activity is easy to follow by a spectrophotometric assay, it is used as a marker protein for trichomonad hydrogenosomes. Because the membrane of hydrogenosome is impermeable to NAD^+, the isolated and osmotically protected hydrogenosomes usually display rather low malic enzyme activity in vitro. Upon detergent treatment or after several cycles of freezing and thawing, the membrane is permeabilized, and the total activity can be measured. Latency, the ratio of total activity to the activity of intact hydrogenosomes, is a measure of the physiological state of the hydrogenosomal membrane.

4.5 NADH Dehydrogenase

NADH dehydrogenase (NADH:ubiquinone oxidoreductase, complex 1, EC 1.6.5.3) is the most complicated multisubunit complex of the respiratory chain. It consists of 13–14 subunits in bacteria and of over 40 subunits in typical mitochondria (Gabaldon et al. 2005). The complex catalyzes the oxidation of NADH coupled to the reduction of membrane-soluble ubiquinone (coenzyme Q). The transfer of electrons along the redox centers (FMN, FeS clusters) within the complex is linked to the extrusion of protons that form an electrochemical gradient across the inner mitochondrial (or bacterial) membrane. For a long time, it was assumed that trichomonad hydrogenosomes did not contain any component of the classical respiratory chain. That two subunits of complex 1 are present in *T. vaginalis* hydrogenosomes was discovered in 2004 (Dyall et al. 2004; Hrdý et al. 2004). These are a 47 kDa protein (homologue of mitochondrial 51 kDa subunit or NuoF in bacterial terminology), with primary structure that is suggestive of the presence of one [4Fe4S] cluster and an FMN cofactor, and a 22 kDa protein (homologue of 24 kDa mitochondrial subunit or NuoE) that likely binds one [2Fe2S] cluster (Dyall et al. 2004; Hrdý et al. 2004). The *T. vaginalis* complex purifies as an active, NADH-oxidizing heterodimer. The genome analysis did not reveal other subunits of complex 1 with significant confidence. Experimental data suggest that the physiological electron acceptor is likely a [2Fe2S] ferredoxin (Hrdý et al. 2004), although the exact mechanism of this reaction, which may involve the so-called electron bifurcation, along with possible association of complex 1 subunits with hydrogenase (Schut and Adams 2009), remains to be elucidated. The trichomonad protein is insensitive to rotenone, a complex 1 inhibitor, which is not surprising as the rotenone-binding subunits are not present. *T. vaginalis* encodes for two similar copies of the 47 kDa subunit, but expression has only been confirmed for one. The 22 kDa subunit is encoded by a single gene (Carlton et al. 2007). No subunits of other respiratory complexes were found in the *T. vaginalis* genome. Similar two-subunit complex 1 remnant was more recently identified in anaerobic/microaerophilic protists *Pygsuia* and *Stygiella* (Stairs et al. 2014; Leger et al. 2016), and somewhat less reduced complex 1 has been observed in hydrogen-producing mitochondria of the ciliate *Nyctotherus ovalis*, with seven identified subunits (Boxma et al. 2005).

4.6 Succinyl-CoA:Acetate CoA Transferase

Succinyl-CoA:acetate CoA transferase (acetate/succinate CoA transferase (ASCT), EC 2.8.3.8) catalyzes the transfer of the CoA moiety between acetate and succinate and releases the hydrogenosomal end product, acetate. This activity was first detected in the hydrogenosomes of *Tritrichomonas foetus* in the mid-1970s (Lindmark 1976) and subsequently in *T. vaginalis* as well (Steinbüchel and Müller 1986). However, during the analysis of the *T. vaginalis* genome, no ASCT gene could be found, but four similar genes were annotated as putative members of the acetyl-CoA hydrolase/transferase family, with easily recognizable hydrogenosomal targeting sequences on their amino-termini. Later, the work of van Grinsven and colleagues has shown that these genes in fact encode the enzyme with ASCT activity (van Grinsven et al. 2008).

All four paralogues were subsequently identified in hydrogenosomes in proteomic studies (Rada et al. 2011; Schneider et al. 2011; Beltrán et al. 2013). The trichomonad enzyme is homologous to ASCT from anaerobic mitochondria of *Fasciola hepatica* (van Grinsven et al. 2009) but structurally unrelated to an isofunctional protein from *Trypanosoma brucei* mitochondria (Rivière et al. 2004).

4.7 Succinyl-CoA Synthetase

Succinyl-CoA synthetase (SCS), also known as succinate thiokinase (STK) or succinate-CoA ligase (EC 6.2.1.4-5), is the only known hydrogenosomal enzyme directly involved in energy conservation. The protein catalyzes the reversible, substrate-level phosphorylation of ADP or GDP to the respective triphosphate at the expense of the high-energy thioester bond of succinyl-CoA. Succinate and CoA are released in the reaction. The *T. vaginalis* enzyme consists of two subunits, α (32.5 kDa) and β (43 kDa), which constitute the active heterotetramer located in the hydrogenosomal matrix. The enzyme is apparently able to utilize both ADP and GDP as phosphate acceptor (Jenkins et al. 1991; Lahti et al. 1992, 1994). Three similar genes encoding either subunit are present in the *T. vaginalis* genome, and all six genes are expressed (Rada et al. 2011; Schneider et al. 2011; Beltrán et al. 2013).

4.8 Adenylate Kinase

Adenylate kinase (ATP:AMP phosphotransferase, EC 2.7.4.3) is a ubiquitous central enzyme of energy metabolism, catalyzing the reversible transfer of a phosphate group between two molecules of ADP, forming ATP and AMP and thus playing an important role in the maintenance of cellular energy homeostasis. The enzyme from *T. vaginalis* hydrogenosomes shares some properties with cytosolic and some with mitochondrial homologues (Declerck and Müller 1987). The *T. vaginalis* genome encodes ten adenylate kinase homologues (Carlton et al. 2007). Based on the presence of a stretch of 27 amino acid residues in the carboxy-terminal half of the proteins, only two of these sequences belong to the longer, mitochondrial-type eukaryotic adenylate kinases, while the rest belong among the shorter, cytosolic-like ones. Both mitochondrial-type enzymes were identified in the hydrogenosomal proteome (Beltrán et al. 2013), but only one of them is equipped with a typical hydrogenosomal targeting signal; this sequence corresponds to the protein that was previously purified and characterized (Lange et al. 1994).

5 Interaction with Oxygen and Reactive Oxygen Species

Trichomonas vaginalis is usually regarded as an anaerobic organism for which oxygen concentrations higher than those encountered in situ in the vagina (above ≈60 μM) are toxic (Ellis et al. 1994b). On the other hand, it was observed that very low oxygen concentrations (less than 0.25 μM) actually can stimulate the growth of trichomonad cultures (manifested as shorter doubling time), and it was therefore suggested to regard *T. vaginalis* as a microaerophile (Paget and Lloyd 1990). This stimulatory effect of traces of oxygen might be possibly explained on the basis of cytosolic redox balance: NADH produced during glycolysis would be reoxidized by the oxygen-reducing NADH oxidase and not through the activity of lactate dehydrogenase, which would allow more pyruvate to enter the hydrogenosome and provide additional ATP.

Despite being fermentative organisms that lack a respiratory chain and therefore the ability to carry out oxidative phosphorylation, trichomonads do take up oxygen at a high rate when it is present. If the oxygen concentration is high enough to saturate the cytosolic oxygen reductases, the gas diffuses into the hydrogenosome, and instead of protons, it assumes the role of terminal electron acceptor in the hydrogenosomal metabolism. Production of hydrogen stops (Lloyd and Kristensen 1985), but the composition of metabolic end products remains largely the same as under anaerobic conditions, only the ratio of hydrogenosomal and cytosolic end products may shift in response to actual oxygen and CO_2 concentrations. It was observed that acetate prevails under microaerobiosis, while glycerol and lactate formed in the cytosol become dominant when, together with traces of oxygen, CO_2 is present at a high (ca 5 mM) concentration (Paget and Lloyd 1990).

Two pyridine nucleotide-specific dehydrogenases are responsible for oxygen reduction in the cytosol: a highly active NADH oxidase that reduces oxygen to water (Tanabe 1979; Linstead and Bradley 1988) and a minor NADPH oxidase that produces hydrogen peroxide (Linstead and Bradley 1988). The latter enzyme was later shown to be in fact flavin reductase that produces hydrogen peroxide indirectly via free flavines (Leitsch et al. 2014).

Oxygen-related biochemistry of hydrogenosomes is rather complex. That hydrogenosomes are able to utilize oxygen as a terminal electron acceptor with high activity was recognized already in the pioneer times in the 1970s. When incubated under aerobic conditions, hydrogenosomes isolated from *T. foetus* respired actively when supplied with α-glycerophosphate, the main hydrogenosomal substrate pyruvate, or malate together with NAD^+ (Čerkasov et al. 1978). While the respiration with α-glycerophosphate was not studied any further, respiration with pyruvate was found to be dependent on the availability of ADP, CoA, and a catalytic amount of succinate, all compounds necessary in the oxidative decarboxylation of pyruvate coupled to substrate-level phosphorylation in the principal hydrogenosomal catabolic pathway. That observed respiration was not due to the activity of the standard respiratory chain was obvious, since rotenone and cyanide, the inhibitors of mitochondrial electron transport and respiration, had no effect. Hydrogenosomal respiration differed from that of mitochondria in yet another important aspect: its rate declined rapidly, and the oxygen uptake soon became irreversibly inhibited, indicating damage to components of the oxygen-reducing pathway (Čerkasov et al. 1978; Müller and Lindmark 1978). The putative terminal oxidase has not been identified, but involvement of a flavoprotein that produces harmful hydrogen peroxide has been proposed (Čerkasov et al. 1978). However, as has already been suggested (Docampo et al. 1987), it is possible that auto-oxidation of iron-sulfur clusters of proteins involved in pyruvate oxidation, namely, PFOR and/or ferredoxin, contributes to the high respiration rate at the atmospheric oxygen concentration. Resulting products of incomplete oxygen reduction, such as superoxide anion, hydrogen peroxide, and the highly toxic hydroxyl radical formed from hydrogen peroxide in the presence of ferrous ions through Fenton chemistry, could account for inactivation of sensitive hydrogenosomal components and could be partly responsible for the toxicity of elevated nonphysiological levels of oxygen to trichomonads.

The key hydrogenosomal enzymes hydrogenase and pyruvate:ferredoxin oxidoreductase are sensitive proteins prone to inactivation by oxygen (Lindmark and Müller 1973; Lloyd and Kristensen 1985). Obviously, hydrogenosomes need to be equipped with defense mechanisms to combat the damage caused by oxygen and reactive oxygen species formed either enzymatically or upon contact of oxygen with reduced flavines and FeS clusters. The almost ubiquitous superoxide dismutase (SOD), which converts two molecules of superoxide radical into hydrogen peroxide and oxygen, is present in the cytosol and, to a lesser extent, in the hydrogenosomes of *Tritrichomonas foetus* (Lindmark and Müller 1974), and the same, albeit lower, activity has been detected (without determination of subcellular localization) also in *T. vaginalis* (Ellis et al. 1994b). Seven genes encoding iron-containing SOD (FeSOD) have been found in *T. vaginalis* by classical molecular methods. The *T. vaginalis* SOD is a dimeric protein composed of 22 kDa subunits, which is similar to the homologues from parasitic protists like *Trypanosoma* or *Entamoeba* and to the proteobacterial sequences (Viscogliosi et al. 1998). The number of SOD genes was confirmed by the *T. vaginalis* genome project (Carlton et al. 2007); however, none of the genes was found to contain typical amino-terminal hydrogenosomal targeting sequence. Nevertheless, an SOD protein spot has been identified by mass spectrometry on 2D gels of carefully purified hydrogenosomes (Pütz et al. 2005), and subsequently two SOD paralogues were found in hydrogenosome proteome (Schneider et al. 2011), confirming the presence of this defense protein in the organelle. The presence of an SOD certainly accounts for the observed hydrogen peroxide formation in the hydrogenosomes under aerobic conditions, possibly together with reduced flavines (Chapman et al. 1999) and the activity of FeS flavoprotein (Smutná et al. 2014, and see below). However, a catalase that removes hydrogen peroxide has been detected in *T. foetus* but not in *T. vaginalis* (Page-Sharp et al. 1996).

Proteomic and more recent biochemical studies of isolated organelles showed the presence of additional enzymatic systems, often not seen in aerobic eukaryotes but common in anaerobic bacteria, which likely play a role in the defense of hydrogenosomes against oxygen and reactive oxygen species and xenobiotics. Some of these, for a eukaryote rather exotic, defense proteins may explain the known relative oxygen tolerance of *T. vaginalis* and could be potentially exploited as novel specific drug targets (Leitsch et al. 2018).

A possible means to cope with deleterious effects of peroxide buildup was proposed following the finding of putative peroxide-reducing enzymes, rubrerythrin and peroxiredoxin, by a proteomic approach in *T. vaginalis* hydrogenosomes (Pütz et al. 2005).

The first of these, rubrerythrin, is a diiron center and rubredoxin-like center-containing protein (Jin et al. 2002) with peroxidase-like activity, so far known only from anaerobic prokaryotes and amoebae *E. histolytica* and *Naegleria gruberi* in which homologous sequences are present (Pütz et al. 2005; https://genome.jgi.doe.gov/Naegr1/Naegr1.home.html). Six rubrerythrin genes have been identified in the *T. vaginalis* genome; however, none of them possesses the typical amino-terminal hydrogenosomal targeting sequence. The amino-terminus of the mature, hydrogenosome-located protein has not been determined, so it is unknown whether it is processed upon import into the organelle. Nevertheless, at least two of the paralogues are localized in the organelle, as verified by immunofluorescence microscopy and proteomic studies (Pütz et al. 2005; Schneider et al. 2011; Beltrán et al. 2013).

The second candidate peroxide-reducing protein of *T. vaginalis* hydrogenosomes is bacterial-type thiol-dependent peroxidase, a member of the ubiquitous peroxiredoxin family of proteins that contain active cysteine residues and participate in peroxide detoxification and sensing (Chae et al. 1994; Wood et al. 2003). They likely play a major role in peroxide protection in anaerobic organisms that lack catalase and glutathione peroxidase, e.g., helminths, trypanosomatids, and *Plasmodium falciparum* (McGonigle et al. 1998; Müller et al. 2003). The function of peroxiredoxin is dependent on its reduction by a small protein thioredoxin that itself is reduced by NADPH-dependent thioredoxin reductase (McGonigle et al. 1998). These three proteins form the typical peroxide detoxifying peroxiredoxin system.

The presence of this system has been demonstrated in the cytosol of *T. vaginalis* (Coombs et al. 2004) and later also in hydrogenosomes, where a thiol peroxidase homologue distinct from the cytoplasmic protein has been identified by mass spectroscopy. Similar to SOD and rubrerythrin, this thiol peroxidase does not feature a characteristic hydrogenosomal target signal (Pütz et al. 2005), nor is it known whether the amino terminus of any of these proteins is processed. The cysteine-regenerating thioredoxin has been identified by mass spectroscopy in the hydrogenosome as well. In this instance, the corresponding sequence does contain the amino-terminal extension with strong similarity to known hydrogenosomal targeting sequences (Pütz et al. 2005). The first protein of the catalytic cascade, the thioredoxin reductase, has been found in hydrogenosomes in 2008 when two out of five paralogues of a low molecular weight-type thioredoxin reductase were shown to partially localize to the organelles (Mentel et al. 2008). One of the proteins possesses amino-terminal extension necessary for organellar import, while the other paralogue was partially imported into hydrogenosomes irrespective of the presence of a substantial stretch of N-terminal amino acids and thus must contain an internal localization signal (Mentel et al. 2008).

Another hydrogensomal protein involved in oxidative stress protection belongs to the family of OsmC (osmotically inducible) and Ohr (organic hydroperoxide resistance) (Gutierrez and Devedjian 1991; Mongkolsuk et al. 1998) cysteine-based dithiol-dependent peroxidases. These proteins were originally believed to exist only in bacteria, but the study of trichomonad protein, belonging to the OsmC subfamily, showed the presence of OsmC/Ohr homologues in a variety of eukaryotes as well (Nývltová et al. 2016). *T. vaginalis* OsmC functions as an effective lipoate-dependent reductase of hydrogen peroxide and organic hydroperoxides and was actually the first hydrogenosomal protein for which peroxidase activity was experimentally demonstrated (Nývltová et al. 2016). Remarkably, the hydrogenosomal OsmC receives electrons from NADH via L and H proteins of glycine decarboxylase complex (see below), constituting a novel peroxidase system not recognized in other organisms so far.

A fourth protein detected in *T. vaginalis* hydrogenosomes and involved in their protection is a member of flavodiiron (FDP) protein superfamily. These proteins are widespread in anaerobic or facultative anaerobic Eubacteria as well as Archaea and occur also in some anaerobic, mostly parasitic protists. They were first proposed to provide protection against oxygen by safely reducing it to water (Chen et al. 1993), but it was later recognized that they also possess considerable nitric oxide reductase activity and therefore are likely important in protection against nitrosative stress (Gomes et al. 2002). These proteins are functional dimers that share a common structural core, consisting of an amino-terminal β-lactamase-like domain with a diiron center and of a carboxy-terminal flavodoxin domain that non-covalently binds FMN. The simplest and most common representatives, designated as Class A flavodiiron proteins, consist of only this two-domain core. Proteins of other currently recognized classes (Class B through H) possess additional domains in the carboxy-terminal part of the molecule that participate in electron transfer (Folgosa et al. 2018). FDP residing in *T. vaginalis* hydrogenosomes belongs to the

Class A subfamily and functions as oxygen (and not nitric oxide) reductase, similar to the homologues from other eukaryotic parasites *Giardia intestinalis* and *Entamoeba histolytica* (Di Matteo et al. 2008; Smutná et al. 2009; Vicente et al. 2012). Like most of its homologues, it is a dimer and binds one FMN per subunit. Its amino-terminus displays a typical hydrogenosomal targeting peptide that is cleaved from the mature protein. The electron donor for *T. vaginalis* FDP is ferredoxin (Smutná et al. 2009).

Another hydrogenosomal enzyme with likely protective function is iron-sulfur flavoprotein (Isf). Proteins of this family bind one FMN and one compact low redox potential [4Fe4S] cluster per subunit and again are common among anaerobic prokaryotes, while the only known eukaryotes possessing them are trichomonads and *Entamoeba* (Zhao, Cruz and Ferry 2001; Smutná et al. 2014). Physiological role of hydrogenosomal Isf is not entirely clear. It reduces oxygen, but unlike its characterized homologue from *Methanosarcina* (Cruz and Ferry 2006), it produces harmful hydrogen peroxide instead of water (Smutná et al. 2014). On the other hand, it is able to reduce metronidazole with even number of electrons, obviating the formation of highly reactive free nitro radical anion, and also reduces and detoxifies chloramphenicol, a broad-spectrum antibiotic with a nitro group. Both ferredoxin and NADH can deliver electrons to hydrogenosomal Isf, and it is unknown what is the preferred reductant in vivo (Smutná et al. 2014). *T. vaginalis* encodes seven Isf paralogues, all featuring characteristic amino-terminal hydrogensomal targeting sequences. At least three paralogues are expressed (Schneider et al. 2011).

6 Iron-Sulfur Cluster Assembly Machinery

Iron-sulfur clusters (ISC) are ubiquitous inorganic cofactors which are required for the biological function of a number of essential proteins in various cell compartments. Although some FeS clusters in apoproteins can be reconstituted chemically, in the living cell this process is mediated by a complex ISC assembly machinery. Eukaryotes inherited this machinery from the bacterial endosymbiont which gave rise to mitochondria and related organelles, such as hydrogenosomes and mitosomes. In fact, formation of FeS clusters is the only common function which is shared by mitochondria, hydrogenosomes, and certain mitosomes, indicating their common origin (see chapter "Mitosomes in Parasitic Protists"). Several reviews summarize the mechanisms of ISC assembly in eukaryotes and bacteria (Johnson and Dean 2004; Lill and Mühlenhoff 2006; Balk and Schaedler 2014; Stehling et al. 2014; Lill et al. 2015). In principle, FeS clusters are first assembled on a scaffold protein, and then they are transferred to acceptor apoproteins. The first step requires the cysteine desulfurase activity of the IscS/Isd11/Acp1 complex. This enzyme provides sulfur for the scaffold protein IscU to form transient [2Fe2S] cluster. Delivery of iron for FeS cluster assembly, as well as for heme synthesis, is most likely facilitated by mitochondrial frataxin. The chaperone system, including Hsp70, J-type cochaperone and ADP/ATP nucleotide exchange factor, was suggested to participate in the second step in which FeS clusters are transferred either directly or via glutaredoxin 5 to apoproteins. The complex IscA1/IscA2/Iba57 serves for maturation of more complex [4Fe4S] clusters. Nfu, Ind1, and BolA family proteins contribute to the specific targeting of the cluster to apoproteins. The assembly of an FeS cluster also requires reducing equivalents which are provided by [2Fe2S] ferredoxin. Genes coding for key components of FeS cluster assembly machinery were identified in the *T. vaginalis* genome (Carlton et al. 2007); however, experimental studies on the hydrogenosomal ISC assembly machinery are rather limited.

Initially, two genes coding for IscS were characterized in *T. vaginalis*, and hydrogenosomal localization of the gene products was predicted based on the presence of amino-terminal targeting presequences (Tachezy et al. 2001). The hydrogenosomal localization of the *iscs* gene product in hydrogenosomes was later confirmed in isolated organelles as well as by immunofluorescence microscopy (Sutak et al. 2004). This finding strongly suggested that ISC assembly operates in hydrogenosomes (Fig. 1). Indeed, the ability of isolated organelles to catalyze assembly and insertion of an FeS cluster into apoprotein was demonstrated (Sutak et al. 2004).

Frataxin is another protein which was shown to be targeted into hydrogenosomes (Doležal et al. 2007). Importantly, *T. vaginalis* frataxin can, in part, functionally replace mitochondrial frataxin as demonstrated by its ability to partially restore defects in ISC assembly in *S. cerevisiae Δyfh1* mutants and frataxin-deficient *T. brucei* (Long et al. 2008). In yeast, *T. vaginalis* frataxin also partially restored defects in heme synthesis, although proteins involved in heme synthesis have not been identified in *T. vaginalis* (Doležal et al. 2007).

Interestingly, the transcription of genes coding for hydrogenosomal ISC assembly components including frataxin was markedly upregulated under iron deficiency, which might reflect an increased demand for the synthesis of new FeS clusters (Horváthová et al. 2012; Beltrán et al. 2013). This is in direct contrast to what was found with yeast mitochondrial frataxin (yfh1), the expression of which is strongly stimulated by iron (Santos et al. 2004). The observed differences may reflect differences in the function of frataxin in various organisms. For example, Gakh et al. (2002) proposed that frataxin may serve as a mitochondrial iron-storage molecule, for which upregulation of transcription by iron could be expected. According to our transcriptomic data, *T. vaginalis* frataxin is unlikely to have this function in hydrogenosomes (Doležal et al. 2007).

As mentioned above, sequencing of the complete *T. vaginalis* genome revealed the presence of seven [2Fe2S] ferredoxin homologues. It is likely that some of them serve as electron donors for FeS cluster biogenesis, while others are required for electron transport associated with the energy metabolism of hydrogenosomes.

Some components required for transfer of FeS clusters to apoproteins are present, including Nfu and Ind1. Ind1 is known to contribute to FeS cluster incorporation into complex 1 subunits. Its presence is consistent with identification of two complex 1 subunits, NuoE and NuoF, in hydrogenosomes. Other components including glutaredoxin 5 and BolA seem to be absent.

The mitochondrial ISC assembly machinery is essential not only for maturation of FeS proteins within mitochondria but also for extramitochondrial FeS proteins (Lill and Kispal 2000). Although the nature of compounds exported from mitochondria is not known, the formation of extramitochondrial FeS clusters is dependent on ISC export machinery, which involves the ABC transporter Atm1, sulfhydryl oxidase Erv1 and glutathione, and nine components of the cytosolic FeS cluster assembly (CIA) system (Paul and Lill 2015). It is noteworthy that no candidates for a member of the ISC export machinery were identified in the *T. vaginalis* genome, although most cytosolic components are present. Thus, it would be interesting to investigate whether, similarly to mitochondria, hydrogenosomes are involved in the maturation of FeS proteins outside of hydrogenosomes and, if so, which membrane components are involved in this process.

7 Amino Acid and Polyamine Metabolism

Analysis of the *T. vaginalis* genome gave a number of unexpected results. One of these was the detection of two putatively hydrogenosomal enzymes with a possible role in amino acid metabolism. Amino acid metabolism in *T. vaginalis* hydrogenosomes was regarded to be essentially restricted to arginine hydrolysis to citrulline and ammonia (Yarlett et al. 1994), and apart from that, the organelles were assumed to have no enzymes involved in amino acid metabolism. Surprisingly, genes encoding proteins involved in the interconversion of serine and glycine, serine hydroxymethyltransferase (SHMT, EC 2.1.2.1), and glycine decarboxylase complex (GDC, also known as glycine cleavage system, EC 1.4.4.2) were detected.

7.1 *Serine Hydroxymethyltransferase and the Case of Glycine Decarboxylase Complex*

In *T. vaginalis*, the single gene coding for SHMT has a presequence that is somewhat different from the typical hydrogenosomal targeting peptides but nevertheless possesses a mitochondrial-like processing site. The closest homologues of *T. vaginalis* SHMT are of the mitochondrial type, but the *T. vaginalis* enzyme was found to be a dimer, based on sucrose density gradient analysis of overexpressed protein, like its bacterial homologues rather than a tetramer as is observed in mitochondria (Mukherjee et al. 2006b). In addition to identifying the presence of overexpressed and tagged SHMT in *T. vaginalis* hydrogenosomes by immunofluorescence, the corresponding (albeit low) enzymatic activity was also detected in the hydrogenosomes of wild-type trichomonads (Mukherjee et al. 2006b).

GDC is in eukaryotes exclusively mitochondrial multienzyme complex composed of four loosely associated component enzymes: the P protein (a homodimer containing pyridoxal phosphate), the H protein (a monomeric lipoamide-containing protein), the T protein (a monomeric protein requiring tetrahydrofolate cofactor), and the L protein, a dihydrolipoamide dehydrogenase (a homodimer containing FAD). GDC catalyzes the oxidative decarboxylation and deamination of glycine, thereby producing CO_2 and NH_3 and concomitantly reducing NAD^+ coenzyme to NADH. The remaining methylene carbon of glycine is transferred to tetrahydrofolate cofactor, forming N^5, N^{10}-methylene tetrahydrofolate. SHMT, a pyridoxal phosphate-dependent enzyme that catalyzes reversible conversion of serine to glycine, uses N^5, N^{10}-methylene tetrahydrofolate as a donor of one-carbon unit in the reaction leading to the formation of serine from another molecule of glycine (Schirch and Szebenyi 2005). SHMT and GDC are thus interconnected through a common glycine substrate and a soluble pool of tetrahydrofolate. Physiological function of these two enzymes is the interconversion of serine and glycine, an essential and ubiquitous step in primary metabolism. It not only produces serine from glycine (and vice versa), both precursors for amino acid and phospholipid biosynthesis, but, most importantly, it provides the activated one-carbon units bound to tetrahydrofolate that are utilized in a number of biosynthetic pathways, such as the biosynthesis of methionine, pyrimidines, and purines. In other words, the role of the GDC-SHMT pathway is to interconnect the metabolism of one-, two-, and three-carbon compounds (Bauwe and Kolukisaoglu 2003).

The discovery of genes coding for GDC proteins in *T. vaginalis* was reported along with the identification of SHMT, and the presence of both enzymes was interpreted as evidence for so far unrecognized and unexpected role of hydrogenosomes in amino acid metabolism (Mukherjee et al. 2006a, b). Two genes coding for the H protein and a single gene for the L protein have been identified in *T. vaginalis* genome, and both H and L proteins were later detected in the hydrogenosome proteome (Schneider et al. 2011; Beltrán et al. 2013).

However, the two remaining and obligate components of GDC, the P protein that is the glycine-decarboxylating enzyme and the T protein that transfers the methylene group to tetrahydrofolate, are absent from *T. vaginalis* genome, meaning that the L and H proteins alone cannot function as glycine decarboxylase. Moreover, the dihydrofolate reductase, the enzyme necessary to convert the imported folate to tetrahydrofolate, the active coenzyme, could not be detected in the *T. vaginalis* genome database either, making also the physiological role of SHMT questionable.

The function of the two hydrogenosomal GDC proteins was thus unknown until it was discovered that the L and H proteins constitute a redox system that can deliver electrons from NADH to OsmC peroxidase and thus provide protection against both organic hydroperoxides and hydrogen peroxide. The L protein functions as NADH-dependent reductase of lipoamide moiety of H protein, which in turn serves as a reductant of dithiol-dependent OsmC (Nývltová et al. 2016). Such functional connection of GDC proteins and OsmC/Ohr peroxidase has not been reported for any other organism so far.

7.2 Polyamine Metabolism

One reaction of arginine metabolism apparently takes place in hydrogenosomes. Arginine is a precursor in the biosynthesis of polyamines, the ubiquitous polycations that participate in pivotal cellular functions such as macromolecule biosynthesis, cell growth, and differentiation (Pegg 1986; Igarashi and Kashiwagi 2000). In trichomonads, *Giardia*, and likely many other representatives of Metamonada (Novák et al. 2016), arginine is metabolized to the polyamine putrescine by a bacterial-type arginine dihydrolase pathway in which arginine is first converted to citrulline by arginine deiminase. Citrulline is then converted to ornithine and carbamylphosphate by catabolic ornithine carbamyl transferase, and putrescine is formed from ornithine by ornithine decarboxylase. Carbamylphosphate is a macroergic molecule that is broken down to bicarbonate and ammonia by carbamate kinase, which concomitantly phosphorylates one molecule of ADP to ATP (Yarlett et al. 1994). This pathway is probably a significant source of ATP to the parasite in situ, as evidenced by rapid replacement of arginine with putrescine in the vaginal fluid of infected patients (Chen et al. 1982), but when abundant glucose is available, as in in vitro cultures, the arginine dihydrolase pathway adds only about 1% of extra ATP to that generated from carbohydrates (Yarlett et al. 1996).

Localization of enzymes of arginine dihydrolase pathway has been studied in *T. vaginalis* as well as in *T. foetus*. Almost half of the arginine deiminase activity was found to localize to the sedimentable fraction, while the other enzymes of the pathway were present predominantly in the cytosol (Yarlett et al. 1994). Analysis of the *T. vaginalis* genome revealed the presence of three arginine deiminase genes, all with amino-termini strongly resembling known hydrogenosomal target signals, and subsequent experiments with hemagglutinin-tagged, episomally overexpressed arginine deiminase confirmed the co-localization of the protein in the organelles labeled with the hydrogenosomal marker protein, malic enzyme (Morada et al. 2011). The association of all three arginine deiminase paralogues with hydrogenosomes was further supported by mass spectroscopic analysis of purified hydrogenosomes (Beltrán et al. 2013). The reason why only the first enzyme of the arginine dihydrolase pathway is localized in the organelle while the other members of the pathway are cytosolic is unknown.

Polyamines are of another significance for the hydrogenosomes: inhibition of putrescine-synthesizing enzyme ornithine decarboxylase by 1,4-diamino-2-butanone (DAB) resulted in markedly decreased levels of polyamines in *T. foetus* cells and ultimately led to reduction of hydrogenosome numbers and to alterations of hydrogenosomal ultrastructure in growth-inhibited cultures (Reis et al. 1999). Since polyamines have been implicated in protection against oxidative stress (Tadolini 1988) and in membrane stability (Tabor and Tabor 1976), it was suggested that damage to hydrogenosomes in DAB-treated and polyamine-depleted trichomonads could be, at least in part, caused by membrane disorganization and by reactive oxygen species generated within hydrogenosomes under aerobic conditions (Reis et al. 1999).

8 Concluding Remarks

While our knowledge of hydrogenosome biochemistry has increased considerably over the last four decades, we are still far from a complete understanding of the workings of the trichomonad hydrogenosome. The core hydrogenosomal pathway involved in ATP synthesis and hydrogen evolution and the role of the organelle in the biosynthesis of iron-sulfur centers are reasonably well defined. Also some of the enzymes participating in oxygen and reactive oxygen species detoxification are now characterized, and their roles were experimentally determined. The *T. vaginalis* genome project and subsequent proteomic studies provided us with a wealth of data that indicate, however, the presence of previously unknown functions. Even a number of relatively reliably identified proteins cannot be accommodated into our current biochemical schemes. Classical biochemistry combined with recombinant molecular methods will be necessary to clarify the roles of different hydrogenosomal hydrogenases, flavoproteins, ATPases, membrane proteins and putative enzymes of amino acid, and carbohydrate metabolism, to name just a few. That there is still a long way to go is evident from the fact that proteins with unknown functions found in hydrogenosome proteome represent a high percentage of all proteins present in the organelle.

Establishing the protein roles in *T. vaginalis* has been hampered by the lack of usable gene silencing methodology. The recent adaptation of CRISPR/Cas9 system for targeted gene knockout in *T. vaginalis* (Janssen et al. 2018) thus holds the promise for the future.

We shall stress that our still limited knowledge of biochemistry of hydrogenosomes as described in this chapter resulted from the work on important parasites that are easily grown in axenic cultures and whose genomes were sequenced. These organisms represent only a tiny fraction of the diversity of (anaerobic) protists that harbor other variants of mitochondrial organelle. Some work on other eukaryotic anaerobes has already been done (Müller et al. 2012), but much more is needed if we want to comprehend the functional variability of protistan organelles. And even more so if we admit that genomic and proteomic data are helpful predictions but true functions are often uncovered only experimentally.

Acknowledgments The excellent technical assistance of Ms. Míša Marcinčiková is gratefully acknowledged.

Part of the original research presented in this chapter was supported by the NIH (grant no. AI 11942 to Miklós Müller), the Ministry of Education, Youth and Sports of the Czech Republic project NPU II (LQ1604), and by ERD Funds project CePaViP (CZ.02.1.01/0.0/0.0/16_019/0000759) to Jan Tachezy.

References

Balk J, Schaedler TA (2014) Iron cofactor assembly in plants. Annu Rev Plant Biol 65:125–153

Bauwe H, Kolukisaoglu U (2003) Genetic manipulation of glycine decarboxylation. J Exp Bot 54:1523–1535

Beltrán NC, Horváthová L, Jedelský PL et al (2013) Iron-induced changes in the proteome of *Trichomonas vaginalis* hydrogenosomes. PLoS One 8:e65148. https://doi.org/10.1371/journal.pone.0065148

Benchimol M, De Souza W (1983) Fine structure and cytochemistry of the hydrogenosome of *Tritrichomonas foetus*. J Protozool 30:422–425

Boxma B, de Graaf RM, van der Staay GWM et al (2005) An anaerobic mitochondrion that produces hydrogen. Nature 434:74–79

Bui ET, Johnson PJ (1996) Identification and characterization of [Fe]-hydrogenases in the hydrogenosome of *Trichomonas vaginalis*. Mol Biochem Parasitol 76:305–310

Carlton JM, Hirt RP, Silva JC et al (2007) Draft genome sequence of the sexually transmitted pathogen *Trichomonas vaginalis*. Science 315:207–212

Čerkasov J, Čerkasovová A, Kulda J, Vilhelmová D (1978) Respiration of hydrogenosomes of *Tritrichomonas foetus*. J Biol Chem 253:1207–1214

Čerkasovová A, Čerkasov J, Kulda J, Reischig J (1976) Circular DNA and cardiolipin in hydrogenosomes, microbody-like organelles of trichomonads. Folia Parasitol Praha 23:33–37

Chabrière E, Charon MH, Volbeda A et al (1999) Crystal structures of the key anaerobic enzyme pyruvate: ferredoxin oxidoreductase, free and in complex with pyruvate. Nat Struct Biol 6:182–190. https://doi.org/10.1038/5870

Chae HZ, Robison K, Poole LB et al (1994) Cloning and sequencing of thiol-specific antioxidant from mammalian brain – alkyl hydroperoxide reductase and thiol-specific antioxidant define a large family of antioxidant enzymes. Proc Natl Acad Sci U S A 91:7017–7021

Chapman A, Linstead DJ, Lloyd D (1999) Hydrogen peroxide is a product of oxygen consumption by *Trichomonas vaginalis*. J Biosci 24:339–344

Chen KC, Amsel R, Eschenbach DA, Holmes KK (1982) Biochemical diagnosis of vaginitis: determination of diamines in vaginal fluid. J Infect Dis 145:337–345

Chen L, Liu MY, LeGall J et al (1993) Rubredoxin oxidase, a new flavo-hemo-protein, is the site of oxygen reduction to water by the strict anaerobe *Desulfovibrio gigas*. Biochem Biophys Res Commun 193:100–00105

Coombs GH, Westrop GD, Suchan P et al (2004) The amitochondriate eukaryote *Trichomonas vaginalis* contains a divergent t00hioredoxin-linked peroxiredoxin antioxidant system. J Biol Chem 279:5249–5256

Crossnoe CR, Germanas JP, LeMagueres P et al (2002) The crystal structure of *Trichomonas vaginalis* ferredoxin provides insight into metronidazole activation. J Mol Biol 318:503–518

Cruz F, Ferry JG (2006) Interaction of iron-sulfur flavoprotein with oxygen and hydrogen peroxide. Biochim Biophys Acta – Gen Subj. https://doi.org/10.1016/j.bbagen.2006.02.016

Čtrnáctá V, Ault JG, Stejskal F, Keithly JS (2006) Localization of pyruvate: NADP(+) oxidoreductase in sporozoites of *Cryptosporidium parvum*. J Eukaryot Microbiol 53:225–231

Declerck PJ, Müller M (1987) Hydrogenosomal ATP: AMP phosphotransferase of *Trichomonas vaginalis*. Comp Biochem Physiol B 88:575–580. https://doi.org/10.1016/0305-0491(87)90347-6

Di Matteo A, Scandurra FM, Testa F et al (2008) The O_2-scavenging flavodiiron protein in the human parasite *Giardia intestinalis*. J Biol Chem 283:4061–4068. https://doi.org/10.1074/jbc.M705605200

Docampo R, Moreno SN, Mason RP (1987) Free radical intermediates in the reaction of pyruvate: ferredoxin oxidoreductase in *Tritrichomonas foetus* hydrogenosomes. J Biol Chem 262:12417–12420

Doležal P, Vaňáčová Š, Tachezy J et al (2004) Malic enzymes of *Trichomonas vaginalis*: two enzyme families, two distinct origins. Gene 329:81–92. https://doi.org/10.1016/j.gene.2003.12.022

Doležal P, Dancis A, Lesuisse E et al (2007) Frataxin, a conserved mitochondrial protein, in the hydrogenosome of *Trichomonas vaginalis*. Eukaryot Cell 6:1431–1438. https://doi.org/10.1128/EC.00027-07

Drmota T, Proost P, Van Ranst M et al (1996) Iron-ascorbate cleavable malic enzyme from hydrogenosomes of *Trichomonas vaginalis*: purification and characterization. Mol Biochem Parasitol 83:221–234

Dyall SD, Yan W, Gadillo-Correa MG et al (2004) Non-mitochondrial complex I proteins in a hydrogenosomal oxidoreductase complex. Nature 431:1103–1107

Ellis JE, Setchell KDR, Kaneshiro ES (1994a) Detection of ubiquinone in parasitic and free-living protozoa, including species devoid of mitochondria. Mol Biochem Parasitol 65:213–224

Ellis JE, Yarlett N, Cole D et al (1994b) Antioxidant defences in the microaerophilic protozoan *Trichomonas vaginalis*: comparison of metronidazole-resistant and sensitive strains. Microbiology 140:2489–2494. https://doi.org/10.1099/13500872-140-9-2489

Folgosa F, Martins MC, Teixeira M (2018) Diversity and complexity of flavodiiron NO/O_2 reductases. FEMS Microbiol Lett 365. https://doi.org/10.1093/femsle/fnx267

Gabaldon T, Rainey D, Huynen MA (2005) Tracing the evolution of a large protein complex in the eukaryotes, NADH: ubiquinone oxidoreductase (Complex I). J Mol Biol 348:857–870

Gakh O, Adamec J, Gacy AM et al (2002) Physical evidence that yeast frataxin is an iron storage protein. Biochemistry 41:6798–6804. https://doi.org/10.1021/bi025566+

Gomes CM, Giuffre A, Forte E et al (2002) A novel type of nitric-oxide reductase. *Escherichia coli* flavorubredoxin 2. J Biol Chem 277:25273–25276

Gorrell TE, Yarlett N, Müller M (1984) Isolation and characterization of *Trichomonas vaginalis* ferredoxin. Carlsberg Res Commun 246:529–536. https://doi.org/10.1007/BF02913954

Guschina IA, Harris KM, Maskrey B et al (2009) The microaerophilic flagellate, *Trichomonas vaginalis*, contains unusual acyl lipids but no detectable cardiolipin. J Eukaryot Microbiol 56:52–57

Gutierrez C, Devedjian JC (1991) Osmotic induction of gene osmC expression in *Escherichia coli* K12. J Mol Biol 220:959–973. https://doi.org/10.1016/0022-2836(91)90366-E

Horváthová L, Šafaříková L, Basler M et al (2012) Transcriptomic identification of iron-regulated and iron-independent gene copies within the heavily duplicated *Trichomonas vaginalis* genome. Genome Biol Evol 4:1017–1029. https://doi.org/10.1093/gbe/evs078

Hrdý I, Müller M (1995) Primary structure and eubacterial relationships of the pyruvate:Ferredoxin oxidoreductase of the amitochondriate eukaryote *Trichomonas vaginalis*. J Mol Evol 41:388–396. https://doi.org/10.1007/BF00186551

Hrdý I, Mertens E, Van Schaftingen E (1993) Identification, purification and separation of different isozymes of NADP-specific malic enzyme from *Tritrichomonas foetus*. Mol Biochem Parasitol 57:253–260

Hrdý I, Hirt RP, Doležal P et al (2004) *Trichomonas* hydrogenosomes contain the NADH dehydrogenase module of mitochondrial complex I. Nature 432:618–622

Hrdý I, Cammack R, Stopka P et al (2005) Alternative pathway of metronidazole activation in *Trichomonas vaginalis* hydrogenosomes. Antimicrob Agents Chemother 49:5033–5036. https://doi.org/10.1128/AAC.49.12.5033-5036.2005

Huang KY, Ku FM, Cheng WH et al (2015) Novel insights into the molecular events linking to cell death induced by tetracycline in the amitochondriate protozoan *Trichomonas vaginalis*. Antimicrob Agents Chemother 59:6891–6903. https://doi.org/10.1128/AAC.01779-15

Huber M, Garfinkel L, Gitler C et al (1988) Nucleotide-sequence analysis of an *Entamoeba histolytica* ferredoxin gene. Mol Biochem Parasitol 31:27–33

Hug LA, Stechmann A, Roger AJ (2010) Phylogenetic distributions and histories of proteins involved in anaerobic pyruvate metabolism in eukaryotes. Mol Biol Evol 27:311–324. https://doi.org/10.1093/molbev/msp237

Igarashi K, Kashiwagi K (2000) Polyamines: mysterious modulators of cellular functions. Biochem Biophys Res Commun 271:559–564

Inui H, Ono K, Miyatake K et al (1987) Purification and characterization of pyruvate - NADP+ oxidoreductase in *Euglena gracilis*. J Biol Chem 262:9130–9135

Janssen BD, Chen YP, Molgora BM et al (2018) CRISPR/Cas9-mediated gene modification and gene knock out in the human-infective parasite *Trichomonas vaginalis*. Sci Rep 8:270. https://doi.org/10.1038/s41598-017-18442-3

Jenkins TM, Gorrell TE, Müller M, Weitzman PD (1991) Hydrogenosomal succinate thiokinase in *Tritrichomonas foetus* and *Trichomonas vaginalis*. Biochem Biophys Res Commun 179:892–896

Jin S, Kurtz DM, Liu ZJ et al (2002) X-ray crystal structures of reduced rubrerythrin and its azide adduct: a structure-based mechanism for a non-heme diiron peroxidase. J Am Chem Soc 124:9845–9855

Johnson D, Dean D (2004) Structure, function, and formation of biological iron-sulfur clusters. Annu Rev Biochem 74:247–281

Johnson PJ, d'Oliveira CE, Gorrell TE et al (1990) Molecular analysis of the hydrogenosomal ferredoxin of the anaerobic protist *Trichomonas vaginalis*. Proc Natl Acad Sci U S A 87:6097–6101. https://doi.org/10.1073/pnas.87.16.6097

Kulda J (1999) Trichomonads, hydrogenosomes and drug resistance. Int J Parasitol 29:199–212

Lahti CJ, d'Oliveira CE, Johnson PJ (1992) Beta-succinyl-coenzyme A synthetase from *Trichomonas vaginalis* is a soluble hydrogenosomal protein with an amino-terminal sequence that resembles mitochondrial presequences. J Bacteriol 174:6822–6830

Lahti CJ, Bradley PJ, Johnson PJ (1994) Molecular characterization of the alpha-subunit of *Trichomonas vaginalis* hydrogenosomal succinyl CoA synthetase. Mol Biochem Parasitol 66:309–318

Land KM, Delgadillo-Correa MG, Tachezy J et al (2004) Targeted gene replacement of a ferredoxin gene in *Trichomonas vaginalis* does not lead to metronidazole resistance. Mol Microbiol 51:115–122

Lange S, Rozario C, Müller M (1994) Primary structure of the hydrogenosomal adenylate kinase of *Trichomonas vaginalis* and its phylogenetic relationships. Mol Biochem Parasitol 66:297–308

Lantsman Y, Tan KSW, Morada M, Yarlett N (2008) Biochemical characterization of amitochondrial-like organelle from *Blastocystis* sp. subtype 7. Microbiology 154 (Pt 9):2757–2766. https://doi.org/10.1099/mic.0.2008/017897-0

Leger MM, Gawryluk RMR, Gray MW, Roger AJ (2013) Evidence for a hydrogenosomal-type anaerobic ATP generation pathway in *Acanthamoeba castellanii*. PLoS One 8:e69532. https://doi.org/10.1371/journal.pone.0069532

Leger MM, Eme L, Hug LA, Roger AJ (2016) Novel hydrogenosomes in the microaerophilic jakobid *Stygiella incarcerata*. Mol Biol Evol 33:2318–2336. https://doi.org/10.1093/molbev/msw103

Leitsch D (2017) A review on metronidazole: an old warhorse in antimicrobial chemotherapy. Parasitology 23:1–12. https://doi.org/10.1017/S0031182017002025

Leitsch D, Kolarich D, Binder M et al (2009) *Trichomonas vaginalis*: metronidazole and other nitroimidazole drugs are reduced by the flavin enzyme thioredoxin reductase and disrupt the cellular redox system. Implications for nitroimidazole toxicity and resistance. Mol Microbiol 72:518–536. https://doi.org/10.1111/j.1365-2958.2009.06675.x

Leitsch D, Janssen BD, Kolarich D et al (2014) *Trichomonas vaginalis* flavin reductase 1 and its role in metronidazole resistance. Mol Microbiol 91:198–208. https://doi.org/10.1111/mmi.12455

Leitsch D, Williams CF, Hrdý I (2018) Redox pathways as drug targets in microaerophilic parasites. Trends Parasitol 34:576–589

Lill R, Kispal G (2000) Maturation of cellular Fe-S proteins: an essential function of mitochondria. Trends Biochem Sci 25:352–356

Lill R, Mühlenhoff U (2006) Iron-sulfur protein biogenesis in eukaryotes: components and mechanisms. Annu Rev Cell Dev Biol 22:457–486

Lill R, Dutkiewicz R, Freibert SA et al (2015) The role of mitochondria and the CIA machinery in the maturation of cytosolic and nuclear iron-sulfur proteins. Eur J Cell Biol 94:280–291

Lindmark DG (1976) Acetate production by *Tritrichomonas foetus*. In: Van den Bossche H (ed) Biochemistry of parasites and host-parasite relationships. Elsevier, Amsterdam, pp 15–21
Lindmark DG, Müller M (1973) Hydrogenosome, a cytoplasmic organelle of the anaerobic flagellate *Tritrichomonas foetus*, and its role in pyruvate metabolism. J Biol Chem 248:7724–7728
Lindmark DG, Müller M (1974) Superoxide dismutase in the anaerobic flagellates, *Tritrichomonas foetus* and *Monocercomonas* sp. J Biol Chem 249:4634–4637
Lindmark DG, Müller M, Shio H (1975) Hydrogenosomes in *Trichomonas vaginalis*. J Parasitol 61:552–554
Linstead DJ, Bradley S (1988) The purification and properties of two soluble reduced nicotinamide: acceptor oxidoreductases from *Trichomonas vaginalis*. Mol Biochem Parasitol 27:125–133
Lloyd D, Kristensen B (1985) Metronidazole inhibition of hydrogen production in vivo in drug-sensitive and resistant strains of *Trichomonas vaginalis*. J Gen Microbiol 131:849–853
Long S, Jirků M, Mach J, Ginger M, Šuťák R, Richardson DR, Tachezy J, Lukeš J (2008) Ancestral roles of eukaryotic frataxin: mitochondrial frataxin function and heterologous expression of hydrogenosomal *Trichomonas* homologues in trypanosomes. Mol Microbiol 69:94–109
Mallo N, Lamas J, Leiro JM (2013) Hydrogenosome metabolism is the key target for antiparasitic activity of resveratrol against *Trichomonas vaginalis*. Antimicrob Agents Chemother 57:2476–2484. https://doi.org/10.1128/AAC.00009-13
Marczak R, Gorrell TE, Müller M (1983) Hydrogenosomal ferredoxin of the anaerobic protozoon, *Tritrichomonas foetus*. J Biol Chem 258:12427–12433
McGonigle S, Dalton JP, James ER (1998) Peroxidoxins: a new antioxidant family. Parasitol Today 14:139–145
Mentel M, Zimorski V, Haferkamp P et al (2008) Protein import into hydrogenosomes of *Trichomonas vaginalis* involves both N-terminal and internal targeting signals: a case study of thioredoxin reductases. Eukaryot Cell 7:1750–1757. https://doi.org/10.1128/EC.00206-08
Mertens E, Van Schaftingen E, Müller M (1992) Pyruvate kinase from *Trichomonas vaginalis*, an allosteric enzyme stimulated by ribose 5-phosphate and glycerate 3-phosphate. Mol Biochem Parasitol 54:13–20
Meyer J (2007) [FeFe] hydrogenases and their evolution: a genomic perspective. Cell Mol Life Sci 64:1063–1084
Mogi T, Kita K (2010) Diversity in mitochondrial metabolic pathways in parasitic protists *Plasmodium* and *Cryptosporidium*. Parasitol Int 59:305–312
Mongkolsuk S, Praituan W, Loprasert S et al (1998) Identification and characterization of a new organic hydroperoxide resistance (ohr) gene with a novel pattern of oxidative stress regulation from *Xanthomonas campestris* pv. *phaseoli*. J Bacteriol 180:2636–2643
Morada M, Smid O, Hampl V et al (2011) Hydrogenosome-localization of arginine deiminase in *Trichomonas vaginalis*. Mol Biochem Parasitol 176:51–54. https://doi.org/10.1016/j.molbiopara.2010.10.004
Mukherjee M, Brown MT, McArthur AG, Johnson PJ (2006a) Proteins of the glycine decarboxylase complex in the hydrogenosome of *Trichomonas vaginalis*. Eukaryot Cell 5:2062–2071
Mukherjee M, Sievers SA, Brown MT, Johnson PJ (2006b) Identification and biochemical characterization of serine hydroxymethyl transferase in the hydrogenosome of *Trichomonas vaginalis*. Eukaryot Cell 5:2072–2078
Müller M (1993) The hydrogenosome. J Gen Microbiol 139:2879–2889. https://doi.org/10.1099/00221287-139-12-2879
Müller M, Lindmark DG (1978) Respiration of hydrogenosomes of *Tritrichomonas foetus*. II. Effect of CoA on pyruvate oxidation. J Biol Chem 253:1215–1218
Müller S, Liebau E, Walter RD, Krauth-Siegel RL (2003) Thiol-based redox metabolism of protozoan parasites. Trends Parasitol 19:320–328
Müller M, Mentel M, van Hellemond JJ et al (2012) Biochemistry and evolution of anaerobic energy metabolism in eukaryotes. Microbiol Mol Biol Rev 76:444–495

Novák L, Zubáčová Z, Karnkowska A et al (2016) Arginine deiminase pathway enzymes: evolutionary history in metamonads and other eukaryotes. BMC Evol Biol 16:197. https://doi.org/10.1186/s12862-016-0771-4
Nývltová E, Smutná T, Tachezy J, Hrdý I (2016) OsmC and incomplete glycine decarboxylase complex mediate reductive detoxification of peroxides in hydrogenosomes of *Trichomonas vaginalis*. Mol Biochem Parasitol 206:29–38
Page-Sharp M, Behm CA, Smith GD et al (1996) *Tritrichomonas foetus* and *Trichomonas vaginalis*: the pattern of inactivation of hydrogenase activity by oxygen and activities of catalase and ascorbate peroxidase. Microbiology 142:207–211
Paget TA, Lloyd D (1990) *Trichomonas vaginalis* requires traces of oxygen and high concentrations of carbon dioxide for optimal growth. Mol Biochem Parasitol 41:65–72
Paltauf F, Meingassner JG (1982) The absence of cardiolipin in hydrogenosomes of *Trichomonas vaginalis* and *Tritrichomonas foetus*. J Parasitol 68:949–950
Paul VD, Lill R (2015) Biogenesis of cytosolic and nuclear iron-sulfur proteins and their role in genome stability. Biochim Biophys Acta – Mol Cell Res 1853:1528–1539
Payne MJ, Chapman A, Cammack R (1993) Evidence for an [Fe]-type hydrogenase in the parasitic protozoan *Trichomonas vaginalis*. FEBS Lett 317:101–104
Pegg AE (1986) Recent advances in the biochemistry of polyamines in eukaryotes. Biochem J 234:249–262. https://doi.org/10.1016/j.eswa.2016.05.028
Peters JW, Schut GJ, Boyd ES et al (2015) [FeFe]- and [NiFe]-hydrogenase diversity, mechanism, and maturation. Biochim Biophys Acta 1853:1350–1369. https://doi.org/10.1016/j.bbamcr.2014.11.021
Pütz S, Gelius-Dietrich G, Piotrowski M, Henze K (2005) Rubrerythrin and peroxiredoxin: two novel putative peroxidases in the hydrogenosomes of the microaerophilic protozoon *Trichomonas vaginalis*. Mol Biochem Parasitol 142:212–223
Pütz S, Doležal P, Gelius-Dietrich G et al (2006) Fe-hydrogenase maturases in the hydrogenosomes of *Trichomonas vaginalis*. Eukaryot Cell 5:579–586
Rada P, Doležal P, Jedelský PL et al (2011) The core components of organelle biogenesis and membrane transport in the hydrogenosomes of *Trichomonas vaginalis*. PLoS One 6:e24428. https://doi.org/10.1371/journal.pone.0024428
Rasoloson D, Vaňáčová Š, Tomková E et al (2002) Mechanisms of in vitro development of resistance to metronidazole in *Trichomonas vaginalis*. Microbiology 148:2467–2477. https://doi.org/10.1099/00221287-148-8-2467
Reeves PR, Guthrie JD, Lobelle-Rich P (1980) *Entamoeba histolytica*: isolation of ferredoxin. Exp Parasitol 49:83–88
Reis IA, Martinez MP, Yarlett N et al (1999) Inhibition of polyamine synthesis arrests trichomonad growth and induces destruction of hydrogenosomes. Antimicrob Agents Chemother 43:1919–1923
Rivière L, Van Weelden SWH, Glass P et al (2004) Acetyl:succinate CoA-transferase in procyclic *Trypanosoma brucei*. Gene identification and role in carbohydrate metabolism. J Biol Chem. https://doi.org/10.1074/jbc.M407513200
Rotte C, Stejskal F, Zhu G et al (2001) Pyruvate: NADP(+) oxidoreductase from the mitochondrion of *Euglena gracilis* and from the apicomplexan *Cryptosporidium parvum*: a biochemical relic linking pyruvate metabolism in mitochondriate and amitochondriate protists. Mol Biol Evol 18:710–720
Rubach JK, Brazzolotto X, Gaillard J, Fontecave M (2005) Biochemical characterization of the HydE and HydG iron-only hydrogenase maturation enzymes from *Thermatoga maritima*. FEBS Lett 579:5055–5060
Santos R, Buisson N, Knight SAB et al (2004) *Candida albicans* lacking the frataxin homologue: a relevant yeast model for studying the role of frataxin. Mol Microbiol 54:507–519. https://doi.org/10.1111/j.1365-2958.2004.04281.x
Schirch V, Szebenyi DME (2005) Serine hydroxymethyltransferase revisited. Curr Opin Chem Biol 9:482–487. https://doi.org/10.1016/j.cbpa.2005.08.017

Schneider RE, Brown MT, Shiflett AM et al (2011) The *Trichomonas vaginalis* hydrogenosome proteome is highly reduced relative to mitochondria, yet complex compared with mitosomes. Int J Parasitol 41:1421–1434
Schut GJ, Adams MWW (2009) The iron-hydrogenase of *Thermotoga maritima* utilizes ferredoxin and NADH synergistically: a new perspective on anaerobic hydrogen production. J Bacteriol 191:4451–4457. https://doi.org/10.1128/JB.01582-08
Slamovits CH, Keeling PJ (2006) Pyruvate-phosphate dikinase of oxymonads and parabasalia and the evolution of pyrophosphate-dependent glycolysis in anaerobic eukaryotes. Eukaryot Cell 5:148–154
Smutná T, Goncalves VL, Saraiva LM et al (2009) Flavodiiron protein from *Trichomonas vaginalis* hydrogenosomes: the terminal oxygen reductase. Eukaryot Cell 8:47–55
Smutná T, Pilařová K, Tarábek J et al (2014) Novel functions of an iron-sulfur flavoprotein from *Trichomonas vaginalis* hydrogenosomes. Antimicrob Agents Chemother 58:3224–3232
Stairs CW, Eme L, Brown MW et al (2014) A SUF Fe-S cluster biogenesis system in the mitochondrion-related organelles of the anaerobic protist *Pygsuia*. Curr Biol 24:1176–1186. https://doi.org/10.1016/j.cub.2014.04.033
Stehling O, Wilbrecht C, Lill R (2014) Mitochondrial iron-sulfur protein biogenesis and human disease. Biochimie 100:61–77
Steinbüchel A, Müller M (1986) Anaerobic pyruvate metabolism of *Tritrichomonas foetus* and *Trichomonas vaginalis* hydrogenosomes. Mol Biochem Parasitol 20:57–65
Sutak R, Doležal P, Fiumera HL et al (2004) Mitochondrial-type assembly of FeS centers in the hydrogenosomes of the amitochondriate eukaryote *Trichomonas vaginalis*. Proc Natl Acad Sci U S A 101:10368–10373
Sutak R, Hrdý I, Doležal P et al (2012) Secondary alcohol dehydrogenase catalyzes the reduction of exogenous acetone to 2-propanol in *Trichomonas vaginalis*. FEBS J 279:2768–2780
Tabor CW, Tabor H (1976) 1,4-Diaminobutane (Putrescine), Spermidine, and Spermine. Annu Rev Biochem 45:285–306
Tachezy J, Sanchez LB, Müller M (2001) Mitochondrial type iron-sulfur cluster assembly in the amitochondriate eukaryotes *Trichomonas vaginalis* and *Giardia intestinalis*, as indicated by the phylogeny of IscS. Mol Biol Evol 18:1919–1928
Tadolini B (1988) Polyamine inhibition of lipoperoxidation – the influence of polyamines on iron oxidation in the presence of compounds mimicking phospholipid polar heads. Biochem J 249:33–36
Tanabe M (1979) *Trichomonas vaginalis* NADH oxidase activity. Exp Parasitol 48:135–143
Thong KW, Coombs GH (1987) *Trichomonas* species: homocysteine desulphurase and serine sulphydrase activities. Exp Parasitol 63:143–151
Townson M, Hanson GR, Upcroft JA, Upcroft P (1994) Purified ferredoxin from *Giardia duodenalis*. Eur J Biochem 220:439–446
van Grinsven KWA, Rosnowsky S, Van Weelden SWH et al (2008) Acetate:succinate CoA-transferase in the hydrogenosomes of *Trichomonas vaginalis*: Identification and characterization. J Biol Chem 283:1411–1418. https://doi.org/10.1074/jbc.M702528200
van Grinsven KWA, van Hellemond JJ, Tielens AGM (2009) Acetate:succinate CoA-transferase in the anaerobic mitochondria of *Fasciola hepatica*. Mol Biochem Parasitol 164:74–79. https://doi.org/10.1016/j.molbiopara.2008.11.008
Vicente JB, Tran V, Pinto L et al (2012) A detoxifying oxygen reductase in the anaerobic protozoan *Entamoeba histolytica*. Eukaryot Cell 11:1112–1118. https://doi.org/10.1128/EC.00149-12
Vignais PM, Billoud B, Meyer J (2001) Classification and phylogeny of hydrogenases. FEMS Microbiol Rev 25:455–501
Vilela R, Menna-Barreto RFS, Benchimol M (2010) Methyl jasmonate induces cell death and loss of hydrogenosomal membrane potential in *Trichomonas vaginalis*. Parasitol Int 59:387–393. https://doi.org/10.1016/j.parint.2010.05.003

Viscogliosi E, Delgado-Viscogliosi P, Gerbod D et al (1998) Cloning and expression of an iron-containing superoxide dismutase in the parasitic protist, *Trichomonas vaginalis*. FEMS Microbiol Lett 161:115–123

Westrop GD, Wang L, Blackburn GJ et al (2017) Metabolomic profiling and stable isotope labelling of *Trichomonas vaginalis* and *Tritrichomonas foetus* reveal major differences in amino acid metabolism including the production of 2-hydroxyisocaproic acid, cystathionine and S-methylcysteine. PLoS One 12:e0189072. https://doi.org/10.1371/journal.pone.0189072

Williams K, Lowe PN, Leadlay PF (1987) Purification and characterization of pyruvate: ferredoxin oxidoreductase from the anaerobic protozoon *Trichomonas vaginalis*. Biochem J 246:529–536

Wood ZA, Schroder E, Harris JR, Poole LB (2003) Structure, mechanism and regulation of peroxiredoxins. Trends Biochem Sci 28:32–40

Yarlett N, Lindmark DG, Goldberg B et al (1994) Subcellular localization of the enzymes of the arginine dihydrolase pathway in *Trichomonas vaginalis* and *Tritrichomonas foetus*. J Eukaryot Microbiol 41:554–559

Yarlett N, Martinez MP, Moharrami MA, Tachezy J (1996) The contribution of the arginine dihydrolase pathway to energy metabolism by *Trichomonas vaginalis*. Mol Biochem Parasitol 78:117–125

Zhao T, Cruz F, Ferry JG (2001) Iron-sulfur flavoprotein (Isf) from *Methanosarcina thermophila* is the prototype of a widely distributed family. J Bacteriol 183:6225–6233

Hydrogenosomes of Anaerobic Fungi: An Alternative Way to Adapt to Anaerobic Environments

Johannes H. P. Hackstein, Scott E. Baker, Jaap J. van Hellemond, and Aloysius G. M. Tielens

Contents

1 Introduction 160
2 Mitochondria Versus Hydrogenosomes 162
3 Anaerobic Fungi Possess Hydrogenosomes and Perform a (Bacterial-Type) Mixed Acid Fermentation 164
4 Hydrogenosomal Metabolism of *Piromyces* and *Neocallimastix* 165
5 The Role of the Hydrogenosomes in the Energy Metabolism of *Piromyces* sp. E2 169
6 The *Mitochondrial* Origin of the Hydrogenosomes in Anaerobic Fungi 170
References 171

Abstract Fungi form a very diverse group of eukaryotes. The majority of investigated fungi contain mitochondria and are capable of oxidative phosphorylation. On the other hand, anaerobically functioning fungi, found as symbionts in the gastrointestinal tract of many herbivorous mammals, contain hydrogenosomes. These organelles of mitochondrial origin are also found in multiple classes of anaerobically functioning protists. Hydrogenosomes produce hydrogen as an end product of a

J. H. P. Hackstein
Faculty of Science, IWWR, Radboud University Nijmegen, Nijmegen, The Netherlands

S. E. Baker
Earth and Biological Sciences Directorate Pacific Northwest National Laboratory, Environmental Molecular Science Division, Biosystems Dynamics and Simulation Group, Richland, WA, USA

J. J. van Hellemond
Department of Medical Microbiology and Infectious Diseases, Erasmus MC University Medical Center Rotterdam, Rotterdam, The Netherlands

A. G. M. Tielens (✉)
Department of Medical Microbiology and Infectious Diseases, Erasmus MC University Medical Center Rotterdam, Rotterdam, The Netherlands

Faculty of Veterinary Medicine, Department of Biochemistry and Cell Biology, Utrecht University, Utrecht, The Netherlands
e-mail: A.G.M.Tielens@uu.nl

J. Tachezy (ed.), *Hydrogenosomes and Mitosomes: Mitochondria of Anaerobic Eukaryotes*, Microbiology Monographs 9,
https://doi.org/10.1007/978-3-030-17941-0_7

fermentative energy metabolism and produce ATP by substrate-level phosphorylation. However, the hydrogenosomes of the anaerobic fungi *Neocallimastix* and *Piromyces* differ from the hydrogenosomes of trichomonads and those of anaerobic ciliates in the way they convert pyruvate to acetyl-CoA. The hydrogenosomes of these anaerobic fungi use pyruvate:formate lyase (PFL), whereas trichomonads use pyruvate-ferredoxin oxidoreductase (PFO), and anaerobic ciliates use pyruvate dehydrogenase (PDH) for the degradation of pyruvate. The characteristics and role of these hydrogenosomes in the energy metabolism of anaerobic fungi are discussed.

1 Introduction

Fungi comprise a monophyletic but morphologically and biochemically extremely diverse eukaryotic kingdom. The taxon *Fungi* includes not only some of the best studied eukaryotic model organisms, i.e. the baker's yeast *Saccharomyces cerevisiae*, the fission yeast *Schizosaccharomyces pombe* and the common bread moulds *Neurospora crassa* and *Aspergillus nidulans*, but also poorly understood, uncultured fungi from soil and aquatic environments, or symbiotic fungi such as mycorrhiza, and elusive anaerobic fungi from the gastrointestinal tract of herbivorous mammals (Trinci et al. 1994; Schadt et al. 2003; Strack et al. 2003; Anderson and Cairney 2004; Luo et al. 2005).

Anaerobic fungi are important symbionts in the gastrointestinal tract of herbivorous mammals. A flagellated rumen-dwelling organism, *Neocallimastix frontalis*, was described in 1975 by Colin Orpin of Australia's CSIRO (Orpin 1975). Two years later, Orpin published a report showing that *N. frontalis* and two other anaerobes had cell walls that contained chitin, indicating that these rumen-dwelling organisms are fungi (Orpin 1977). Over the next three decades, a body of research was generated covering the physiology, ultrastructure and taxonomy, ecology, cell biology, biochemistry and metabolism and genomics of anaerobic fungi.

Monocentric anaerobic fungi are defined by a single thallus (zoospore containing structure) and centre of growth from which rhizoids, fine anucleate filaments involved in substrate anchorage and nutrient absorption, radiate. In contrast, polycentric anaerobic fungi have multiple thalli and centres of growth. Zoospores of anaerobic fungi can have one or multiple flagella depending on the species. Taxonomy, life cycle, ecology and biotechnological potential of anaerobic fungi are reviewed by Gruninger et al. (2014). Anaerobic fungi belong to the phylum *Neocallimastigomycota*, an early diverging lineage in the kingdom *Fungi*. They are closely related to the chytrids (phylum *Chytridiomycota*) and earlier were even a group in the same phylum, but the anaerobic fungi now form a separate phylum, *Neocallimastigomycota* (Hibbett et al. 2007). Nine genera of anaerobic fungi have been described so far in the *Neocallimastigomycota* phylum, but it is expected that more will follow (Edwards et al. 2017).

The genomes of mammalian herbivores do not encode enzymes that are capable to degrade the ingested plant material and rely on symbiotic interactions with microorganisms such as bacteria, protozoa and anaerobic fungi that live in their gut. In this microbial world, the anaerobic fungi are especially known for their capacity to degrade lignocellulosic plant fibres that form the cell walls of plants. Lignocellulosic biomass refers to plant biomass that is mainly composed of cellulose (an insoluble polysaccharide comprising long linear chains of glucose molecules), hemicellulose (a mixture of branched polymers of various 5- and 6-carbon sugars, such as xylose, mannose and galactose) and lignin (non-polysaccharide aromatic cross-linked polymers). A select number of anaerobic bacteria produce cellulosomes, multi-protein complexes that serve to improve the degradation of lignocellulose biomass (Artzi et al. 2017). It was known already for a long time that the anaerobic fungi known to degrade lignocellulose also contain extracellular cellulosome-like structures (Wilson and Wood 1992). Recently it was shown that these anaerobic fungi produce a large array of biomass-degrading enzymes that are assembled into these cellulosomes and synergistically degrade crude, untreated plant biomass, which makes these fungal enzymes an interesting option for industrial applications (Solomon et al. 2016). A comparative genomic study revealed that the scaffolding proteins (or "scaffoldins") for fungal cellulosomes are quite different from their bacterial counterparts and contain proteins unique to anaerobic fungi (Haitjema et al. 2017).

Anaerobic fungi are highly effective fibre-degrading microorganisms in the gut of mammalian herbivores. Their diversity is large, and they are found in the gastrointestinal tract of nearly all large herbivores, ranging from ruminants such as cattle, sheep, goat, deer and antelopes to the foregut-fermenting marsupials and camelids on the one hand and hindgut-fermenting species such as horse, elephant, rhinoceros, mara (Patagonian hare) and capybara ("water pig", the world's largest rodent) on the other hand. Anaerobic fungi can be isolated from rumen fluid or faeces, and they are maintained in anaerobic culture, most of them as pure axenic cultures. In the rumen of cattle or sheep, these anaerobic fungi can be as frequent as 7.6×10^8 thallus-forming units; in the faeces there are still 4.2×10^4 units per g dry weight (Trinci et al. 1994). Anaerobic fungi are not truly host-specific since it is possible to transfaunate various host animals with isolates from different hosts. On the other hand, the various isolates are not the same, even if collected from the same host species and assigned to the same species. The patterns of utilization of substrates and the metabolic properties are different from isolate to isolate (Trinci et al. 1994).

We now know that the anaerobic fungi comprise many species that are integral in the rumen ecosystem and crucial in the digestion of plant material to metabolizable substrates such as simple sugars. Moreover, they produce hydrogen needed for the growth of methanogenic bacteria (reviewed in Williams et al. 1994). With regard to degradation of plant material, the rumen anaerobic fungi produce a wide variety of glycosyl hydrolases needed for the breakdown of plant biomass (reviewed in Chen et al. 1995). Because of their intimate association with the plant biomass being degraded, anaerobic fungi were often discarded with the solid gut material leaving a paucity of culturable material behind, which kept them from being more thoroughly characterized earlier (Bauchop 1979). Later on they were frequently studied, mainly for biotechnological applications of their cellulolytic enzymes.

From both an applied and basic science perspective, anaerobic fungi are interesting and important organisms. For that reason several genome sequencing projects were performed. An early study of the DNA content of a *Neocallimastix* species found that it has an "extreme" genome which is highly A+T rich (up to 82%, Brownlee 1989). Subsequent studies on other anaerobic fungi produced similar results (Brownlee 1994). These results were later confirmed with a limited gene sequencing study in *Orpinomyces* (Nicholson et al. 2005). It has been further speculated that rDNA constitutes up to 25% of the genome (Brownlee 1989, 1994; Nicholson et al. 2005). A number of studies in which genes have been cloned from a variety of anaerobic fungal species further have shown that the protein-coding sequences contain more "normal" levels of A+T (40–60%) (see e.g. Nicholson et al. 2005). Thus, it is likely that introns and intragenic spaces are responsible for the A+T richness of anaerobic fungal genomes.

Thanks to improved techniques in long-read sequencing methods, five genomes of anaerobic fungi are published up to now: *Piromyces* sp. E2, *Pecoramyces ruminantium* C1A (formerly *Orpinomyces* sp. C1A), *Neocallimastix californiae*, *Piromyces finnis* and *Anaeromyces robustus* (see Haitjema et al. 2017). Several efforts to use these methods to sequence the genomes of other anaerobic fungi failed however. Practical considerations and current challenges when generating and analysing anaerobic fungal genomes are reviewed by Edwards et al. (2017) and Peng et al. (2018).

2 Mitochondria Versus Hydrogenosomes

The majority of the cultured fungi contain mitochondria. These mitochondria host a genome of varying sizes, which characteristically encodes only a handful of proteins (Bullerwell and Lang 2005). This implies that the vast majority of the 700–800 mitochondrial proteins (Sickmann et al. 2003) are nuclear encoded, synthesized in the cytoplasm and imported into the organelles. Interestingly, certain cultivars of mitochondriate species are able to maintain mitochondria in the absence of a mitochondrial genome. Such yeasts are known as "petites"; they are viable but respiration deficient and incapable of growing on non-fermentable substrates (Contamine and Picard 2000; Lipinski et al. 2010). In this respect they are similar to the genome-less hydrogenosomes of anaerobic fungi to be discussed in detail below.

On the other hand, two natural isolates of fission yeasts, *Schizosaccharomyces japonicus* var. *japonicus* and *S. japonicus* var. *versatilis*, lack detectable cytochromes and are respiration deficient, but DNA sequencing has revealed that they nevertheless have apparently fully functional mtDNA (Bullerwell and Lang 2005). These fission yeasts are considered to be an intermediate evolutionary stage in between respiratory-competent fungi and those that completely lack mitochondrial DNA. The mitochondria of these yeast species might be similar to the "hydrogenosomes" of the ciliate *Nyctotherus ovalis* described in chapter "Hydrogenosomes of Anaerobic Ciliates" and, therefore, might represent an evolutionary intermediate between the classical mitochondria of most fungi and the "hydrogenosomes" of the anaerobic fungi described below. Notably, the members of the phylum *Neocallimastigomycota*, such as *Piromyces* (Fig. 1) and *Neocallimastix*, which possess hydrogenosomes,

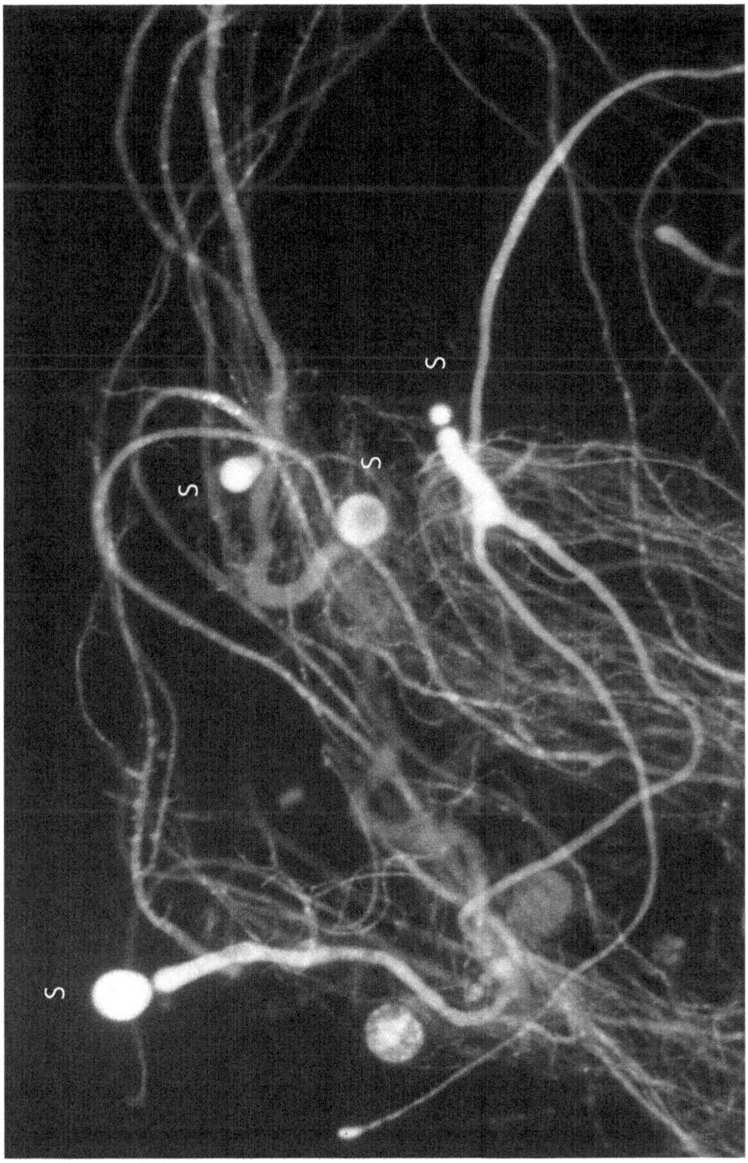

Fig. 1 Epifluorescence micrograph of *Piromyces* sp. E2 originally isolated from the faeces of an Indian elephant. Magnification about ×400. The organism was vitally stained with rhodamine 123. S: young sporangia

lack both mitochondria and a mitochondrial genome (van der Giezen et al. 1997). These hydrogenosomes of anaerobic fungi are membrane-bounded compartments up to 1 μm in size that produce ATP by substrate-level phosphorylation together with hydrogen as an end product allowing growth under anoxic conditions (Marvin-Sikkema et al. 1993, 1994; Hackstein et al. 2001; Voncken et al. 2002b). We will discuss here that these "hydrogenosomes" are specialized mitochondria, just like all other hydrogenosomes, but that these fungal hydrogenosomes are nevertheless different from those of trichomonads (see chapter "Metabolism of Trichomonad Hydrogenosomes") as well as from those of anaerobic ciliates (see chapter "Hydrogenosomes of Anaerobic Ciliates").

3 Anaerobic Fungi Possess Hydrogenosomes and Perform a (Bacterial-Type) Mixed Acid Fermentation

As explained in the Introduction, anaerobic fungi in the gut of herbivores digest various plant carbohydrates using large extracellular enzyme complexes known as cellulosomes which consist of many cellulolytic and hemicellulolytic enzymes bound to scaffolding proteins (Wang et al. 2014; Solomon et al. 2016; Haitjema et al. 2017; Kameshwar and Qin 2018). The hydrolysis of cellulose by these cellulosomes results in the liberation of glucose molecules only, while hemicellulose hydrolysis yields also pentoses such as xylose and arabinose. All these monosaccharides are then fermented by the anaerobic fungi. The fermentative metabolism by anaerobic fungi has not been studied in great detail, but it is known that they produce, besides hydrogen and carbon dioxide, also formate, acetate, lactate, ethanol and succinate when growing on cellulose, glucose or fructose as a carbon source (Julliand et al. 1998; Edwards et al. 2017). Such a mixed acid fermentation is very similar to bacterial mixed acid fermentations that are, for example, well known for facultative anaerobic enteric bacteria, such as *Escherichia coli*.

Many bacterial mixed acid fermentations do not produce ethanol from pyruvate via pyruvate decarboxylase and alcohol dehydrogenase as in the alcoholic fermentation of yeast but by the successive action of pyruvate:formate lyase (PFL) and alcohol dehydrogenase E (ADHE). The latter enzyme combines aldehyde dehydrogenase (ALDH) and alcohol dehydrogenase (ADH) activities, and acetyl-CoA is the substrate (Kessler et al. 1991; Arnau et al. 1998; Fontaine et al. 2002). Until now, ADHE has been found exclusively in bacteria, with the remarkable exception of certain eukaryotic anaerobes such as *Giardia*, *Spironucleus*, *Entamoeba* and *Mastigamoeba*, which do not possess mitochondria or hydrogenosomes but possess mitosomes (Bruchhaus and Tannich 1994; Sánchez 1998; Dan and Wang 2000; Field et al. 2000; Andersson et al. 2003; Müller et al. 2012). However, these eukaryotic anaerobes do *not* perform a bacterial-type mixed acid fermentation, as they do *not* exhibit PFL activity and, notably, they *lack* ATP-producing organelles such as mitochondria or hydrogenosomes (Müller 1998; Müller et al. 2012).

4 Hydrogenosomal Metabolism of *Piromyces* and *Neocallimastix*

The hydrogenosomal metabolism has been studied in more detail in the anaerobic fungi *Piromyces* and *Neocallimastix*. As discussed earlier, the lignocellulosic plant material is first degraded to fermentable monosaccharides which are then degraded in the cytosol and hydrogenosomes of the anaerobic fungi. Glucose and xylose are the most abundant carbohydrates in lignocellulotic hydrolysates. The glucose is degraded via a mixed acid fermentation, introduced in paragraph 3, and for *Piromyces* further discussed in detail below. The pentose xylose first has to be converted to intermediates of glycolysis, which will then follow the same pathways of mixed acid fermentation. It was shown that in contrast to most yeasts and fungi, *Piromyces* converts xylose via the combined action of xylose isomerase and a xylulokinase to xylulose-5-phosphate, which is the bacterial pathway (Harhangi et al. 2003). This xylulose-5-phosphate is then converted to ribulose-5-phosphate and via the final reaction of the oxidative phase of the pentose phosphate pathway (PPP) to ribose-5-phosphate (Fig. 2). Three molecules of ribose-5-phosphate can then be converted to two molecules of fructose-6-phosphate and one molecule of glyceraldehyde-3-phosphate via the nonoxidative phase of the PPP. These compounds are intermediates in the mixed acid fermentation pathways and will be further metabolized, resulting in the formation of the same end products as glucose metabolism (Fig. 2).

Notably, the hydrogenosomes of anaerobic fungi are structurally and metabolically clearly different from those known of trichomonads and anaerobic ciliates. Most importantly, the hydrogenosomes of *Neocallimastix* sp. L2 and *Piromyces* sp. E2 contain PFL as key enzyme and not PDH (as in *N. ovalis*) or PFO (as in *Trichomonas vaginalis*) (Fig. 2).

As discussed above, these anaerobic fungi perform a mixed acid fermentation, where carbohydrate degradation results in the production of hydrogen and carbon dioxide plus formate, acetate, lactate, ethanol and succinate (Fig. 2). However, the ratio of these excreted end products is not constant, as it was shown that growth of *Piromyces* sp. E2 in the presence of increasing concentrations of fructose is accompanied by changes in the fermentation pattern (Boxma et al. 2004). Increasing the fructose concentration from 0.1 to 0.5% resulted in a threefold increase in degradation of this substrate to end products. It is remarkable that the relative fluxes of fructose degradation through the various pathways were not constant during changing fructose concentrations (Fig. 3). Although the absolute amounts of hydrogen formed in the incubations during growth at these increasing concentrations of fructose remained constant, the relative flux of malate into the hydrogenosomes

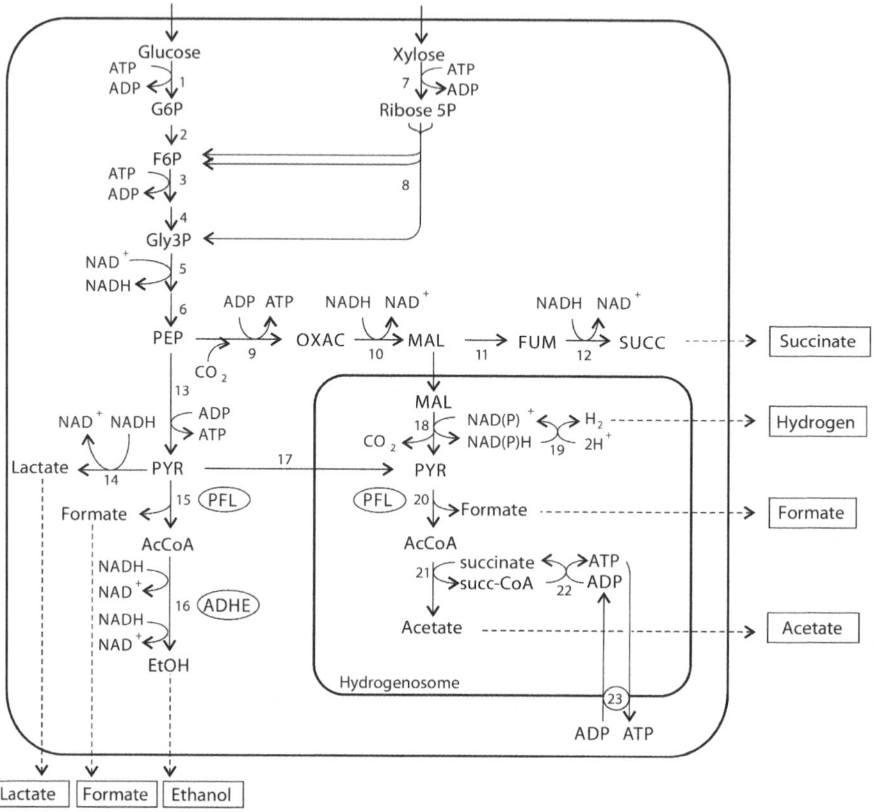

Fig. 2 Energy metabolism of *Piromyces* sp. E2. Shown is a scheme of the metabolic pathways involved in the production of the major end products. The numbers indicate the following enzymes: (1) hexokinase; (2) glucose-6-phosphate isomerase; (3) phosphofructokinase; (4) aldolase and triose phosphate isomerase; (5) glyceraldehyde 3-phosphate dehydrogenase; (6) phosphoglycerate kinase, phosphoglycerate mutase and enolase; (7) oxidative phase of pentose phosphate pathway; (8) nonoxidative phase of pentose phosphate pathway; (9) phosphoenolpyruvate carboxykinase; (10) malate dehydrogenase; (11) fumarase; (12) fumarate reductase; (13) pyruvate kinase; (14) lactate dehydrogenase; (15) cytosolic pyruvate:formate lyase; (16) alcohol dehydrogenase E; (17) pyruvate import into hydrogenosomes; (18) malic enzyme; (19) hydrogenase; (20) hydrogenosomal pyruvate:formate lyase; (21) acetate:succinate CoA-transferase; (22) succinyl-CoA synthetase; and (23) ADP/ATP carrier. Abbreviations: *AcCoA* acetyl-CoA, *EtOH* ethanol, *FUM* fumarate, *G3P* glyceraldehyde-3-phosphate, *MAL* malate, *OXAC* oxaloacetate, *PEP* phosphoenolpyruvate, *PYR* pyruvate, *SUCC* succinate (modified after Boxma et al. 2004)

and hence the relative flux to hydrogen decreased from 47 to 15% (Fig. 3, Boxma et al. 2004). In contrast, the relative fluxes in the formation of the cytosolic end-products lactate, ethanol and succinate increased severalfold. These observations show that increasing amounts of a fermentable carbon source result in an increased metabolism without an increased production of hydrogen and cause a relative shift from hydrogenosomal carbon metabolism to a cytosolic one (Fig. 3).

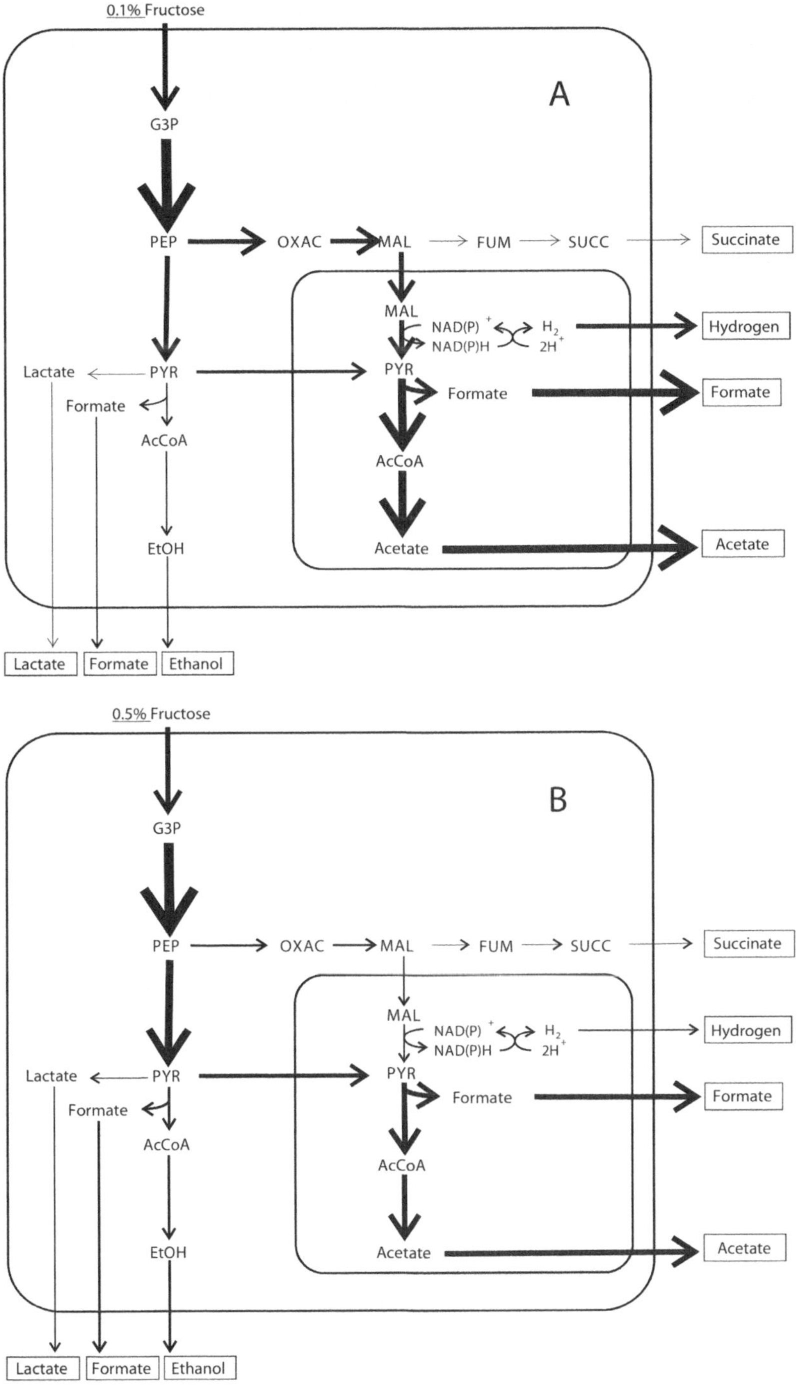

Fig. 3 Changes in metabolic fluxes in the energy metabolism of *Piromyces* induced by changes in the fructose concentration in the medium. The thickness of the arrows is proportional to the

Experiments using [6-^{14}C]-glucose and [U-^{14}C]-glucose indicated that an incomplete Krebs cycle operates in the reductive mode allowing the formation of succinate from oxaloacetate via a malate intermediate (Fig. 2). As no significant amounts of labelled CO_2 were produced, while formate and acetate plus ethanol were formed in a 1:1 ratio, it must be concluded that PFL and not pyruvate:ferredoxin oxidoreductase (PFO) or pyruvate dehydrogenase (PDH) play the central role in the hydrogenosomal metabolism. The activity of the latter enzymes would have generated one molecule of labelled carbon dioxide per molecule of pyruvate degraded. However, less than 1% of the expected amount was detected (Boxma et al. 2004).

ADHE activity, characteristic for bacterial mixed acid fermentations, is present in *Piromyces* sp. E2 and *Neocallimastix* sp. L2. In *E. coli*, where PFL is only expressed under anaerobic growth conditions, ADHE can act as a PFL inactivase, thereby protecting PFL against irreversible damage by oxygen (Kessler et al. 1991, 1992; Sawers and Watson 1998). The situation is likely to be comparable in anaerobic fungi, and, notably, anaerobic fungi also possess a typical PFL-activating enzyme (Gelius-Dietrich and Henze 2004).

There has been a controversy about the presence or activity of pyruvate:ferredoxin oxidoreductase (PFO) in anaerobic fungi: in a few cases, PFO activity instead of PFL activity has been reported to be responsible for the hydrogenosomal pyruvate degradation. A low enzymatic activity attributed to PFO has been measured in *Neocallimastix* sp. L2 and *N. patriciarum* (Yarlett et al. 1986; Marvin-Sikkema et al. 1993, 1994), while in another species, *N. frontalis*, the absence has been reported of any detectable PFO activity, measured as pyruvate synthase using ferredoxin as electron acceptor (O'Fallon et al. 1991). Notably, the observation of putative PFO activity in anaerobic fungi has not been substantiated by characterization of the enzyme nor by identification of a PFO gene in these organisms. Furthermore, the observed formate production in all metabolic studies strongly suggests the participation of an active PFL in the metabolism of anaerobic fungi, as there is no other fermentative pathway known to result in formate production. On the other hand, PFO is not completely unknown in the fungal world, as it has been shown that *Saccharomyces* and other fungi contain enzymes involved in methionine biosynthesis that are fusion proteins of PFO domains and fragments of redox enzymes (Horner et al. 1999).

In the gastrointestinal tract of herbivores, the anaerobic fungi live together with methanogens, archaea that can grow autotrophically on carbon dioxide and hydrogen, producing methane. The methanogens use end products of the fermentative metabolism of the anaerobic fungi as substrate. The hydrogen and formate produced by the anaerobic fungi are used by the archaea for the production of methane, a process that yields ATP. This further use of the fungal end products by the methanogens results in changes in metabolism in the anaerobic fungi. Coculturing

Fig. 3 (continued) calculated fluxes in the presence of 0.1% fructose (Panel **a**) or 0.5% fructose (Panel **b**). Abbreviations are as in Fig. 2 (calculated from the data in Boxma et al. 2004)

of anaerobic fungi and methanogens increased the rate of substrate fermentation and growth of the fungi, together with a shift in metabolism (Marvin-Sikkema et al. 1990; Li et al. 2017; Wei et al. 2017; Wilken et al. 2018). The production of hydrogen is energetically unfavourable at higher hydrogen concentration. Therefore the use of hydrogen by the methanogens results in low hydrogen levels, facilitating the hydrogenase reaction which results in a shift of metabolism away from the cytosolic reactions and towards the malate-pyruvate pathway (Fig. 2).

5 The Role of the Hydrogenosomes in the Energy Metabolism of *Piromyces* sp. E2

The observation that the hydrogenosomal PFL and the cytoplasmic ADHE are the key enzymes in the degradation of carbohydrates by anaerobic fungi reveals that the metabolism of these hydrogenosomes is fundamentally different from the hydrogenosomal metabolism in both trichomonads and *N. ovalis*-like ciliates. Obviously, anaerobic fungi chose their own way to adapt to anaerobic environments by evolutionary tinkering. The metabolic scheme displayed in Fig. 2 shows a generalized metabolism. A quantitative analysis revealed that (1) PFL must be present and (2) that under certain conditions, hydrogen formation can become marginal (Boxma et al. 2004). The evolutionary strategy of anaerobic fungi apparently tends to avoid the formation of reduction equivalents by using PFL instead of PFO or PDH (Akhmanova et al. 1999; Hackstein et al. 1999, 2006; Voncken 2001).

At higher fructose concentrations, the major flow through the fungal hydrogenosome involves pyruvate, which is split by PFL into formate and acetate without the generation of reduction equivalents (Fig. 2). Thus, these reactions do not contribute to hydrogen formation by the [FeFe] hydrogenase (Davidson et al. 2002; Voncken 2001; Voncken et al. 2002b). Hydrogen formation depends on the import of malate into the hydrogenosome, where malate is decarboxylated by malic enzyme, which provides the electrons for the reduction of H^+ to hydrogen (Marvin-Sikkema et al. 1994). However, the latter reaction seems to represent only a minor pathway of the anaerobic energy metabolism in this type of hydrogenosome and might indicate a role in controlling or fine-tuning of the intra-hydrogenosomal environment.

The major role of the fungal hydrogenosomes seems to be the generation of ATP by substrate-level phosphorylation. The presence of PFL in the absence of hydrogenosomal ADHE most probably directs all organellar pyruvate into substrate-level ATP formation. A possible presence of ADHE inside the hydrogenosomes would compromise this function of the hydrogenosome as an ATP-generating organelle. In the cytoplasm, however, ADHE might allow regulation of PFL activity, thus saving pyruvate (and its metabolites) for anapleurotic pathways. A partial Krebs cycle with links to the anapleurotic pathways operates in the cytoplasm (Akhmanova et al. 1998). This hypothesis is supported by the observation that several mitochondrial enzymes, which are involved in anabolic reactions, e.g. malate dehydrogenase, aconitase, isocitrate dehydrogenase and acetohydroacid reductoisomerase, have been retargeted to the cytoplasm in *Piromyces* sp. E2 (Akhmanova et al. 1998; Hackstein et al. 1999). Consequently, compartmentalization of the energy metabolism seems to enhance the possibilities for regulation of the metabolic pathways of this organism.

6 The *Mitochondrial* Origin of the Hydrogenosomes in Anaerobic Fungi

It is now generally accepted that different types of mitochondrion-related organelles (MROs) evolved from an ancestral endosymbiont. Furthermore, all eukaryotes contain MROs, and all MROs are descendants of one ancestral endosymbiont (Embley and Martin 2006; Hackstein et al. 2006; Tielens and Van Hellemond 2007; Howe 2008; van der Giezen 2009; Müller et al. 2012). By differential loss and gain of metabolic functions, the organellar evolution resulted in a mosaic of mitochondrial functions in the various eukaryotic lineages (Hjort et al. 2010; Maguire and Richards 2014; Makiuchi and Nozaki 2014; Stairs et al. 2015).

Anaerobic fungi are no exception; all data are consistent with a mitochondrial origin of the hydrogenosomes of anaerobic fungi: their morphology, the mitochondrial-type targeting signals that are used to import proteins, the ADP/ATP carriers and mitochondrial-type chaperones and all this supplemented by genomic analyses of separate hydrogenosomal proteins.

Using 18S rDNA phylogenies, or the phylogenies of mitochondrial genes from aerobic fungi, a monophyletic origin of all anaerobic fungi becomes evident (Bullerwell and Lang 2005). An analysis of biochemical and morphological traits consistently establishes a close relationship between anaerobic and aerobic fungi (Ragan and Chapman 1978), and Akhmanova et al. (1998) demonstrated that several enzymes of mitochondrial origin, which lack putative targeting signals, were retargeted to the cytoplasm (in active form) in the hydrogenosome-bearing anaerobic fungus *Piromyces*.

Hydrogenosomes of anaerobic fungi look rather different from the pictures of mitochondria in textbooks (Fig. 4). However, serial sectioning followed by electron microscopical analysis revealed a structure resembling the ultrastructure of mitochondria from particular diseased human patients (Frey and Mannella 2000; Hackstein et al. 2001; Voncken et al. 2002a). Apparently, also in these cases, the inner membrane undergoes a derangement in the mechanism that normally stabilizes the crista junctions (Mannella 2006).

From the metabolic and phylogenetic studies, it is clear that the hydrogenosomes of anaerobic fungi are substantially different from the hydrogenosomes of trichomonads (chapter "Metabolism of Trichomonad Hydrogenosomes" of this volume) and also from those of ciliates (chapter "Hydrogenosomes of Anaerobic Ciliates" of

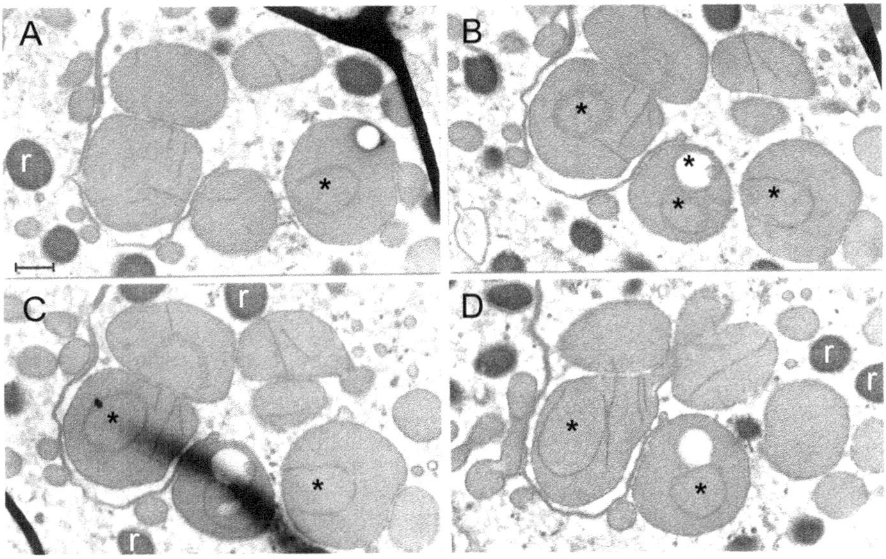

Fig. 4 Serial sectioning of the hydrogenosomes of *Neocallimastix* sp. L2. (**a**)–(**d**) Bar = 0.5 μm; *h* hydrogenosomes, *r* ribosome globules (Munn et al. 1988). Asterisk, internal vesicular structures. From Voncken et al. (2002a), modified

this volume). From a metabolic point of view, the main difference is probably their pyruvate metabolism. Whereas anaerobic fungi use PFL activity to produce acetyl-CoA, in the hydrogenosomes of trichomonads, acetyl-CoA is produced by PFO, and in the hydrogenosomes of the anaerobic ciliate *N. ovalis*, this is performed by the pyruvate dehydrogenase complex (PDH). In contrast to PFO and PDH, the PFL reaction does not generate reduced equivalents, and therefore, the role of the hydrogenase in the hydrogenosomes of anaerobic fungi is smaller than in those two other types of hydrogenosomes.

References

Akhmanova A, Voncken FGJ, Harhangi H, Hosea KM, Vogels GD, Hackstein JHP (1998) Cytosolic enzymes with a mitochondrial ancestry from the anaerobic chytrid *Piromyces* sp. E2. Mol Microbiol 30(5):1017–1027

Akhmanova A, Voncken FGJ, Hosea KM, Harhangi H, Keltjens JT, den Camp HJMO, Vogels GD, Hackstein JHP (1999) A hydrogenosome with pyruvate formate-lyase: anaerobic chytrid fungi use an alternative route for pyruvate catabolism. Mol Microbiol 32(5):1103–1114

Anderson IC, Cairney JWG (2004) Diversity and ecology of soil fungal communities: increased understanding through the application of molecular techniques. Environ Microbiol 6 (8):769–779

Andersson JO, Sjogren AM, Davis LAM, Embley TM, Roger AJ (2003) Phylogenetic analyses of diplomonad genes reveal frequent lateral gene transfers affecting eukaryotes. Curr Biol 13 (2):94–104

Arnau J, Jorgensen F, Madsen SM, Vrang A, Israelsen H (1998) Cloning of the *Lactococcus lactis* adhE gene, encoding a multifunctional alcohol dehydrogenase, by complementation of a fermentative mutant of *Escherichia coli*. J Bacteriol 180(12):3049–3055

Artzi L, Bayer EA, Moraïs S (2017) Cellulosomes: bacterial nanomachines for dismantling plant polysaccharides. Nat Rev Microbiol 15:83–95

Bauchop T (1979) Rumen anaerobic fungi of cattle and sheep. Appl Environ Microbiol 38 (1):148–158

Boxma B, Voncken F, Jannink S, van Alen T, Akhmanova A, van Weelden SWH, van Hellemond JJ, Ricard G, Huynen M, Tielens AGM, Hackstein JHP (2004) The anaerobic chytridiomycete fungus *Piromyces* sp E2 produces ethanol via pyruvate: formate lyase and an alcohol dehydrogenase E. Mol Microbiol 51(5):1389–1399

Brownlee AG (1989) Remarkably at-rich genomic dna from the anaerobic fungus *Neocallimastix*. Nucleic Acids Res 17(4):1327–1335

Brownlee AG (1994) The nucleic acids of anaerobic fungi. In: Mountfort DO, Orpin CG (eds) Anaerobic fungi. Biology, ecology, and function. Marcel Dekker, New York, pp 241–256

Bruchhaus I, Tannich E (1994) Purification and molecular characterization of the NAD(+)-dependent acetaldehyde alcohol-dehydrogenase from *Entamoeba histolytica*. Biochem J 303:743–748

Bullerwell CE, Lang BF (2005) Fungal evolution: the case of the vanishing mitochondrion. Curr Opin Microbiol 8(4):362–369

Chen HZ, Li XL, Ljungdahl LG (1995) Biomass degrading enzymes from anaerobic rumen fungi. SAAS Bull Biochem Biotechnol 8:1–6

Contamine V, Picard M (2000) Maintenance and integrity of the mitochondrial genome: a plethora of nuclear genes in the budding yeast. Microbiol Mol Biol Rev 64(2):281–315

Dan MX, Wang CC (2000) Role of alcohol dehydrogenase E (ADHE) in the energy metabolism of *Giardia lamblia*. Mol Biochem Parasitol 109(1):25–36

Davidson EA, van der Giezen M, Horner DS, Embley TM, Howe CJ (2002) An [Fe] hydrogenase from the anaerobic hydrogenosome-containing fungus *Neocallimastix frontalis* L2. Gene 296 (1–2):45–52

Edwards JE, Forster RJ, Callaghan TM, Dollhofer V, Dagar SS, Cheng Y, Chang J, Kittelmann S, Fliegerova K, Puniya AK, Henske JK, Gilmore SP, O'Malley MA, Griffith GW, Smidt H (2017) PCR and omics based techniques to study the diversity, ecology and biology of anaerobic fungi: insights, challenges and opportunities. Front Microbiol 8:1657

Embley TM, Martin W (2006) Eukaryotic evolution, changes and challenges. Nature 440 (7084):623–630

Field J, Rosenthal B, Samuelson J (2000) Early lateral transfer of genes encoding malic enzyme, acetyl-CoA synthetase and alcohol dehydrogenases from anaerobic prokaryotes to *Entamoeba histolytica*. Mol Microbiol 38(3):446–455

Fontaine L, Meynial-Salles I, Girbal L, Yang XH, Croux C, Soucaille P (2002) Molecular characterization and transcriptional analysis of adhE2, the gene encoding the NADH-dependent aldehyde/alcohol dehydrogenase responsible for butanol production in alcohologenic cultures of *Clostridium acetobutylicum* ATCC 824. J Bacteriol 184(3):821–830

Frey TG, Mannella CA (2000) The internal structure of mitochondria. Trends Biochem Sci 25 (7):319–324

Gelius-Dietrich G, Henze K (2004) Pyruvate formate lyase (PFL) and PFL activating enzyme in the chytrid fungus *Neocallimastix frontalis*: a free-radical enzyme system conserved across divergent eukaryotic lineages. J Eukaryot Microbiol 51(4):456–463

Gruninger RJ, Puniya AK, Callaghan TM, Edwards JE, Youssef N, Dagar SS, Fliegerova K, Griffith GW, Forster R, Tsang A, McAllister T, Elshahed MS (2014) Anaerobic fungi (phylum Neocallimastigomycota): advances in understanding their taxonomy, life cycle, ecology, role and biotechnological potential. FEMS Microbiol Ecol 90:1–17

Hackstein JHP, Akhmanova A, Boxma B, Harhangi HR, Voncken FGJ (1999) Hydrogenosomes: eukaryotic adaptations to anaerobic environments. Trends Microbiol 7(11):441–447

Hackstein JHP, Akhmanova A, Voncken F, van Hoek A, van Alen T, Boxma B, Moon-van der Staay SY, van der Staay G, Leunissen J, Huynen M, Rosenberg J, Veenhuis M (2001) Hydrogenosomes: convergent adaptations of mitochondria to anaerobic environments. Zool Anal Complex Syst 104(3–4):290–302

Hackstein JHP, Tjaden J, Huynen M (2006) Mitochondria, hydrogenosomes and mitosomes: products of evolutionary tinkering! Curr Genet 50(4):225–245

Haitjema CH, Gilmore SP, Henske JK, Solomon KV, de Groot R, Kuo A, Mondo SJ, Salamov AA, LaButti K, Zhao Z, Chiniquy J, Barry K, Brewer HM, Purvine SO, Wright AT, Hainaut M, Boxma B, van Alen T, Hackstein JHP, Henrissat B, Baker SE, Grigoriev IV, O'Malley MA (2017) A parts list for fungal cellulosomes revealed by comparative genomics. Nat Microbiol 2:17087

Harhangi HR, Akhmanova AS, Emmens R, van der Drift C, de Laat WT, van Dijken JP, Jetten MS, Pronk JT, Op den Camp HJ (2003) Xylose metabolism in the anaerobic fungus *Piromyces* sp. strain E2 follows the bacterial pathway. Arch Microbiol 180:134–141

Hibbett DS, Binder M, Bischoff JF, Blackwell M, Cannon PF, Eriksson OE, Huhndorf S, James T, Kirk PM et al (2007) A higher-level phylogenetic classification of the fungi. Mycol Res 111:509–547

Hjort K, Goldberg AV, Tsaousis AD, Hirt RP, Embley TM (2010) Diversity and reductive evolution of mitochondria among microbial eukaryotes. Philos Trans R Soc B Lond B Biol Sci 365:713–727

Horner DS, Hirt RP, Embley TM (1999) A single eubacterial origin of eukaryotic pyruvate: ferredoxin oxidoreductase genes: implications for the evolution of anaerobic eukaryotes. Mol Biol Evol 16(9):1280–1291

Howe CJ (2008) Cellular evolution: what's in a mitochondrion? Curr Biol 18:R429–R431

Julliand V, Riondet C, de Vaux A, Alcaraz G, Fonty G (1998) Comparison of metabolic activities between *Piromyces citronii*, and equine fungal species, and *Piromyces communis*, a ruminal species. Anim Feed Sci Technol 70(1–2):161–168

Kameshwar AKS, Qin W (2018) Genome wide analysis reveals the extrinsic cellulolytic and biohydrogen generating abilities of *Neocallimastigomycota* fungi. J Genomics 6:74–87

Kessler D, Leibrecht I, Knappe J (1991) Pyruvate-formate-lyase-deactivase and acetyl-coa reductase activities of *Escherichia coli* reside on a polymeric protein particle encoded by adhe. FEBS Lett 281(1–2):59–63

Kessler D, Herth W, Knappe J (1992) Ultrastructure and pyruvate formate-lyase radical quenching property of the multienzymatic ADHE protein of *Escherichia coli*. J Biol Chem 267 (25):18073–18079

Li Y, Jin W, Mu C, Cheng Y, Zhu W (2017) Indigenously associated methanogens intensified the metabolism in hydrogenosomes of anaerobic fungi with xylose as substrate. J Basic Microbiol 57:933–940

Lipinski KA, Kaniak-Golik A, Golik P (2010) Maintenance and expression of the *S. cerevisiae* mitochondrial genome–from genetics to evolution and systems biology. Biochim Biophys Acta 1797:1086–1098

Luo QW, Krumholz LR, Najar FZ, Peacock AD, Roe BA, White DC, Elshahed MS (2005) Diversity of the microeukaryotic community in sulfide-rich Zodletone spring (Oklahoma) Appl. Environ Microbiol 71(10):6175–6184

Maguire F, Richards TA (2014) Organelle evolution: a mosaic of 'mitochondrial' functions. Curr Biol 24:R518–R520

Makiuchi T, Nozaki T (2014) Highly divergent mitochondrion-related organelles in anaerobic parasitic protozoa. Biochimie 100:3–17

Mannella CA (2006) The relevance of mitochondrial membrane topology to mitochondrial function. Biochim Biophys Acta-Mol Basis Dis 1762(2):140–147

Marvin-Sikkema FD, Richardson AJ, Stewart CS, Gottschal JC, Prins RA (1990) Influence of hydrogen-consuming bacteria on cellulose degradation by anaerobic fungi. Appl Environ Microbiol 56:3793–3797

Marvin-Sikkema FD, Gomes TMP, Grivet JP, Gottschal JC, Prins RA (1993) Characterization of hydrogenosomes and their role in glucose-metabolism of *Neocallimastix* sp L2. Arch Microbiol 160(5):388–396

Marvin-Sikkema FD, Driessen AJM, Gottschal JC, Prins RA (1994) Metabolic energy generation in hydrogenosomes of the anaerobic fungus *Neocallimastix* – evidence for a functional-relationship with mitochondria. Mycol Res 98:205–212

Müller M (1998) Enzymes and compartmentation of core energy metabolism of anaerobic protists – a special case in eukaryotic evolution? In: Coombs GH, Vickerman K, Sleigh MA, Warren A (eds) Evolutionary relationships among protozoa. The Systematics Association, Special Volume Series 56. Kluwer Academic, Dordrecht, pp 109–132

Müller M, Mentel M, van Hellemond JJ, Henze K, Woehle C, Gould SB, Yu RY, van der Giezen M, Tielens AGM, Martin WF (2012) Biochemistry and evolution of anaerobic energy metabolism in eukaryotes. Microbiol Mol Biol Rev 76:444–495

Munn EA, Orpin CG, Greenwood CA (1988) The ultrastructure and possible relationships of 4 obligate anaerobic chytridiomycete fungi from the rumen of sheep. Biosystems 22:67–81

Nicholson MJ, Theodorou MK, Brookman JL (2005) Molecular analysis of the anaerobic rumen fungus *Orpinomyces* – insights into an AT-rich genome. Microbiology (UK) 151:121–133

O'Fallon JV, Wright RW, Calza RE (1991) Glucose metabolic pathways in the anaerobic rumen fungus *Neocallimastix frontalis* EB188. Biochem J 274:595–599

Orpin CG (1975) Studies on rumen flagellate *Neocallimastix frontalis*. J Gen Microbiol 91:249–262

Orpin CG (1977) Occurrence of chitin in cell-walls of rumen organisms *Neocallimastix frontalis, Piromonas communis* and *Sphaeromonas communis*. J Gen Microbiol 99:215–218

Peng X, Swift CL, Theodorou MK, O'Malley MA (2018) Methods for genomic characterization and maintenance of anaerobic fungi. In: Fungal genomics. Humana, New York, pp 53–67

Ragan MA, Chapman DJ (1978) A biochemical phylogeny of the protists. Academic, New York

Sánchez LB (1998) Aldehyde dehydrogenase (CoA-acetylating) and the mechanism of ethanol formation in the amitochondriate protist, *Giardia lamblia*. Arch Biochem Biophys 354:57–64

Sawers G, Watson G (1998) A glycyl radical solution: oxygen-dependent interconversion of pyruvate formate-lyase. Mol Microbiol 29(4):945–954

Schadt CW, Martin AP, Lipson DA, Schmidt SK (2003) Seasonal dynamics of previously unknown fungal lineages in tundra soils. Science 301(5638):1359–1361

Sickmann A, Reinders J, Wagner Y, Joppich C, Zahedi R, Meyer HE, Schonfisch B, Perschil I, Chacinska A, Guiard B, Rehling P, Pfanner N, Meisinger C (2003) The proteome of *Saccharomyces cerevisiae* mitochondria. Proc Natl Acad Sci U S A 100(23):13207–13212

Solomon KV, Haitjema CH, Henske JK, Gilmore SP, Borges-Rivera D, Lipzen A, Brewer HM, Purvine SO, Wright AT, Theodorou MK, Grigoriev IV, Regev A, Thompson DA, O'Malley MA (2016) Early-branching gut fungi possess a large, comprehensive array of biomass-degrading enzymes. Science 351:1192–1195

Stairs CW, Leger MM, Roger AJ (2015) Diversity and origins of anaerobic metabolism in mitochondria and related organelles. Philos Trans R Soc Lond B Biol Sci 370:20140326

Strack D, Fester T, Hause B, Schliemann W, Walter MH (2003) Arbuscular mycorrhiza: biological, chemical, and molecular aspects. J Chem Ecol 29(9):1955–1979

Tielens AGM, van Hellemond JJ (2007) Anaerobic mitochondria: properties and origins. In: Martin WF, Müller M (eds) Origin of mitochondria and hydrogenosomes. Springer, Berlin, pp 85–103

Trinci APJ, Davies DR, Gull K, Lawrence MI, Nielsen BB, Rickers A, Theodorou MK (1994) Anaerobic fungi in herbivorous animals. Mycol Res 98:129–152. Part 2

van der Giezen M (2009) Hydrogenosomes and mitosomes: conservation and evolution of functions. J Eukaryot Microbiol 56:221–231

Van der Giezen M, Sjollema KA, Artz RRE, Alkema W, Prins RA (1997) Hydrogenosomes in the anaerobic fungus *Neocallimastix frontalis* have a double membrane but lack an associated organelle genome. FEBS Lett 408(2):147–150

Voncken F (2001) Hydrogenosomes: eukaryotic adaptations to anaerobic environments. PhD Thesis, University of Nijmegen. ISBN 90-9014868-x

Voncken F, Boxma B, Tjaden J, Akhmanova A, Huynen M, Verbeek F, Tielens AGM, Haferkamp I, Neuhaus HE, Vogels G, Veenhuis M, Hackstein JHP (2002a) Multiple origins of hydrogenosomes: functional and phylogenetic evidence from the ADP/ATP carrier of the anaerobic chytrid *Neocallimastix* sp. Mol Microbiol 44(6):1441–1454

Voncken FGJ, Boxma B, van Hoek AHAM, Akhmanova AS, Vogels GD, Huynen M, Veenhuis M, Hackstein JHP (2002b) A hydrogenosomal [Fe]-hydrogenase from the anaerobic chytrid *Neocallimastix* sp L2. Gene 284(1–2):103–112

Wang HC, Chen YC, Hseu RS (2014) Purification and characterization of a cellulolytic multienzyme complex produced by *Neocallimastix patriciarum* J11. Biochem Biophys Res Commun 45:190–195

Wei YQ, Yang HJ, Long RJ, Wang ZY, Cao BB, Ren QC, Wu TT (2017) Characterization of natural co-cultures of *Piromyces* with *Methanobrevibacter ruminantium* from yaks grazing on the Qinghai-Tibetan Plateau: a microbial consortium with high potential in plant biomass degradation. AMB Exp 7:160

Wilken SE, Saxena M, Petzold LR, O'Malley MA (2018) In silico identification of microbial partners to form consortia with anaerobic fungi. Processes 6(1):7

Williams AG, Withers SE, Naylor GE, Joblin KN (1994) Interactions between the rumen chytrid fungi and other microorganisms. In: Mountfort DO, Orpin CG (eds) Anaerobic fungi. Biology, ecology, and function. Marcel Dekker, New York, pp 191–227

Wilson CA, Wood TM (1992) The anaerobic fungus *Neocallimastix frontalis*: isolation and properties of a cellulosome-type enzyme fraction with the capacity to solubilize hydrogen-bond-ordered cellulose. Appl Microbiol Biotechnol 37:125–129

Yarlett N, Orpin CG, Munn EA, Yarlett NC, Greenwood CA (1986) Hydrogenosomes in the rumen fungus *Neocallimastix patriciarum*. Biochem J 236(3):729–739

The Proteome of *T. vaginalis* Hydrogenosomes

Petr Rada and Jan Tachezy

Contents

1 Introduction 178
2 Experimental Procedures 179
3 In Silico Predictions 180
4 The Proteome of *Trichomonas vaginalis* Hydrogenosomes 181
4.1 Energy Metabolism 182
4.2 Iron-Sulfur Cluster Assembly 187
4.3 Reactive Oxygen Species Defense 188
4.4 Amino Acid Metabolism 189
4.5 Protein Import 190
4.6 Carriers of the Hydrogenosomal Inner Membrane 193
4.7 C-Tail-Anchored Proteins 193
4.8 Contaminants or Surface-Associated Proteins? 194
5 Quantitative Analysis of the Proteome: Iron-Induced Changes 195
6 The Proteome of *Pentatrichomonas hominis* Hydrogenosome 198
7 Concluding Remarks 198
References 199

Abstract *Trichomonas vaginalis* and related organisms of Parabasalia group possess anaerobic forms of mitochondria named hydrogenosomes. These organelles lost most of mitochondrial pathways during adaptation to anaerobic lifestyle. Proteomic analysis of hydrogenosomes purified from *T. vaginalis* and more recently from *Pentatrichomonas hominis* revealed that trichomonad hydrogenosomes consist of about 600 proteins, which is about half of proteins present in yeast mitochondria. Matrix proteins of known function that account for about 19% of the proteome include enzymes of energy metabolism, components of Fe-S cluster assembly, oxygen and xenobiotic stress response, amino acid metabolism, proteolytic enzymes, and chaperones. Approximately 11% of the proteome represent membrane

P. Rada · J. Tachezy (✉)
Faculty of Science, Department of Parasitology, BIOCEV, Charles University, Vestec, Czech Republic
e-mail: tachezy@natur.cuni.cz

J. Tachezy (ed.), *Hydrogenosomes and Mitosomes: Mitochondria of Anaerobic Eukaryotes*, Microbiology Monographs 9,
https://doi.org/10.1007/978-3-030-17941-0_8

components of protein import machinery and other membrane proteins. Three large protein groups consist of hypothetical proteins (20%), miscellaneous proteins (16%), and proteins that might be associated with outer hydrogenosomal membrane (34%). The proteome is highly redundant, which reflects the presence of multiple gene copies for most of proteins in trichomonad genome. Interestingly, most of paralogues are expressed and identified in the proteome. However, the protein level of individual paralogues is rather variable, and iron availability appears to be an important factor in regulation of the proteome composition.

1 Introduction

Hydrogenosomes are anaerobic relatives of mitochondria, which are believed to evolve from a common proteobacterial endosymbiont (Martijn et al. 2018). Their main functions are the production of ATP by substrate-level phosphorylation, connected with evolution of molecular hydrogen, and the biosynthesis of Fe-S clusters. Except for plants, hydrogenosomes and other mitochondria-related organelles (MROs) are found among all major supergroups of eukaryotes and most likely represent only a specific part of a very large diversity of mitochondrial relatives (Keeling et al. 2005; Stairs et al. 2015). Hydrogenosomes found in various lineages of protists and fungi differ in shape, size, membrane composition, and metabolic pathways, suggesting that hydrogenosomes evolved from a mitochondrial ancestor multiple times as a response to oxygen-poor environment. The first hydrogenosomes were discovered in parabasalids *Tritrichomonas foetus* and *Trichomonas vaginalis*, the causative agents of urogenital trichomoniasis of cattle and of humans, respectively (Lindmark and Müller 1973; Lindmark et al. 1975; Cerkasovova et al. 1973). Later, hydrogenosomes have been found in various parasitic and free-living protists, such as marine free-living ciliates *Plagiopyla frontata* (Fenchel and Finlay 1991), *Metopus concordus* (van Bruggen et al. 1984), and *Sonderia* sp. (Dyer 1989); freshwater free-living ciliates *Trimyema* sp. (Broers et al. 1991); rumen-dwelling ciliates *Dasytrichia ruminantium* (Yarlett et al. 1981), *Isotricha* sp., *Dasytrichia* sp., and *Entodinium* sp. (Lloyd et al. 1989; Yarlett et al. 1983; Paul et al. 1990); ciliate *Nyctotherus ovalis* from the alimentary tract of cockroaches (Hackstein et al. 1999; Boxma et al. 2005); rumen-dwelling chytrid fungi *Neocallimastix frontalis* (O'Fallon et al. 1991; Marvin-Sikkema et al. 1994) and *Piromyces* sp. (Akhmanova et al. 1999); and heteroloboseans *Psalteriomonas lanterna* (Brul et al. 1994) and *Sawyeria marylandensis* (Barberà et al. 2010). Organelles with the features corresponding to hydrogenosomes have been investigated in the free-living amoeba *Mastigamoeba balamuthi* (Gill et al. 2007; Nývltová et al. 2013, 2015), diplomonad *Spironucleus salmonicida* (Jerlström-Hultqvist et al. 2013), and jakobid *Stygiella incarcerata* (Leger et al. 2016).

Hydrogenosomes were mostly subjected to small-scale studies, which were focused on the investigations of selected proteins or pathways. Large-scale projects focused on the whole hydrogenosomal proteome are essential to expand our knowledge about the organelle biogenesis and biochemistry and to help us better understand its evolutionary connections to other mitochondria-related organelles across various eukaryotic lineages. However, the proteomic studies are still scarce due to technical limitations. There are three main conditions that typically need to be fulfilled: (1) hydrogenosome-harboring organism can be cultivated in vitro in axenic cultures, (2) suitable method for purification of hydrogenosomes exists, and (3) genomic or transcriptomic data are available for identification of protein sequences. Due to these limitations, the proteome of the hydrogenosomes has been investigated only in the parasitic protist *Trichomonas vaginalis* and its relative *Pentatrichomonas hominis* thus far.

2 Experimental Procedures

The prerequisite for the proteome analysis is the effective separation of hydrogenosomes of the highest possible purity from other cellular structures. Hydrogenosomes of trichomonads could be purified from cell homogenate by sequential differential and isopycnic centrifugation using continuous Percoll gradient and discontinuous OptiPrep gradient (Bradley et al. 1997; Sutak et al. 2004; Beltrán et al. 2013). There are various strategies for further processing of hydrogenosomal proteins. In initial studies, purified hydrogenosomes were subjected to two-dimensional (2D) SDS PAGE, and protein spots were excised from the gel and analyzed by mass spectrometry (MS) (Pütz et al. 2005; Henze 2007). This method led to successful identification of 61 hydrogenosomal proteins (Henze 2007). However, there are several limitations of this traditional method, particularly the difficulties in separation of hydrophobic and highly charged proteins (Bunai and Yamane 2005; Rabilloud and Lelong 2011; Magdeldin et al. 2014). Therefore, several different approaches were used for more efficient protein extraction of soluble, hydrophobic, and membrane proteins such as combination of sodium carbonate and sodium hydroxide extraction (Schneider et al. 2011), zinc chelating column chromatography (Schneider et al. 2011), or the Triton X-114 extraction (Bordier 1981; Rada et al. 2011). Extracted proteins were then separated by 1D SDS electrophoresis, excised from the gel and digested by trypsin. The tryptic peptides were separated by nano-liquid chromatography (nano-LC) and analyzed by mass spectrometry (MS) (Rada et al. 2011; Schneider et al. 2011). Beltrán et al. (2013) and Fang et al. (2016) analyzed protein sequences using gel-free approaches. Beltrán et al. (2013) precipitated proteins from purified hydrogenosomes by acetone, the proteins were then digested with trypsin, and tryptic peptides were separated into two steps by isoelectric focusing and nano-LC. Fang et al. (2016) dissolved hydrogenosomal extract in SDS buffer, and separated tryptic peptides using 2D LC. Protein sequences that were determined by MS were automatically annotated using available *T. vaginalis* protein database (UniProt, TrichoDB, NCBI) and manually curated using a range of available bioinformatic tools (e.g., see https://

www.expasy.org/tools/). When possible, the protein localization of selected proteins was experimentally validated using transfection systems (Delgadillo et al. 1997; Hrdy et al. 2004).

3 In Silico Predictions

The hydrogenosomal genome was completely lost during the mitochondrion-to-hydrogenosome transition. This loss is likely related to the loss of respiratory complexes that are partially encoded by mitochondrial DNA (Gray 2012). Consequently, all hydrogenosomal proteins are encoded in the nucleus and imported posttranslationally. Proteins transported to the hydrogenosomal membranes or into the hydrogenosomal matrix possess targeting signals that are recognized at the organelle surface. Significant portion of proteins that are targeted to the hydrogenosomal matrix contain N-terminal targeting sequences (NTS) that are upon transport cleaved from the mature protein by hydrogenosomal processing peptidase (Carlton et al. 2007; Brown et al. 2007; Smíd et al. 2008; Shiflett and Johnson 2010). In silico identification of genes coding for hydrogenosomal NTS in *T. vaginalis* genome led to the first estimations of the size of the hydrogenosomal proteome. Carlton et al. (2007) predicted 138 and Smíd et al. (2008) 226 hydrogenosomal proteins containing NTS. The latter predictions were based on two main parameters: (1) the NTS start motif was defined as ML(STACGR) or MTL or MSL, and (2) the cleavage site motif was specified as RXF/(ILFSAGQ) or R (FNESG)/(ILFSAGQ) (the slash indicates the cleavage site and brackets mean one residue position). The maximum NTS length was set up to 25 residues (Smíd et al. 2008). However, the actual size of the hydrogenosomal proteome was estimated to be larger, considering proteins which are targeted to the organelle independently of the cleavable NTS (Mentel et al. 2008; Burstein et al. 2012; Zimorski et al. 2013; Garg et al. 2015; Rada et al. 2015). These include (1) matrix proteins with non-cleavable N-terminal targeting signals, (2) matrix proteins that possess internal targeting signals (ITS), or (3) membrane proteins and proteins of the intermembrane space with ITSs. Advanced machine-learning approach was used for predictions of hydrogenosomal proteins that employed, in addition to prediction of NTS, various genomic, biochemical, and evolutionary traits of the proteins (Burstein et al. 2012). However, reliability of this approach for prediction of hydrogenosomal proteins without the canonical NTS is still limited. Therefore, proteomic analysis of hydrogenosomes using mass spectrometry in combination with advanced bioinformatics is necessary to obtain considerably more complete picture of the hydrogenosomal proteome.

4 The Proteome of *Trichomonas vaginalis* Hydrogenosomes

The first partial proteome of the hydrogenosome was investigated in *Trichomonas vaginalis* and *Tritrichomonas foetus* and was based on 2D SDS PAGE protein separation (Pütz et al. 2005; Henze 2007). The proteome pointed out the importance of NTS-independent protein import into the organelle (Pütz et al. 2005; Henze 2007). In the *T. vaginalis* hydrogenosome, altogether 61 proteins were identified of which 34 did not contain NTS. The partial proteome included 55 proteins that were involved in pyruvate fermentative metabolism, iron-sulfur cluster (ISC) assembly machinery, amino acid metabolism, oxidative stress protection, and 6 proteins with unknown function (Pütz et al. 2005; Henze 2007). In this proteome, only a single membrane protein, the putative pore-forming Hmp35 (hydrogenosomal membrane protein 35 kDa, Dyall et al. 2003), was identified, although components of the protein import machinery and various metabolite transporters were expected to function in hydrogenosomal membranes (Dyall and Dolezal 2007; Henze 2007). Novel approaches based on improved protein extraction and more sensitive nano-LC MS/MS analysis have been performed to obtain more complete hydrogenosomal proteome. Based on several current studies, the proteome of *T. vaginalis* hydrogenosomes consists of almost 600 proteins (Rada et al. 2011; Schneider et al. 2011; Beltrán et al. 2013). Compared to more complex mitochondria that according to current estimations contain 1100–1900 proteins (Calvo et al. 2016; Smith and Robinson 2016), the lower number of identified hydrogenosomal proteins apparently reflects the reduction of most mitochondrial pathways associated with oxygen metabolism, including oxidative phosphorylation, inner membrane-associated electron transport chain, tricarboxylic acid cycle, β-oxidation of fatty acids, as well as biosynthesis of phospholipids, heme, and biotine.

The proteome of *T. vaginalis* hydrogenosome consists mainly of proteins that are involved in energy metabolism, amino acid metabolism, biosynthesis of Fe-S clusters, oxygen and reactive oxygen species detoxification, and protein maturation, membrane proteins involved in protein import and metabolite exchange, and proteins with other or unknown function (Fig. 1) (Rada et al. 2011; Schneider et al. 2011) (chapter "Metabolism of Trichomonad Hydrogenosomes"). The proteome is highly redundant, reflecting the nature of the *T. vaginalis* genome (Carlton et al. 2007). Most of proteins are encoded by multiple gene copies and most of copies are expressed. For example, α- and β-subunits of succinyl-CoA synthetase (SCS) (also known as succinate thiokinase, STK) are each coded by three genes, and all six protein sequences were identified in proteomes (Rada et al. 2011; Schneider et al. 2011; Beltrán et al. 2013). However, the level of expression is affected by various environmental conditions such as availability of iron. The differential expression of hydrogenosomal proteins affected by iron availability has been studied using quantitative MS (Beltrán et al. 2013). This study revealed 58 differentially expressed protein in response to iron level, including protein paralogues. Products of some genes have not been detected in any proteomic study. For example, pyruvate/ferredoxin oxidoreductase (PFO) is coded by seven genes, but only six paralogues

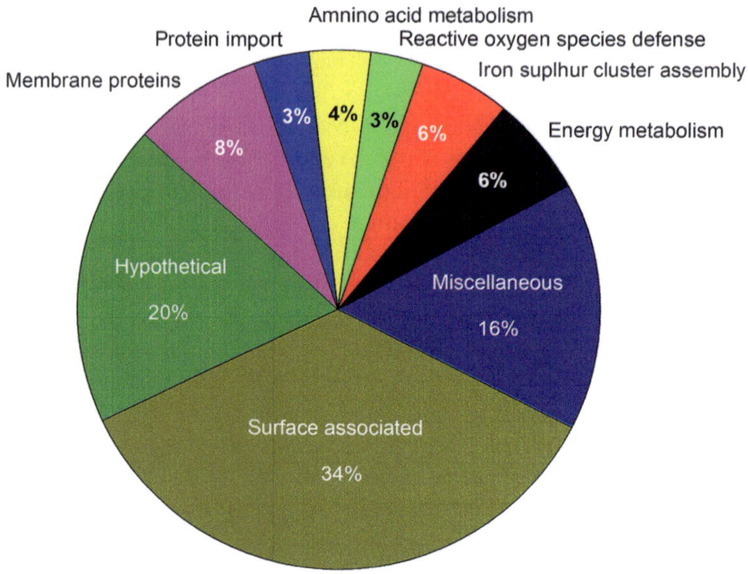

Fig. 1 Functional groups of proteins identified in proteomic analyses of *T. vaginalis* hydrogenosomes. Proteins were assembled into functional groups based on proteomes published by Rada et al. (2011), Schneider et al. (2011), and Beltrán et al. (2013)

were identified in the proteome (Rada et al. 2011; Schneider et al. 2011; Beltrán et al. 2013). Whether genes coding for proteins that could not been detected represent pseudogenes and are permanently silenced or their expression requires specific environmental factors remains to be established (Horváthová et al. 2012; Huang et al. 2012; Beltrán et al. 2013; Gould et al. 2013; Woehle et al. 2014).

4.1 Energy Metabolism

Hydrogenosomes oxidize pyruvate to acetate, CO_2, and molecular hydrogen. This metabolic activity is coupled with production of ATP by substrate-level phosphorylation. Proteins involved in energy metabolism of the hydrogenosome are malic enzyme, PFO, [2Fe-2S] ferredoxin, [FeFe] hydrogenase, SCS, acetate/succinate-CoA transferase (ASCT), and adenylate kinase (AK). NADH resulting from malic enzyme reaction is regenerated by the residual subunits of the complex 1, Tvh47 (NuoF), and Tvh22 (NuoE) (Hrdý et al. 2004; chapter "Metabolism of Trichomonad Hydrogenosomes"). All these proteins were identified in at least one of the three investigated hydrogenosomal proteomes, several of them in multiple paralogues (Table 1) (Rada et al. 2011; Schneider et al. 2011; Beltrán et al. 2013). *Trichomonas* genome codes for a diverse group of iron hydrogenases that contain, in addition to conserved H cluster, the site of hydrogen synthesis, also several N-terminally located

Table 1 *T. vaginalis* components of major hydrogenosomal pathways determined by three independent proteomic studies

	Rada et al.	Schneider et al.	Beltrán et al.
Energy metabolism			
Hydrogenosomal malic enzyme	TVAG_068130; TVAG_183790; TVAG_238830; TVAG_267870; TVAG_340290; TVAG_412220	TVAG_068130; TVAG_183790; TVAG_238830; TVAG_267870; TVAG_340290; TVAG_412220; TVAG_416100	TVAG_068130; TVAG_183790; TVAG_238830; TVAG_267870; TVAG_340290; TVAG_412220; TVAG_416100
[2Fe-2S] ferredoxin	TVAG_003900	TVAG_003900; TVAG_292710; TVAG_399860	TVAG_003900; TVAG_292710; TVAG_399860; TVAG_078730; TVAG_213140; TVAG_251200
[FeFe] hydrogenase	TVAG_037570; TVAG_361590; TVAG_182620; TVAG_310050	TVAG_037570; TVAG_361590; TVAG_182620; TVAG_310050	TVAG_037570; TVAG_182620; TVAG_310050
[FeFe] hydrogenase-related proteins		TVAG_160930; TVAG_255980	
Pyruvate ferredoxin oxidoreductase	TVAG_198110; TVAG_230580; TVAG_242960; TVAG_254890	TVAG_198110; TVAG_230580; TVAG_242960; TVAG_254890; TVAG_105770; TVAG_466790	TVAG_198110; TVAG_230580; TVAG_242960; TVAG_254890; TVAG_105770; TVAG_466790
Succinyl-CoA synthetase α-subunit	TVAG_165340; TVAG_318670	TVAG_165340; TVAG_318670; TVAG_047890	TVAG_165340; TVAG_318670; TVAG_047890
Succinyl-CoA synthetase β-subunit	TVAG_144730; TVAG_183500; TVAG_259190	TVAG_144730; TVAG_183500; TVAG_259190	TVAG_144730; TVAG_183500; TVAG_259190
Adenylate kinase	TVAG_489800	TVAG_489800	TVAG_489800; TVAG_072700
Acetyl/succinate-CoA transferase	TVAG_164890; TVAG_113870	TVAG_164890; TVAG_395550; TVAG_113870	TVAG_164890; TVAG_395550; TVAG_393850
NuoE	TVAG_133030	TVAG_133030	TVAG_133030
NuoF	TVAG_296220	TVAG_296220	TVAG_296220
Iron-sulfur cluster assembly			
IscS	TVAG_239660	TVAG_239660	TVAG_239660
IscU	TVAG_432650	TVAG_432650	TVAG_432650
Isd11			TVAG_242760; TVAG_215080
GrpE		TVAG_297650	TVAG_182340; TVAG_297650

(continued)

Table 1 (continued)

	Rada et al.	Schneider et al.	Beltrán et al.
Hsp70	TVAG_381470; TVAG_237140; TVAG_340390; TVAG_433130	TVAG_340390; TVAG_433130	TVAG_381470; TVAG_100490; TVAG_253630; TVAG_237140; TVAG_340390; TVAG_433130
Jac1		TVAG_032090	TVAG_305730; TVAG_019190; TVAG_032090
Nfu	TVAG_008840		TVAG_008840; TVAG_044500; TVAG_146780; TVAG_451860
IscA	TVAG_456770	TVAG_361540; TVAG_456770	TVAG_055320; TVAG_361540; TVAG_456770
Frataxin			TVAG_114560; TVAG_182150
Ind	TVAG_027170; TVAG_076230; TVAG_217870	TVAG_076230; TVAG_217870	TVAG_027170; TVAG_076230; TVAG_217870; TVAG_277380
Maturases of [FeFe] hydrogenase			
Hyd E			TVAG_027060; TVAG_329200
Hyd F	TVAG_205390	TVAG_205390	TVAG_205390
Hyd G	TVAG_257780	TVAG_257780	TVAG_257780; TVAG_469020
Reactive oxygen species defense			
Thioredoxin	TVAG_385350	TVAG_385350	TVAG_086470; TVAG_125500; TVAG_385350; TVAG_418970
Superoxide dismutase		TVAG_039980; TVAG_049140	TVAG_049140
Thioredoxin peroxidase (peroxiredoxin)	TVAG_055200; TVAG_455310	TVAG_055200; TVAG_114310	TVAG_055200; TVAG_038090; TVAG_114310; TVAG_455310
Thioredoxin reductase			TVAG_474980
Rubrerythrin	TVAG_064490; TVAG_275660	TVAG_064490	TVAG_064490; TVAG_275660
OsmC		TVAG_410350	TVAG_410350; TVAG_412560
FDP	TVAG_036010	TVAG_036010	TVAG_036010

(continued)

Table 1 (continued)

	Rada et al.	Schneider et al.	Beltrán et al.
Isf		TVAG_040030; TVAG_154730; TVAG_327760	TVAG_040030; TVAG_154730
HCP	TVAG_206500	TVAG_206500; TVAG_336320	TVAG_206500; TVAG_336320
Amino acid metabolism			
H protein	TVAG_177600	TVAG_100550; TVAG_177600	TVAG_177600
L protein		TVAG_272760	TVAG_272760
Serine hydroxymethyltransferase		TVAG_109540	TVAG_109540
PSAT		TVAG_183300	
MGL		TVAG_147790	
Cysteine synthase		TVAG_547520	
Alanine aminotransferase	TVAG_074600; TVAG_088220; TVAG_132440; TVAG_379550	TVAG_074600; TVAG_088220; TVAG_098820; TVAG_379550	TVAG_074600; TVAG_088220; TVAG_098820; TVAG_132440; TVAG_379550
Arginine deiminase	TVAG_183850; TVAG_344520	TVAG_183850; TVAG_344520	TVAG_183850; TVAG_344520; TVAG_467820
Aspartate aminotransferase	TVAG_088220	TVAG_088220; TVAG_268020; TVAG_419720	TVAG_088220
Glutamate dehydrogenase		TVAG_025980	
IscS	TVAG_239660	TVAG_239660	TVAG_239660
Protein import			
Tom40	TVAG_123100; TVAG_195900; TVAG_332970; TVAG_341190; TVAG_399510; TVAG_450220	TVAG_123100; TVAG_195900; TVAG_332970; TVAG_341190; TVAG_399510; TVAG_450220	TVAG_195900; TVAG_399510; TVAG_450220
Sam50	TVAG_178100	TVAG_178100	TVAG_178100
Small Tims	TVAG_026080; TVAG_287510		TVAG_026080; TVAG_287510
Tim17/22/23	TVAG_061900 (A); TVAG_198350 (B); TVAG_370860 (C); TVAG_379950 (D)	TVAG_198350 (B); TVAG_370860 (C)	TVAG_061900 (A); TVAG_370860 (C)
Tim17 like	TVAG_447580	TVAG_447580	TVAG_447580
Tim44	TVAG_008790	TVAG_008790	TVAG_008790
Pam18	TVAG_436580		
Pam16	TVAG_470110	TVAG_470110	TVAG_470110
Tom36	TVAG_277930	TVAG_277930	TVAG_277930

(continued)

Table 1 (continued)

	Rada et al.	Schneider et al.	Beltrán et al.
Tom46	TVAG_137270	TVAG_137270	TVAG_137270
Homp19	TVAG_283120	TVAG_283120	TVAG_283120
Homp38	TVAG_190830		TVAG_190830
Membrane proteins			
Porins	TVAG_146920; TVAG_340380	TVAG_146920; TVAG_340380	TVAG_146920
Hmp35	TVAG_104250; TVAG_590550	TVAG_104250	TVAG_104250
Hmp36	TVAG_031860; TVAG_216170	TVAG_031860; TVAG_216170	TVAG_031860; TVAG_216170
ADP/ATP carriers	TVAG_237680 (Hmp-31); TVAG_051820; TVAG_164560; TVAG_196220; TVAG_262210	TVAG_237680 (Hmp-31); TVAG_051820; TVAG_164560; TVAG_196220; TVAG_262210	TVAG_237680 (Hmp-31); TVAG_051820; TVAG_262210
C-tail-anchored proteins	TVAG_090120; TVAG_211970; TVAG_272350; TVAG_369980; TVAG_458060; TVAG_174010; TVAG_240680; Tom36; Tom46; Homp19; Homp38	TVAG_130330; TVAG_211970; TVAG_272350; TVAG_369980; TVAG_458060; TVAG_174010; TVAG_214300; Tom36; Tom46; Homp19	TVAG_090120; TVAG_211970; TVAG_272350; TVAG_369980; TVAG_174010; Tom39; Tom46; Homp19; Homp38
Integral polytopic	TVAG_039960; TVAG_455090; TVAG_489980; TVAG_127990; TVAG_440200; TVAG_136450; TVAG_192370	TVAG_455090; TVAG_440200; TVAG_136450; TVAG_192370	TVAG_455090; TVAG_440200; TVAG_136450
Integral single spanning	TVAG_032990; TVAG_080160; TVAG_094480; TVAG_152710; TVAG_178320; TVAG_182990; TVAG_210010; TVAG_218130; TVAG_225560; TVAG_251750; TVAG_252220; TVAG_295140; TVAG_331680; TVAG_333160; TVAG_337270; TVAG_341690;	TVAG_423530	TVAG_225560

(continued)

Table 1 (continued)

	Rada et al.	Schneider et al.	Beltrán et al.
	TVAG_370950; TVAG_403380; TVAG_413430; TVAG_425430; TVAG_423530		

Fe-S clusters (Carlton et al. 2007). Proteomic analysis identified in the hydrogenosome four iron hydrogenases with a conserved H cluster-binding motive and two hydrogenase-related proteins that possess an incomplete H cluster-binding motive with some key amino acid residues missing. Interestingly, Beltrán et al. (2013) identified six out of seven ferredoxins that were differentially expressed in response to iron. Three paralogues were upregulated, and three were downregulated upon iron-limited conditions. The ferredoxin TVAG_068150 is missing in all examined proteomes. The absence of this paralogue in proteomes is possibly due to very low level of its transcription (Gould et al. 2013). The most abundant hydrogenosomal enzymes were identified in each of proteomes published by Schneider et al. (2011) and Beltrán et al. (2013), namely, PFO in six paralogues, malic enzyme in eight, and each of α- and β-SCS in three (Table 1).

4.2 *Iron-Sulfur Cluster Assembly*

The Fe-S proteins, which occur within hydrogenosomes, cytoplasm, and nucleus, require Fe-S cluster biosynthesis for their maturation. Immunolocalization and phylogenetic analysis of key enzymes involved in Fe-S cluster formation revealed that Fe-S cluster biosynthesis is localized within hydrogenosomes and is homologous to the mitochondrial *i*ron-*s*ulfur *c*luster (ISC) biosynthesis pathway (Tachezy et al. 2001; Šuťák et al. 2004; chapter "Metabolism of Trichomonad Hydrogenosomes"). Altogether genes coding for ten components of the ISC pathway have been identified in the *Trichomonas* genome, some of them in multiple copies including cysteine desulfurase IscS and its partner protein Isd11, scaffold proteins IscU and IscA, electron carrier [2Fe-2S] ferredoxin, putative iron donor frataxin, and factors involved in transfer of Fe-S cluster to apoproteins, the chaperones Jac1 (Hsc20), nucleotide exchange factor GrpE (Mge1), Hsp70, Nfu, and Ind1 (Rada et al. 2011; Schneider et al. 2011; Beltrán et al. 2013). There are also three maturases of [FeFe] hydrogenases HydE, HydF, and HydG, which are required for assembly of catalytic H cluster. Some of these proteins are present in multiple paralogues, namely, frataxin, HydE, and HydG, Isd11 in two, IscA in three, Jac1, Nfu, and Ind1, each in four paralogues. Cysteine desulfurase IscS is coded by two genes, IscS-1 (TVAG_365590) and IscS-2 (TVAG_239660). IscS-2 contains cleavable N-terminal targeting signal, and it was localized into the organelle by immunofluorescence assay and immunoblotting (Tachezy et al. 2001; Sutak et al. 2004). In contrast, IscS-1 does not contain cleavable NTS and has been suggested to be a pseudogene (Tachezy et al. 2001; Sutak et al. 2004). However, extremely low level of transcription for IscS-1 was observed by Gould et al. (2013). Not surprisingly, only IscS-2 was identified in all three proteomes (Rada et al. 2011; Schneider et al. 2011; Beltrán et al. 2013).

4.3 Reactive Oxygen Species Defense

Anaerobic protists have to cope with a low oxygen concentration in their natural environment. Oxygen and reactive oxygen species (ROS), in particular superoxide radical anion (O_2^-) and hydrogen peroxide (H_2O_2), damage Fe-S clusters of various enzymes such as PFO and [FeFe] hydrogenase and oxidize other molecules with low redox potential and therefore are highly toxic for the cell (Lloyd and Williams 2015; Leitsch et al. 2018). In hydrogenosomes, oxygen and ROS are detoxified by a set of enzymes and molecules that include superoxide dismutase (SOD), proteins that belong to the thioredoxin-linked peroxiredoxin system, flavodiiron protein (FDP), Fe-S flavoprotein (Isf), rubrerythrin peroxidase (Rbr), and OsmC system (Viscogliosi et al. 1996, 1998; Coombs et al. 2004; Pütz et al. 2005; Mentel et al. 2008; Smutna et al. 2009, 2014; Schneider et al. 2011). Iron-containing superoxide dismutase (SOD) transforms O_2^- into hydrogen peroxide and O_2. Peroxide is removed by the activity of thioredoxin-linked peroxiredoxin system, which includes thioredoxin, thioredoxin reductase, and thioredoxin peroxidase (Leitsch et al. 2018). These proteins are coded in the *Trichomonas* genome by multiple copies; however, cellular distribution of gene products is difficult to predict. For example, there are five genes for thioredoxin reductase, of which one is coding for a protein with NTS that is targeted to hydrogenosomes (TrxRh1). However, another paralogue TrxRh2 has been experimentally shown to target to hydrogenosomes without NTS, whereas the other three are cytosolic. TrxRh2 was later identified in the proteome by Beltrán et al. (2013). Thioredoxins are coded in the *Trichomonas* genome by 32 gene copies, but only 1–4 paralogues were identified in the hydrogenosome proteome (Table 1). SOD is coded by 7 genes, of which 1–2 were found in proteomes (Schneider et al. 2011; Beltrán et al. 2013). Thioredoxin peroxidases (peroxiredoxins) require for their function the presence of redox-active cysteine in the active site, which is oxidized by the peroxide substrate to a sulfenic acid. In class 2-Cys peroxiredoxins, sulfenic acid is recycled by thioredoxin. *Trichomonas* genome is coding for nine genes of 2-Cys peroxiredoxins. Proteomics detected in the hydrogenosome only five paralogues (Pütz et al. 2005; Rada et al. 2011; Schneider et al. 2011; Beltrán et al. 2013). In addition to the thioredoxin-linked peroxiredoxin system, two other antioxidant proteins were identified in the proteome of hydrogenosomes, namely, peroxidase rubrerythrin and lipoate-dependent peroxidase OsmC. Rubrerythrin (TVAG_064490) was first identified using 2D gel separation of hydrogenosomal proteins by Pütz et al. (2005). Beltrán et al. (2013) identified two paralogues of rubrerythrin (TVAG_064490, TVAG_275660). However, there are altogether six rubrerythrin paralogues in *T. vaginalis* genome and none of them possesses NTS, including those two identified in the hydrogenosome proteome. The cellular localization of other four paralogues is unknown. OsmC is a lipoate-dependent peroxidase that in hydrogenosomes catalyzes reductive detoxification of peroxides (Nývltová et al. 2016). Rubrerythrin and OsmC are related to prokaryotic peroxidases that are found in anaerobic and facultatively anaerobic eubacteria, but their presence in eukaryotes is rather limited (rubrerythrin was so far identified only in trichomonads, *Entamoeba*, and *Naegleria*) (Zysset-Burri et al. 2014; Cabeza et al. 2015).

In *Trichomonas* hydrogenosomes, oxygen is reduced to water by flavodiiron protein (FDP) or to hydrogen peroxide by iron-sulfur flavoprotein (Isf) (Smutna et al. 2009, 2014). Single FDP and three Isf paralogues were identified in the hydrogenosomal proteome (Rada et al. 2011; Schneider et al. 2011; Beltrán et al. 2013). Homologues of FDP can be found among other anaerobic protists such as *Giardia*, *Entamoeba*, or *Mastigamoeba*, while Isf has been found, in addition to *Trichomonas*, only in *Entamoeba* to date (Smutna et al. 2009, 2014; Leitsch et al. 2018). Hybrid cluster protein (HCP) is another example of typically bacterial protein in hydrogenosomes (Schneider et al. 2011; Leitsch et al. 2018). HCP contains hybrid iron-sulfur-oxygen cluster (Arendsen et al. 1998). *Trichomonas* hydrogenosomal proteome contains two paralogues of HCP (TvHCP-1 and TvHCP-2). Localization of TvHCP-1 (TVAG_206500) in hydrogenosomes was verified by immunofluorescence microscopy (Rada et al. 2011; Schneider et al. 2011). In bacteria, HCP is likely involved in detoxification of NO, hydroxylamine, and H_2O_2 (Wang et al. 2016). However, the function of HCP in *T. vaginalis* remains unknown.

4.4 Amino Acid Metabolism

Metabolism of amino acids is another common feature between hydrogenosomes and mitochondria; however, the spectrum of metabolized amino acids in hydrogenosomes is considerably reduced. The proteome of the *T. vaginalis* hydrogenosome contains 11 enzymes (Table 1) that are involved in the metabolism of only 7 amino acids (Ala, Cys, Asp, Glu, Gly, Arg, and Ser) (Rada et al. 2011; Schneider et al. 2011; Beltrán et al. 2013). In contrast, mitochondria usually metabolize 17 amino acids, of which 11 are newly synthesized (Guda et al. 2007).

Cysteine is likely the only amino acid that might be synthesized in hydrogenosomes as predicted by proteomics (Schneider et al. 2011). *Trichomonas* generates cysteine from phosphohydroxypyruvate and homocysteine by three enzymes, phosphoserine aminotransferase (PSAT), methionine-γ-lyase (MGL), and cysteine synthase (CS) (Westrop et al. 2006). Schneider et al. (2011) identified all these three enzymes in the hydrogenosome proteome and confirmed the presence of PSAT in hydrogenosomes by immunofluorescence microscopy.

The proteome of the hydrogenosome includes three paralogues of arginine deiminase (ADI1, TVAG_467820; ADI2, TVAG_344520; and ADH3, TVAG_183850). ADI converts arginine to citrulline, which is the first step of arginine dihydrolase (ADH) pathway. Interestingly, the other components including ornithine carbamyltransferase and carbamate kinase are present exclusively in the cytosol (Morada et al. 2011). The reason why ADI is compartmentalized to the hydrogenosomes remains unclear.

Several enzymes are involved in amino acid interconversion. Aspartate aminotransferase (AAT) and glutamate dehydrogenase (GDH) catalyze reversible conversion of aspartate, glutamate, oxoglutarate, and oxaloacetate. Alanine is generated by the activity of cysteine desulfurase (IscS) and further could be converted to glutamate by alanine aminotransferase (ALT). Five paralogues of alanine aminotransferases were identified in the proteome (Rada et al. 2011; Schneider et al. 2011; Beltrán et al. 2013).

Hydrogenosomes possess two proteins of glycine cleavage system (GCS, also known as glycine decarboxylase complex (GDC)), dihydrolipoil dehydrogenase (L protein) and lipoamide-binding protein (H protein), that were found in the proteome (Rada et al. 2011; Schneider et al. 2011; Beltrán et al. 2013). Presence of both proteins in hydrogenosomes was also observed by immunofluorescence microscopy (Mukherjee et al. 2006). However, complete GCS consists of four proteins that are all present in mitochondria and conserved in their various reduced forms (Leger et al. 2017). In addition to L and H proteins, GCS consists of glycine dehydrogenase (decarboxylating), (P protein), and aminomethyltransferase (T protein). All four proteins are required for catalysis of oxidative decarboxylation and deimination of glycine. Although possible involvement of reduced hydrogenosomal GCS in amino acid metabolism has been suggested (Mukherjee et al. 2006), it has been shown that its function is different. The reduced hydrogenosomal GCS represents a unique system for reductive detoxification of peroxides via OsmC peroxidase (Nývltová et al. 2016; chapter "Metabolism of Trichomonad Hydrogenosomes"). Interestingly, serine hydroxymethyltransferase that is functionally linked with GCS to convert serine to glycine is present in hydrogenosomes; however, its physiological role is unclear.

4.5 Protein Import

The majority of proteins directed into mitochondria are transported through the translocase of the outer membrane (TOM), which serves as the general gate into the organelle. Imported proteins are subsequently sorted according to their final destination to TIM23 complex that facilitates import into the matrix or to Tim22 translocase, which imports proteins into the inner membrane. β-Barrel-containing proteins are integrated into the outer membrane by the β-barrel sorting and assembly SAM complex (Wiedemann and Pfanner 2017).

Tom40 is a β-barrel protein that is a core component of TOM complex. Like VDAC (*v*oltage-*d*ependent *a*nion *c*hannel), both proteins are assembled from 19 beta-strands and contain a porin-3 motif, suggesting that Tom40 and VDACs represent ancient paralogues and belong to a common mitochondrial porin family (Pfam family PF01459). Although mitochondrial porins were not initially identified in the *Trichomonas* genome (Carlton et al. 2007), following proteomic studies focused on extraction of membrane proteins and bioinformatic analyses based on hidden Markov model searches identified eight proteins of the porin protein family (Rada et al. 2011; Schneider et al. 2011). All of them possess 19 β-strands and contain β-signal motif, which is recognized by SAM complex and is required for their insertion into the outer membrane of the organelle (Kutik et al. 2008). Six paralogues were predicted to represent Tom40; however, their sequences were highly divergent from known Tom40 orthologues, making it difficult to clearly distinguish between Tom40 and VDAC without experimental verification (Rada et al. 2011). *T. vaginalis* paralogue Tom40-3 was localized to hydrogenosomes and purified by gel filtration within the ~230 kDa and ~590 kDa complexes. The presence of Tom40-3 in high molecular weight complexes is consistent with mitochondrial Tom40 that forms multisubunit complexes of similar weight (Rada et al. 2011). More recent study of Tom40-2 confirmed the function of Tom40 in hydrogenosomes, and proteomic analysis of isolated Tom40-2 complex revealed the presence of new TOM receptors named Tom46 and Tom36 and two other membrane proteins of unknown function (Homp38, Homp19). Moreover, the hydrogenosomal TOM complex seems to be associated with another β-barrel protein Sam50 (Makki et al. 2019; chapter "Protein Import into Hydrogenosomes and Mitosomes"). Finally, there are few other β-barrel proteins, two paralogues of hydrogenosomal 35 kDa membrane protein (Hmp35) (Dyall et al. 2003), and two other related proteins Hmp36 (Rada et al. 2011; Schneider et al. 2011). Hmp35 possesses cysteine-rich C-terminal domain, which contains metal-binding motif (Dyall et al. 2003), and their last β-strand contains putative β-signal. The cysteine-rich C-terminal domain and putative β-signal are absent in both Hmp36 (Rada et al. 2011).

Multiple spanning proteins such as members of mitochondrial carrier protein family (MCF) are imported into the inner membrane of the organelle by the activity of Tim22 translocase. Import of proteins that are addressed for the matrix, as well as insertion of single-spanning proteins into the inner membrane, is facilitated by a hetero-oligomeric complex TIM23. The core components of TIM23 complex are the Tim23 channel and its associated subunit Tim17. TIM22 and TIM23 belong to the Tim17/22/23 protein family (Chacinska et al. 2009; Wiedemann and Pfanner 2017) that typically possesses four transmembrane domains and PRAT motif (Rassow et al. 1999). Searches within the proteome identified two hydrogenosomal proteins (TvTim17/22/23-A and TvTim17/22/23-B) that are homologous to the Tim17/22/23 family. Three other proteins matched with some overall sequence similarity to this protein family and were referred to as TvTim17/22/23-C, TvTim17/22/23-D, and TvTim-like (Rada et al. 2011). The PRAT motif was partially conserved in both TvTim17/22/23-A and B and less in TvTim17/22/23-C and TvTim17/22/23-D, and it was absent in Tim-like protein (Rada et al. 2011). TvTim17/22/23-A also contains features of the Tim23 translocase, which are positively charged residues at its C-terminus (Lys130 and Lys131) and negatively charged residues at its N-terminus (Glu26 and Glu27) (Murcha et al. 2003; Rada et al. 2011). Investigations of TvTim17/22/23-A using isolated *Saccharomyces cerevisiae* mitochondria showed that TvTim17/22/23-A is translocated across the outer membrane and incorporated into the inner membrane. Its incorporation is dependent on membrane potential, a feature that is known for the mitochondrial orthologues (Wiedemann and Pfanner 2017). Based on these observations, TvTim17/22/23-A and B were suggested as candidates for the inner membrane translocase of the hydrogenosome. However, their sequences are too divergent to assign them to each of the Tim17, Tim22, or Tim23 protein subfamilies (Rada et al. 2011).

Import of proteins into the matrix of mitochondria is co-facilitated by the TIM23-associated molecular motor, called PAM complex. In yeast, PAM complex consists of three membrane-associated components Tim44, Pam18, and Pam16, a soluble matrix chaperone mtHsp70, and a nucleotide exchange factor Mge1 (Wiedemann and Pfanner 2017). Except for Pam16, all these components were identified in the *Trichomonas* genome, suggesting that a PAM system may function in protein import into the hydrogenosome (chapter "Protein Import into Hydrogenosomes and Mitosomes").

Precursors of mitochondrial and bacterial proteins are sorted within the intermembrane space with the assistance of small chaperones (Tims) that prevent their misfolding and non-specific aggregation. In yeast and humans, five small Tims (Tim8, Tim9, Tim10, Tim12, and Tim13) form two functional pairs Tim9/10 and Tim8/13 that assemble into a hexameric complex ($\alpha_3\beta_3$) (Gentle et al. 2007). Tim12 acts as a transient peripheral subunit of the Tim22. The small Tim complex ferries hydrophobic precursors to the inner membrane and guides β-barrels to the outer membrane of mitochondria (Chacinska et al. 2009).

It is expected that various hydrogenosomal membrane proteins depend on small Tim-like chaperones which should assist their insertion into the hydrogenosomal membranes. Previous bioinformatic searches failed to recognize small Tims in *Trichomonas* genome (Dyall and Dolezal 2007). However, in the proteome of *Trichomonas* hydrogenosome two small Tims named TvTim9-10a and TvTim9-10b were identified with limited similarity to mitochondrial Tim9 and Tim10. Hydrogenosomal localization was confirmed for TvTim9-10a by immunofluorescence microscopy (Rada et al. 2011). The particular divergence of TvTim9-10a-b from other homologues is in highly conserved twin cysteine motif (CX3CXnCX3C). Both TvTim9-10a and b possess only the second conserved cysteine residue. Since all four cysteine residues are crucial for formation of Tim9/10 complex and are required for import of precursor protein, it is puzzling how TvTim9-10a and b function with the degenerate twin cysteine motif.

4.6 Carriers of the Hydrogenosomal Inner Membrane

Hydrogenosomes and mitochondria produce various metabolites including ATP that are transported across their membranes. In mitochondria, exchange of metabolites across the inner membrane is facilitated by up to 55 distinct carriers that belong to the large group of MCF. MCF members include ATP/ADP carriers (ACCs), amino acid carriers, and transporters of various cofactors and inorganic ions (Kunji and Crichton 2010; Kunji et al. 2016). The *Trichomonas* genome is coding for five genes with overall similarity to MCF members. All five MCF proteins were identified in the proteome of the hydrogenosome (Rada et al. 2011; Schneider et al. 2011). Each of them contains characteristic six transmembrane α-helixes and specific signature motifs which are crucial for the formation of the carrier pore and its nucleotide exchange activity (Kunji et al. 2016). One of these, called hydrogenosomal 31 kDa membrane protein (Hmp31), represents one of the most abundant hydrogenosomal membrane proteins (Dyall et al. 2000). The other four have no obvious homologues in other organisms. Hmp31 is phylogenetically related to the mitochondrial ATP/ADP carriers and was successfully targeted to the inner membrane of yeast mitochondria (Dyall et al. 2000). Hmp31 ATP/ADP nucleotide exchange activity has not been investigated in *T. vaginalis* hydrogenosome so far; however, this activity was experimentally confirmed for the Hmp31 orthologue in *Trichomonas gallinae* (Tjaden et al. 2004). Mitochondrial-type ACCs were also reported in hydrogenosomes of the ciliate *Nyctotherus ovalis* and the fungus *Neocallimastix frontalis* (Voncken et al. 2002; de Graaf et al. 2011).

4.7 C-Tail-Anchored Proteins

Rada et al. (2011) and Schneider et al. (2011) identified in the membranes of *T. vaginalis* hydrogenosome 33 integral membrane proteins, and of these 21 possess transmembrane domain (TMD) in the middle of the protein (so-called integral single-spanning proteins) and 12 contain TMD close to the C-terminus (Rada et al. 2011). Surprisingly, both types of these membrane proteins have no homologues in any other organism. The latter type displays structural characteristics of C-tail-anchored proteins. The α-helical C-tail-anchored proteins (CTAs) represent a heterogeneous group of membrane proteins with a large functional N-terminal domain exposed to the cytosol and a short membrane insertion at their C-terminus. CTAs represent 3–5% of the eukaryotic membrane proteome and play a key role in various cell processes such as lipid trafficking, vesicle targeting/fusion, organelle division, and apoptosis (Borgese and Fasana 2011; Hegde and Keenan 2011; Rao et al. 2016; Costello et al. 2017; Chio et al. 2017). Subcellular targeting of C-tail-anchored proteins is defined by the hydrophobicity and length of the TMD together with the positive net charge in TMD flanking regions (Rao et al. 2016; Chio et al. 2017; Costello et al. 2017). Predicted transmembrane domains of *Trichomonas* CTAs are between 18 and 22 amino acid residues in length, somewhat longer than in their mitochondrial counterparts. This feature may reflect the thickness and variable composition of the outer membrane of hydrogenosomes and mitochondria. The C-terminal segments that follow the transmembrane domain are 3–16 amino acid residues in length and contain dibasic motif R/K-R/K or longer stretches of up to 4 basic amino acids (i.e., KRRK, RKKK) (Rada et al. 2018). Representative *T. vaginalis* C-tail-anchored proteins TVAG_272350, TVAG_277930, TVAG_283120, TVAG_369980, and TVAG_069740 were localized to hydrogenosomes by immunofluorescent microscopy, and their topology was confirmed by protein protection assay (Rada et al. 2018).

4.8 Contaminants or Surface-Associated Proteins?

In addition to matrix and membrane proteins, all proteomic studies of highly purified hydrogenosomes identified a large group of proteins that are known to reside in the cytosol. It is rather difficult to determine whether they represent contamination of the organellar fraction or whether they are functionally associated with the outer hydrogenosomal membrane. In particular, these proteins include various GTPases (137), glycolytic enzymes (10), and ribosomal proteins (60).

Interestingly, ribosome-like particles have been observed on the outer membrane of hydrogenosomes (Benchimol 2001) and mitochondria by transmission electron microscopy (Kellems et al. 1975). There is a growing evidence for translation of nucleus-coded proteins near to mitochondria and/or direct co-translational protein import into mitochondria (Eliyahu et al. 2010; Gao et al. 2014; Williams et al. 2014; Saint-Georges et al. 2008). Whether co-translational import operates also into the hydrogenosome is not known. However, the identifications of a large set of ribosomal proteins in the proteome of hydrogenosomes should stimulate investigation in this direction.

The proteome includes ten enzymes of the glycolytic pathway (Rada et al. 2011; Schneider et al. 2011; Beltrán et al. 2013) including pyrophosphate- and ATP-dependent phosphofructokinases (PFK). It has been speculated that identified glycolytic enzymes may form functional complexes at the surface of hydrogenosomes as observed for mitochondria (Treberg et al. 2007). However, the investigation of ATP-dependent PFK revealed that this protein is targeted to the hydrogenosomal matrix and is not associated with the surface (Rada et al. 2015). Association of other glycolytic enzymes with outer hydrogenosomal membrane remains to be investigated; however, available cell fractionation studies do not support their organellar association (Markos et al. 1993; Rada et al. 2015).

5 Quantitative Analysis of the Proteome: Iron-Induced Changes

Quantitative mass spectrometry allows to investigate effect of various environmental factors on the proteome of the whole cell or separated organelles. This approach is particularly important for *T. vaginalis* that is living in vivo in highly dynamic environment and needs to modulate gene expression and consequently proteome to optimize its physiological status according to changing conditions. Iron belongs among the most important nutritional factors for *T. vaginalis*, and its level has strong regulatory effect on trichomonad transcriptome, especially on transcription of genes coding for hydrogenosomal proteins (Peterson and Alderete 1984; Vanacova et al. 2001; Horváthová et al. 2012). Mass-spectrometry-based quantification method was used to investigate effect of iron on hydrogenosomes at protein level in the study by Beltrán et al. (2013). *T. vaginalis* cells were grown in vitro under iron-rich and iron-restricted conditions before isolation of hydrogenosomes. The organellar protein samples from both cell lines were digested and peptides labeled with isobaric tags for relative and absolute quantification (iTRAQ).

Altogether, 179 identified proteins were considered to be hydrogenosomal using stringent criteria to eliminate most of contaminants. Of these, 58 were differentially expressed with a fold change of at least 2.0.; 31 of these proteins were more abundant, and 27 were less abundant under iron-deficient conditions (Table 2). Iron-restricted conditions particularly downregulated hydrogenosomal enzymes of pyruvate and malate catabolism that includes PFO, malic enzyme, [2Fe-2S] ferredoxin, iron hydrogenase, and both subunits of rudimental complex I (NuoE and NuoF) (Beltrán et al. 2013). In contrast, all members involved in Fe-S cluster assembly were significantly upregulated at iron-depleted conditions (Table 2). We can speculate that the pathway of pyruvate/malate catabolism, in which abundant Fe-S proteins operate, is downregulated to lower the need for iron supply, while ISC system is strengthened to increase the rate of Fe-S cluster formation, which is required for vital housekeeping proteins. Interestingly, there are six paralogues of ferredoxin, of which three were upregulated and three were downregulated. Such a differential expression of paralogues may reflect their different functions in hydrogenosomes. [4Fe-4S] ferredoxin serves as an electron carrier between PFO and hydrogenase, and downregulated paralogues are likely members of this pathway. The second function of the ferredoxin is to provide electrons to ISC machinery, and thus upregulated ferredoxin paralogues are most likely linked with Fe-S cluster assembly. However, there are more examples of differential expression between protein paralogues. For example, only one of three detected paralogues of acetate/succinate-CoA transferase and α-SCS subunits was significantly downregulated under iron-restricted conditions, whereas two others were upregulated. The differential expression of individual paralogues seems to be a unique mode of trichomonad response to environmental changes such as availability of key nutrients and may provide explanation for massive retention of numerous paralogues in the *Trichomonas* genome.

Table 2 Iron regulated proteins in iron depleted conditions (Beltrán et al. 2013)

Downregulated			Upregulated		
Accession no.	Annotation	Fold change	Accession no.	Annotation	Fold change
TVAG_230580	Pyruvate/ferredoxin oxidoreductase BI	−37.4	TVAG_469020	HydG-1	10.4
TVAG_254890	Pyruvate/ferredoxin oxidoreductase E	−16.9	TVAG_146780	Nfu-4	8.7
TVAG_242960	Pyruvate/ferredoxin oxidoreductase BII	−13.5	TVAG_479680	2-Nitropropane dioxygenase precursor	5.4
TVAG_310050	[Fe] hydrogenase-3	−9.0	TVAG_451860	Nfu-3	4.7
TVAG_265760	FAD/FMN-binding family protein-4	−8.8	TVAG_253630	Hsp70 mitochondrial type-3	3.5
TVAG_292710	Ferredoxin 4	−6.6	TVAG_008840	Nfu-2	3.2
TVAG_064490	Rubrerythrin-1	−6.4	TVAG_055320	IscA2-2	3.2
TVAG_154730	Iron-sulfur flavoprotein (Isf3)	−6.3	TVAG_044500	Nfu-1	3.1
TVAG_399860	Ferredoxin 2	−5.9	TVAG_361540	IscA2-3	2.8
TVAG_198110	Pyruvate/ferredoxin oxidoreductase A	−5.4	TVAG_329200	HydE-2	2.7
TVAG_049140	Superoxide dismutase [Fe], putative	−5.3	TVAG_282580	Conserved unknown protein	2.6
TVAG_183790	Malic enzyme F	−4.8	TVAG_412560	OsmC-2	2.5
TVAG_076510	Serine palmitoyltransferase	−4.2	TVAG_048590	Thioesterase family protein	2.5
TVAG_412220	Malic enzyme D	−3.8	TVAG_344280	Conserved unknown protein	2.5
TVAG_296220	Complex 1, Tvh21	−3.6	TVAG_342900	Flavin reductase	2.4
TVAG_037570	[Fe] hydrogenase-2 (64 kDa)	−3.3	TVAG_060450	Acetyltransferase-1	2.3
TVAG_395550	Acetate/succinate-CoA transferase-3	−3.2	TVAG_385350	Thioredoxin	2.3
TVAG_466790	Pyruvate/ferredoxin oxidoreductase F	−3.0	TVAG_456770	IscA2-1	2.3
TVAG_181350	Conserved unknown protein	−2.9	TVAG_242760	Isd11-1	2.2
TVAG_133030	Complex 1, Tvh47	−2.9	TVAG_182340	Mge (GrpE) protein-1	2.2
TVAG_003900	Ferredoxin 1	−2.8	TVAG_183850	Arginine deiminase-3	2.1
TVAG_416100	Malic enzyme C	−2.4	TVAG_370860	Tim17/22/23C	2.1
TVAG_047890	Succinyl-CoA synthetase, α-subunit	−2.3	TVAG_177600	Glycine cleavage system H protein	2.1
TVAG_182620	[Fe] Hydrogenase-1 (50 kDa)	−2.3	TVAG_381290	Hsp20-3	2.0
TVAG_340290	Malic enzyme H	−2.1	TVAG_277380	Ind-4 (P-Loop ATPase)	2.0

TVAG_104250	Hmp35-2	−2.1	TVAG_205390	HydF	2.0
TVAG_351540	FAD/FMN-binding family protein-3	−2.0	TVAG_086470	Thioredoxin	2.0
			TVAG_433130	Hsp70 mitochondrial type-6	2.0
			TVAG_467820	Arginine deiminase-1	2.0
			TVAG_318670	Succinyl-CoA synthetase, alpha	2.0
			TVAG_109540	Serine hydroxymethyltransferase	2.0

6 The Proteome of *Pentatrichomonas hominis* Hydrogenosome

The hydrogenosomes of *Pentatrichomonas hominis* are the only hydrogenosomes beside *Trichomonas vaginalis*, for which the proteome was analyzed (Fang et al. 2016). *P. hominis* is an anaerobic protist that colonizes the large intestine of humans and various mammals, including cats, dogs, goats, pigs, and primates (Maritz et al. 2014). The genome of *P. hominis* has not been decoded yet to provide a reference database for analysis of MS spectra. Therefore, Fang et al. used an alternative data set based on a sequencing of total RNA isolated from the *P. hominis* axenic culture. The protein analysis of purified hydrogenosomes resulted in the set of 442 hydrogenosomal proteins. The functional categories of proteins were very similar to these in *T. vaginalis*, including components of energy metabolism, amino acid metabolism, oxygen stress response, chaperones, peptidases, protein import components, and large set of hypothetical proteins. Some expected proteins were not identified, such as ISC assembly protein Isd11 and membrane-associated proteins Pam16, Pam18, and Hmp36. The possible explanation is that their mRNA expression was extremely low under experimental conditions, resulting in missing these transcripts in the RNA-seq data set.

As in *T. vaginalis*, *P. hominis* hydrogenosomes contain multiple paralogues from each protein family. For example, there are three PFO, four ferredoxins, nine iron hydrogenases, four ADP/ATP carriers, and five Tom40 paralogues. Interestingly, comparison of hydrogenosomal proteomes between these two trichomonads revealed only 107 common proteins, whereas 270 were *P. hominis* specific and 206 *T. vaginalis* specific. It seems that *Pentatrichomonas* and *Trichomonas* express different paralogues of the same multiple gene family. The largest difference was in the category of hypothetical proteins. The research strategy by Feng et al. (2014) using RNA sequencing and proteomics proved to be valid alternative for investigations of understudied organisms without known genome sequence.

7 Concluding Remarks

Research of hydrogenosomal proteomes made a significant progress in the past decade. Pilot studies have been performed based on laborious separation of proteins by 2D SDS electrophoresis that revealed over 60 hydrogenosomal proteins (Pütz et al. 2005; Henze 2007). New gel-free techniques for protein separation and new generation of mass spectrometry instruments with high-resolution power and sensitivity dramatically increased the number of identified proteins in complex protein samples. For example, Feng et al. (2014) identified 2934 proteins in the hydrogenosomal preparation. This high number reflects a high sensitivity of new instruments, in this case mass spectrometer based on an ion trap mass analyzer. However, such a sensitivity provides not only more correct identifications but also more cellular contaminations that need to be carefully filtered. Indeed, the raw data set of 2934 proteins was filtered to the set of "only" 442 proteins using elaborated bioinformatic tools. Thus, with current sensitivity of mass spectrometers, the most challenging issues are correct evaluation of raw data and development of high-throughput methods for validation of cell localization of detected proteins. Such methods are currently available, e.g., for *Trypanosoma brucei* (Dyer et al. 2016).

However, the most challenging issue is proteomics of hydrogenosomes and other MROs in protists that are difficult to cultivate in vitro to obtain sufficient amounts of cells required for current proteomic methods. The analysis of their proteomes awaits development of new technologies for a single-cell proteomics in the future (Doerr 2019).

Acknowledgments Research in the J.T. laboratory is supported by the Ministry of Education, Youth and Sports of the Czech Republic project NPU II (LQ1604) and by ERD Funds project CePaViP (CZ.02.1.01/0.0/0.0/16_019/0000759).

References

Akhmanova A, Voncken FGJ, Hosea KM et al (1999) A hydrogenosome with pyruvate formate-lyase: anaerobic chytrid fungi use an alternative route for pyruvate catabolism. Mol Microbiol 32:1103–1114

Arendsen AF, Hadden J, Card G et al (1998) The "prismane" protein resolved: X-ray structure at 1.7 Å and multiple spectroscopy of two novel 4Fe clusters. J Biol Inorg Chem 3:81–95

Barberà MJ, Ruiz-Trillo I, Tufts JYA et al (2010) *Sawyeria marylandensis* (Heterolobosea) has a hydrogenosome with novel metabolic properties. Eukaryot Cell 9:1913–1924

Beltrán NC, Horváthová L, Jedelský PL et al (2013) Iron-induced changes in the proteome of *Trichomonas vaginalis* hydrogenosomes. PLoS One 8:e65148

Benchimol M (2001) Hydrogenosome morphological variation induced by fibronectin and other drugs in *Trichomonas vaginalis* and *Tritrichomonas foetus*. Parasitol Res 87:215–222

Bordier C (1981) Phase separation of integral membrane proteins in Triton X-114 solution. J Biol Chem 256:1604–1607

Borgese N, Fasana E (2011) Biochimica et Biophysica Acta Targeting pathways of C-tail-anchored proteins. Biochim Biophys Acta 1808:937–946

Boxma B, de Graaf RM, van der Staay GWM et al (2005) An anaerobic mitochondrion that produces hydrogen. Nature 434:74–79

Bradley PJ, Lahti CJ, Plümper E, Johnson PJ (1997) Targeting and translocation of proteins into the hydrogenosome of the protist *Trichomonas*: similarities with mitochondrial protein import. EMBO J 16:3484–3493

Broers C, Stumm CK, Vogels GD (1991) Axenic cultivation of the anaerobic Free-Living Ciliate *Trimyema compressum*. J Protozool 38:507–511

Brown MT, Goldstone HMH, Bastida-Corcuera F et al (2007) A functionally divergent hydrogenosomal peptidase with protomitochondrial ancestry. Mol Microbiol 64:1154–1163

Brul S, Veltman RH, Lombardo MCP, Vogels GD (1994) Molecular cloning of hydrogenosomal ferredoxin cDNA from the anaerobic amoeboflagellate *Psalteriomonas lanterna*. Biochim Biophys Acta Bioenerg 1183:544–546

Bunai K, Yamane K (2005) Effectiveness and limitation of two-dimensional gel electrophoresis in bacterial membrane protein proteomics and perspectives. J Chromatogr B 815:227–236

Burstein D, Gould SB, Zimorski V et al (2012) A machine learning approach to identify hydrogenosomal proteins in *Trichomonas vaginalis*. Eukaryot Cell 11:217–228

Cabeza MS, Guerrero SA, Iglesias AA, Arias DG (2015) New enzymatic pathways for the reduction of reactive oxygen species in *Entamoeba histolytica*. Biochim Biophys Acta Gen Subj 1850:1233–1244

Calvo SE, Clauser KR, Mootha VK (2016) MitoCarta2.0: an updated inventory of mammalian mitochondrial proteins. Nucleic Acids Res 44:D1251–D1257

Carlton JM, Hirt RP, Silva JC et al (2007) Draft genome sequence of the sexually transmitted pathogen *Trichomonas vaginalis*. Science 315:207–212

Cerkasovova A, Lukasova G, Cerkasov J, Kulda J (1973) Biochemical characterization of large granule fraction of *Tritrichomonas foetus* (KV1 strain). J Protozool 20:537

Chacinska A, Koehler CM, Milenkovic D et al (2009) Importing mitochondrial proteins: machineries and mechanisms. Cell 138:628–644

Chio US, Cho H, Shan S-O (2017) Mechanisms of tail-anchored membrane protein targeting and insertion. Annu Rev Cell Dev Biol 33:417–438

Coombs GH, Westrop GD, Suchan P et al (2004) The amitochondriate eukaryote *Trichomonas vaginalis* contains a divergent thioredoxin-linked peroxiredoxin antioxidant system. J Biol Chem 279:5249–5256

Costello JL, Castro IG, Camões F et al (2017) Predicting the targeting of tail-anchored proteins to subcellular compartments in mammalian cells. J Cell 130:1675–1687

de Graaf RM, Ricard G, van Alen TA et al (2011) The organellar genome and metabolic potential of the hydrogen-producing mitochondrion of *Nyctotherus ovalis*. Mol Biol Evol 28:2379–2391

Delgadillo MG, Liston DR, Niazi K, Johnson PJ (1997) Transient and selectable transformation of the parasitic protist *Trichomonas vaginalis*. Proc Natl Acad Sci U S A 94:4716–4720

Doerr A (2019) Single-cell proteomics. Nat Methods 16:20–20

Dyall SD, Dolezal P (2007) Protein import into hydrogenosomes and mitosomes. In: Tachezy J (ed) Hydrogenosomes and mitosomes: mitochondria of anaerobic eukaryotes, 1st edn. Springer, Berlin, pp 21–73

Dyall SD, Koehler CM, Delgadillo-Correa MG et al (2000) Presence of a member of the mitochondrial carrier family in hydrogenosomes: conservation of membrane-targeting pathways between hydrogenosomes and mitochondria. Mol Cell Biol 20:2488–2497

Dyall SD, Lester DC, Schneider RE et al (2003) *Trichomonas vaginalis* Hmp35, a putative pore-forming hydrogenosomal membrane protein, can form a complex in yeast mitochondria. J Biol Chem 278:30548–30561

Dyer BD (1989) *Metopus*, *Cyclidium* and *Sonderia*: ciliates enriched and cultured from sulfureta of a microbial mat community. Biosystems 23:41–51

Dyer P, Dean S, Sunter J (2016) High-throughput gene tagging in *Trypanosoma brucei*. J Vis Exp 114:e54342

Eliyahu E, Pnueli L, Melamed D et al (2010) Tom20 mediates localization of mRNAs to mitochondria in a translation-dependent manner. Mol Cell Biol 30:284–294

Fang Y-K, Chien K-Y, Huang K-Y et al (2016) Responding to a zoonotic emergency with multi-omics research: *Pentatrichomonas hominis* hydrogenosomal protein characterization with use of RNA sequencing and proteomics. Omi A J Integr Biol 20:662–669

Fenchel T, Finlay BJ (1991) The biology of free-living anaerobic ciliates. Eur J Protistol 26:201–215

Gao J, Schatton D, Martinelli P et al (2014) CLUH regulates mitochondrial biogenesis by binding mRNAs of nuclear-encoded mitochondrial proteins. J Cell Biol 207:213–223

Garg S, Stölting J, Zimorski V et al (2015) Conservation of transit peptide-independent protein import into the mitochondrial and hydrogenosomal matrix. Genome Biol Evol 7:2716–2726

Gentle IE, Perry AJ, Alcock FH et al (2007) Conserved motifs reveal details of ancestry and structure in the small Tim chaperones of the mitochondrial intermembrane space. Mol Biol Evol 24:1149–1160

Gill EE, Diaz-Triviño S, Barberà MJ et al (2007) Novel mitochondrion-related organelles in the anaerobic amoeba *Mastigamoeba balamuthi*. Mol Microbiol 66:1306–1320
Gould SB, Woehle C, Kusdian G et al (2013) Deep sequencing of *Trichomonas vaginalis* during the early infection of vaginal epithelial cells and amoeboid transition. Int J Parasitol 43:707–719
Gray MW (2012) Mitochondrial evolution. Cold Spring Harb Perspect Biol 4:a011403
Guda P, Guda C, Subramaniam S (2007) Reconstruction of pathways associated with amino acid metabolism in human mitochondria. Genomics Proteomics Bioinformatics 5:166–176
Hackstein JH, Akhmanova A, Boxma B et al (1999) Hydrogenosomes: eukaryotic adaptations to anaerobic environments. Trends Microbiol 7:441–447
Hegde RS, Keenan RJ (2011) Tail-anchored membrane protein insertion into the endoplasmic reticulum. Nat Rev Mol Cell Biol 12:787–798
Henze K (2007) The proteome of *T. vaginalis* hydrogenosomes. In: Tachezy J (ed) Hydrogenosomes and mitosomes: mitochondria of anaerobic eukaryotes, 1st edn. Springer, Berlin, pp 163–178
Horváthová L, Šafaříková L, Basler M et al (2012) Transcriptomic identification of iron-regulated and iron-independent gene copies within the heavily duplicated *Trichomonas vaginalis* genome. Genome Biol Evol 4:1017–1029
Hrdy I, Hirt RP, Dolezal P et al (2004) *Trichomonas* hydrogenosomes contain the NADH dehydrogenase module of mitochondrial complex I. Nature 432:618–622
Huang K-Y, Huang P-J, Ku F-M et al (2012) Comparative transcriptomic and proteomic analyses of *Trichomonas vaginalis* following adherence to fibronectin. Infect Immun 80:3900–3911
Jerlström-Hultqvist J, Einarsson E, Xu F et al (2013) Hydrogenosomes in the diplomonad *Spironucleus salmonicida*. Nat Commun 4:2493
Keeling PJ, Burger G, Durnford DG et al (2005) The tree of eukaryotes. Trends Ecol Evol 20:670–676
Kellems RE, Allison VF, Butow RA (1975) Cytoplasmic type 80S ribosomes associated with yeast mitochondria. IV. Attachment of ribosomes to the outer membrane of isolated mitochondria. J Cell Biol 65:1–14
Kunji ERS, Crichton PG (2010) Mitochondrial carriers function as monomers. Biochim Biophys Acta Bioenerg 1797:817–831
Kunji ERS, Aleksandrova A, King MS et al (2016) The transport mechanism of the mitochondrial ADP/ATP carrier. Biochim Biophys Acta Mol Cell Res 1863:2379–2393
Kutik S, Stojanovski D, Becker L et al (2008) Dissecting membrane insertion of mitochondrial β-barrel proteins. Cell 132:1011–1024
Leger MM, Eme L, Hug LA, Roger AJ (2016) Novel hydrogenosomes in the microaerophilic jakobid *Stygiella incarcerata*. Mol Biol Evol 33:2318–2336
Leger MM, Kolisko M, Kamikawa R et al (2017) Organelles that illuminate the origins of *Trichomonas* hydrogenosomes and *Giardia* mitosomes. Nat Ecol Evol 1:0092
Leitsch D, Williams CF, Hrdý I (2018) Redox pathways as drug targets in microaerophilic parasites. Trends Parasitol 34:576–589
Lindmark DG, Müller M (1973) Hydrogenosome, a cytoplasmic organelle of the anaerobic flagellate *Tritrichomonas foetus*, and its role in pyruvate metabolism. J Biol Chem 248:7724–7728
Lindmark DG, Müller M, Shio H (1975) Hydrogenosomes in *Trichomonas vaginalis*. J Parasitol 61:552. https://doi.org/10.2307/3279345
Lloyd D, Williams CF (2015) Avoid excessive oxygen levels in experiments with organisms, tissues and cells. Adv Microb Physiol 67:293–314
Lloyd D, Hillman K, Yarlett N, Williams AG (1989) Hydrogen production by rumen holotrich protozoa: effects of oxygen and implications for metabolic control by *in situ* conditions. J Protozool 36:205–213
Magdeldin S, Enany S, Yoshida Y et al (2014) Basics and recent advances of two dimensional-polyacrylamide gel electrophoresis. Clin Proteomics 11:16

Makki A, Rada P, Žárský V et al (2019) Triplet-pore structure of a highly divergent TOM complex of hydrogenosomes in *Trichomonas vaginalis*. PLoS Biol 17:e3000098
Maritz JM, Land KM, Carlton JM, Hirt RP (2014) What is the importance of zoonotic trichomonads for human health? Trends Parasitol 30:333–341
Markos A, Miretsky A, Müller M (1993) A glyceraldehyde-3-phosphate dehydrogenase with eubacterial features in the amitochondriate eukaryote, *Trichomonas vaginalis*. J Mol Evol 37:631–643
Martijn J, Vosseberg J, Guy L et al (2018) Deep mitochondrial origin outside the sampled alphaproteobacteria. Nature 557:101–105
Marvin-Sikkema FD, Driessen AJM, Gottschal JC, Prins RA (1994) Metabolic energy generation in hydrogenosomes of the anaerobic fungus *Neocallimastix*: evidence for a functional relationship with mitochondria. Mycol Res 98:205–212
Mentel M, Zimorski V, Haferkamp P et al (2008) Protein import into hydrogenosomes of *Trichomonas vaginalis* involves both N-terminal and internal targeting signals: a case study of thioredoxin reductases. Eukaryot Cell 7:1750–1757
Morada M, Smid O, Hampl V et al (2011) Hydrogenosome-localization of arginine deiminase in *Trichomonas vaginalis*. Mol Biochem Parasitol 176:51–54
Mukherjee M, Brown MT, McArthur AG, Johnson PJ (2006) Proteins of the glycine decarboxylase complex in the hydrogenosome of *Trichomonas vaginalis*. Eukaryot Cell 5:2062–2071
Murcha MW, Lister R, Ho AYY, Whelan J (2003) Identification, expression, and import of components 17 and 23 of the inner mitochondrial membrane translocase from Arabidopsis. Plant Physiol 131:1737–1747
Nývltová E, Šuták R, Harant K et al (2013) NIF-type iron-sulfur cluster assembly system is duplicated and distributed in the mitochondria and cytosol of *Mastigamoeba balamuthi*. Proc Natl Acad Sci U S A 110:7371–7376
Nývltová E, Stairs CW, Hrdý I et al (2015) Lateral gene transfer and gene duplication played a key role in the evolution of *Mastigamoeba balamuthi* hydrogenosomes. Mol Biol Evol 32:1039–1055
Nývltová E, Smutná T, Tachezy J, Hrdý I (2016) OsmC and incomplete glycine decarboxylase complex mediate reductive detoxification of peroxides in hydrogenosomes of *Trichomonas vaginalis*. Mol Biochem Parasitol 206:29–38
O'Fallon JV, Wright RW, Calza RE, Calza RE (1991) Glucose metabolic pathways in the anaerobic rumen fungus *Neocallimastix frontalis* EB188. Biochem J 274(Pt 2):595–599
Paul RG, Williams AG, Butler RD (1990) Hydrogenosomes in the rumen entodiniomorphid ciliate *Polyplastron multivesiculatum*. J Gen Microbiol 136:1981–1989
Peterson KM, Alderete JF (1984) Iron uptake and increased intracellular enzyme activity follow host lactoferrin binding by *Trichomonas vaginalis* receptors. J Exp Med 160:398–410
Pütz S, Gelius-Dietrich G, Piotrowski M, Henze K (2005) Rubrerythrin and peroxiredoxin: two novel putative peroxidases in the hydrogenosomes of the microaerophilic protozoon *Trichomonas vaginalis*. Mol Biochem Parasitol 142:212–223
Rabilloud T, Lelong C (2011) Two-dimensional gel electrophoresis in proteomics: a tutorial. J Proteomics 74:1829–1841
Rada P, Doležal P, Jedelský PL et al (2011) The core components of organelle biogenesis and membrane transport in the hydrogenosomes of *Trichomonas vaginalis*. PLoS One 6:e24428
Rada P, Makki AR, Zimorski V et al (2015) N-terminal presequence-independent import of phosphofructokinase into hydrogenosomes of *Trichomonas vaginalis*. Eukaryot Cell 14:1264–1275
Rada P, Makki A, Žárský V, Tachezy J (2018) Targeting of tail-anchored proteins to *Trichomonas vaginalis* hydrogenosomes. Mol Microbiol. https://doi.org/10.1111/mmi.14175
Rao M, Okreglak V, Chio US et al (2016) Multiple selection filters ensure accurate tail-anchored membrane protein targeting. Elife. https://doi.org/10.7554/eLife.21301
Rassow J, Dekker PJ, van Wilpe S et al (1999) The preprotein translocase of the mitochondrial inner membrane: function and evolution. J Mol Biol 286:105–120

Saint-Georges Y, Garcia M, Delaveau T et al (2008) Yeast mitochondrial biogenesis: a role for the PUF RNA-binding protein Puf3p in mRNA localization. PLoS One 3:e2293
Schneider RE, Brown MT, Shiflett AM et al (2011) The *Trichomonas vaginalis* hydrogenosome proteome is highly reduced relative to mitochondria, yet complex compared with mitosomes. Int J Parasitol 41:1421–1434
Shiflett AM, Johnson PJ (2010) Mitochondrion-related organelles in eukaryotic protists. Annu Rev Microbiol 64:409–429
Smíd O, Matusková A, Harris SR et al (2008) Reductive evolution of the mitochondrial processing peptidases of the unicellular parasites *Trichomonas vaginalis* and *Giardia intestinalis*. PLoS Pathog 4:e1000243
Smith AC, Robinson AJ (2016) MitoMiner v3.1, an update on the mitochondrial proteomics database. Nucleic Acids Res 44:D1258–D1261
Smutna T, Goncalves VL, Saraiva LM et al (2009) Flavodiiron protein from *Trichomonas vaginalis* hydrogenosomes: the terminal oxygen reductase. Eukaryot Cell 8:47–55
Smutna T, Pilarová K, Tarábek J et al (2014) Novel functions of an iron-sulfur flavoprotein from *Trichomonas vaginalis* hydrogenosomes. Antimicrob Agents Chemother 58:3224–3232
Stairs CW, Leger MM, Roger AJ (2015) Diversity and origins of anaerobic metabolism in mitochondria and related organelles. Philos Trans R Soc Lond B Biol Sci 370:20140326
Sutak R, Dolezal P, Fiumera HL et al (2004) Mitochondrial-type assembly of FeS centers in the hydrogenosomes of the amitochondriate eukaryote *Trichomonas vaginalis*. Proc Natl Acad Sci U S A 101:10368–10373
Tachezy J, Sánchez LB, Müller M (2001) Mitochondrial type iron-sulfur cluster assembly in the amitochondriate eukaryotes *Trichomonas vaginalis* and *Giardia intestinalis*, as indicated by the phylogeny of IscS. Mol Biol Evol 18:1919–1928
Tjaden J, Haferkamp I, Boxma B et al (2004) A divergent ADP/ATP carrier in the hydrogenosomes of *Trichomonas gallinae* argues for an independent origin of these organelles. Mol Microbiol 51:1439–1446
Treberg JR, MacCormack TJ, Lewis JM et al (2007) Intracellular glucose and binding of hexokinase and phosphofructokinase to particulate fractions increase under hypoxia in heart of the amazonian armored catfish (*Liposarcus pardalis*). Physiol Biochem Zool 80:542–550
van Bruggen JJA, Zwart KB, van Assema RM et al (1984) *Methanobacterium formicicum*, an endosymbiont of the anaerobic ciliate *Metopus striatus* McMurrich. Arch Microbiol 139:1–7
Vanacova S, Rasoloson D, Rázga J, Hrdý I, Kulda J, Tachezy J (2001) Iron-induced changes in pyruvate metabolism of *Tritrichomonas foetus* and involvement of iron in expression of hydrogenosomal proteins. Microbiology 147:53–62
Viscogliosi E, Durieux I, Delgado-Viscogliosi P et al (1996) Phylogenetic implication of iron-containing superoxide dismutase genes from trichomonad species. Mol Biochem Parasitol 80:209–214
Viscogliosi E, Delgado-Viscogliosi P, Gerbod D et al (1998) Cloning and expression of an iron-containing superoxide dismutase in the parasitic protist, *Trichomonas vaginalis*. FEMS Microbiol Lett 161:115–123
Voncken F, Boxma B, Tjaden J et al (2002) Multiple origins of hydrogenosomes: functional and phylogenetic evidence from the ADP/ATP carrier of the anaerobic chytrid *Neocallimastix* sp. Mol Microbiol 44:1441–1454
Wang J, Vine CE, Balasiny BK et al (2016) The roles of the hybrid cluster protein, Hcp and its reductase, Hcr, in high affinity nitric oxide reduction that protects anaerobic cultures of *Escherichia coli* against nitrosative stress. Mol Microbiol 100:877–892
Westrop GD, Goodall G, Mottram JC, Coombs GH (2006) Cysteine biosynthesis in *Trichomonas vaginalis* involves cysteine synthase utilizing *O* -Phosphoserine. J Biol Chem 281:25062–25075
Wiedemann N, Pfanner N (2017) Mitochondrial machineries for protein import and assembly. Annu Rev Biochem 86:685–714

Williams CC, Jan CH, Weissman JS (2014) Targeting and plasticity of mitochondrial proteins revealed by proximity-specific ribosome profiling. Science 346:748–751

Woehle C, Kusdian G, Radine C et al (2014) The parasite *Trichomonas vaginalis* expresses thousands of pseudogenes and long non-coding RNAs independently from functional neighbouring genes. BMC Genomics 15:906

Yarlett N, Hann AC, Lloyd D, Williams A (1981) Hydrogenosomes in the rumen protozoon *Dasytricha ruminantium* Schuberg. Biochem J 200:365–372

Yarlett N, Hann AC, Lloyd D, Williams AG (1983) Hydrogenosomes in a mixed isolate of *Isotricha prostoma* and *Isotricha intestinalis* from ovine rumen contents. Comp Biochem Physiol Part B Comp Biochem 74:357–364

Zimorski V, Major P, Hoffmann K et al (2013) The N-Terminal sequences of four major hydrogenosomal proteins are not essential for import into hydrogenosomes of *Trichomonas vaginalis*. J Eukaryot Microbiol 60:89–97

Zysset-Burri DC, Müller N, Beuret C et al (2014) Genome-wide identification of pathogenicity factors of the free-living amoeba *Naegleria fowleri*. BMC Genomics 15:496

Mitosomes in Parasitic Protists

Jan Tachezy and Ondřej Šmíd

Contents

1 Introduction 206
2 Discovery of Mitosomes 208
3 Morphology 210
4 Biogenesis 213
4.1 The Genome 213
4.2 Protein Targeting, Translocation, and Maturation 214
4.3 Replication 217
5 Physiological Functions 221
5.1 Iron-Sulfur Cluster Assembly Machinery 221
5.2 Requirements for ATP, Membrane Potential and Electron Transport 229
5.3 Sulfate Activation and Other Mitosomal Functions 231
6 Perspectives 232
References 233

Abstract Mitosomes are highly reduced forms of mitochondria, which were found in several parasitic protists of various eukaryotic lineages, including the human parasites *Entamoeba histolytica*, *Giardia intestinalis*, *Cryptosporidium* spp., *Mikrocytos mackini*, and microsporidians. Although all these organisms underwent different evolutionary histories, they arrived at common life strategies for which oxygen-dependent ATP synthesis is not required: they inhabit either an oxygen-poor environment such as the intestinal tract of their hosts or they are adapted to intracellular parasitism. Consequently, the majority of their mitochondrial functions were permanently lost including ATP synthesis with concomitant loss of the organellar genome. The common features of mitosomes, which were retained and pointed to their mitochondrial origin, are a double membrane surrounding the organellar matrix, conserved mechanisms of protein import and processing, and

J. Tachezy (✉) · O. Šmíd
Faculty of Science, BIOCEV, Department of Parasitology, Charles University, Vestec, Czech Republic
e-mail: tachezy@natur.cuni.cz

J. Tachezy (ed.), *Hydrogenosomes and Mitosomes: Mitochondria of Anaerobic Eukaryotes*, Microbiology Monographs 9,
https://doi.org/10.1007/978-3-030-17941-0_9

the biosynthesis of iron-sulfur (Fe-S) clusters. Finding the latter function in mitosomes supports the notion that Fe-S cluster assembly is the only essential function of mitochondria necessary for the maturation of cellular Fe-S proteins. Only in the mitosomes of *Entamoeba histolytica* the mitochondrion type of Fe-S cluster assembly machinery was not conserved, and these organelles gained a unique sulfate activation pathway. In spite of a great progress in elucidation of the evolutionary paths leading to the formation of mitosomes, cellular functions of mitosomes are still poorly understood.

1 Introduction

A parasitic lifestyle leads to the evolutionary adaptation of the parasitic organisms to the new niches of the host environment. As a result, the parasites can regulate their biochemical capabilities during their often complicated life cycles to optimize their metabolism within different hosts and host tissues. Such changes lead to various functional and consequently morphological modifications of the parasitic cells, including alterations in the functions and structure of mitochondria, the principal ATP-generating organelles of eukaryotes. A textbook example is the mitochondrion of *Trypanosoma brucei*. This organelle undergoes dramatic metabolic and structural changes during the trypanosome switch between the blood of the mammalian host and the digestive tract of the tsetse fly. An excess of glucose in the mammalian host permits the bloodstream stage to mainly employ glycolysis for ATP generation. Consequently, the mitochondrial energy metabolism is repressed (van Dooren et al. 2005), lacking the ability to produce ATP by oxidative phosphorylation (Clayton and Michels 1996), and on the contrary ATP is hydrolyzed by reverse activity of F_0F_1-ATPase to maintain a membrane potential (Schnaufer et al. 2005). In the insect vector, trypanosomes encounter a nutrient-poor environment, where they can only survive upon the activation of the mitochondrion. Structurally, these biochemical changes are reflected by the tubular cristae mitochondrion in its bloodstream forms, which transforms to the platelike cristae-containing mitochondrion in its vector stages (Vickerman 1985) (Fig. 1). Some other parasites, including certain parasitic helminths, remain within the same host; however, they migrate through different host tissues, for part of their life cycle is entering the oxygen-poor environment of the host's intestinal tract. These organisms adapted their mitochondria for either aerobic or anaerobic respiration according to the availability of oxygen (Komuniecki et al. 1993; Müller et al. 2012). The function of mitochondria is often altered in intracellular parasites adapted to life within the host cell, and their bioenergetics rely mainly on cytosolic glycolysis similarly to anaerobes. For example, *Plasmodium* mitochondria lack pyruvate dehydrogenase to feed the tricarboxylic acid (TCA) cycle with acetyl-CoA; thus, TCA is unlikely to perform a cycle: they lost respiratory

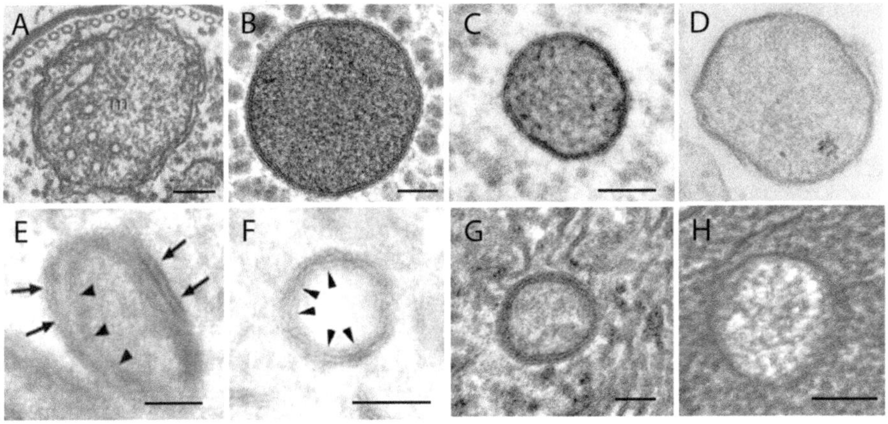

Fig. 1 Morphology of mitochondria, hydrogenosomes, and mitosomes in parasitic protists. (**a**) Mitochondrion of bloodstream stage of *Trypanosoma brucei* with tubular cristae. (**b**, **c**) Hydrogenosomes from (**b**) *Monocercomonas ruminatium* and (**c**) *Mastigamoeba balamuthi*. (**d–h**) Mitosomes from (**d**) *Entamoeba histolytica*, (**e**) *Giardia intestinalis*, (**f**) *Trachipleistophora hominis*, (**g**) *Spironucleus vortens*, and (**h**) *Globulispora mitoportans*. Arrows and arrowheads indicate outer and inner membranes, respectively. (**g**) and (**h**) kindly provided by J. Kulda and J. Vávra, respectively. Figures reproduced with permission from the International Journal for Parasitology, Elsevier Limited (**a**), the American Society for Microbiology (**d**), Nature Publishing Group, Macmillan Limited (**e**, **f**). Scale bars: (**a**)–(**e**), (**g**) 100 nm; (**e**), (**f**), and (**h**) 50 nm

complex I and possess unconventional ATP synthase (van Dooren et al. 2006; Vaidya and Mather 2009).

In more extreme cases, the adaptive and reductive course of evolution led to permanent rejection of a number of metabolic pathways, including those residing in mitochondria, and consequently to the transformation of mitochondria to organelles known as hydrogenosomes and mitosomes. Hydrogenosomes have been found in unicellular eukaryotes (protists), parasites, or commensals, typically inhabiting the intestinal or urogenital tract, as well as in some free-living protists inhabiting oxygen-poor environments. They evolved repeatedly, being found in phylogenetically distant protists, including trichomonads, diplomonads, jakobids, fungi, heteroloboseans, and archamoebae (Müller et al. 2012; Nývltová et al. 2015; Jerlstrom-Hultqvist et al. 2013; de Graaf et al. 2009; chapter "Metabolism of Trichomonad Hydrogenosomes"). The common characteristics of hydrogenosomes include the production of molecular hydrogen, which is coupled with ATP synthesis by substrate-level phosphorylation, lack of a membrane-associated electron transport chain and consequently ATP synthesis by oxidative phosphorylation, and absence of the genome (Müller et al. 2012). A number of mitochondrial functions have been lost, which is reflected by reduced hydrogenosome proteome (Schneider et al. 2011; chapter "The Proteome of *T. vaginalis* Hydrogenosomes").

Yet the most reduced mitochondria-derived organelles are mitosomes, in which mostly a single biosynthetic pathway mediating the formation of Fe-S clusters is retained. Exceptionally, mitosomes may harbor a sulfate activation pathway that has been found in *Entamoeba histolytica* (Mi-ichi et al. 2009). This chapter reviews the current status of knowledge in mitosome research comparing the morphology, biogenesis, and function of mitosomes in various lineages of parasitic protists.

2 Discovery of Mitosomes

Mitosomes have been found in parasitic protists employing two different living strategies: (1) anaerobic parasites invading the intestinal tract—*Giardia intestinalis* (Tovar et al. 2003), *Entamoeba histolytica* (Mai et al. 1999; Tovar et al. 1999)—and (2) intracellular parasite invading various cell tissues, apicomplexan *Cryptosporidium* (Riordan et al. 1999), microsporidia (Williams et al. 2002), and rhizarian *Mikrocytos mackini* (Burki et al. 2013). Though all these organisms underwent different evolutionary paths, they arrived at common life strategies, for which oxygen-dependent ATP synthesis is not required. Consequently, the majority of mitochondrial functions were permanently lost. Previously, the absence of "canonical" mitochondrial pathways and cellular structures conforming to typical mitochondria in several groups of protists was interpreted as the absence of mitochondria, and such organisms were labeled "amitochondriates." The lack of mitochondria was considered to be a result of two possible scenarios: (1) According to the Archezoa hypothesis, contemporary "amitochondriates" were the direct descendants of primitive amitochondrial proto-eukaryotes, which separated them from the main eukaryotic trunk before the endosymbiotic acquisition of mitochondria (Cavalier-Smith 1987a, b). Based on this attractive hypothesis, four amitochondrial groups including Archamoebae (e.g., *Entamoeba histolytica*), metamonads (*Giardia intestinalis*), microsporidia (*Encephalitozoon cuniculi*), and Parabasalia (*Trichomonas vaginalis*) were incorporated into the taxon named Archezoa to indicate their ancient origin. Two types of amitochondriates were later distinguished: Type I amitochondriates, in which energy metabolism is not compartmentalized (*Giardia*, *Entamoeba*), and Type II amitochondriates possessing ATP-generating hydrogenosomes (trichomonads) (Martin and Müller 1998). (2) According to the second scenario, the amitochondriates lost mitochondria secondarily due to their specific adaptation to anaerobic or oxygen-restricted conditions. The second scenario obviously explains their amitochondrial status in organisms, which belong to the monophyletic taxons together with organisms possessing typical mitochondria such as the apicomplexan *Cryptosporidium*. Indirect evidence supporting the secondary loss of mitochondria also in *Entamoeba histolytica* came from the phylogenetic analysis of its small subunit ribosomal RNA, which places this parasite after the branch of mitochondrion-harboring protists such as kinetoplastids (Sogin 1991). Similarly, a phylogenetic analysis of tubulins (Keeling and Doolittle 1996) and the largest subunit of the RNA polymerase II (Hirt et al. 1999) of microsporidians revealed the affinity of these protists to fungi, which typically possess mitochondria.

It is generally accepted that the majority of genes of the endosymbiotic ancestor, which gave rise to mitochondria, was either lost or transferred into the host nucleus.

In order to provide more direct evidence of the secondary loss of mitochondria, further investigations were focused on uncovering genes of mitochondrial origin, which would have been transferred from the mitochondrial to the nuclear genome if mitochondria were once present and which might be retained in contemporary "amitochondriates." Indeed, various genes regarded as mitochondrial in origin, including mitochondrial-type heat shock protein 70 (mitHSP 70) (Arisue et al. 2002; Bui et al. 1996; Germot et al. 1996, 1997; Morrison et al. 2001), chaperonin 60 (CPN60) (Bui et al. 1996; Clark and Roger 1995; Horner et al. 1996; Roger et al. 1998), valyl-tRNA synthetase (Hashimoto 1998), pyridine-nucleotide transhydrogenase (PNT) (Clark and Roger 1995), cysteine desulfurase IscS (Tachezy et al. 2001), and other components of the Fe-S cluster assembly machinery (Katinka et al. 2001) were identified in the genomes of all members of Archezoa groups.

The next logical step, which led to the discovery of mitosomes, was to determine the cellular localization of proteins coded by genes of mitochondrial origin. The first mitosomes were described in *Entamoeba histolytica* (Mai et al. 1999; Tovar et al. 1999). Two research groups showed in parallel studies that amebic CPN60 was present in microsomal cellular fractions. The organellar localization of CPN60 was confirmed by immunofluorescence microscopy in wild-type cells as well as in *Entamoeba* expressing recombinant CPN60 with *c-myc* tag. Subsequently, transmission electron microscopy of the mitosome-enriched fraction showed organelles bound by a double membrane (Ghosh et al. 2000).

Analysis of the *Encephalitozoon cuniculi* genome sequence, which revealed the presence of several genes coding for proteins with putative mitochondrial function, led to a prediction of mitosome existence in microsporidia (Katinka et al. 2001). Direct evidence came from studies of *Trachipleistophora hominis*, an intracellular parasite causing severe myositis in patients with AIDS. Williams et al. (2002) demonstrated the localization of *Trachipleistophora* mitHSP70 to a number of double-membrane-bound organelles by means of immunofluorescence and immunoelectron microscopy.

Preliminary evidence of a mitosome (relict mitochondrion) in *Cryptosporidium parvum* was provided by an electron-microscopic study that revealed the presence of a single double-membrane-bound organelle (Riordan et al. 1999). These observations were supported by *C. parvum* genome analysis that identified several putative mitochondrial proteins such as a mitochondrial protein-import apparatus and chaperones (Abrahamsen et al. 2004). Subsequently, CPN60 (Riordan et al. 2003) and HSP70 (Slapeta and Keithly 2004) were detected within this organelle by immunomicroscopy, and detail structure of the mitosome was reconstructed by electron tomography (Keithly et al. 2005).

The discovery of the Fe-S cluster assembly as an essential mitochondrial function provided new markers for tracing mitochondrion-derived organelles such as cysteine desulfurase IscS and the molecular scaffold protein IscU (Tachezy et al. 2001). These two markers were used for the identification of mitosomes in *Giardia intestinalis* (Tovar et al. 2003). Specific antibodies against *Giardia* IscS and IscU uncovered their localization in double-membrane-bound organelles using immunomicroscopy and cell fractionation. In subsequent studies, *Giardia* CPN60 and mitHSP70 were detected within the same cellular compartment as IscS and IscU (Regoes et al. 2005).

Giardia intestinalis was for a while considered one of the most solid candidates to represent the deepest diverging lineages of "amitochondrial" eukaryotes (Adam 2001; Best et al. 2004). Thus, the finding that *Giardia* also possesses a mitochondrion (mitosome) strongly contributed to the corrosion of the Archezoa concept. The finding of mitosomes in "amitochondrial" organisms that belong to various eukaryotic lineages let to the notion that all contemporary eukaryotes once possess mitochondria and that mitosomes represent the most reduced form of mitochondria that underwent the deepest reduction of the mitochondrial metabolic functions. However, the reductive evolution can proceed exceptionally even deeper. Current investigations of *Monocercomonoides exilis* revealed that this intestinal endosymbiont completely lost mitochondria due to replacement of the mitochondrial Fe-S cluster assembly machinery with the system of bacterial origin operating in the cytosol (see chapter "Protists Without Mitochondria, How It May Happen?"). *M. exilis* and most likely related organisms of Preaxostyla clade represent the truly amitochondrial eukaryotes. However, it needs to be pointed out that amitochondrial status of *M. exilis* is a result of reductive evolution not due to primitive absence of mitochondria as it has been proposed for Archezoa.

3 Morphology

Mitosomes are tiny ovoid or elongated organelles that measure less than 0.5 μm in length. In electron micrographs, the organelles appeared to be limited by two membranes, which envelop a granular electron-dense matrix. Mitosomes do not form cristae, the inner membrane foldings, which are typically present in respiring mitochondria. Morphologically, mitosomes are reminiscent of the mitochondria found in yeast cells grown anaerobically, "acristae" mitochondria in *Plasmodium falciparum* merozoites (Fry and Beesley 1991) as well as hydrogenosomes found in some anaerobic protists (Fig. 1). The identification of the double-limiting membrane in mitosomes was critical to the argument for their relationship with mitochondria (Ghosh et al. 2000; Riordan et al. 1999; Tovar et al. 2003; Williams et al. 2002). The detailed morphology of mitosomal double membranes, however, is not identical for the mitosomes observed in different species.

Electron micrographs of the mitosomal membranes in *Entamoeba histolytica* and *Trachipleistophora hominis* showed that they are formed by two closely opposed membranes without an apparent intramembrane space (Fig. 1). The membranes are visible as three lipidic layers: the outer layer of the outer membrane, the inner layer of the inner membrane, and the central lipidic bilayer. This arrangement is very similar to that observed in hydrogenosomes. In these organelles, the central bilayer separates around calcium-rich intramembrane vesicles, which are surrounded by two complete membrane units (see chapter "Structure of the Hydrogenosome").

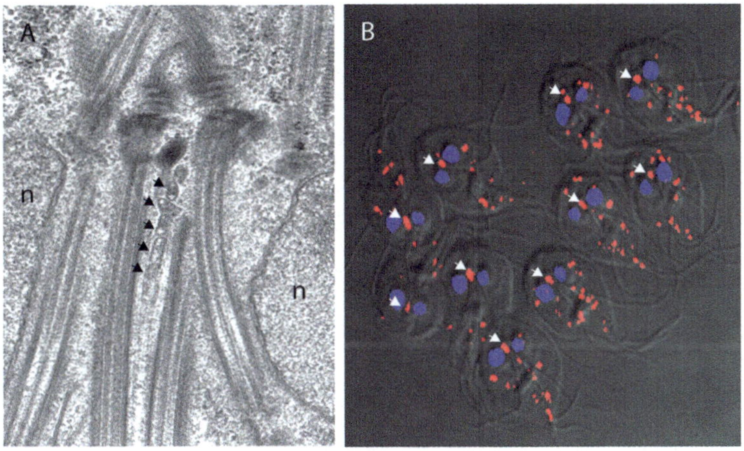

Fig. 2 Mitosomes of *Giardia intestinalis*. (**a**) Central mitosomes localized between axonemes of caudal flagella close to basal bodies (*n* nucleus). Scale bar = 500 nm. Kindly provided by E. Nohýnková, Charles University in Prague, Czech Republic. (**b**) Central and peripheral mitosomes visualized by immunofluorescence microscopy using anti GiIscU antibody (red). Arrows pointed to central mitosomes

However, such intramembrane vesicles have not been observed in mitosomes so far. An intramembrane space reminiscent of mitochondria was observed in *Giardia* mitosomes and mitosome-like organelles in some microsporidians. Also, the size and number of mitosomes varied in different cells. In initial studies of *Entamoeba histolytica*, a single mitosome was observed in most cells using immunofluorescence microscopy, but a few cells contained two and rarely three organelles (Mai et al. 1999; Tovar et al. 1999). The size of mitosomes was estimated to be 0.5–1.0 μm based on electron microscopy micrographs. However, in a later study, confocal microscopy images revealed that mitosomes are rather abundant organelles of over 150 mitosomes per *Entamoeba* trophozoite with an estimated size of less than 0.5 μm for most organelles (Leon-Avila and Tovar 2004). *Giardia* mitosomes are also abundant structures, ranging from 25 to 100 per cell (Doležal et al. 2005; Regoes et al. 2005; Tovar et al. 2003). However, they are considerably smaller, with a size of less than 0.2 μm (an average of 184 × 140 nm). The majority of *Giardia* mitosomes are randomly distributed throughout the cytoplasm (peripheral mitosomes), often in lateral and posterior regions of the trophozoites. Central mitosomes form a distinct rodlike structure between the two *Giardia* nuclei in close proximity to the basal bodies. This structure, which is invariably present in all cells, consists of several attached mitosomes organized between the axonemes of caudal flagella (Fig. 2). The microsporidian *Trachipleistophora hominis* contains between 7 and 47 mitosomes with an average of 28 mitosomes per cell observed throughout the cytoplasm. With a size of 50 nm by 90 nm estimated by electron microscopy, these structures are the smallest mitosomes to have been observed thus far (Williams et al. 2002). Similar structures have been observed in other

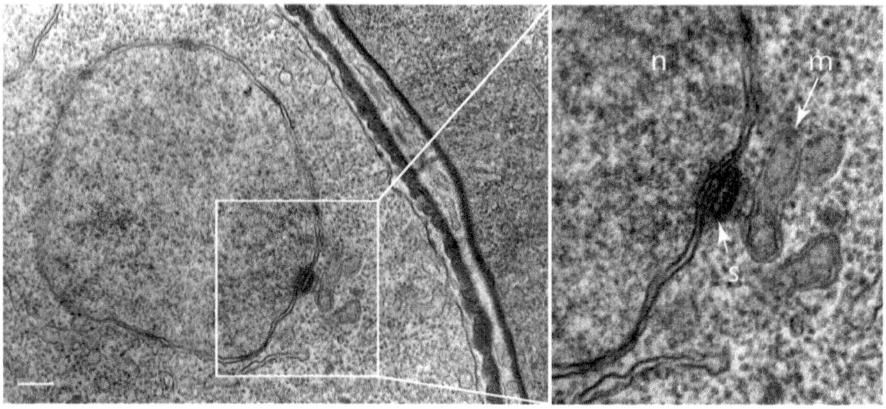

Fig. 3 Mitosome of *Vavraia culicis* associated with a spindle plaque (*n* nucleus, *s* spindle plaque, *m* mitosome). Scale bar = 200 nm. Kindly provided by J. Vávra, Charles University in Prague, Czech Republic

microsporidian species, including the meronts and sporonts of *Vavraia culicis* from *Spodoptera exigua*, *Amblyospora* sp. from *Cyclops strenuus*, *Vairimorpha* sp. from *Lymantria dispar*, *Marssoniella elegans* from *Cyclops vicinus* (Vávra 2005), and *Globulispora mitoportans* parasitizing daphnids (Vávra et al. 2016). Unlike in *Trachipleistophora*, only a few mitosome-like organelles are freely scattered in the cytoplasm of the other microsporidia. Typically, they form a group of organelles which are associated with a spindle plaque, situated in a depression of the nuclear membrane (Fig. 3). The mitosome-like organelles associated with a spindle plaque probably correspond to the "polar vesicles," previously reported in various ultrastructural studies of microsporidia (Vávra 1976; Youssef and Hammond 1971). The mitosomes in these species are oval, about 0.15–0.5 × 0.1–0.25 μm in size. Different morphology was observed for "mitosome-like vesicles" in *Globulispora mitoportans*. These organelles are considerably larger (0.3–1.5 × 0.08–0.1 μm) and contain a loose web of filamentous material (Vávra et al. 2016). The apicomplexan *Cryptosporidium parvum* contains a single mitosome (relict mitochondrion), which is invariably localized to the posterior part of the cell between the nucleus and crystalloid body. The size of the organelle estimated by electron microscopy ranges between 0.2 and 0.5 μm in diameter (Putignani et al. 2004; Riordan et al. 2003). The organelle is enveloped by a rough endoplasmic reticulum that extends from the outer nuclear membrane. A tomographic reconstruction showed that the inner mitosomal membrane of *Cryptosporidium* is highly folded without a cristae-like connection, or it forms independent vesicles (chapter "The Mitochondrion-Related Organelles of *Cryptosporidium* Species").

4 Biogenesis

Mitosomal biogenesis, as well as the biogenesis of mitochondria, is a complex process of maintenance and multiplication of the organelle in the cell. It also includes segregation of the organelles during the cell cycle.

4.1 The Genome

No evidence has been found for the mitosomal genome in any organism so far. An earlier study on *Entamoeba histolytica* reported the presence of DNA in *Entamoeba* mitosomes (Ghosh et al. 2000) and synonymized these organelles with DNA-containing structures named kinetoplast-like organelles (EhKO) (Orozco et al. 1997). However, the work by Leon-Avila and Tovar later showed that EhKO and mitosomes are not related structures. Moreover, in situ nick translation coupled to immunofluorescence microscopy failed to detect the presence of DNA in *Entamoeba* mitosomes (Leon-Avila and Tovar 2004).

In contemporary mitochondria, the genome of the original endosymbiotic ancestor is retained to various extents coding from 100 genes in the protist *Andalucia godoyi* (Burger et al. 2013) to 2 genes in *Chromera velia* (Flegontov et al. 2015). Several hypotheses have been proposed to explain the reason why the genome is retained: (1) The import of highly hydrophobic proteins from cytosol to mitochondria is not efficient. (2) Certain proteins coded in the mitochondrial genome are toxic when present in the cytosol. (3) The retention of genes, whose products are involved in electron transport and energy metabolism, may reflect the need for precise regulation of their expression by the redox state of the mitochondrion. (4) In some eukaryotes, mitochondrial genes may be trapped in the mitochondrial genome by a nonstandard genetic code usage in the organelle (Adams and Palmer 2003). Interestingly, two genes, coding for cytochrome *b* (cyt b) and subunit I of cytochrome *c* oxidase (cox1), are present in all mitochondrial genomes known to date. These genes may represent core components of the mitochondrial genome whose transfer to the nucleus is not possible, for any of the reasons mentioned above. Both cyt b and cox1 are components of the respiratory chain, which is linked to ATP production by oxidative phosphorylation. Although these pathways are typically present in mitochondria, they are absent in hydrogenosomes and mitosomes. Thus, the loss of a cytochrome-dependent respiratory chain might be the reason why the mitosomal genome was not required any more for the functioning of the organelles, and consequently the genome was completely eliminated.

4.2 Protein Targeting, Translocation, and Maturation

The majority or virtually all proteins targeted to mitochondria or mitosomes, respectively, are encoded in the nucleus, translated in the cytosol on free ribosomes, and posttranslationally imported to their final destination. However, this classical view was challenged by proximity-specific ribosome profiling that provided evidence for alternative protein synthesis at the mitochondrial outer membrane (Ingolia et al. 2009; Ingolia 2014).

The current knowledge on protein targeting and translocation to mitochondria is summarized in several excellent reviews (Pfanner et al. 2019; Straub et al. 2016; Wiedemann and Pfanner 2017). Briefly, the information for the import of mitochondrial proteins is carried either within their N-terminal extension (targeting presequence of matrix proteins, NTS) or within the polypeptide chain (internal targeting signals of membrane proteins, ITS). The primary structure of the targeting signals is not conserved. However, the secondary structure of NTSs exhibits a consensus conformation of an amphipathic alpha-helix with positively charged and apolar sides. Mitochondrial import signals are recognized by receptors on the outer mitochondrial membrane. These receptors are clustered together with a general import pore to the complex called translocase of the outer mitochondrial membrane (TOM). TOM complex consists of at least three highly conserved core components (Tom40, Tom22, and Tom7) and various peripheral components, namely, pre-protein receptors (Fukasawa et al. 2017). Once proteins pass through the outer membrane, they follow one of three major import pathways: (1) Presequence-carrying proteins are directed to the translocase of the inner membrane called the TIM23 complex. The pore within TIM23 cooperates with mitochondrial HSP70 and other proteins that form the presequence translocase-associated motor (PAM) that pulls the imported protein to the matrix. Once the protein reaches the matrix, the targeting sequence is cleaved off by a matrix-localized mitochondrial processing peptidase (MPP). (2) Proteins of the inner membrane are guided by small Tim chaperones (Tim8/13/9/10) through the intermembrane space to the TIM22 complex. The pore of TIM22 inserts these pre-proteins into the mitochondrial inner membrane. The core components of TIM23 (Tim17 and Tim23) and TIM22 (Tim22) are paralogs and belong to the same Tim17/22/23 protein family. Only a few proteins follow an alternative route, being exported to the inner membrane from the mitochondrial matrix by the OXA complex. (3) Outer membrane proteins are guided via small Tims to the sorting and assembly machinery (SAM) complex (Wiedemann et al. 2004).

Mitosomes of *Giardia intestinalis* have been demonstrated to share a common mode of protein targeting with hydrogenosomes and mitochondria (Doležal et al. 2005; Regoes et al. 2005) (chapter "Protein Import into Hydrogenosomes and Mitosomes"). Targeting sequences resembling the N-terminal extensions of mitochondrial proteins are required for the efficient import of some proteins into the *Giardia* mitosome. Namely, the role of the targeting sequences of the iron-sulfur cluster assembly protein IscU and mitosomal ferredoxin has been demonstrated.

Moreover, the IscU and ferredoxin targeting sequences were able to direct proteins into the hydrogenosomes of *Trichomonas vaginalis* (Doležal et al. 2005) and the mitochondria of human kidney cells (Regoes et al. 2005). However, majority of the mitosomal matrix proteins including IscS, Cpn60, and mtHsp70 lack N-terminal targeting sequences, and the import into the mitosomes is driven by ITS (Šmíd et al. 2008).

A characterization of the translocase complexes responsible for mitosomal protein import in *Giardia* is underway. Hidden Markov model searches in the *Giardia* genome revealed a Tom40 homologue that is a core subunit of protein translocation channel (Dagley et al. 2009). However, no other TOM subunits have been found so far. Bioinformatics and following proteomic studies identified components of PAM including Pam18, Pam16, and Hsp70 as well as Tim44 that provides an interface between TIM channel and PAM motor (Jedelsky et al. 2011; Martincová et al. 2015). A multiple experimental and bioinformatic attempts have failed to identify core components of TIM complexes in *Giardia*. Recently, Tim17-like proteins has been identified in several other metamonads, and implementations of these sequences to metamonad-specific hidden Markov model led to identification of putative Tim17 in *Giardia*, whereas Tim23 and Tim22 seem to be absent (Pyrihová et al. 2018). As a classical TIM23 channel comprise of Tim17 and Tim23, more experimental works are required to elucidate whether and how the single Tim17 forms the protein-conducting channel in mitosomes without Tim23.

After translocation into the mitosome, the N-terminal presequences of mitosomal proteins are cleaved off, by a peptidase that is homologous to MPP. The MPP is a matrix-localized metallopeptidase that consist of α-subunit with a flexible glycine-rich loop for the substrate recognition and β-subunit that possesses a catalytic zinc-binding motif His-X-X-Glu-His (Gakh et al. 2002). A presence of putative β-MPP subunit in *G. intestinalis* and its mitosomal localization has been initially demonstrated by Doležal et al. (2005). Later investigations revealed that unlike all eukaryotic MPPs, the *Giardia* processing peptidase (GPP) consists of only β-subunit, whereas α-subunit is absent (Šmíd et al. 2008). The unique simplicity of GPP most likely reflect highly reduced function of mitosomes and consequently the paucity of proteins that are targeted to this organelle as well as lack of NTSs in the majority of mitosomal proteins.

Similarly, a reduced mitosomal protein-import machinery has been described in microsporidia (Burri et al. 2006). Some mitosomal proteins of microsporidia carry a short NTS. In some cases, when expressed in the yeast *S. cerevisiae*, these presequences are able to direct proteins into mitochondria. Other proteins possess an ITS or both types of targeting signals. Only two components of the TOM complex have been identified, Tom40 and ITS receptor Tom70. TIM components consist of Tim50 receptor and Tim22, whereas Tim17 and Tim23 are absent. Identification of only a single member of Tim17/22/23 family suggests that microsporidia may form a single type of TIM channel in inner mitosomal membrane as *Giardia*; however, microsporidian TIM channel is derived from different Tim17/22/23 paralog (Pyrihová et al. 2018). PAM components include Pam16 and Pam18 (Waller et al. 2009). The mitosomal presequences of matrix proteins are not processed by a homologue of MPP that is absent in microsporidia. As these presequences are short, they probably do not interfere with the function of matrix proteins. Intermembrane space proteins require the processing of another type of protease, the inner membrane peptidase. A homologue of this enzyme has been demonstrated in the microsporidia *Antonospora locustae*, while it is missing in *Encephalitozoon cuniculi*. It has been hypothesized that the mitosome of the latter organism represents a further stage of reductive adaptation of the pristine fungal mitochondrion (Burri et al. 2006).

The mitosome of *Cryptosporidium parvum* represents another example of simplified TOM and TIM complexes (Alcock et al. 2012). Similarly to mitosomes of previous organisms, only Tom40 is known in *Cryptosporidium parvum*, whereas Tom7 and Tom22 were not identified. TIM channel is probably based on a Tim17 member of Tim17/22/23 family like in *Giardia*. The other identified components of the import machinery include Tim44, Pam18, and Hsp70. The remarkable difference from other mitosomes is identification of a single type of small Tim named *Cp*TimS with strong similarity to Tim8 and some similarity to Tim13 and identification of core components of SAM (Sam50) (Alcock et al. 2012). These findings suggest that more sophisticated protein sorting takes place in the intermembrane space of *Cryptosporidium parvum* in comparison to *Giardia* and microsporidia in which neither small Tims nor Sam50 are present.

Overall reduction of protein-import machinery in mitosomes, changes in properties of NTSs of pre-proteins, and particularly modification of TIM complex apparently reflect functional minimalism of mitosomes. In comparison to mitochondria that contain over 1000 proteins, considerably less proteins are targeted to mitosomes (Jedelsky et al. 2011). More importantly, the mitosomes lack a membrane potential that provides a main driving force for protein translocation across inner mitochondrial membrane. Mitochondrial NTSs are mostly of 15–55 amino acid long peptides with a net positive charge (+3 to +6 in *Saccharomyces cerevisiae*) that can form amphipathic α-helices (von Heijne 1986). These properties allow both hydrophobic and electrostatic interactions between NTS and components of the import machinery during the translocation of pre-proteins and their maturation. NTSs are enriched in arginine, serine, alanine, and leucine, while acidic amino acids are rare. Typically, there is an arginine at the cleavage motif for MPP and several more positively charged residuals N-terminally distal from the cleavage site (Fukasawa et al. 2015). NTS is initially recognized via the hydrophobic interaction with Tom20 receptor that faces to the cytosol (Saitoh et al. 2007). Then NTS is transferred to negatively charged N-terminal domain of Tom22 receptor and Tom5 that facilitate pre-protein entry to the TOM gate. Subsequently, NTS interacts with acidic patches within Tom40 channel and C-terminal domain of Tom22 when NTS appears in the intermembrane space (Komiya et al. 1998; Schatz 1997). The sequential binding and movement of the positively charged NTS over the membrane components was termed the "acid chain mechanism" (Voos et al. 1999). NTS is then bound to Tim50 and N-terminal domain of Tim23 for translocation across inner membrane (Yamamoto et al. 2002). Most importantly, the net positive charge is required for NTS translocation through the TIM23 channel by electrophoretic force. This step is dependent on the inner membrane potential (ΔΨ) that is generated by the electron transport chain. The ΔΨ-dependent step is followed by protein translocation via hydrophobic interaction with PAM machinery that is driven at the expense of ATP (Pfanner et al. 2019; Schulz et al. 2015). Finally, electrostatic interactions facilitate accommodation of NTS within negatively charged cavity of MPP to be cleaved off (Gakh et al. 2002). In mitosomes and similarly hydrogenosomes, mitochondrial electron transport chain is absent; consequently ΔΨ is low or absent in these organelles, and the electrophoretic step cannot operate in the protein import. The lack of ΔΨ most likely released a selection pressure that keeps net positive charge at NTS, and its role in protein import declined. Indeed, mitosomal NTSs are short, for example, NTSs in *Giardia* are 10–18 amino acids with 1–2 arginines only at the cleavage site motif. However, most of mitosomal pre-proteins entirely lack NTSs (Šmíd et al. 2008). Changes in NTS properties of pre-proteins and/or their elimination due to lack of ΔΨ seem to go with modifications or elimination of the import components involved in interactions with NTS such as Tom22, Tom5, Tim50, Tim23, and MPP. For example, putative Tom22 was recently identified as a component of hydrogenosomal TOM complex in *T. vaginalis*; however, it consists of only a transmembrane domain for interaction with Tom40, whereas charged N- and C-terminal domains are absent (Makki et al. 2019). Modifications of the GPP adapted for processing of mitosomal NTS represent another example. In addition to absence of α-subunit, βGPP forms a cavity with acid residues only close to the active site to interact with arginine residue of the NTS cleavage site motif, whereas the rest of the predicted βGPP cavity is, unlike βMPP, positively charged (Šmíd et al. 2008). Mechanisms of import of pre-proteins with mitosomal NTS as well as pre-proteins without NTS are currently unclear. Experimental data provided evidences that these proteins possess ITSs; however, there is a little information about the ITS characters. Collectively, it is not clear what drives import of mitosomal pre-proteins across mitosomal membrane. However, it is likely that the final step of the pre-protein translocation to the mitosomal matrix is mediated by ATP-dependent PAM machinery that is conserved in all known mitosomes.

4.3 Replication

Structures of mitosomes resembling dividing mitochondria have been occasionally observed on some electron micrographs (Keithly et al. 2005) and recently studied to some details in *Giardia* (Voleman et al. 2017). Interestingly, unlike mitochondria, no fusion of these organelles has been observed so far.

4.3.1 Division

Bacteria, the progenitors of mitosomes, mitochondria, and hydrogenosomes, employ the ring-forming protein FtsZ in their division. Some eukaryotes retained FtsZ-based machinery (*Cyanidioschyzon merolae* of Rhodophyta, *Cyanophora paradoxa* of Glaucophyta, *Mallomonas splendens* of Heterokontophyta, and *Dictyostelium discoideum* of Amoebozoa) to be employed in the division of mitochondria (Beech et al. 2000). However, in most eukaryotic organisms including parasitic protists, FtsZ has been lost, and mitochondrial fission depends on a soluble cytosolic dynamin-related protein Drp1/Dnm1 (Bleazard et al. 1999). Drp1/Dnm1 is recruited to the mitochondria by adaptor proteins such as Mff, Mid49, and Mid51 that are anchored in the outer mitochondrial membrane (Otera et al. 2010; Palmer et al. 2011). Dynamins are GTPases with various functions that include the formation of clathrin-coated endocytic vesicles. Interestingly, *T. brucei*, a kinetoplastid protist, exploits a single dynamin-related protein for both mitochondrial division and formation of endocytic vesicles (Benz et al. 2017; Chanez et al. 2006). The only dynamin-related protein of *Giardia intestinalis* is involved in the scission of budding vesicles but not in the division of mitosomes (Gaechter et al. 2008; Voleman et al. 2017). Thus, the molecular mechanisms of *Giardia* mitosome division remain unknown. Similarly, *Spironucleus salmonicida*, a diplomonad protist related to *Giardia intestinalis*, and *Encephalitozoon cuniculi* of Microsporidia possess a single dynamin-related protein (Andersson et al. 2007). However, whether this protein is involved in mitosome division remains to be elucidated.

The mitochondria of mammals and fungi are known as the dynamic organelles that continually undergo fission and fusion to maintain homeostasis of mitochondrial mass through mitochondrial biogenesis and mitophagy, organelle homogeneity, and changes in energy needs (Liesa and Shirihai 2013; Mishra and Chan 2014). Although there is little information about dynamics of mitosomes, recent investigations of *Giardia* suggest that dynamics of mitosomes is remarkably different in comparison to the mitochondria of classical models. Life-cell imaging of *Giardia* with Halo-labeled mitosomal marker revealed no fusion of mitosomes, and fission was observed only during mitosis that was followed by either cytokinesis or encystation. Two types of mitosomes could be distinguished in *Giardia* according to the cell localization: (1) the central mitosomes that are invariably present between two nuclei in contact with the basal bodies of flagellar axonemes and (2) the peripheral mitosomes scattered within the cytosol but preferentially at the lateral and posterior regions. Interestingly, investigation of particular phase of mitosis, at which mitosomes divide, revealed that peripheral mitosomes undergo division during all mitosis with increasing number toward telophase, whereas the central mitosomes divide exclusively during prophase. Morphology of dividing mitosomes ranged from dumbbell-shaped to threadlike vesicles. Interestingly, co-labeling of mitosomes and endoplasmic reticulum (ER) revealed close association between these two types of organelles (Fig. 4). Particularly, the dividing mitosomes formed ringlike structure around the ER tubes (Voleman et al. 2017). Previously, association of ER with mitochondria has been shown to stimulate mitochondrial fission in human cells (Korobova et al. 2013). Therefore, it is likely that observed tight association between dividing mitosomes and the ER is related to the mechanism of the mitosome division.

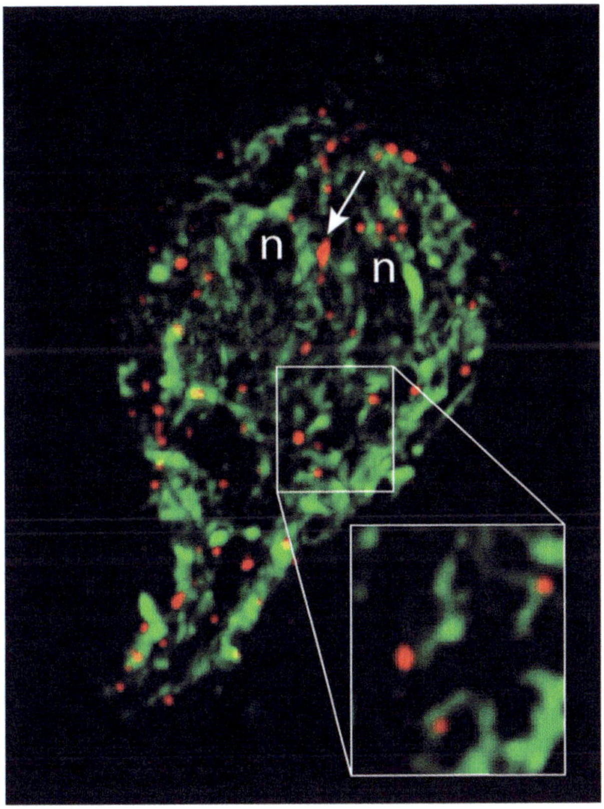

Fig. 4 Super-resolution structured illumination microscopy (SIM) of *Giardia intestinalis* showing association between mitosomes and structures of endoplasmic reticulum. Mitosomes were detected by an antibody raised against GL50803_9296 protein (red) (Martincová et al. 2015), and ER was visualized by the antibody against protein disulfide isomerase PDI2 (green). Arrow indicates central mitosomes; *n* nucleus. Kindly provided by L. Voleman, Charles University in Prague, Czech Republic

4.3.2 Segregation

In the highly organized eukaryotic cell, organelles including mitochondria are not randomly distributed within the cytosol, but their position is determined by specific interactions with one or more of the three cytoskeletal systems (microtubules, microfilaments, and intermediate filaments). Mechanisms that control the movement of mitochondria are highly diverse, even within a group of closely related organisms. For example, the migration of mitochondria in the cells of *Saccharomyces cerevisiae* depends on actin filaments (Boldogh et al. 2005), while the related fungus *Schizosaccharomyces pombe* uses microtubule tracks to control mitochondrial motility (Yaffe et al. 2003). The microtubule-dependent movement is facilitated by molecular motors, kinesin, and dynein (Saxton and Hollenbeck 2012; Tanaka et al. 1998). The association between mitochondria and the molecular motors is mediated by a protein complex that involves mitochondrial Rho GTPases 1 and 2 (Miro1/2) and TRAK1/2 (Glater et al. 2006; Guo et al. 2005; Schuler et al. 2017). The same cytoskeletal components that are involved in mitochondrial movement during interphase mediate the segregation of mitochondria into daughter cells during cytokinesis

(Boldogh et al. 2005; Yaffe et al. 2003). Precise segregation of mitochondria is essential, since the organelles cannot be formed de novo.

Considering the origin of mitosomes, the mitosomal segregation during cytokinesis may be analogous to the segregation of mitochondria. For example, electron-microscopic study of Microsporidia showed the association of mitosomes with electron-dense depressions of the nuclear membrane. Since these structures are mitotic spindle organization centers of microsporidia, they may be responsible for the distribution of mitosomes into daughter cells (Vávra 2005). Indeed, in *Globulispora mitoportans*, mitosomes were observed to divide when nucleus divides (Vávra et al. 2016).

Much less is known about the movement and segregation of mitochondria in the cells of other protists. As for the Excavata group, the protist *T. brucei* harbors a single mitochondrion. The organelle is immobile, being connected with the flagellar basal body by filaments that span both mitochondrial membranes and bind mitochondrial DNA. Replication of the basal body and its segregation into daughter cells is responsible for the precise segregation of mitochondria during cell division (Ogbadoyi et al. 2003). In another excavate protist, *Giardia intestinalis*, the central mitosomes seem to have a prominent role in the organelle segregation. These mitosomes were reported to divide together with flagellar basal bodies before mitosis. A connection between daughter basal bodies and mitosomes has been proposed to be responsible for the segregation of mitosomes into daughter cells (Regoes et al. 2005). However, the basal bodies of the eight flagella of *G. intestinalis* do not replicate before the start of mitosis as previously thought but during the complex process of parental flagella migration (Nohýnkova et al. 2006). Therefore, precise mitosomal segregation must be achieved by a different mechanism. According to our observations, the central mitosomes divide during prophase and then migrate to the opposite spindle poles in association with the axonemes of caudal flagella. During cytokinesis, these are the only flagella that neither change their function nor position in daughter cells and may facilitate segregation of the central mitosomes into the daughter cells (Fig. 4).

Cryptosporidium parvum is a protist related to the apicomplexans *Plasmodium falciparum* and *Toxoplasma gondii*. The latter parasites possess the apicoplast, a remnant plastid acquired by secondary symbiosis. It has been shown that the *Toxoplasma* apicoplast is segregated into daughter cells by microtubules of mitotic spindle poles (Striepen et al. 2000). As the apicoplast is in contact with the mitochondrion, it has been hypothesized that inheritance of a single mitochondrion that is tightly coupled with cell division cycle is mediated indirectly via the apicoplast (van Dooren et al. 2005). However, later observations of living cells by time-lapse microscopy using differentially labeled mitochondrion and apicoplast revealed that apicoplast enters daughter cells in early phase of daughter cell development, whereas segregation of mitochondrion appeared at last possible moment (Nishi et al. 2008). It is not yet known how mitosomal inheritance is controlled in *Cryptosporidium parvum*. However, as *Cryptosporidium parvum* is the apicomplexan, which lacks an apicoplast, direct contact between the mitosomes and mitotic spindle or other microtubular scaffolding might be expected.

There is still no information available on the movement and segregation of the mitosomes of *E. histolytica*.

5 Physiological Functions

The selection pressure orchestrating the mitochondria-to-mitosome transition was different in different organisms. Nevertheless, it resulted in the formation of similar, albeit not identical organelles, which completely lost their energy metabolism and the majority of other mitochondrial functions over the course of reductive evolution. Unlike mitochondria and hydrogenosomes, mitosomes are not able to synthesize ATP. There is no evidence of the pyruvate dehydrogenase complex producing acetyl-CoA, components of the tricarboxylic acid cycle generating electrons or of the cytochrome-dependent respiratory chain required for oxidative phosphorylation. Pyruvate/ferredoxin oxidoreductase, which generates acetyl-CoA for substrate-level phosphorylation in hydrogenosomes as well as hydrogenase, seems to be also absent in mitosomes. The activity of the former enzyme was detected in non-sedimentable fractions of mitosome-bearing protists (Lindmark 1980; Reeves et al. 1977), although its association with some cellular structures cannot be excluded (Ellis et al. 1993; Rodriguez et al. 1998). There is no evidence of mitochondrial biosynthetic pathways, including the biosynthesis of heme, biotin, and cardiolipin, as well as for the urea cycle, fatty acid degradation, metabolism of amino acids, and nucleotides. The only known mitochondrion function to be retained in the vast majority of mitosomes is the biosynthesis of Fe-S clusters (Tachezy and Doležal 2007).

5.1 Iron-Sulfur Cluster Assembly Machinery

Fe-S clusters represent one of the most ancient ubiquitous prosthetic groups and are essential to the function of a number of proteins involved in numerous distinct metabolic pathways (Johnson and Dean 2004). Proteins containing Fe-S clusters in their active sites (Fe-S proteins) mediate electron transport, oxygen sensing, iron homeostasis, and various enzymatic catalyses. In bacteria, there are at least three independent systems responsible for Fe-S cluster assembly. The ISC (**i**ron-**s**ulfur **c**luster) assembly machinery provides Fe-S clusters for the maturation of various "housekeeping" Fe-S proteins (Zheng et al. 1998). The NIF (nitrogen-fixing) system is primarily dedicated to form specific Fe-S clusters in the nitrogenase of nitrogen-fixing bacteria (Kennedy and Dean 1992). Finally, the SUF (sulfur-mobilization) system repairs Fe-S clusters under oxidative stress and iron-restricted conditions (Takahashi and Tokumoto 2002). In eukaryotes, Fe-S cluster assembly is considered to be the only essential function of mitochondria (Lill et al. 1999). The machinery mediating this process was shown to be of an ISC type. Phylogenetic analyses indicate that the mitochondrial ISC machinery was inherited from the proteobacterial endosymbiont, which is consistent with the proposed origin of mitochondria (Tachezy et al. 2001). Components of the SUF system were found mostly in plastids and in the apicoplast of apicomplexan such as *P. falciparum* (Wilson et al. 2003; Pala et al. 2018). The SUF machinery was most likely inherited from cyanobacteria, the ancestors of plastids (Tachezy et al. 2001). However, some components of SUF system were acquired by various anaerobic protists that lack plastids via lateral gene transfer (LGT) (Karnkowska et al. 2016). Archamoebae are the only eukaryotes, which replaced mitochondrial ISC system by bacterial NIF system (Ali et al. 2004; van der Giezen et al. 2004; Nývltová et al. 2013).

5.1.1 The Mitochondrial Model

The current model for Fe-S cluster formation is based mainly on studies of *Saccharomyces cerevisiae* and human cells (reviewed in Lill and Mühlenhoff 2006; Lill et al. 2015; Stehling et al. 2014). The pyridoxal phosphate-dependent cysteine desulfurase IscS (Nfs1 in yeast) together with two partner proteins Isd11 and Acp1 forms a central complex catalyzing the release of sulfur from L-cysteine and formation of IscS-bound persulfide (-SSH) (Cory et al. 2017; Herrera et al. 2018; Li et al. 1999; Schwartz et al. 2000). Sulfur is then transferred via still poorly understand process to IscU, which constitutes a scaffold for the formation of a transient [2Fe-2S] cluster (Agar et al. 2000). This steps requires electrons provided by the NADPH-dependent ferredoxin reductase (FdxR) and [2Fe-2S] ferredoxin (Webert et al. 2014). The source of iron for the Fe-S cluster formation is not known. Frataxin, a mitochondrial protein whose impaired expression is associated with Friedreich ataxia in humans, is able to bind iron and interacts with the ISC assembly complex (Gerber et al. 2003; Shan et al. 2007). However, the role of frataxin in iron delivery is still controversial (Campuzano et al. 1996). In the next step, the [2Fe-2S] cluster is transferred with assistance of the Hsp70 (Ssq1) chaperone, J-type (Jac1) co-chaperone, and the nucleotide exchange factor Mge1 directly or via the monothiol glutaredoxin Grx5 to [2Fe-2S]apoproteins (Dutkiewicz et al. 2003). Grx5 binds the [2Fe-2S] cluster either as the homodimer or as the heterodimer with BolA family proteins (Bol1and Bol3) (Banci et al. 2014; Nasta et al. 2017). Formation of [4Fe-4S] cluster is dependent on coupling of two [2Fe-2S] clusters delivered by Grx5 on the complex IscA1-IscA2-Iba57 (Brancaccio et al. 2014; Sheftel et al. 2012). Transfer of [4Fe-4S] clusters from IscA2-IscA2-Iba57 complex to specific apoproteins is facilitated by Ind1 and Nfu (Py et al. 2012; Sheftel et al. 2009) that can coordinate transient [4Fe-4S] cluster and BolA family proteins that also contribute to the specific targeting of the cluster (Uzarska et al. 2016).

5.1.2 The Role of Mitochondria in the Maturation of Extramitochondrial Fe-S Proteins

Experiments on *S. cerevisiae* mutant strains, human cells as well as *T. brucei* deficient in mitochondrial Fe-S cluster assemblies indicated that the mitochondrial pathway is crucial for the biogenesis of cytosolic and nuclear Fe-S proteins (Kispal et al. 1999; Šmíd et al. 2006). Three mitochondrial membrane components were found to be essential for the biogenesis of extramitochondrial Fe-S proteins: Atm1, an inner mitochondrial membrane transporter (Kispal et al. 1999); Erv1, a sulfhydryl oxidase present in the intermembrane space of mitochondria (Lange et al. 2001); and the tripeptide glutathione, which represents a major free thiol pool (Sipos et al. 2002). The nature of the compound exported from mitochondria by Atm1 is not known; however, it most likely contains sulfur (compound X-S). Investigations of plant ortholog of Atm1 suggested that a compound X-S might be related to glutathione polysulfide (Schaedler et al. 2014).

Until now, nine cytosolic CIA components were characterized that contribute to the maturation of extramitochondrial Fe-S proteins (Paul and Lill 2015). Cfd1 (Roy et al. 2003) and Nbp35 (Hausmann et al. 2005) are essential soluble P-loop NTPases. They form a complex which acts as a scaffold for the formation of a transient [4Fe-4S] clusters (Netz et al. 2007). These proteins are paralogs and evolved most likely by gene duplication. Nbp35 possesses N-terminal Fe-S cluster that lack Cfd1 and is present in all eukaryotes, whereas patchy distribution in eukaryotic lineages was observed for Cfd1 (Tsaousis et al. 2014). Similarly to mitochondria, assembly of a Fe-S cluster in the cytosol is dependent on the electron supply that is facilitated by the NADPH-dependent diflavin protein Tah18 and Fe-S protein Dre2 (Netz et al. 2010). A WD40 repeat protein named Cia1, Cia2 with DUF59 domain, and HEAT repeat protein MMS19 then facilitates the incorporation of pre-assembled Fe-S clusters into their target proteins (Gari et al. 2012; Srinivasan et al. 2007; Stehling et al. 2018), which is dependent on interaction with Nar1 (human IOP1), a homologue of [FeFe] hydrogenases (Balk et al. 2005; Horner et al. 2000).

5.1.3 Fe-S Cluster Assembly in Mitosomes

The central role of mitochondria in mediating the biosynthesis of Fe-S clusters in unicellular eukaryotes is noteworthy. It is this physiological function which may provide a critical selective pressure to retain highly reduced mitochondria over the course of the protist's adaptation to oxygen-poor niches and a parasitic lifestyle.

Giardia intestinalis and Other Diplomonads

The identification of the IscS gene in *Giardia intestinalis* provided the first evidence for the presence of the ISC machinery in this parasite (Tachezy et al. 2001). A phylogenetic reconstruction revealed that *Giardia* IscS is related to its mitochondrial homologues (Emelyanov 2003; Tachezy et al. 2001). Following cell localization of IscU and IscS by immunofluorescence and immunoelectron microscopy led to the discovery of *Giardia* mitosomes (Tovar et al. 2003). The presence of [2Fe2S] ferredoxin in the same compartment was later demonstrated by expression of the recombinant protein in *Giardia* (Doležal et al. 2005). Gene-mining and a proteomic analysis identified the presence of Grx5 in mitosome-enriched cell fraction together with IscS, IscU, IscA, Nfu, Hsp70, and Jac1 (Jedelsky et al. 2011). It has been demonstrated that mitosomal Grx5 in the form of homodimer is able to coordinate the [2Fe-2S] cluster, which is consistent with its role in mitochondria (Rada et al. 2009). Recent identification of BolA-like proteins in *Giardia* mitosomes rises a question whether the mitosomal Grx5 can form a heterodimer with BolA-like protein to coordinate [2Fe-2S] clusters as in yeast and human mitochondria (Stairs et al. 2018). Thus, the *Giardia* mitosomes contain key components of the Fe-S cluster assembly machinery (Table 1), and, importantly, the ability of the mitosome-rich fraction to form a Fe-S cluster in apoprotein was demonstrated (Tovar et al. 2003). However, some important components were not found in *Giardia* such as IscS partner proteins Isd11 and Acp1 and frataxin. An intriguing question is identifying the source of ATP and reducing equivalents, which are required for Fe-S cluster assembly, since no ATP-generating pathway that could supply the ISC assembly machinery with electrons has so far been found in the mitosomes. Although the electron carrier [2Fe-2S] ferredoxin is in mitosomes, FdxR that provides electron to ferredoxin in mitochondria is absent. A novel mitosomal NADPH-dependent diflavin oxidoreductase GiOR-1 has been identified in *Giardia*. This protein is able to utilize NADPH for reduction of various electron acceptors such as cytochrome b_5 however not ferredoxin (Jedelsky et al. 2011). Therefore, a physiological function of GiOR-1 is currently unknown. Collectively, the source of ATP that is most likely imported from the cytosol to mitosomes and the character of electron transport remains to be elucidated.

Orthologs of the mitochondrial membrane export system (Atm1, Erv1) have not been identified in the *Giardia* genome. The formation of glutathione was not detected in this organism (Brown et al. 1993) although its presence cannot be excluded as genes coding two enzymes involved in glutathione synthesis, glutamate cysteine ligase, and glutathione synthase have been found in the genome (Rada et al. 2009). Although it is likely that the mitosomal Fe-S cluster assembly is required for the maturation of extramitosmal Fe-S proteins, it is not clear which components of mitosomal membranes are involved in the export of X-S compound of the mitosomal Fe-S cluster assembly machinery for CIA system. Interestingly, cell localization studies of CIA components revealed dual localization for two of three Nbp35 paralogs (Nbp35-1 and Nbp35-2) in the cytosol and in association with outer mitosomal membrane. Moreover, Cia2 has been found mostly in the mitosomal intermembrane space (Pyrih et al. 2016). Based on these studies, it has been proposed that Cia2 and Nbp35 proteins in *Giardia* might represent a novel connect between ISC and CIA pathways.

Table 1 Components of Fe-S cluster assembly in "amitochondriate" eukaryotes

S. cerevisiae	*T. vaginalis*	*G .intestinalis*	*E. histolytica*	*E. cuniculi*	*C. parvum*	*M. mackini*
Mitochondrial type of iron-sulphur cluster assembly machinery						
IscS (Nfs1)	+	+	NIFS	+	+	+
Isd11	+	-	-	-	-	*
ACP1	-	-	-	-	-	
IscU (Isu1/Isu2)	+	+	NIFU	+	+	+
IscA (Isa1/Isa2)	+	+	-	-	-	*
Iba57	-	-	-		-	*
Ferredoxin (Yah1)	+	+	+	+	+	*
Ferredoxin reductase (Arh1)	-	-	-	+	+	+
Frataxin (Yfh1)	+	-	-	+	+	*
Nfu	+	+	-	-	-	*
Ind1	+	-	-		-	*
BolA	-	+	-	-	-	*
Grx5	-	+	-	-	+	*
mitHsp70 (Ssq1)	+	+	+	+	+	+
Jac1	+	+	-	+	+	*
Mge1	+	+	-	-	+	*
Export machinery in mitochondrial membrane						
Atm1	-	-	-	?	?	*
Erv1	-	-	-	-	-	*
Cytosolic components						
Cfd1	+	+	+	+	-	*
Nbp35	+	+	+	+	+	*
Nar1	+	-	+	+	+	*
Cia1	+	+	+	+	+	*
Cia2	+	+	+	+	+	*
MMS19	-	-	+	-	-	*
Tah18	-	-	-	+	+	*
Dre2	-	-	-	+	+	*

mitHsp70 (Ssq1) proteins cannot be distinguished from other mitHsp70 isoforms
+ high sequence similarity, ? limited sequence similarity, * partial transcriptomic data, – no similarity found

In addition to *Giardia*, it is likely that mitosomes are present in related diplomonad the fish parasite *Spironucleus vortens*. A recent genomic survey of this organism indicated the presence of IscS, IscU, [2Fe-2S] ferredoxin, and frataxin in this parasite, which might be localized into putative mitosomes (Stoklasa, 2019, Mitochondrial-related organelles in diplomonads, Diploma thesis, Charles University, Czech Republic).

Microsporidia

Analysis of the *Encephalitozoon cuniculi* and *Trachipleistophora hominis* genome sequences (Katinka et al. 2001; Heinz et al. 2012) revealed that most of ISC proteins are conserved in these microsporidian parasites including putative components of the export machinery (Table 1) (Freibert et al. 2017). The missing components are proteins required for formation of [4Fe-4S] clusters, which is consistent with absence of target [4Fe-4S]proteins such as aconitase, respiratory complexes as well as pyruvate/ferredoxin oxidoreductase in mitosomes. Because of lack of systems for genetic manipulation of microsporidia, *Saccharomyces cerevisiae* cells were used for functional investigations of microsporidian ISC components. Initially, several putative ISC proteins were expressed in yeast to investigate their targeting into mitochondria. However, only few proteins appeared in mitochondria. Specifically, the scaffold protein IscU of *Encephalitozoon cuniculi* and three ISC components, [2Fe2S] ferredoxin, FdxR, and frataxin of *Antonospora locustae*, were correctly targeted to yeast mitochondria, whereas *Encephalitozoon cuniculi* Atm1 and Erv1 and *Antonospora locustae* Hsp70, IscS, and IscU were mislocalized (Burri et al. 2006). These results most likely reflect differences in targeting signals. Most of microsporidian proteins lack predictable N-terminal targeting signal and possibly contain poorly understood internal signals, of these only some are recognized by yeast import machinery. Presence of IscS, IscU, and frataxin in mitosomes of *Encephalitozoon cuniculi* was later confirmed using homologous antibodies. Curiously, parallel experiments with *Trachipleistophora hominis* suggested that IscS is in mitosomes as expected; however, IscU and frataxin were localized to the cytosol (Goldberg et al. 2008). The unusual distribution of key ISC subunits led to the suggestion that persulfide compound generated by IscS in mitosomes might be exported and delivered to the cytosolic IscU in this microsporidia. However, these observations were not confirmed by later studies that revealed mitosomal localization of *Trachipleistophora* IscU and frataxin using immunoelectron microscopy. The mitosomal localization of ISC proteins was further confirmed by biochemical analysis of mitosome-enriched fraction that was isolated from *Trachipleistophora*-infected rabbit kidney cells (Freibert et al. 2017). For further functional studies in yeasts, the microsporidian ISC proteins were expressed with fungal targeting presequences to be efficiently delivered to the mitochondria (Goldberg et al. 2008). Expression of *Trachipleistophora* IscU and Isd11 in yeast mutants with impaired endogenous IscU1/2 and Isd11, respectively, fully rescued the growth defect of the mutant strain and maturation of yeast Fe-S proteins. By contrast,

Trachipleistophora IscS, [2Fe-2S] ferredoxin, FdxR, and frataxin did not complement the corresponding yeast homologues and expression of *Trachipleistophora* Hsp70 which reduced the growth rate of Ssq1 (yeast Hsp70 member of ISC machinery)-depleted yeast cells. Parallel experiments with *Encephalitozoon* ISC components showed that frataxin and Grx5 can replace the yeast homologues, whereas IscU and IscS did not complement function of corresponding yeast proteins (Goldberg et al. 2008). The reason for inability of some microsporidian components to rescue yeast mutants is likely due to high divergency of these proteins and consequently incompatibility with endogenous yeast ISC proteins to form functional complexes. The direct proof for function of *Trachipleistophora* ISC components provided in vitro reconstitution test. Five recombinant *Trachipleistophora* ISC proteins (IscS-Isd11, IscU, ferredoxin, and frataxin) together with human recombinant FdxR in the presence of reduced iron, cysteine, and NADPH supported formation of a [2Fe-2S] cluster on IscU dimmer. Reconstitution of [2Fe-2S] decreased when any of the ISC components was omitted (Freibert et al. 2017).

All these data confirmed that microsporidian mitosomes are involved in Fe-S cluster assembly. However, the role of predicted components for the export of X-S compound to the cytosol is still unclear. Searches in *Trachipleistophora hominis* genome identified three candidates for Atm1, sulfhydryl oxidase as well as two genes for enzymes of glutathione synthesis pathway (γ-glutamylcysteine synthase and glutathione synthase). Nevertheless, none of Atm1 candidates complemented the yeast Atm1 and predicted sulfhydryl oxidase appeared to be an enzyme of endoplasmic reticulum (Erv2), whereas mitosomal paralog Erv1 seems to be absent (Freibert et al. 2017). Thus, connection between mitosomal ISC machinery and the CIA remains to be elucidated.

Cryptosporidium spp.

Cryptosporidium spp. provide a nice example of closely related organisms that underwent reductive evolution of mitochondria to various extents as predicted based on genome analyses (Liu et al. 2016). Gastric species *Cryptosporidium andersoni* and *Cryptosporidium muris* possess mitochondria with pathways typical for aerobic metabolism including TCA cycle and membrane-associated respiratory chain, although it lacks cytochromes and oxidative phosphorylation. However, there is no evidence that these organelles possess a mitochondrial genome. Mitochondria of *Cryptosporidium parvum* and *Cryptosporidium hominis* are highly reduced and most likely lost energy metabolism, a feature corresponding to mitosomes. The possible ATP generation is discussed elsewhere (chapter "The Mitochondrion-Related Organelles of *Cryptosporidium* Species"). In addition to Fe-S cluster assembly, it has been predicted that *Cryptosporidium parvum* and *Cryptosporidium hominis* mitosomes possess at least two more pathways, ubiquinone and polyisoprenoid biosynthesis (Liu et al. 2016). The most reduced mitochondria have been predicted in *Cryptosporidium ubiquitum*. These organelles contain Fe-S cluster assembly as the only complete mitosomal pathway (Liu et al. 2016).

Fe-S assembly machinery has been investigated in some details in *Cryptosporidium parvum*. Protein sequence analyses revealed that ISC components possess predictable N-terminal targeting presequences that are consistent with the presence of MPP (Henriquez et al. 2005). To test their function, the presequences of IscS and IscU were fused at C-terminus with green fluorescence protein and expressed in yeast (LaGier et al. 2003). Both presequences delivered GFP to yeast mitochondria that strongly suggest their ability to deliver protein also into mitosomes (LaGier et al. 2003). Localization of IscS, IscU, and frataxin in *Cryptosporidium parvum* mitosomes were confirmed in various life cycle stages using specific antibodies. Moreover, it has been shown that *Cryptosporidium parvum* IscS can functionally substitute the yeast homologue (chapter "The Mitochondrion-Related Organelles of *Cryptosporidium* Species"; Miller et al. 2018).

Mikrocytos mackini

Mikrocytos mackini is a microcell parasite of oysters that belong to the group Rhizaria. Ultrastructural studies did not reveal a presence of mitochondria, and consequently amitochondrial status of this organism has been suggested (Hine et al. 2001). However, sequencing of *Mikrocytos mackini* transcriptome identified four ISC candidates, IscU, IscS, mtHsp70, and FdxR (Burki et al. 2013). All these proteins possess predictable N-terminal presequences. Other mitochondrial components were not identified. Based on this analysis of partial transcriptome, it has been predicted that *M. mackini* possess mitosomes. However, more sequencing and experimental data are required to confirm this prediction and to characterize putative *M. mackini* mitosome.

Entamoeba histolytica and Related Organisms

Archamoebae is the eukaryotic lineage in which homologues of ISC machinery have not been found. Instead of an ISC system, these organisms including *E. histolytica* possess components of a nonredundant NIF system. Two components, NifS and NifU, were shown to be expressed and active in this parasite (Ali et al. 2004). In addition to *Entamoeba histolytica*, homologues of NIF components are present in the genomes of other species of the *Entamoeba* genus such as the reptile parasite *Entamoeba invadens*, mostly free-living *Entamoeba moshkovskii*, gastrointestinal pathogen of nonhuman primates *Entamoeba nuttalli* (http://amoebadb.org/amoeba/) as well as in the related free-living protist *Mastigamoeba balamuthi* (Nývltová et al. 2013). A phylogenetic reconstruction of *E. histolytica* and *M. balamuthi* NifS and NifU indicated that both genes were acquired by horizontal gene transfer from the group of most likely environmental ε-proteobacteria (Gill et al. 2007). When expressed in a strain of *Escherichia coli* with deletions of both the ISC and SUF systems, NifU and NifS of *E. histolytica* complement the growth of the mutant bacteria (Ali et al. 2004). This result suggests that NIF components can catalyze general Fe-S cluster assembly in *Entamoeba histolytica*, as was shown for some non-diazotrophic bacteria (Olson et al. 2000). Importantly, the analysis of entamoebid NifS and NifU protein sequences did not reveal the presence of N-terminal targeting presequences that would suggest their localization within the mitosomes (van der Giezen et al. 2004). Interestingly, *M. balamuthi* has two gene copies for each NifS and NifU gene; one copy encoded a paralog with and the second copy without the targeting presequence. The presence and the absence of *M. balamuthi* targeting sequences correspond with the organellar and the cytosolic localization of corresponding proteins, respectively (Nývltová et al. 2013). Mitosomal localization of NifS and NifU in *E. histolytica* is unlikely although dual localization in the cytosol and mitosomes have been suggested (Maralikova et al. 2009). However, mitosomal localization of the NIF machinery has not been confirmed by neither proteomic analysis (Mi-ichi et al. 2009) nor following cell localization studies (Doležal et al. 2010; Nývltová et al. 2013). Moreover, the *E. histolytica* NIF components were found mainly in the cytosol and partially attached to organellar membranes facing the cytosol upon expression in *Saccharomyces cerevisiae* (Nývltová et al. 2013). Therefore, the entamoebid mitosomes are the only mitochondria-derived organelles so far, which entirely lost the ability to synthesize Fe-S clusters.

5.2 Requirements for ATP, Membrane Potential and Electron Transport

In mitosome-harboring organisms, the generation of ATP is exclusively dependent on cytosolic energy metabolism. Mitosomes are neither able to synthesize ATP by oxidative nor substrate-level phosphorylation. However, ATP hydrolysis is required for the functioning of mtHSP70 and CPN60, which are present in mitochondria as well as all derived organelles, including mitosomes. In mitochondria, they are involved in protein import, protein maturation as well as the formation of Fe-S clusters. It is likely that similar demands for ATP are required for corresponding functions in mitosomes. If so, ATP has to be imported from the cytosol to the organelles to satisfy mitosomal needs. Indeed, genes for the ATP/ADP transporter have been identified in *Cryptosporidium parvum* (Abrahamsen et al. 2004; Katinka et al. 2001). An unusual member of the mitochondrial carrier family (MCF) has been found to mediate ADP/ATP transport in *Entamoeba histolytica* (Chan et al. 2005). Unlike mitochondrial carriers, the *Entamoeba* carrier is not sensitive to the classic inhibitors carboxyatractyloside and bongkrekic acid and also exhibited differences in its substrate preferences. The transport is mediated via a specific mechanism that is not dependent on a membrane potential. In addition to ATP and ADP, which is transported by mitochondrial carriers, the *Entamoeba* carrier also transports AMP. A unique bacterial-like ADP/ATP nucleotide transporters (NTT) were characterized in the intracellular parasite *Encephalitozoon cuniculi*. Four NTT genes were identified, of which three NTTs were localized at the parasite surface that most likely allows to steal ATP directly from the cytosol of the host cell. The forth gene was localized in the mitosome, and its involvement in ATP import to these organelles has been suggested (Tsaousis et al. 2008). Neither genome analysis nor proteomic studies reveal any candidate for ATP/ADP transport in *Giardia intestinalis* mitosomes.

Mitochondrial protein import is dependent on the membrane potential, which is mainly generated by cytochrome-dependent electron transport. In mitosome-carrying protists, however, neither cytochromes nor other components of the membrane-associated respiratory chain have been found, and it is not clear whether and if so how the mitosomal membrane potential is generated. The only candidate that might be involved in mitosomal transmembrane proton translocation is pyridine-nucleotide transhydrogenase (PNT). This proton pump, typically present in the inner mitochondrial membrane, catalyzes direct transfer of the hydride ion between NAD (H) and NADP(H) providing intramitochondrial NADPH (Hatefi and Yamaguchi 1996). In helminths such as *Hymenolepis diminuta* that enter the anaerobic phase of their life cycle, PNT is the principle proton pump generating a membrane potential. It operates in the reverse direction, translocating protons from the matrix to the intramembrane mitochondrial space (Mercer et al. 1999). The genes coding for PNT have been found in the genomes of *Entamoeba histolytica* and *Cryptosporidium parvum* (Putignani et al. 2004; Yu and Samuelson 1994). Catalytic activity was demonstrated in the former organisms. Interestingly, PNT activity was associated with a sedimentable cellular fraction. Together with the observation that the amino terminus of *E. histolytica* PNT possesses an extension similar to targeting presequences, these findings suggested that PNT may operate in mitosomes. However, later localization studies revealed that PNT is in organelles other than mitosomes such as lysosomes and phagosomes (Yousuf et al. 2010).

In *Giardia*, the presence of membrane-bound structures displaying membrane potential and electron transport was tested using the cationic dye rhodamine 123 and 5-cyano-2,3-ditolyl tetrazolium chloride (CTC), respectively (Lloyd et al. 2002). Rhodamine 123 was accumulated in spherical and oval structures adjacent to the inner face of the *Giardia* plasma membrane. CTC is an artificial electron acceptor, which reduces to fluorescent insoluble formazan. The formation of formazan was observed in several membrane structures under the *Giardia* plasma membrane. However, the CTC-labeled structures were visibly distinct from those labeled with rhodamine 123. Although both types of labeling were considered evidence of the presence of the membrane system with mitochondrion-like functions in *Giardia*, the specificity of the labeling in *Giardia* and the character of the observed structures need to be validated. First, it is clear that the rhodamine 123-labeled structures are distinct from mitosomes: they are considerably larger than mitosomes (1.3–2 μm), and their number (4–5 per cell) and cellular distribution are different. The specificity of CTC labeling is questionable, as the reduction of tetrazolium salts is mediated by a number of non-mitochondrial enzymes and electron carriers, and formazan might be also unspecifically formed by superoxide radicals (Bernas and Dobrucki 2000; Breeuwer and Abee 2000). Moreover, neither rhodamine 123 nor CTC labeled the mitosomal rodlike structure, which is invariantly observed between *Giardia* nuclei

(Doležal et al. 2005; Regoes et al. 2005; Tovar et al. 2003). Thus, co-localization studies with specific markers are required to clarify the relationship between rhodamine 123- and CTC-labeled structures and mitosomes.

Indirect evidence of some electron transport in mitosomes indicates the presence of [2Fe2S] ferredoxin. This electron carrier is typically present in mitochondria and hydrogenosomes, providing electrons for various pathways including Fe-S cluster formations. The mitochondrial type of ferredoxin has been found in the genomes of *Encephalitozoon cuniculi* and *Cryptosporidium parvum* together with partner protein FdxR that can transfer electrons from NADPH (Lei et al. 2010). In the former organism, the mitosomal localization of ferredoxin and FdxR has been experimentally verified (Freibert et al. 2017), whereas cellular localization in *Cryptosporidium parvum* remains to be determined. *Giardia* mitosomes possess [2Fe2S] ferredoxin; however, there is not known any electron donor. Generally, the source of reducing equivalents for electron transport via ferredoxins within mitosomes is unknown. Interestingly, [2Fe2S] ferredoxin is absent in *Entamoeba histolytica*, which is consistent with the replacement of ISC by the NIF type of Fe-S cluster assembly machinery in this parasite. In the NIF machinery, the function of [2Fe-2S] ferredoxin is likely substituted by a ferredoxin-like module of NifU (Agar et al. 2000).

5.3 Sulfate Activation and Other Mitosomal Functions

More information about the functions of mitosomes was expected to be obtained from an analysis of the complete genomes of mitosome-bearing protists and following proteomic studies. The most unexpected discovery was the identification of a sulfate activation pathway in mitosome of *E. histolytica* (Mi-ichi et al. 2009). Proteomic analysis of partially purified mitosome identified 95 putative mitosomal proteins including three components of a sulfate activation, namely, ATP sulfurylase (AS) that converts inorganic sulfate and ATP to adenosine-5′-phosphosulfate (APS) and pyrophosphate, APS kinase (APSK) that coverts APS to 3′ phosphoadenosinc-5′-phosphosulfate (PAPS), and inorganic pyrophosphatase (IPP) to break down pyrophosphate to phosphates. This pathway is present also in *Mastigamoeba balamuthi* hydrogenosomes suggesting that enzymes for sulfate activation have been gained by a common ancestor after split of Archamoebae and Mycetozoa (Nývltová et al. 2013). PAPS synthesized in mitosomes is then exported to the cytosol and may serve as a sulfuryl donor for various acceptors such as lipids, proteins, and small organic molecules. The export of PAPS is likely mediated by MCF that has the capacity to exchange PAPS with ATP (Mi-ichi et al. 2015b). In the cytosol, PAPS is used as a substrate by sulfotransferases (SULTs). There are ten homologous genes for SULTs in the *Entamoeba histolytica* of which SULT6 was shown to catalyze formation of cholesteryl sulfate (Mi-ichi et al. 2015a). Synthesis of this sulfolipid was shown to be associated with the encystation of *Entamoeba invadens*, which is an essential process for the pathogen transmission. These findings underscore an important function of mitosome for vital function of *Entamoeba* parasite. Interestingly, acquisition of SULTs was suggested to be specific for parasitic *Entamoeba* species. However, our recent analysis of *Mastigamoeba balamuthi* genome revealed presence of at least four genes for SULTs in these free-living protists (our unpublished data). What is a role of *Mastigamoeba* SULTs remains to be elucidated.

The genomes of *Cryptosporidium parvum*, *Cryptosporidium hominis*, and *Cryptosporidium ubiquitum* revealed the presence of genes coding for putative components associated with mitochondrial membrane electron transport and ATP synthesis. There are gene coding for mitosomal alternative NADH dehydrogenase, which replaced respiratory complex I in most apicomplexans, two subunits (α and β) of multisubunit F-type ATP synthase, and malate:quinone oxidoreductase. What is a role of these isolated components of various biochemical pathways in *Cryptosporidium ubiquitum* is currently unknown. However, *Cryptosporidium parvum* and *Cryptosporidium hominis* possess in addition to alternative NADH dehydrogenase also membrane-associated alternative oxidase (AOX) and enzymes for ubiquinone biosynthesis. These predictions suggested that in mitosomes, a short membrane-associated electron chain may operate including electron transfer from NADH via NADH dehydrogenase and ubiquinone to AOX. Then AOX can transfer electrons to molecular oxygen and may contribute to the scavenging of toxic oxygen species (Putignani et al. 2004; Liu et al. 2016).

6 Perspectives

Since 1999 when the first mitosome was discovered in *Entamoeba histolytica*, a number of genomic, proteomic, and biochemical studies have been performed to characterize these mysterious organelles. The most exciting question was and still is what the essential function is that forces organisms in various eukaryotic lineages to retain mitosomes over the course of their reductive evolution. Such an essential function might not be identical in all mitosomes as documented by unexpected discovery of unique sulfate activation pathway in *Entamoeba histolytica* mitosomes. The formation of Fe-S clusters might be the essential function, common to mitosomes in *Giardia intestinalis*, *Spironucleus vortens*, microsporidians, and *Cryptosporidium* spp. It has been speculated that the compartmentalization of this function might be essential to avoid the toxicity of ferrous iron and sulfide, which are needed for Fe-S cluster assembly (Tachezy and Doležal 2007). However, absence of Fe-S cluster assembly in *Entamoeba histolytica*, which synthesizes Fe-S clusters in the cytosol, as well as complete absence of mitochondria in *Monocercomonoides exilis* challenged this hypothesis. Interestingly both entamoebids and *Monocercomonoides exilis* lost the mitochondrial ISC machinery, which was replaced either by simple NIF or SUF components in the cytosol, respectively. These cytosolic systems need somehow to deal with the toxicity of ferrous iron and sulfide in the cytosol; however, the mechanism is not known. Moreover the cytosolic NIF and SUF components most likely interact directly with CIA machinery to form Fe-S clusters and thus replace the function of mitochondrial ISC machinery in providing still unknown substrate X-S directly to CIA system. In mitosomes with ISC machinery, it is still not clear whether and how these organelles are involved in export of X-S compound for CIA machinery. Although involvement of mitosomes in the cytosolic Fe-S cluster formation is likely, components of the mitochondrial export machinery are mostly absent in mitosomes. Another open question is what the source of electrons is and which components of expected electron transport provide reductive equivalents for the mitosomal Fe-S cluster assembly. Moreover, because of absence of organellar genome, mitosomes need to import all matrix protein for Fe-S assembly and other pathways via TOM/TIM complexes to cross both mitosomal membranes. However, what the main force for protein import is upon absence of the membrane potential remains unknown. Many more functional studies are required to clarify these questions and to decipher mitosomal functions that are still at the beginning of our understanding.

Acknowledgments Research in the JT laboratory is supported by the Ministry of Education, Youth and Sports of the Czech Republic project NPU II (LQ1604) and by ERD Funds project CePaViP (CZ.02.1.01/0.0/0.0/16_019/0000759).

References

Abrahamsen MS, Templeton TJ, Enomoto S, Abrahante JE, Zhu G, Lancto CA, Deng M, Liu C, Widmer G, Tzipori S, Buck GA, Xu P, Bankier AT, Dear PH, Konfortov BA, Spriggs HF, Iyer L, Anantharaman V, Aravind L, Kapur V (2004) Complete genome sequence of the apicomplexan, *Cryptosporidium parvum*. Science 304:441–445

Adam RD (2001) Biology of *Giardia lamblia*. Clin Microbiol Rev 14:447–475

Adams KL, Palmer JD (2003) Evolution of mitochondrial gene content: gene loss and transfer to the nucleus. Mol Phylogenet Evol 29:380–395

Agar JN, Yuvaniyama P, Jack RF, Cash VL, Smith AD, Dean DR, Johnson MK (2000) Modular organization and identification of a mononuclear iron-binding site within the NifU protein. J Biol Inorg Chem 5:167–177

Alcock F, Webb CT, Dolezal P, Hewitt V, Shingu-Vasquez M, Likic VA, Traven A, Lithgow T (2012) A small Tim homohexamer in the relict mitochondrion of *Cryptosporidium*. Mol Biol Evol 29:113–122

Ali V, Shigeta Y, Tokumoto U, Takahashi Y, Nozaki T (2004) An intestinal parasitic protist, *Entamoeba histolytica*, possesses a non-redundant nitrogen fixation-like system for iron-sulfur cluster assembly under anaerobic conditions. J Biol Chem 279:16863–16874

Andersson JO, Sjogren AM, Horner DS, Murphy CA, Dyal PL, Svard SG, Logsdon JM Jr, Ragan MA, Hirt RP, Roger AJ (2007) A genomic survey of the fish parasite *Spironucleus salmonicida* indicates genomic plasticity among diplomonads and significant lateral gene transfer in eukaryote genome evolution. BMC Genomics 8:51

Arisue N, Sachez LB, Weiss LM, Müller M, Hashimoto T (2002) Mitochondrial-type hsp70 genes of the amitochondriate protists, *Giardia intestinalis*, *Entamoeba histolytica* and two microsporidians. Parasitol Int 51:9–16

Balk J, Pierik AJ, Aguilar Netz DJ, Mühlenhoff U, Lill R (2005) Nar1p, a conserved eukaryotic protein with similarity to Fe-only hydrogenases, functions in cytosolic iron-sulphur protein biogenesis. Biochem Soc Trans 33:86–89

Banci L, Brancaccio D, Ciofi-Baffoni S, Del CR, Gadepalli R, Mikolajczyk M, Neri S, Piccioli M, Winkelmann J (2014) [2Fe-2S] cluster transfer in iron-sulfur protein biogenesis. Proc Natl Acad Sci U S A 111:6203–6208
Beech PL, Nheu T, Schultz T, Herbert S, Lithgow T, Gilson PR, McFadden GI (2000) Mitochondrial FtsZ in a chromophyte alga. Science 287:1276–1279
Benz C, Stribrna E, Hashimi H, Lukeš J (2017) Dynamin-like proteins in *Trypanosoma brucei*: a division of labour between two paralogs? PLoS One 12:e0177200
Bernas T, Dobrucki JW (2000) The role of plasma membrane in bioreduction of two tetrazolium salts, MTT, and CTC. Arch Biochem Biophys 380:108–116
Best AA, Morrison HG, McArthur AG, Sogin ML, Olsen GJ (2004) Evolution of eukaryotic transcription: insights from the genome of *Giardia lamblia*. Genome Res 14:1537–1547
Bleazard W, McCaffery JM, King EJ, Bale S, Mozdy A, Tieu Q, Nunnari J, Shaw JM (1999) The dynamin-related GTPase Dnm1 regulates mitochondrial fission in yeast. Nat Cell Biol 1:298–304
Boldogh IR, Fehrenbacher KL, Yang HC, Pon LA (2005) Mitochondrial movement and inheritance in budding yeast. Gene 354:28–36
Brancaccio D, Gallo A, Mikolajczyk M, Zovo K, Palumaa P, Novellino E, Piccioli M, Ciofi-Baffoni S, Banci L (2014) Formation of [4Fe-4S] clusters in the mitochondrial iron-sulfur cluster assembly machinery. J Am Chem Soc 136:16240–16250
Breeuwer P, Abee T (2000) Assessment of viability of microorganisms employing fluorescence techniques. Int J Food Microbiol 55:193–200
Brown DM, Upcroft JA, Upcroft P (1993) Cysteine is the major low-molecular weight thiol in *Giardia duodenalis*. Mol Biochem Parasitol 61:155–158
Bui ET, Bradley PJ, Johnson PJ (1996) A common evolutionary origin for mitochondria and hydrogenosomes. Proc Natl Acad Sci U S A 93:9651–9656
Burger G, Gray MW, Forget L, Lang BF (2013) Strikingly bacteria-like and gene-rich mitochondrial genomes throughout jakobid protists. Genome Biol Evol 5:418–438
Burki F, Corradi N, Sierra R, Pawlowski J, Meyer GR, Abbott CL, Keeling PJ (2013) Phylogenomics of the intracellular parasite *Mikrocytos mackini* reveals evidence for a mitosome in rhizaria. Curr Biol 23:1541–1547
Burri L, Williams BA, Bursac D, Lithgow T, Keeling PJ (2006) Microsporidian mitosomes retain elements of the general mitochondrial targeting system. Proc Natl Acad Sci U S A 103:15916–15920
Campuzano V, Montermini L, Molto MD, Pianese L, Cossee M, Cavalcanti F, Monros E, Rodius F, Duclos F, Monticelli A, Zara F, Canizares J, Koutnikova H, Bidichandani SI, Gellera C, Brice A, Trouillas P, DeMichele G, Filla A, DeFrutos R, Palau F, Patel PI, DiDonato S, Mandel JL, Cocozza S, Koenig M, Pandolfo M (1996) Friedreich's ataxia: autosomal recessive disease caused by an intronic GAA triplet repeat expansion. Science 271:1423–1427
Cavalier-Smith T (1987a) Eukaryotes with no mitochondria. Nature 326:332–333
Cavalier-Smith T (1987b) The origin of eukaryotic and archaebacterial cells. Ann N Y Acad Sci 503:17–54
Chan KW, Slotboom DJ, Cox S, Embley TM, Fabre O, van der Giezen M, Harding M, Horner DS, Kunji ERS, Leon-Avila G, Tovar J (2005) A novel ADP/ATP transporter in the mitosome of the microaerophilic human parasite *Entamoeba histolytica*. Curr Biol 15:737–742
Chanez AL, Hehl AB, Engstler M, Schneider A (2006) Ablation of the single dynamin of *T. brucei* blocks mitochondrial fission and endocytosis and leads to a precise cytokinesis arrest. J Cell Sci 119:2968–2974
Clark CG, Roger AJ (1995) Direct evidence for secondary loss of mitochondria in *Entamoeba histolytica*. Proc Natl Acad Sci U S A 92:6518–6521
Clayton CE, Michels P (1996) Metabolic compartmentation in African trypanosomes. Parasit Today 12:465–471

Cory SA, Van Vranken JG, Brignole EJ, Patra S, Winge DR, Drennan CL, Rutter J, Barondeau DP (2017) Structure of human Fe-S assembly subcomplex reveals unexpected cysteine desulfurase architecture and acyl-ACP-ISD11 interactions. Proc Natl Acad Sci U S A 114:E5325–E5334

Dagley MJ, Dolezal P, Likic VA, Smid O, Purcell AW, Buchanan SK, Tachezy J, Lithgow T (2009) The protein import channel in the outer mitosomal membrane of *Giardia intestinalis*. Mol Biol Evol 26:1941–1947

de Graaf RM, Duarte I, van Alen TA, Kuiper JW, Schotanus K, Rosenberg J, Huynen MA, Hackstein JH (2009) The hydrogenosomes of *Psalteriomonas lanterna*. BMC Evol Biol 9:287

Doležal P, Šmíd O, Rada P, Zubáčová Z, Bursac D, Sutak R, Nebesářová J, Lithgow T, Tachezy J (2005) *Giardia* mitosomes and trichomonad hydrogenosomes share a common mode of protein targeting. Proc Natl Acad Sci U S A 102:10924–10929

Doležal P, Dagley MJ, Kono M, Wolynec P, Likic VA, Foo JH, Šedinová M, Tachezy J, Bachmann A, Bruchhaus I, Lithgow T (2010) The essentials of protein import in the degenerate mitochondrion of *Entamoeba histolytica*. PLoS Pathog 6:e1000812

Dutkiewicz R, Schilke B, Knieszner H, Walter W, Craig EA, Marszalek J (2003) Ssq1, a mitochondrial Hsp70 involved in iron-sulfur (Fe/S) center biogenesis – similarities to and differences from its bacterial counterpart. J Biol Chem 278:29719–29727

Ellis JE, Williams R, Cole D, Cammack R, Lloyd D (1993) Electron transport components of the parasitic protozoon *Giardia lamblia*. FEBS Lett 325:196–200

Emelyanov VV (2003) Phylogenetic affinity of a *Giardia lamblia* cysteine desulfurase conforms to canonical pattern of mitochondrial ancestry. FEMS Microbiol Lett 226:257–266

Flegontov P, Michalek J, Janouskovec J, Lai DH, Jirku M, Hajduskova E, Tomcala A, Otto TD, Keeling PJ, Pain A, Obornik M, Lukeš J (2015) Divergent mitochondrial respiratory chains in phototrophic relatives of apicomplexan parasites. Mol Biol Evol 32:1115–1131

Freibert SA, Goldberg AV, Hacker C, Molik S, Dean P, Williams TA, Nakjang S, Long S, Sendra K, Bill E, Heinz E, Hirt RP, Lucocq JM, Embley TM, Lill R (2017) Evolutionary conservation and in vitro reconstitution of microsporidian iron-sulfur cluster biosynthesis. Nat Commun 8:13932

Fry M, Beesley JE (1991) Mitochondria of mammalian *Plasmodium* spp. Parasitology 102:17–26

Fukasawa Y, Tsuji J, Fu SC, Tomii K, Horton P, Imai K (2015) MitoFates: improved prediction of mitochondrial targeting sequences and their cleavage sites. Mol Cell Proteomics 14:1113–1126

Fukasawa Y, Oda T, Tomii K, Imai K (2017) Origin and evolutionary alteration of the mitochondrial import system in eukaryotic lineages. Mol Biol Evol 34:1574–1586

Gaechter V, Schraner E, Wild P, Hehl AB (2008) The single dynamin family protein in the primitive protozoan *Giardia lamblia* is essential for stage conversion and endocytic transport. Traffic 9:57–71

Gakh O, Cavadini P, Isaya G (2002) Mitochondrial processing peptidases. Biochim Biophys Acta 1592:63–77

Gari K, Leon Ortiz AM, Borel V, Flynn H, Skehel JM, Boulton SJ (2012) MMS19 links cytoplasmic iron-sulfur cluster assembly to DNA metabolism. Science 337:243–245

Gerber J, Mühlenhoff U, Lill R (2003) An interaction between frataxin and Isu1/Nfs1 that is crucial for Fe/S cluster synthesis on Isu1. EMBO Rep 4:906–911

Germot A, Philippe H, Le Guyader H (1996) Presence of a mitochondrial-type 70-kDa heat shock protein in *Trichomonas vaginalis* suggests a very early mitochondrial endosymbiosis in eukaryotes. Proc Natl Acad Sci U S A 93:14614–14617

Germot A, Philippe H, Le Guyader H (1997) Evidence for loss of mitochondria in microsporidia from a mitochondrial-type HSP70 in *Nosema locustae*. Mol Biochem Parasitol 87:159–168

Ghosh S, Field J, Rogers R, Hickman M, Samuelson J (2000) The *Entamoeba histolytica* mitochondrion-derived organelle (crypton) contains double-stranded DNA and appears to be bound by a double membrane. Infect Immun 68:4319–4322

Gill EE, Diaz-Trivino S, Barbera MJ, Silberman JD, Stechmann A, Gaston D, Tamas I, Roger AJ (2007) Novel mitochondrion-related organelles in the anaerobic amoeba *Mastigamoeba balamuthi*. Mol Microbiol 66:1306–1320

Glater EE, Megeath LJ, Stowers RS, Schwarz TL (2006) Axonal transport of mitochondria requires milton to recruit kinesin heavy chain and is light chain independent. J Cell Biol 173:545–557
Goldberg AV, Molik S, Tsaousis AD, Neumann K, Kuhnke G, Delbac F, Vivares CP, Hirt RP, Lill R, Embley TM (2008) Localization and functionality of microsporidian iron-sulphur cluster assembly proteins. Nature 452:624–628
Guo X, Macleod GT, Wellington A, Hu F, Panchumarthi S, Schoenfield M, Marin L, Charlton MP, Atwood HL, Zinsmaier KE (2005) The GTPase dMiro is required for axonal transport of mitochondria to *Drosophila* synapses. Neuron 47:379–393
Hashimoto T (1998) Secondary absence of mitochondria in *Giardia lamblia* and *Trichomonas vaginalis* revealed by valyl-tRNA synthetase phylogeny. Proc Natl Acad Sci U S A 95:6860–6865
Hatefi Y, Yamaguchi M (1996) Nicotinamide nucleotide transhydrogenase: a model for utilization of substrate binding energy for proton translocation. FASEB J 10:444–452
Hausmann A, Aguilar Netz DJ, Balk J, Pierik AJ, Mühlenhoff U, Lill R (2005) The eukaryotic P loop NTPase Nbp35: an essential component of the cytosolic and nuclear iron-sulfur protein assembly machinery. Proc Natl Acad Sci U S A 102:3266–3271
Heinz E, Williams TA, Nakjang S, Noel CJ, Swan DC, Goldberg AV, Harris SR, Weinmaier T, Markert S, Becher D, Bernhardt J, Dagan T, Hacker C, Lucocq JM, Schweder T, Rattei T, Hall N, Hirt RP, Embley TM (2012) The genome of the obligate intracellular parasite *Trachipleistophora hominis*: new insights into microsporidian genome dynamics and reductive evolution. PLoS Pathog 8:e1002979
Henriquez FL, Richards TA, Roberts F, McLeod R, Roberts CW (2005) The unusual mitochondrial compartment of *Cryptosporidium parvum*. Trends Parasitol 21:68–74
Herrera MG, Pignataro MF, Noguera ME, Cruz KM, Santos J (2018) Rescuing the rescuer: on the protein complex between the human mitochondrial acyl carrier protein and ISD11. ACS Chem Biol 13:1455–1462
Hine PM, Bower SM, Meyer GR, Cochennec-Laureau N, Berthe FC (2001) Ultrastructure of *Mikrocytos mackini*, the cause of Denman Island disease in oysters *Crassostrea* spp. and *Ostrea* spp. in British Columbia, Canada. Dis Aquat Organ 45:215–227
Hirt RP, Logsdon JM, Healy B, Dorey MW, Doolittle WF, Embley TM (1999) Microsporidia are related to fungi: evidence from the largest subunit of RNA polymerase II and other proteins. Proc Natl Acad Sci U S A 96:580–585
Horner DS, Hirt RP, Kilvington S, Lloyd D, Embley TM (1996) Molecular data suggest an early acquisition of the mitochondrion endosymbiont. Proc R Soc Lond Ser B Biol Sci 263:1053–1059
Horner DS, Foster PG, Embley TM (2000) Iron hydrogenases and the evolution of anaerobic eukaryotes. Mol Biol Evol 17:1695–1709
Ingolia NT (2014) Ribosome profiling: new views of translation, from single codons to genome scale. Nat Rev Genet 15:205–213
Ingolia NT, Ghaemmaghami S, Newman JR, Weissman JS (2009) Genome-wide analysis in vivo of translation with nucleotide resolution using ribosome profiling. Science 324:218–223
Jedelsky PL, Dolezal P, Rada P, Pyrih J, Smid O, Hrdy I, Sedinova M, Marcincikova M, Voleman L, Perry AJ, Beltran NC, Lithgow T, Tachezy J (2011) The minimal proteome in the reduced mitochondrion of the parasitic protist *Giardia intestinalis*. PLoS One 6:e17285
Jerlstrom-Hultqvist J, Einarsson E, Xu F, Hjort K, Ek B, Steinhauf D, Hultenby K, Bergquist J, Andersson JO, Svard SG (2013) Hydrogenosomes in the diplomonad *Spironucleus salmonicida*. Nat Commun 4:2493
Johnson D, Dean D (2004) Structure, function, and formation of biological iron-sulfur clusters. Annu Rev Biochem 74:247–281
Karnkowska A, Vacek V, Zubáčová Z, Treitli SC, Petrzelkova R, Eme L, Novák L, Žárský V, Barlow LD, Herman EK, Soukal P, Hroudova M, Doležal P, Stairs CW, Roger AJ, Elias M, Dacks JB, Vlček C, Hampl V (2016) A eukaryote without a mitochondrial organelle. Curr Biol 26:1274–1284

Katinka MD, Duprat S, Cornillot E, Metenier G, Thomarat F, Prensier G, Barbe V, Peyretaillade E, Brottier P, Wincker P, Delbac F, El Alaoui H, Peyret P, Saurin W, Gouy M, Weissenbach J, Vivares CP (2001) Genome sequence and gene compaction of the eukaryote parasite *Encephalitozoon cuniculi*. Nature 414:450–453

Keeling PJ, Doolittle WF (1996) Alpha-tubulin from early-diverging eukaryotic lineages and the evolution of the tubulin family. Mol Biol Evol 13:1297–1305

Keithly JS, Langreth SG, Buttle KF, Mannella CA (2005) Electron tomographic and ultrastructural analysis of the *Cryptosporidium parvum* relict mitochondrion, its associated membranes, and organelles. J Eukaryot Microbiol 52:132–140

Kennedy C, Dean D (1992) The nifU, nifS and nifV gene products are required for activity of all three nitrogenases of *Azotobacter vinelandii*. Mol Gen Genet 231:494–498

Kispal G, Csere P, Prohl C, Lill R (1999) The mitochondrial proteins Atm1p and Nfs1p are essential for biogenesis of cytosolic Fe/S proteins. EMBO J 18:3981–3989

Komiya T, Rospert S, Koehler C, Looser R, Schatz G, Mihara K (1998) Interaction of mitochondrial targeting signals with acidic receptor domains along the protein import pathway: evidence for the 'acid chain' hypothesis. EMBO J 17:3886–3898

Komuniecki PR, Johnson J, Kamhawi M, Komuniecki R (1993) Mitochondrial heterogeneity in the parasitic nematode, *Ascaris suum*. Exp Parasitol 76:424–437

Korobova F, Ramabhadran V, Higgs HN (2013) An actin-dependent step in mitochondrial fission mediated by the ER-associated formin INF2. Science 339:464–467

LaGier MJ, Tachezy J, Stejskal F, Kutišová K, Keithly JS (2003) Mitochondrial-type iron-sulfur cluster biosynthesis genes (IscS and IscU) in the apicomplexan *Cryptosporidium parvum*. Microbiology 149:3519–3530

Lange H, Lisowsky T, Gerber J, Mühlenhoff U, Kispal G, Lill R (2001) An essential function of the mitochondrial sulfhydryl oxidase Erv1p/ALR in the maturation of cytosolic Fe/S proteins. EMBO Rep 2:715–720

Lei C, Rider SD Jr, Wang C, Zhang H, Tan X, Zhu G (2010) The apicomplexan *Cryptosporidium parvum* possesses a single mitochondrial-type ferredoxin and ferredoxin:NADP+ reductase system. Protein Sci 19:2073–2084

Leon-Avila G, Tovar J (2004) Mitosomes of *Entamoeba histolytica* are abundant mitochondrion-related remnant organelles that lack a detectable organellar genome. Microbiology 150:1245–1250

Li J, Kogan M, Knight SA, Pain D, Dancis A (1999) Yeast mitochondrial protein, Nfs1p, coordinately regulates iron-sulfur cluster proteins, cellular iron uptake, and iron distribution. J Biol Chem 274:33025–33034

Liesa M, Shirihai OS (2013) Mitochondrial dynamics in the regulation of nutrient utilization and energy expenditure. Cell Metab 17:491–506

Lill R, Mühlenhoff U (2006) Iron-sulfur protein biogenesis in eukaryotes: components and mechanisms. Ann Rev Cell Dev Biol 22:457–486

Lill R, Diekert K, Kaut A, Lange H, Pelzer W, Prohl C, Kispal G (1999) The essential role of mitochondria in the biogenesis of cellular iron-sulfur proteins. Biol Chem 380:1157–1166

Lill R, Dutkiewicz R, Freibert SA, Heidenreich T, Mascarenhas J, Netz DJ, Paul VD, Pierik AJ, Richter N, Stumpfig M, Srinivasan V, Stehling O, Mühlenhoff U (2015) The role of mitochondria and the CIA machinery in the maturation of cytosolic and nuclear iron-sulfur proteins. Eur J Cell Biol 94:280–291

Lindmark DG (1980) Energy metabolism of the anaerobic protozoon *Giardia lamblia*. Mol Biochem Parasitol 1:1–12

Liu S, Roellig DM, Guo Y, Li N, Frace MA, Tang K, Zhang L, Feng Y, Xiao L (2016) Evolution of mitosome metabolism and invasion-related proteins in *Cryptosporidium*. BMC Genomics 17:1006

Lloyd D, Ralphs JR, Harris JC (2002) *Giardia intestinalis*, a eukaryote without hydrogenosomes, produces hydrogen. Microbiology 148:727–733

Mai Z, Ghosh S, Frisardi M, Rosenthal B, Rogers R, Samuelson J (1999) Hsp60 is targeted to a cryptic mitochondrion-derived organelle ("crypton") in the microaerophilic protozoan parasite *Entamoeba histolytica*. Mol Cell Biol 19:2198–2205

Makki A, Rada P, Zarsky V, Kereiche S, Kovacik L, Novotny M, Jores T, Rapaport D, Tachezy J (2019) Triplet-pore structure of a highly divergent TOM complex of hydrogenosomes in *Trichomonas vaginalis*. PLoS Biol 17:e3000098

Maralikova B, Ali V, Nakada-Tsukui K, Nozaki T, van der Giezen M, Henze K, Tovar J (2009) Bacterial-type oxygen detoxification and iron-sulfur cluster assembly in amoebal relict mitochondria. Cell Microbiol 12:331–342

Martin W, Müller M (1998) The hydrogen hypothesis for the first eukaryote. Nature 392:37–41

Martincová E, Voleman L, Pyrih J, Žárský V, Vondračková P, Kolísko M, Tachezy J, Doležal P (2015) Probing the biology of *Giardia intestinalis* mitosomes using *in vivo* enzymatic tagging. Mol Cell Biol 35:2864–2874

Mercer NA, McKelvey JR, Fioravanti CF (1999) *Hymenolepis diminuta*: catalysis of transmembrane proton translocation by mitochondrial NADPH->NAD transhydrogenase. Exp Parasitol 91:52–58

Mi-ichi F, Abu YM, Nakada-Tsukui K, Nozaki T (2009) Mitosomes in *Entamoeba histolytica* contain a sulfate activation pathway. Proc Natl Acad Sci U S A 106:21731–21736

Mi-ichi F, Miyamoto T, Takao S, Jeelani G, Hashimoto T, Hara H, Nozaki T, Yoshida H (2015a) *Entamoeba* mitosomes play an important role in encystation by association with cholesteryl sulfate synthesis. Proc Natl Acad Sci U S A 112:E2884–E2890

Mi-ichi F, Nozawa A, Yoshida H, Tozawa Y, Nozaki T (2015b) Evidence that the *Entamoeba histolytica* mitochondrial carrier family links mitosomal and cytosolic pathways through exchange of 3′-phosphoadenosine 5′-phosphosulfate and ATP. Eukaryot Cell 14:1144–1150

Miller CN, Josse L, Tsaousis AD (2018) Localization of Fe-S biosynthesis machinery in *Cryptosporidium parvum* mitosome. J Eukaryot Microbiol 65:913–922

Mishra P, Chan DC (2014) Mitochondrial dynamics and inheritance during cell division, development and disease. Nat Rev Mol Cell Biol 15:634–646

Morrison HG, Roger AJ, Nystul TG, Gillin FD, Sogin ML (2001) *Giardia lamblia* expresses a proteobacterial-like DnaK homolog. Mol Biol Evol 18:530–541

Müller M, Mentel M, van Hellemond JJ, Henze K, Woehle C, Gould SB, Yu RY, van der Giezen M, Tielens AG, Martin WF (2012) Biochemistry and evolution of anaerobic energy metabolism in eukaryotes. Microbiol Mol Biol Rev 76:444–495

Nasta V, Giachetti A, Ciofi-Baffoni S, Banci L (2017) Structural insights into the molecular function of human [2Fe-2S] BOLA1-GRX5 and [2Fe-2S] BOLA3-GRX5 complexes. Biochim Biophys Acta Gen Subj 1861:2119–2131

Netz DJA, Pierik AJ, Stuempfig M, Mühlenhoff U, Lill R (2007) The Cfd1-Nbp35 complex acts as a scaffold for iron-sulfur protein assembly in the yeast cytosol. Nat Chem Biol 3:278–286

Netz DJ, Stumpfig M, Dore C, Mühlenhoff U, Pierik AJ, Lill R (2010) Tah18 transfers electrons to Dre2 in cytosolic iron-sulfur protein biogenesis. Nat Chem Biol 6:758–765

Nishi M, Hu K, Murray JM, Roos DS (2008) Organellar dynamics during the cell cycle of *Toxoplasma gondii*. J Cell Sci 121:1559–1568

Nohýnkova E, Tůmova P, Kulda J (2006) Cell division of *Giardia intestinalis*: flagellar developmental cycle involves transformation and exchange of flagella between mastigonts of a diplomonad cell. Eukaryot Cell 5:753–761

Nývltová E, Sutak R, Harant K, Šedinová M, Hrdy I, Pačes J, Vlček C, Tachezy J (2013) NIF-type iron-sulfur cluster assembly system is duplicated and distributed in the mitochondria and cytosol of *Mastigamoeba balamuthi*. Proc Natl Acad Sci U S A 110:7371–7376

Nývltová E, Stairs CW, Hrdý I, Ridl J, Mach J, Pačes J, Roger AJ, Tachezy J (2015) Lateral gene transfer and gene duplication played a key role in the evolution of *Mastigamoeba balamuthi* hydrogenosomes. Mol Biol Evol 32:1039–1055

Ogbadoyi EO, Robinson DR, Gull K (2003) A high-order trans-membrane structural linkage is responsible for mitochondrial genome positioning and segregation by flagellar basal bodies in trypanosomes. Mol Biol Cell 14:1769–1779

Olson JW, Agar JN, Johnson MK, Maier RJ (2000) Characterization of the NifU and NifS Fe-S cluster formation proteins essential for viability in *Helicobacter pylori*. Biochemistry 39:16213–16219

Orozco E, Baez-Camargo M, Riveron AM, Gharibeh R, Gariglio P, de la Cruz HF, Chavez P (1997) A model for unscheduled DNA replication in *Entamoeba histolytica* trophozoites. Arch Med Res 28 Spec No:24–26

Otera H, Wang C, Cleland MM, Setoguchi K, Yokota S, Youle RJ, Mihara K (2010) Mff is an essential factor for mitochondrial recruitment of Drp1 during mitochondrial fission in mammalian cells. J Cell Biol 191:1141–1158

Pala ZR, Saxena V, Saggu GS, Garg S (2018) Recent advances in the [Fe-S] cluster biogenesis (SUF) pathway functional in the apicoplast of *Plasmodium*. Trends Parasitol 34:800–809

Palmer CS, Osellame LD, Laine D, Koutsopoulos OS, Frazier AE, Ryan MT (2011) MiD49 and MiD51, new components of the mitochondrial fission machinery. EMBO Rep 12:565–573

Paul VD, Lill R (2015) Biogenesis of cytosolic and nuclear iron-sulfur proteins and their role in genome stability. Biochim Biophys Acta 1853:1528–1539

Pfanner N, Warscheid B, Wiedemann N (2019) Mitochondrial proteins: from biogenesis to functional networks. Nat Rev Mol Cell Biol 20:267–284

Putignani L, Tait A, Smith HV, Horner D, Tovar J, Tetley L, Wastling JM (2004) Characterization of a mitochondrion-like organelle in *Cryptosporidium parvum*. Parasitology 129:1–18

Py B, Gerez C, Angelini S, Planel R, Vinella D, Loiseau L, Talla E, Brochier-Armanet C, Garcia SR, Latour JM, Ollagnier-de CS, Fontecave M, Barras F (2012) Molecular organization, biochemical function, cellular role and evolution of NfuA, an atypical Fe-S carrier. Mol Microbiol 86:155–171

Pyrih J, Pyrihová E, Kolísko M, Stojanovová D, Basu S, Harant K, Haindrich AC, Doležal P, Lukeš J, Roger A, Tachezy J (2016) Minimal cytosolic iron-sulfur cluster assembly machinery of *Giardia intestinalis* is partially associated with mitosomes. Mol Microbiol 102:701–714

Pyrihová E, Motyčková A, Voleman L, Wandyszewska N, Fiser R, Seydlova G, Roger A, Kolísko M, Doležal P (2018) A Single Tim translocase in the mitosomes of *Giardia intestinalis* illustrates convergence of protein import machines in anaerobic eukaryotes. Genome Biol Evol 10:2813–2822

Rada P, Šmíd O, Sutak R, Doležal P, Pyrih J, Žárský V, Montagne JJ, Hrdý I, Camadro JM, Tachezy J (2009) The monothiol single-domain glutaredoxin is conserved in the highly reduced mitochondria of *Giardia intestinalis*. Eukaryot Cell 8:1584–1591

Reeves RE, Warren LG, Susskind B, Lo H-S (1977) An energy conserving pyruvate-to-acetate pathway in *Entamoeba histolytica*. Pyruvate synthase and a new acetate thiokinase. J Biol Chem 252:726–731

Regoes A, Zourmpanou D, Leon-Avila G, van der Giezen M, Tovar J, Hehl AB (2005) Protein import, replication and inheritance of a vestigial mitochondrion. J Biol Chem 280:30557–30563

Riordan CE, Langreth SG, Sanchez LB, Kayser O, Keithly JS (1999) Preliminary evidence for a mitochondrion in *Cryptosporidium parvum*: phylogenetic and therapeutic implications. J Euk Microbiol 46:52S–55S

Riordan CE, Ault JG, Langreth SG, Keithly JS (2003) *Cryptosporidium parvum* Cpn60 targets a relict organelle. Curr Genet 44:138–147

Rodriguez MA, Garcia-Perez RM, Mendoza L, Sanchez T, Guillen N, Orozco E (1998) The pyruvate:ferredoxin oxidoreductase enzyme is located in the plasma membrane and in a cytoplasmic structure in *Entamoeba*. Microb Pathogen 25:1–10

Roger AJ, Svard SG, Tovar J, Clark CG, Smith MW, Gillin FD, Sogin ML (1998) A mitochondrial-like chaperonin 60 gene in *Giardia lamblia*: evidence that diplomonads once harbored an endosymbiont related to the progenitor of mitochondria. Proc Natl Acad Sci U S A 95:229–234

Roy A, Solodovnikova N, Nicholson T, Antholine W, Walden WE (2003) A novel eukaryotic factor for cytosolic Fe-S cluster assembly. EMBO J 22:4826–4835

Saitoh T, Igura M, Obita T, Ose T, Kojima R, Maenaka K, Endo T, Kohda D (2007) Tom20 recognizes mitochondrial presequences through dynamic equilibrium among multiple bound states. EMBO J 26:4777–4787

Saxton WM, Hollenbeck PJ (2012) The axonal transport of mitochondria. J Cell Sci 125:2095–2104

Schaedler TA, Thornton JD, Kruse I, Schwarzlander M, Meyer AJ, van Veen HW, Balk J (2014) A conserved mitochondrial ATP-binding cassette transporter exports glutathione polysulfide for cytosolic metal cofactor assembly. J Biol Chem 289:23264–23274

Schatz G (1997) Just follow the acid chain. Nature 388:121–122

Schnaufer A, Clark-Walker GD, Steinberg AG, Stuart K (2005) The F1-ATP synthase complex in bloodstream stage trypanosomes has an unusual and essential function. EMBO J 24:4029–4040

Schneider RE, Brown MT, Shiflett AM, Dyall SD, Hayes RD, Xie Y, Loo JA, Johnson PJ (2011) The *Trichomonas vaginalis* hydrogenosome proteome is highly reduced relative to mitochondria, yet complex compared with mitosomes. Int J Parasitol 41:1421–1434. https://doi.org/10.1016/j.ijpara.2011.10.001

Schuler MH, Lewandowska A, Caprio GD, Skillern W, Upadhyayula S, Kirchhausen T, Shaw JM, Cunniff B (2017) Miro1-mediated mitochondrial positioning shapes intracellular energy gradients required for cell migration. Mol Biol Cell 28:2159–2169

Schulz C, Schendzielorz A, Rehling P (2015) Unlocking the presequence import pathway. Trends Cell Biol 25:265–275

Schwartz CJ, Djaman O, Imlay JA, Kiley PJ (2000) The cysteine desulfurase, IscS, has a major role in in vivo Fe-S cluster formation in *Escherichia coli*. Proc Natl Acad Sci U S A 97:9009–9014

Shan Y, Napoli E, Cortopassi G (2007) Mitochondrial frataxin interacts with ISD11 of the NFS1/ISCU complex and multiple mitochondrial chaperones. Hum Mol Genet 16:929–941

Sheftel AD, Stehling O, Pierik AJ, Netz DJ, Kerscher S, Elsasser HP, Wittig I, Balk J, Brandt U, Lill R (2009) Human ind1, an iron-sulfur cluster assembly factor for respiratory complex I. Mol Cell Biol 29:6059–6073

Sheftel AD, Wilbrecht C, Stehling O, Niggemeyer B, Elsasser HP, Mühlenhoff U, Lill R (2012) The human mitochondrial ISCA1, ISCA2, and IBA57 proteins are required for [4Fe-4S] protein maturation. Mol Biol Cell 23:1157–1166

Sipos K, Lange H, Fekete Z, Ullmann P, Lill R, Kispal G (2002) Maturation of cytosolic iron-sulfur proteins requires glutathione. J Biol Chem 277:26944–26949

Slapeta J, Keithly JS (2004) *Cryptosporidium parvum* mitochondrial-type HSP70 targets homologous and heterologous mitochondria. Euk Cell 3:483–494

Šmíd O, Horáková E, Vilímová V, Hrdý I, Cammack R, Horvath A, Lukeš J, Tachezy J (2006) Knock-downs of iron-sulfur cluster assembly proteins IscS and IscU down-regulate the active mitochondrion of procyclic *Trypanosoma brucei*. J Biol Chem 281:28679–28686

Šmíd O, Matušková A, Harris SR, Kučera T, Novotný M, Horváthová L, Hrdý I, Kutějová E, Hirt RP, Embley TM, Janata J, Tachezy J (2008) Reductive evolution of the mitochondrial processing peptidases of the unicellular parasites *Trichomonas vaginalis* and *Giardia intestinalis*. PLoS Pathog 4:e1000243

Sogin ML (1991) Early evolution and the origin of eukaryotes. Curr Biol 1:457–463

Srinivasan V, Netz DJ, Webert H, Mascarenhas J, Pierik AJ, Michel H, Lill R (2007) Structure of the yeast WD40 domain protein Cia1, a component acting late in iron-sulfur protein biogenesis. Structure 15:1246–1257

Stairs CW, Najdrová V, Krupičková A, Voleman L, Svard SG, Ettema TJG, Doležal P (2018) Identification of a BolA-like protein in *Giardia intestinalis*: a new Fe-S cluster assembly factor in mitomes. Paper presented at ISEP 2018, Droushia, Cyprus, 27 May–1 June 2018

Stehling O, Wilbrecht C, Lill R (2014) Mitochondrial iron-sulfur protein biogenesis and human disease. Biochimie 100:61–77

Stehling O, Mascarenhas J, Vashisht AA, Sheftel AD, Niggemeyer B, Rosser R, Pierik AJ, Wohlschlegel JA, Lill R (2018) Human CIA2A-FAM96A and CIA2B-FAM96B integrate iron homeostasis and maturation of different subsets of cytosolic-nuclear iron-sulfur proteins. Cell Metab 27:263

Straub SP, Stiller SB, Wiedemann N, Pfanner N (2016) Dynamic organization of the mitochondrial protein import machinery. Biol Chem 397:1097–1114

Striepen B, Crawford MJ, Shaw MK, Tilney LG, Seeber F, Roos DS (2000) The plastid of *Toxoplasma gondii* is divided by association with the centrosomes. J Cell Biol 151:1423–1434

Tachezy J, Doležal P (2007) Iron-sulfur proteins and iron-sulfur cluster assembly in organisms with hydrogenosomes and mitosomes. 1:105–133

Tachezy J, Sanchez LB, Müller M (2001) Mitochondrial type iron-sulfur cluster assembly in the amitochondriate eukaryotes *Trichomonas vaginalis* and *Giardia intestinalis*, as indicated by the phylogeny of IscS. Mol Biol Evol 18:1919–1928

Takahashi Y, Tokumoto U (2002) A third bacterial system for the assembly of iron-sulfur clusters with homologs in archaea and plastids. J Biol Chem 277:28380–28383

Tanaka Y, Kanai Y, Okada Y, Nonaka S, Takeda S, Harada A, Hirokawa N (1998) Targeted disruption of mouse conventional kinesin heavy chain, kif5B, results in abnormal perinuclear clustering of mitochondria. Cell 93:1147–1158

Tovar J, Fischer A, Clark CG (1999) The mitosome, a novel organelle related to mitochondria in the amitochondrial parasite *Entamoeba histolytica*. Mol Microbiol 32:1013–1021

Tovar J, Leon-Avila G, Sanchez LB, Sutak R, Tachezy J, van der Giezen M, Hernandez M, Müller M, Lucocq JM (2003) Mitochondrial remnant organelles of *Giardia* function in iron-sulphur protein maturation. Nature 426:172–176

Tsaousis AD, Kunji ER, Goldberg AV, Lucocq JM, Hirt RP, Embley TM (2008) A novel route for ATP acquisition by the remnant mitochondria of *Encephalitozoon cuniculi*. Nature 453:553–556

Tsaousis AD, Gentekaki E, Eme L, Gaston D, Roger AJ (2014) Evolution of the cytosolic iron-sulfur cluster assembly machinery in *Blastocystis* species and other microbial eukaryotes. Euk Cell 13:143–153

Uzarska MA, Nasta V, Weiler BD, Spantgar F, Ciofi-Baffoni S, Saviello MR, Gonnelli L, Mühlenhoff U, Banci L, Lill R (2016) Mitochondrial Bol1 and Bol3 function as assembly factors for specific iron-sulfur proteins. Elife 5:e16673

Vaidya AB, Mather MW (2009) Mitochondrial evolution and functions in malaria parasites. Annu Rev Microbiol 63:249–267

van der Giezen M, Cox S, Tovar J (2004) The iron-sulfur cluster assembly genes *iscS* and *iscU* of *Entamoeba histolytica* were acquired by horizontal gene transfer. BMC Evol Biol 4:7

van Dooren GG, Marti M, Tonkin CJ, Stimmler LM, Cowman AF, McFadden GI (2005) Development of the endoplasmic reticulum, mitochondrion and apicoplast during the asexual life cycle of *Plasmodium falciparum*. Mol Microbiol 57:405–419

van Dooren GG, Stimmler LM, McFadden GI (2006) Metabolic maps and functions of the *Plasmodium* mitochondrion. FEMS Microbiol Rev 30:596–630

Vávra J (1976) Structure of the microsporidia. In: Bulla LA, Cheng TC (eds) Biology of microsporidia. Springer, Boston, pp 1–86

Vávra J (2005) "Polar vesicles" of microsporidia are mitochondrial remnants ("mitosomes")? Folia Parasitol 52:193–195

Vávra J, Hyliš M, Fiala I, Nebesářová J (2016) *Globulispora mitoportans* n. g., n. sp., (*Opisthosporidia*: *Microsporidia*) a microsporidian parasite of daphnids with unusual spore organization and prominent mitosome-like vesicles. J Invertebr Pathol 135:43–52

Vickerman K (1985) Development cycles and biology of pathogenic trypanosomes. Br Med Bull 41:105–114

Voleman L, Najdrová V, Astvaldsson A, Tůmová P, Einarsson E, Svindrych Z, Hagen GM, Tachezy J, Svard SG, Doležal P (2017) *Giardia intestinalis* mitosomes undergo synchronized fission but not fusion and are constitutively associated with the endoplasmic reticulum. BMC Biol 15:27

von Heijne G (1986) Mitochondrial targeting sequences may form amphiphilic helices. EMBO J 5:1335–1342

Voos W, Martin H, Krimmer T, Pfanner N (1999) Mechanisms of protein translocation into mitochondria. Biochim Biophys Acta 1422:235–254

Waller RF, Jabbour C, Chan NC, Celik N, Likic VA, Mulhern TD, Lithgow T (2009) Evidence of a reduced and modified mitochondrial protein import apparatus in microsporidian mitosomes. Eukaryot Cell 8:19–26

Webert H, Freibert SA, Gallo A, Heidenreich T, Linne U, Amlacher S, Hurt E, Mühlenhoff U, Banci L, Lill R (2014) Functional reconstitution of mitochondrial Fe/S cluster synthesis on Isu1 reveals the involvement of ferredoxin. Nat Commun 5:5013

Wiedemann N, Pfanner N (2017) Mitochondrial machineries for protein import and assembly. Annu Rev Biochem 86:685–714

Wiedemann N, Frazier AE, Pfanner N (2004) The protein import machinery of mitochondria. J Biol Chem 279:14473–14476

Williams BAP, Hirt RP, Lucocq JM, Embley TM (2002) A mitochondrial remnant in the microsporidian *Trachipleistophora hominis*. Nature 418:865–869

Wilson RJM, Rangachari K, Saldanha JW, Rickman L, Buxton RS, Eccleston JF (2003) Parasite plastids: maintenance and functions. Phil Trans R Soc Lond Ser B Biol Sci 358:155–162

Yaffe MP, Stuurman N, Vale RD (2003) Mitochondrial positioning in fission yeast is driven by association with dynamic microtubules and mitotic spindle poles. Proc Natl Acad Sci U S A 100:11424–11428

Yamamoto H, Esaki M, Kanamori T, Tamura Y, Nishikawa S, Endo T (2002) Tim50 is a subunit of the TIM23 complex that links protein translocation across the outer and inner mitochondrial membranes. Cell 111:519–528

Youssef NN, Hammond D (1971) The fine structure of the developmental stages of the microsporidian *Nosema apis* Zander. Tissue Cell 3:283–294

Yousuf MA, Mi-ichi F, Nakada-Tsukui K, Nozaki T (2010) Localization and targeting of an unusual pyridine nucleotide transhydrogenase in *Entamoeba histolytica*. Eukaryot Cell 9:926–933

Yu Y, Samuelson J (1994) Primary structure of an *Entamoeba histolytica* nicotinamide nucleotide transhydrogenase. Mol Biochem Parasitol 68:323–328

Zheng L, Cash VL, Flint DH, Dean DR (1998) Assembly of iron-sulfur clusters. Identification of an iscSUA-hscBA-fdx gene cluster from *Azotobacter vinelandii*. J Biol Chem 273:13264–13272

The Mitochondrion-Related Organelles of *Cryptosporidium* Species

Anastasios D. Tsaousis and Janet S. Keithly

Contents

1 Introduction 244
2 Ultrastructure Morphology 246
  2.1 Intracellular Location 246
  2.2 Subcellular (Internal) Organization 249
  2.3 Variations in Cristae 249
3 Mitochondrial Cell Biology 250
  3.1 Association of the Mitochondrion with the RER 250
  3.2 Iron-Sulphur Cluster (ISC) Biosynthesis 250
  3.3 Mitochondrial Protein Import 253
4 Carbohydrate Metabolism 254
  4.1 Pyruvate:NADP+ Oxidoreductase (PNO) 254
  4.2 Predicted End Products of Glycolysis 255
5 Energy Metabolism 256
6 The Crystalloid Body 257
  6.1 Ultrastructural Morphology 257
  6.2 Putative Functions 259
7 Diversity of Mitochondria Within the Clade 260
8 Concluding Remarks and Future Perspectives 260
References 261

Abstract Among all apicomplexans, the mitochondrion-related organelle of *Cryptosporidium* species is the least studied. To date, most of our knowledge on this tiny

A. D. Tsaousis (✉)
Laboratory for Evolutionary and Molecular Parasitology, RAPID Group, School of Biosciences, University of Kent, Canterbury, Kent, UK
e-mail: a.tsaousis@kent.ac.uk

J. S. Keithly
Division of Infectious Diseases/Office of Research and Technology, New York State Department of Health, The Wadsworth Center, Albany, NY, USA

Department of Biomedical Sciences, School of Public Health, SUNY-Albany, Albany, NY, USA

J. Tachezy (ed.), *Hydrogenosomes and Mitosomes: Mitochondria of Anaerobic Eukaryotes*, Microbiology Monographs 9,
https://doi.org/10.1007/978-3-030-17941-0_10

organelle stems from observations on the remnant mitochondrion, mitosome, of *Cryptosporidium parvum*. In *C. parvum* the mitosome is structurally distinguished from the hydrogenosomes and mitosomes of other anaerobic protists by its (1) close association with the crystalloid body, an organelle unique to this apicomplexan and the function of which is currently unknown; (2) close association with the outer nuclear membrane and possibly nuclear pores; (3) envelopment by rough endoplasmic reticulum and in some cases an apparent direct tethering to ribosomes; and (4) atypical internal membranous compartments that lack well-defined crista junctions with the mitochondrial inner membrane, a characteristic that defines most aerobic mitochondria. Like most hydrogenosome- and other mitosome-bearing anaerobic protists, however, *C. parvum* lacks a mitochondrial genome, i.e. proteins are encoded by the nucleus and targeted back to the mitosome. As a consequence of this reductive evolution, there are no genes for electron transport or oxidative phosphorylation, and the only function so far ascribed to this tiny organelle is one common to all eukaryotic mitochondria, the assembly and maturation of iron-sulphur clusters. The ultrastructure and tomography of the *C. parvum* mitosome and crystalloid body, as well as the probable functions of these organelles, are the primary topics herein. An overview of iron-sulphur cluster biosynthesis, likely mechanisms for import into and export from the mitosome, as well as core carbohydrate and energy metabolism is also discussed. Similarities and differences in the structure and function of both organelles in the genus *Cryptosporidium*, with anaerobic protists in general, and with other apicomplexans specifically, are described.

1 Introduction

Cryptosporidium belongs to the diverse group of apicomplexans that includes species of medical (*Babesia microti*, *Plasmodium falciparum*, *Toxoplasma gondii*) and veterinary (*Eimeria tenella*, *Neospora caninum*) importance (Thompson et al. 2005; Fayer and Xiao 2007). Using extensive biochemical, genomic, microscopic and molecular data, the genus *Cryptosporidium* has very recently been transferred from class Coccidiomorphea (subclass Coccidia) to subclass Cryptogregaria within the new class Gregarinomorphea (Cavalier-Smith 2014). The genus *Cryptosporidium* is the sole member of this subclass and is defined as an epicellular parasite of vertebrates lacking an apicoplast but possessing a gregarine-like feeder organelle (Cavalier-Smith 2014) [reviewed in Ryan et al. (2016)]. Key similarities supporting this new *Cryptosporidium*/gregarine lineage [reviewed in Liu et al. (2016), Ryan et al. (2016)] include completing host-free life cycles, exhibiting large extracellular gamonts, pairing reproductively end to end (syzygy) and changing cell architecture to adapt to diverse environments, e.g. biofilms, coelom, intestines, soil and water (Ryan et al. 2016). While a single mitochondrion is present in most Apicomplexa

(Henriquez et al. 2005), its existence in *Cryptosporidium parvum* (Aji et al. 1991; Beyer et al. 2000; Fayer and Xiao 2007; Riordan et al. 1999; Tetley et al. 1998), and in other species of the genus (Uni et al. 1987; Alvarez-Pellitero et al. 2004), was, for many years, an open question. The evidence for a mitochondrion-like organelle (mitosome) in *C. parvum* sporozoites was first confirmed by ultrastructural and phylogenetic analyses (Riordan et al. 1999). These initial data showed that the life stages of *C. parvum* possessed a double membrane-bounded single organelle enveloped by rough endoplasmic reticulum (RER) to which *C. parvum* Cpn60 (*Cp*Cpn60) and mitochondrial (mt) Hsp70 (*Cp*mtHsp70) were immunolocalized (Slapeta and Keithly 2004; Riordan et al. 2003). This organelle was also in intimate contact with the nucleus and sandwiched between it and the crystalloid body (CB), a *Cryptosporidium*-specific organelle of unknown function. Mitochondrion-like structures had already been seen in *C. parvum* (Beyer et al. 2000) and *C. muris* (Uni et al. 1987) from infected rats and mice, respectively.

Initially, the ultrastructure of the organelle in sporozoites of *C. parvum* observed posterior to the nucleus was thought to more closely resemble the apicoplast, a chloroplast remnant observed in most apicomplexans (*Eimeria*, *Neospora*, *Plasmodium*, *Toxoplasma*), rather than a mitochondrion (Tetley et al. 1998). Upon further investigation, this supposition could not be sustained. Both transmission electron microscopy (TEM) and electron tomography showed that the organelle in *C. parvum* was bounded by two membranes (Keithly et al. 2005; Slapeta and Keithly 2004; Riordan et al. 2003; Putignani et al. 2004), rather than the three to four observed in apicoplasts [reviewed in Williams and Keeling (2003), Henriquez et al. (2005)]. In addition, tomography indicated that the inner membrane (IM) of the *C. parvum* organelle formed membranous subcompartments (Keithly et al. 2005) and was therefore quite distinct from the infoldings of plastid inner membranes (Henriquez et al. 2005). Indeed, although there did not appear to be "crista junctions" within the IM of the organelle, the internal morphology of the mitochondrion-like organelle was quite similar to that observed for yeast and mammalian mitochondria in a highly "condensed" state when the matrix is contracted [Mannella et al. 2001; reviewed in Frey et al. (2006), Mannella (2006)]. Finally, subsequent ultrastructural studies have indicated that *C. parvum* lacks an apicoplast (Keithly et al. 2005; Putignani et al. 2004; Riordan et al. 1999; Slapeta and Keithly 2004) and that both human species of *Cryptosporidium* (*C. parvum* and *C. hominis*) lack an apicoplast genome (Abrahamsen et al. 2004; Cai et al. 2003; Xu et al. 2004). Taken together, these data indicated that the double membrane-bounded organelle in *C. parvum* was not an apicoplast but rather, an unusually compacted organelle of reduced size that most closely resembled the mitochondria (Krungkrai et al. 2000; Melo et al. 2000), mitosomes (Clark and Roger 1995; Leon-Avila and Tovar 2004; Tovar et al. 1999) and hydrogenosomes (van der Giezen et al. 2005; van Hoek et al. 2000; Muller et al. 2012; Richardson et al. 2015) of other anaerobic parasitic protists. Essentially, it is a relict mitochondrion.

2 Ultrastructure Morphology

Cryptosporidium parvum is distinguished from other mitosome- and hydrogenosome-bearing protists by being an intracellular but extracytoplasmic [or epicellular; see Ryan et al. (2016)] parasite that harbours a remnant mitochondrion, the mitosome, that is drastically reduced in size and structure and that lacks a mitochondrial genome. *Plasmodium* and *Toxoplasma*, its nearest apicomplexan relatives, are also intracellular parasites with mitochondria reduced in structure and function. Unlike *C. parvum*, however, the mitochondria of *P. falciparum* and *T. gondii* still possess small mitochondrial genomes, which permit them to express, during certain parts of the life cycle, cytochrome oxidase b and subunits I and III of the cytochrome c oxidase, the universal mitochondrial genes retained in the mitochondrial genome of all respiring eukaryotes (Kita et al. 2002; Mi-Ichi et al. 2003; Mogi and Kita 2010). In addition, the nuclear genome of the sexual stages of the malarial parasite encodes proteins that enable the transfer of reducing equivalents to oxygen via classical respiratory pathways and to fumarate using a complex II quinol-fumarate reductase. Interestingly, recent genome analyses revealed a progressive reduction of mitochondrial metabolism within the genus *Cryptosporidium* (Liu et al. 2016). For example, genetically related *C. muris* and *C. andersoni* both use the TCA cycle and glycolysis for energy metabolism and possess a classic, nearly complete oxidative phosphorylation. On the other hand, *C. parvum* and *C. hominis* lack the TCA cycle and instead possess an alternative oxidase (AOX) for phosphorylation (Mogi and Kita 2010; Suzuki et al. 2004). At the far end of this spectrum is *C. ubiquitum,* which has lost the entire electron transport system (ETC), and with it, ubiquinone and polyisoprenoid biosynthesis. These genomic data further reveal that differences in energy metabolism within the newly defined *Cryptosporidium*/gregarine lineage (subclass Cryptogregaria) from other apicomplexans are primarily due to the absence of an apicoplast and presence of an alternative electron transport system. In other lineages of apicomplexans, energy metabolism and the ETC are generally conserved (Liu et al. 2016). Thus, species of *Cryptosporidium* are at the far end of the apicomplexan spectrum having lost the apicoplast completely and retained just a replica of the mitochondrion without a genome. It is still thought that the morphological changes in mitosome structure are probably due to adaptations following the loss of organellar metabolism-related genes and adaptations to different parasitic life styles. Furthermore, the presence of the TCA cycle and a conventional ETC in gregarines suggests that they are probably ancestral in the newly defined *Cryptosporidium*/gregarine lineage.

2.1 Intracellular Location

A defining feature of almost all *Cryptosporidium* species is a complex life cycle, and most previous studies have focused on sporozoites, the infectious stage of the

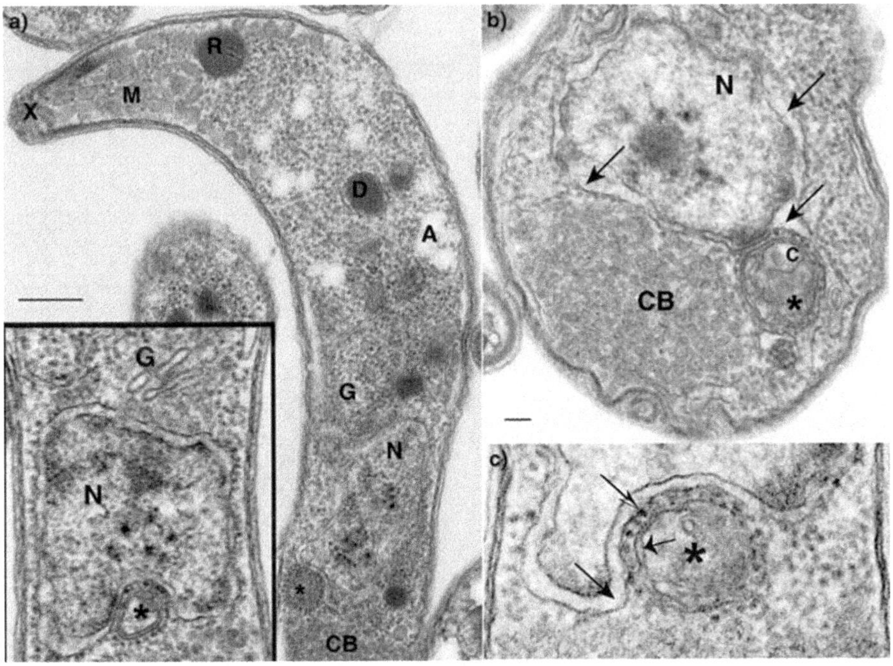

Fig. 1 Transmission electron microscopy of *Cryptosporidium parvum* sporozoite organelles (reprinted from Figs. 1–3 of Keithly et al. 2005 with permission of the publishers). (**a**) Longitudinal section of sporozoite at low magnification. The mitochondrion-related organelle (asterisk) is between the nucleus (*N*) and the crystalloid body (CB). A typical apical complex is evident (*X*), and the Golgi (*G*) is anterior to the nucleus, shown in detail in the inset. Other organelles include the micronemes (*M*), single rhoptry (*R*) and dense granules (*D*) for host cell invasion and amylopectin (*A*), a plantlike storage granule (bar = 0.5 μm). (**b**) Higher magnification showing the intimate relationship between the nucleus, its membranes (arrows), the CB and the mitochondrial remnant (asterisk). The mitochondrion-related organelle is spherical, 150–300 nm in diameter, contains internal crista-like membranes (*C*) and is enveloped by rough endoplasmic reticulum (RER). The vesicular CB is closely apposed to the mitochondrion (bar = 0.1 μm). (**c**) A higher magnification of the nucleus, its membranes, the CB and the relic mitochondrion. A larger arrow denotes the lumen of the nuclear envelope, the outer membrane of which is studded with ribosomes (black/white arrow). The mitochondrion-related organelle is surrounded by two limiting membranes (small arrow). Scale bar = 0.2 μm

parasite capable of invading host cells. Initial TEM and tomographic reconstruction of chemically fixed sporozoites of *C. parvum* have revealed a complex arrangement of membranes both around this organelle and within it (Figs. 1 and 2). Specifically, the mitosome is enveloped by multiple segments of RER that appear to extend from the outer nuclear envelope [Fig. 1b, c; see Figs. 4–9 in Keithly et al. (2005)]. In tomographic reconstructions of fixed sporozoites, a single, highly folded inner membrane (Fig. 2) or multiple internal subcompartments [see Figs. 12–15 in Keithly et al. (2005)] were visualized within the organelle. As mentioned before, the infoldings of the IM appeared to lack tubular crista junctions typical for metazoan,

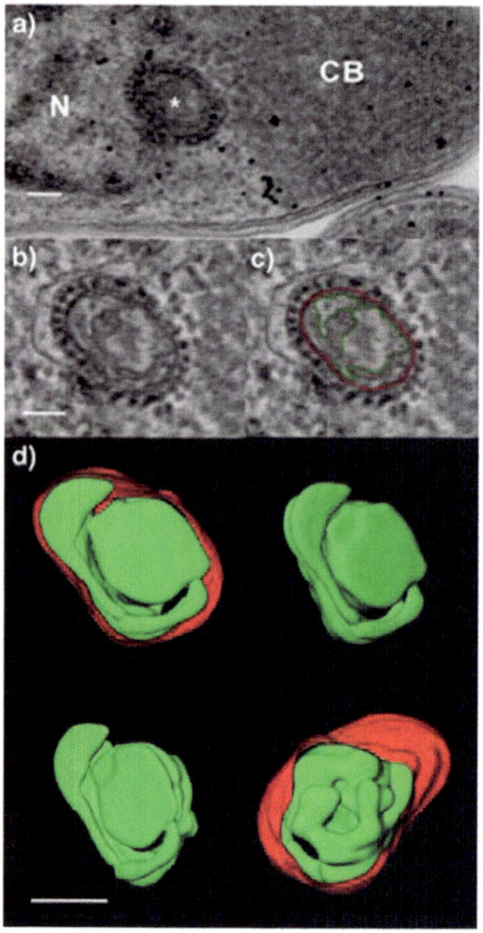

Fig. 2 Electron tomographic reconstruction of a *Cryptosporidium parvum* mitochondrion-related organelle (reprinted from Figs. 16–19 of Keithly et al. 2005 with permission of the publishers). (**a**) Projection image of the untitled specimen showing the nucleus (*N*), crystalloid body (CB), mitochondrion-related organelle and the colloidal gold particles used for alignment. Original section thickness approximately 140 nm (scale bars = 0.1 μm). (**b**) Slice (3.2 nm thick) from the centre of the tomogram, in the plane normal to the electron beam showing the mitochondrion-like organelle surrounded by the prominent RER. (**c**) Identical to b, but contours have been added to show the outer (red) and inner membrane profiles (green). (**d**) Four views of a three-dimensional model of the membrane surfaces in the reconstructed mitochondrion-related organelle showing the outer (red) and one large inner membrane compartment (green). The model represents a volume approximately 130 nm thick at the centre of the reconstruction

fungal and other protists' mitochondrion-related organelles. These observations for the internal membranes of the *C. parvum* mitosome are consistent with the loss, through reductive evolution, of the normal oxidative phosphorylation machinery in *C. parvum* and the loss of its mitochondrial genome (Abrahamsen et al. 2004; Keeling 2004).

2.2 Subcellular (Internal) Organization

The organization of the IM of the *C. parvum* mitosome is quite distinct from other previously considered amitochondriate protists. With the exception of the anaerobic free-living ciliates, e.g. *Metopus* and *Plagiopyla* (Fenchel et al. 1977), the parasitic ciliate *Nyctotherus ovalis* (Boxma et al. 2005) and the stramenopile (syn. heterokont) *Blastocystis* sp. (Nasirudeen and Tan 2004; Tan 2008; Lantsman et al. 2008) and all hydrogenosome- and mitosome-bearing anaerobic protists and fungi lack mitochondrial cristae. These include the parabasalids *Trichomonas vaginalis* and *Tritrichomonas foetus* [reviewed in Muller et al. (2012)]; archaeamoebid *Entamoeba histolytica* (Clark and Roger 1995; Tovar et al. 1999); diplomonads *Giardia intestinalis* and *Spironucleus* sp. (Tovar et al. 2003; Jerlstrom-Hultqvist et al. 2013); and chytrid and microsporidian fungi *Neocallimastix* and *Piromyces* species [reviewed in Williams and Keeling (2003)], *Encephalitozoon cuniculi* (Tsaousis et al. 2008) and *Trachipleistophora hominis* (Williams et al. 2002).

The infoldings of the mitochondrial IM also differ from those of most Apicomplexa, in which the type and number of cristae vary considerably during the life cycle, ranging from tubular in *T. gondii* tachyzoites (Melo et al. 2000) and *P. falciparum* gametocytes to "acristate" in asexual *P. falciparum* and *P. knowlesi* (Fry and Beesley 1991; Krungkrai et al. 1999, 2000). Ultrastructural analyses of the *C. parvum* mitosome indicate that although it has an outer and an inner boundary membrane, the presence of membranous subcompartments sets it apart from these near relatives (Riordan et al. 1999, 2003). Indeed, tomographic reconstructions of fixed sporozoites suggest that junctions, which typically connect cristae to the mitochondrial IM (Mannella et al. 2001), are absent (or greatly reduced) in the mitosome of *C. parvum* (Keithly et al. 2005).

2.3 Variations in Cristae

It is reasonably well established that the morphology of cristae can vary as a result of osmotic adjustments in the matrix volume (Frey et al. 2006; Lucic et al. 2005; Mannella 2006). Most mammalian mitochondria observed in situ are in an "orthodox" conformation, with a large matrix volume that pushes the inner boundary membrane against the outer membrane leaving an 8 nm space between them. Nevertheless, under conditions of stress or loss of genetic function, the IM of mammalian and yeast mitochondria becomes condensed (matrix contracted) [reviewed in Frey et al. (2006)]. In this state, the number of cristae is reduced and their size expands. This is congruent with tomographic reconstructions in chemically fixed cells of the internal membrane organization of the *C. parvum* mitosome, i.e. subcompartments are few and expanded (Fig. 2d). These observations on the lack of typical internal compartmentation in the *C. parvum* mitosome are consistent with a structural reflection of the secondary loss of oxidative phosphorylation during the reductive evolution of this organelle (Richardson et al. 2015; Keeling 2004; Williams and Keeling 2003).

3 Mitochondrial Cell Biology

3.1 Association of the Mitochondrion with the RER

The origin of all mitochondrion-related organelles is generally accepted to be the endosymbiosis of an α-proteobacterium [reviewed in Brown and Doolittle (1997), van der Giezen (2005), van der Giezen and Tovar (2005), Muller et al. (2012), Richardson et al. (2015)], whereas the nucleus, endoplasmic reticulum, lysosomes and apical organelles of the apicomplexan endomembrane system are thought to be derived from the infoldings of the ancestral plasma membrane (Brown and Doolittle 1997). As previously mentioned, TEM confirmed the presence of a roughly spherical 150–300 nm organelle with two closely apposed boundary membranes, the outer membrane of which is closely associated with the RER and which appears to originate from the outer envelope of the nearby nucleus (Fig. 1). In fact, the relic mitochondrion is almost completely enveloped by protrusions of the RER (Fig. 1c), an association that was dramatically visualized by tomographic reconstruction [see Fig. 10 in Keithly et al. (2005)]. These initial TEM and tomographic observations have been supported using cryo-electron tomography (Csordas et al. 2006) of *C. parvum* sporozoites visualized in their "native state", which not only shows envelopment by the RER but the possible intimate association of the RER with nuclear pores (C. Hsieh, M. Marko, J.S. Keithly, C.A. Mannella, unpublished results). The reason behind this close association of the organelles is unknown, but it could be correlated with a potential biochemical function of the mitosomes [e.g. iron-sulphur cluster (ISC) biosynthesis (see below)] that requires immediate access of products necessary for the function of proteins in the RER or in the nucleus.

3.2 Iron-Sulphur Cluster (ISC) Biosynthesis

An important function for all the mitochondrion-related organelles investigated so far is the ISC assembly machinery. Homologues of these machineries have been localized in other mitosomes including *Entamoeba* (Maralikova et al. 2010), *Giardia* (Tovar et al. 2003) and *Microsporidia* (Goldberg et al. 2008; Freibert et al. 2017). The typical mitochondrial ISC machinery is mainly constituted by a cysteine desulphurase (IscS; Nsf1 in yeast), an iron donor (frataxin; Yfh1 in yeast), a scaffold protein (IscU; Isu in yeast) and many other proteins involved in intraorganellar trafficking, transfer and targeting [for review see Braymer and Lill (2017); Fig. 3]. Most of the components of this machinery are encoded by the *Cryptosporidium* genomes (Miller et al. 2018b), and previous reports have demonstrated that in

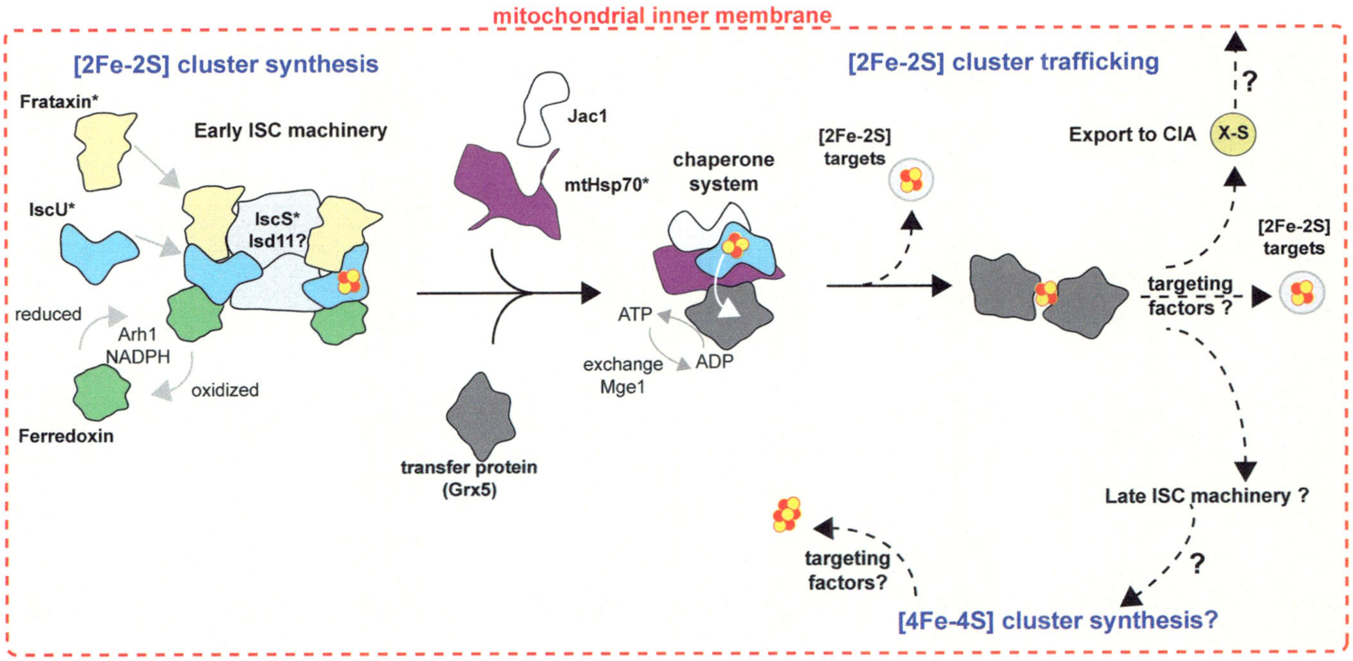

Fig. 3 Cartoon model demonstrating the in silico-predicted Fe-S cluster biosynthetic pathway in *C. parvum* mitosome. The diagram is based upon previous bioinformatics' analyses (Miller et al. 2018b) and the model from Braymer and Lill (2017). Proteins with an asterisk have already been shown experimentally to localize within the *C. parvum* mitosome (Slapeta and Keithly 2004; Miller et al. 2018b). The ISC machinery is the only biosynthetic pathway to be localized in the *C. parvum* mitosome. As shown in the diagram, an assembly of ISC proteins is essential for the de novo synthesis of [2Fe-2S] and [4Fe-4S] clusters, as well as proper trafficking of apoproteins into mitochondria. Firstly, an [2Fe-2S] cluster is synthesized by the early ISC machinery, consisting of the IscU scaffold protein, which also requires a sulphide from the cysteine desulphurase complex IscS-Isd11 (to date, Isd11 has not been identified in any *Cryptosporidium* genome), along with the electrons from the transfer chain NADPH-Arh1 and ferredoxin and the regulator and/or iron donor frataxin. The IscU-bound [2Fe-2S] cluster is subsequently delivered to the monothiol glutaredoxin Grx5, a reaction accomplished by the mitochondrial Hsp70 (mtHsp70) chaperone with the

C. parvum, the presequences of both *Cp*IscS and *Cp*IscU, two genes associated with mitochondrial Fe-S cluster biogenesis, can target GFP to the mitochondrial network of the yeast *Saccharomyces cerevisiae* (LaGier et al. 2003). Recently, a study from the Tsaousis lab using infected COLO-680N in vitro cultured cells (Miller et al. 2018b) and homologous antibodies against *C. parvum* proteins has demonstrated that IscU, IscS and frataxin are localized within the mitosome of *C. parvum* sporozoites and merozoites, whereas localization within the intracellular oocysts needs further investigation (Miller et al. 2018b). Interestingly, until this report, no investigator had shown the fate of the mitosomes during the life cycle of *Cryptosporidium* (Miller et al. 2018b).

Briefly, using long-term in vitro cultivation of *C. parvum* in COLO-680N cells (Miller et al. 2018a) and western blotting, these authors have shown notable expression of the iron-sulphur cluster (ISC) proteins *Cp*IscS, *Cp*IscU and *Cp*frataxin in extracellular sporozoites and merozoites, whereas no expression was detected in host cells infected with oocysts (Miller et al. 2018b). In contrast, immunofluorescent assays showed that all life cycle stages contain these ISC proteins. Furthermore, in sporozoites, direct localization of all three ISC proteins to a 500 nm structure consistent with a mitosome (Slapeta and Keithly 2004) was confirmed by high-resolution confocal microscopy (Miller et al. 2018b). These data are the first to show the direct localization of a biosynthetic pathway (ISC) to the mitosome and that this organelle is present in life cycle stages other than sporozoites. Surprisingly, however, during the life cycle, the localization of immunofluorescence was highly variable, i.e. in some cases it appeared punctate (mitosomal) and in others cytosolic (Miller et al. 2018b).

Two explanations for variations in immunolocalization were proposed by Miller et al. (2018b): first, either that the ISC machinery was translocated to the cytosol during some life cycle stages or, secondly, that late-stage morphological changes, like those observed in the apicomplexans *Plasmodium* when the single mitochondrion becomes branched (van Dooren et al. 2005) or in *Toxoplasma* tachyzoites when the single tubular mitochondrion at the periphery of the parasite begins to concentrate in patches upon transiting from the intra- to extracellular environment (Ovciarikova et al. 2017). Of these two, there is supporting TEM evidence that in *C. parvum*, the latter is more likely [see Fig. 5 in Keithly et al. (2005)]. This figure

Fig. 3 (continued) assistance of the J-type co-chaperone Jac1. This reaction is dependent on ATP hydrolysis by mtHsp70. The exchange factor Mge1 facilitates the exchange of ADP for ATP. The resulting bridging [2Fe-2S] cluster on a Grx5 dimer is inserted directly into [2Fe-2S] recipient apoproteins or trafficked to the late ISC machinery for [4Fe-4S] cluster biogenesis (constituted by Isa1, Isa2 and iBa57 in yeast and currently unidentified in *Cryptosporidium*). The early ISC machinery, including the chaperones and Grx5, is also responsible for generating the component X-S for transport of sulphur out of the mitochondria to the cytosolic ion-sulphur cluster (CIA) machinery for cytosolic-nuclear ISC protein biogenesis. The late ISC machinery consists of the yet structurally and functionally uncharacterized Isa1-Isa2-Iba57 complex, which is needed for the generation of [4Fe-4S] clusters. Dashed arrows indicate biochemical steps that remain poorly elucidated in all organisms investigated so far (Braymer and Lill 2017) clearly shows that the relict mitochondrion and its attendant ribosomes "is either dividing or elongate and tubular as seen in other Apicomplexa (Melo et al. 2000)". Certainly, as Miller et al. (2018b) suggest, this is worth exploring further. The recently developed CRISPR/Cas9 system for *C. parvum* (Vinayak et al. 2015) combined with the use of *Cryptosporidium*-specific fluorescent mitosomal markers (Miller et al. 2018b) using the COLO-680N cultivation of *C. parvum* (Miller et al. 2018a) should permit monitoring these proteins in real time throughout the life cycle. Significantly, the research of Miller et al. (2018b) also demonstrates that *C. parvum* IscS can functionally replace the essential yeast Nfs1 homologue, suggesting that the ISC pathway seen in *C. parvum* is similar to that conserved among other eukaryotes. As stated previously, this is the first localization of proteins involved in a biosynthetic pathway in *C. parvum*, providing a first step into further investigations into the function(s) of this peculiar mitochondrion.

3.3 Mitochondrial Protein Import

In order for proteins to be targeted into the mitosome, a dedicated import machinery is required. This mitochondrial protein import machinery has not only been very well characterized in yeast and mammalian cells but also in many mitochondrion-related organelles of anaerobic protists including *Trichomonas vaginalis*, *Giardia intestinalis* (Dolezal et al. 2005, 2006), and *Blastocystis* sp. (Tsaousis et al. 2011), among others (see other chapters in this book). This machinery is comprised of five complexes (Wiedemann and Pfanner 2017), in which each is responsible for targeting proteins into intraorganellar spaces, namely, the (1) translocase of the outer membrane (TOM), which recognizes nuclear-encoded proteins targeting the mitochondrial IM and OM; (2) translocases of the inner membrane, i.e. TIM22 (proteins targeted to the matrix) with the assistance; (3) of a presequence translocase-associated motor (PAM); (4) TIM23 (proteins targeted to the inner membrane); and (5) sorting assembly machinery (SAM complex) for proteins targeted to the outer membrane. Proteins to be translocated across the OM have nascent membrane proteins that are bound by small Tim chaperones, which assist in delivering them to either the aforementioned Tim22, TIM23 or SAM complex. Like other mitosome-bearing protists (Dolezal et al. 2010; Lithgow and Schneider 2010; Waller et al. 2009), *Cryptosporidium* genomes encode only a few, modified import machinery proteins. In fact, the *C. parvum* genome lacks import proteins usually found in other apicomplexans and instead encodes just one small, atypical Tim protein *Cp*TimS (Alcock et al. 2012). The genome reduction in *C. parvum* to only one TIM complex is thought to reflect an ancestral state, observed previously in trypanosomes (Gentle et al. 2007; Schneider et al. 2008). In these Haematozoa, the authors show that the single complex drives both protein translocation and membrane protein assembly (Schneider et al. 2008). It is as yet unknown whether CpTimS can dock with the single TIM complex in the *C. parvum* mitosome. Using complementation, Alcock et al. (2012) have demonstrated that although

*Cp*TimS could be imported into yeast mitochondria and subsequently become homohexameric, it could not form a stable bond with yeast small Tims (Alcock et al. 2012). This observation suggests either that this protein may function on its own to drive both protein translocation and membrane protein assembly, as trypanosomes accomplish (Schneider et al. 2008), or may simply be a non-functional relic. While the *Cryptosporidium* mitochondrial protein import machinery is reduced, it appears to have adapted its proteins/complexes to accommodate the need of importing proteins into organelles. The single *C. parvum* Tim chaperone has also provided some insight into how the evolution of protein transferred from the TOM complexes to those of TIM and SAM might have occurred [see Alcock et al. (2012)].

4 Carbohydrate Metabolism

Although the carbohydrate metabolism of *C. parvum* is only partially understood, this apicomplexan primarily relies upon glycolysis for energy production (Abrahamsen et al. 2004; Entrala and Mascaro 1997; Liu et al. 2016). In fact, its core metabolism differs significantly from its near relatives *Plasmodium* and *Toxoplasma*, resembling instead that of the microsporidian *Encephalitozoon cuniculi* (Katinka et al. 2001; Keeling and Fast 2002; Tsaousis et al. 2008) and the microaerophilic protists *Giardia intestinalis* and *Entamoeba histolytica* [reviewed in Muller et al. (2012)]. All of these anaerobic protists lack both a functional TCA cycle and oxidative phosphorylation.

4.1 Pyruvate:NADP+ Oxidoreductase (PNO)

In contrast to some species of *Cryptosporidium*, the human species *C. parvum* and *C. hominis* do not have a TCA cycle (Liu et al. 2016). In *C. parvum*, pyruvate is converted to acetyl-CoA by a novel pyruvate:NADP+ oxidoreductase (*Cp*PNO) (Rotte et al. 2001) containing two distinct domains that are fused together: an N-terminal pyruvate:ferredoxin oxidoreductase (PFO), similar to that found in the anaerobic protists *Entamoeba histolytica*, *Giardia intestinalis* and *Trichomonas vaginalis* and a C-terminal NADPH-cytochrome P450 reductase (CPR). A nearly identical PFO-CPR fused protein has also been described from the euglenozoan *Euglena gracilis* (Inui et al. 1987; Nakazawa et al. 2000; Rotte et al. 2001), a distant relative of the *Alveolata* to which all apicomplexans belong. A similar fusion has been identified within the genome of the stramenopile, *Thalassiosira pseudonana*, a marine centric diatom (cited in (Ctrnacta et al. 2006).

Unlike Euglena PNO, which is a mitochondrial protein (Rotte et al. 2001), *Cp*PNO lacks a mitochondrial targeting presequence and does not localize within the relic organelle (Ctrnacta et al. 2006). In fact, sporozoites of *C. parvum* visualized both by confocal immunofluorescent microscopy and immunogold-labelled TEM confirm that *Cp*PNO has a unique compartmentalization: firstly within the cytosol as expected, but secondly within the crystalloid body, a *Cryptosporidium*-specific organelle posterior to the nucleus that is intimately associated with the mitosome (Keithly et al. 2005). Although the role of *Cp*PNO in carbohydrate metabolism is still unknown, a Narf-like hydrogenase has been identified in both *C. parvum* (Stejskal et al. 2003; Abrahamsen et al. 2004) and *C. hominis* (Xu et al. 2004), which may function to oxidize the NADPH produced by PNO during pyruvate decarboxylation. Unlike other previously considered amitochondriate protists (*Entamoeba*, *Giardia*, *Trichomonas*), neither of these human cryptosporidia possesses an [FeFe]-hydrogenase capable of transferring electrons produced during the oxidation of PFO (Horner et al. 2000). It is proposed that the acetyl-CoA resulting from the decarboxylation of pyruvate in *C. parvum* may be converted to malonyl-CoA (Templeton et al. 2004), an important precursor for fatty acid and polyketide biosynthesis (Zhu et al. 2002; Zhu 2004), both of which play an essential role in core fatty acid metabolism of this human pathogen [reviewed in Zhu (2004)]. However, it is unlikely that fatty acids serve as an energy source in *C. parvum* because so far only one enzyme of the fatty acid oxidative pathway, 3-hydroxyacyl-CoA dehydrogenase, has been identified in this apicomplexan (Cai et al. 2005). Interestingly, whole genome sequencing of the gregarine *Ascogregarina taiwanensis* shows that this species shares with *Cryptosporidium* type I fatty acid synthase that is thought to be a polyketide synthase (Templeton et al. 2010). In this paper, the authors have shown many similarities in core metabolic pathways of the *Cryptosporidium*/gregarine clade. On the other hand, there are a number of features absent in *Cryptosporidium* that *A. taiwanensis* shares with other apicomplexans, including the mannitol cycle, shikimate pathway, synthesis of select amino acids and de novo pyrimidine synthesis. These observations further support the idea that while *Cryptosporidium* underwent widespread gene loss during reductive evolution of its core metabolism, the genus has still maintained its metabolic capacities and lineage-specific affinities with gregarines at the base of the apicomplexan clade (Templeton et al. 2010).

4.2 Predicted End Products of Glycolysis

Substrate-level phosphorylation in *C. parvum* predicts that several organic end products, including acetate, lactate, malate and ethanol, will be produced and that, as in other anaerobic protists, one molecule of ATP will result when acetate is generated from acetyl-CoA (Crawford et al. 2003; Muller et al. 2012). The enzymes essential for yielding these products have been identified in the genomes of both *C. parvum* and *C. hominis* (Abrahamsen et al. 2004; Xu et al. 2004), and several, including alcohol-, lactate- and malate-dehydrogenases, have been biochemically and phylogenetically analysed (Madern et al. 2004). In addition, *C. parvum* lactate dehydrogenase (LDH) has been crystallized both in the absence and presence of its substrates and cofactors, whereas pyruvate kinase and glyceraldehyde 3-phosphate dehydrogenase have been crystallized only as apoproteins (Senkovich et al. 2005). These studies confirm the essential role of these enzymes in the core carbohydrate metabolism of *C. parvum*.

5 Energy Metabolism

As mentioned in the previous section, the core energy metabolism of *C. parvum* is primarily fermentative (Abrahamsen et al. 2004; Entrala and Mascaro 1997), resembling that of the microaerophilic protists *G. intestinalis*, *E. histolytica*, *Nyctotherus ovalis* and *T. vaginalis* [reviewed in Muller et al. (2012)]. Although *C. parvum* lacks a mitochondrial genome (Abrahamsen et al. 2004; Keeling 2004; Templeton et al. 2004) and oxidative phosphorylation typical of aerobic mitochondria, this apicomplexan does possess a mitochondrial-derived, cyanide-insensitive alternative oxidase (AOX) which has been proposed to transfer electrons from complex I through ubiquinone (Ellis et al. 1994; Roberts et al. 2004; Suzuki et al. 2004). Phylogenetic analyses place the AOX of *Cryptosporidium* with that of other eukaryotes and α-proteobacteria (Suzuki et al. 2004; Roberts et al. 2004). Although the *C. parvum* AOX targeting presequence predicts that this protein may serve as an alternative terminal electron acceptor for respiration by the organelle (Henriquez et al. 2005; Roberts et al. 2004), this hypothesis remains to be tested. The presence of AOX in *C. parvum* is thought to be an evolutionary adaptation for coping with oxidative stress in the host intestine (Putignani et al. 2004).

Furthermore, although ATP synthesis is thought to occur primarily by substrate-level phosphorylation in the cytosol of *C. parvum*, both α and β subunits of an F_0F_1 ATP synthase (complex V) containing mitochondrial targeting peptides have been found in this parasite (Henriquez et al. 2005; Putignani et al. 2004; Abrahamsen et al. 2004) as well as two genes encoding the ATP/ADP transporter pyridine nucleotide transhydrogenase (PNT) (Abrahamsen et al. 2004). Therefore, ATP may also be generated by proton-gradient-coupled oxidative phosphorylation in the *C. parvum* mitosome. In fact, in both *Plasmodium* and *Cryptosporidium*, it has been proposed that a membrane potential could be generated through a combined action of the matrix-located F_0F_1 of ATP synthase and one (or both) of these membrane-located ATP/ADP transporters (Painter et al. 2007). Briefly, since only a few ions are required for movement across the membrane, ATP might be hydrolysed by the F_0F_1 ATP synthase, and then the ADP generated by it exchanged by one of the two PNT transporters (Painter et al. 2007). Since this remnant organelle lacks both mitochondrial DNA and an electron transport chain, this would certainly help answer the question of how *C. parvum* generates a membrane potential for exporting Fe-S cluster component from the mitosome into the cytoplasm for the cytosolic iron-sulphur cluster assembly (Braymer and Lill 2017; Lill et al. 2015). Interestingly, PNT transporters are also present in the mitosomes of the microaerophilic protist *E. histolytica* (Chan et al. 2005) and microsporidians *Encephalitozoon cuniculi* (Tsaousis et al. 2008) and *Trachipleistophora hominis* (Dean et al. 2018).

As mentioned previously, genome sequencing and comparative analysis (Liu et al. 2016) show that there has been reductive evolution of mitosome metabolism in the genus *Cryptosporidium*. This includes energy production, the ETC and associated pathways. At one end of the spectrum are *C. muris* and *C. andersoni*, which possess a complete TCA cycle, nearly complete ET system and reliance upon glycolysis for energy metabolism (Fig. 4). The human cryptosporidia *C. parvum* and *C. hominis*, on the other hand, lack a TCA cycle and depend upon an AOX for phosphorylation. At the far end of the spectrum is *C. ubiquitum*, which has lost the entire ETC and with it, an ability to synthesize ubiquinone and polyisoprenoids (Fig. 4). This loss in *C. ubiquitum* supports the idea of reductive evolution of energy metabolism and ET systems in the *Cryptosporidium*/gregarine lineage. Since gregarines have retained the TCA cycle and a conventional ETC, they are likely the ancestors of this lineage. Within the genus *Cryptosporidium*, *C. andersoni* appears to be the earliest member, whereas *C. ubiquitum* is likely the most recent one (Liu et al. 2016).

These genome analyses also have implications for the development of therapeutic agents against species of *Cryptosporidium*, since intestinal *C. parvum* and *C. hominis* depend upon glycolysis for energy metabolism and AOX for electron transport, whereas gastric *C. andersoni* and *C. muris* have traditional TCA and ET chains. Intestinal *C. ubiquitum* lacks an AOX, suggesting that inhibitors of this enzyme will not affect its metabolism. Improved long-term in vitro cultivation of all life cycle stages (Miller et al. 2018a) and the development of genetic tools like CRISPR-Cas9 (Vinayak et al. 2015) could provide a way to improve intervention strategies against these pathogens.

6 The Crystalloid Body

6.1 Ultrastructural Morphology

One of the most puzzling structures within *C. parvum* sporozoites is the crystalloid body (CB), which is also in intimate contact with the mitosome, outer nuclear membrane and RER (Fig. 1) (Keithly et al. 2005; Miller et al. 2018a). Although it is still unclear whether this organelle is surrounded by a limiting membrane, or is simply a complex of closely packed membrane-bounded vesicles, it has been shown that like the mitosome (Ctrnacta et al. 2006), the CB takes up MitoTracker dyes (Ctrnacta et al. 2006; Kayser et al. 2002; Keithly et al. 2005). This observation suggests that the CB may have a mechanism for generating a membrane potential (see Sect. 5). Furthermore, in sporozoites its total volume is at least equal to that of the nucleus, which it appears to wrap around (Fig. 1) (Keithly et al. 2005; Ctrnacta et al. 2006). Together these data suggest that the CB may play an essential role during the life cycle of this parasite.

Ultrastructural data have indicated that the CB of *C. parvum* differs markedly from the paracrystalline CB seen in fish coccidians (Lukes 1992) and the randomly

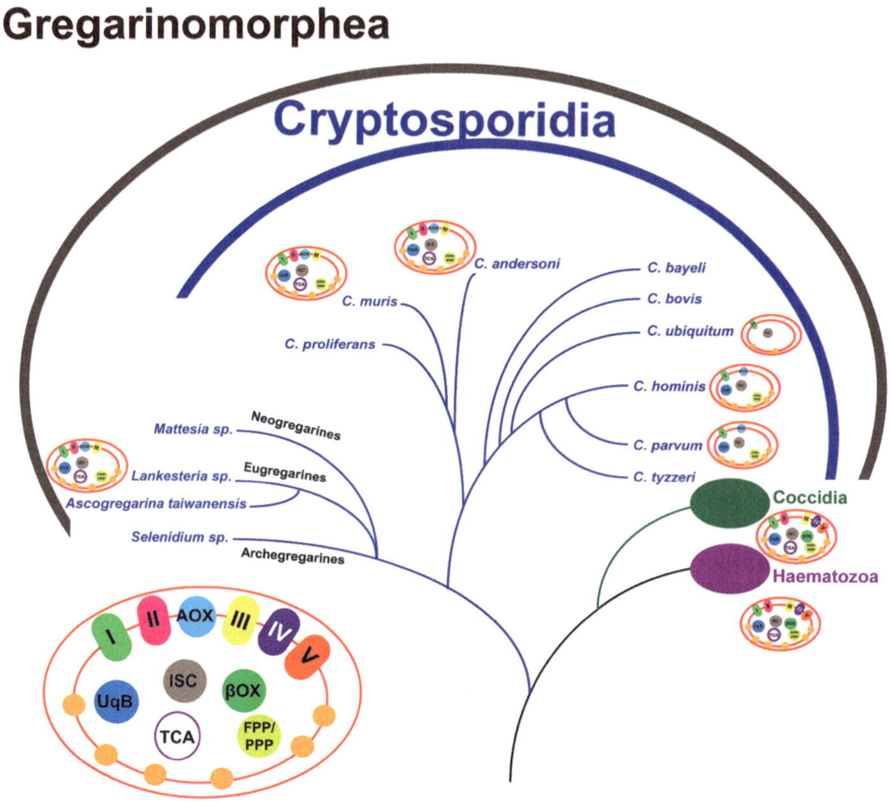

Fig. 4 Synthetic 18S phylogenetic tree of the reductive evolution of predicted mitochondrial-localizing proteins among *Cryptosporidium*/gregarine lineages. Although similarities between species of *Cryptosporidium* and gregarines have been well documented [reviewed in Ryan et al. (2016)], until recently, this relationship was not formally acknowledged. In 2014, however, the genus *Cryptosporidium* was formally reclassified within the phylum Apicomplexa into a new class Gregarinomorphea (subclass Cryptogregaria) which includes epicellular parasites of vertebrates possessing a gregarine-like feeder organelle and lacking an apicoplast (Cavalier-Smith 2014). This figure graphically represents the reductive evolution of energy metabolism within this new class. *Cryptosporidium* species and gregarines differ in energy metabolism from other Apicomplexa (Coccidia, Haematozoa) because they lack the apicoplast and have an AOX system. Species of *Cryptosporidium* have the most reduced organelle-derived metabolism with a total loss of the apicoplast and only a structurally relict mitochondrion lacking a genome. Gregarines are probably ancestral within the *Cryptosporidium*/gregarine lineage because they still have a TCA cycle and conventional ET chain. Within the genus *Cryptosporidium*, *C. andersoni* is probably an earlier member using both a TCA cycle and glycolysis for energy metabolism, with an almost conventional oxidative phosphorylation system, whereas *C. ubiquitum* is a more recent member of the genus having lost the entire ETC and associated biosynthesis of ubiquinone and polyisoprenoids. As seen on the diagram, this species retains only mitosomal ISC biosynthesis. Basic tree colours denote coccidian- and haematozoan-sequenced parasites (green and purple, respectively), species tackled within this investigation (red) and transcriptomic/genomic sequences (blue). Mitosome-labelled abbreviations: AOX, alternative oxidase; βOx, β-oxidation; I, II, II, IV and V, electron transport chain complexes; FPP, farnesyl diphosphate; ISC, iron-sulphur cluster biosynthesis; potential mitochondrial carriers, orange solid circles; PPP, polyprenyl diphosphate; TCA, tricarboxylic acid; UqB, ubiquinone biosynthesis

or regularly arrayed non-membranous particles observed among certain haemogregarines (Siddall and Desser 1992), haemoprotozoa, e.g. *Leucocytozoon* (Trefiak and Desser 1973), and archigregarines [reviewed in Desportes and Schrével (2013)]. Furthermore, although the *C. parvum* CB showed some indication of long-range order in computed diffraction patterns (i.e. Fourier power spectra), these were much less obvious than the cubic paracrystalline patterns of mitochondrial IM observed in starved free-living amoebae (Deng et al. 1999). The vesicular nature of the CB in sporozoites visualized in situ (Keithly et al. 2005) was nearly identical to that of subcellular fractions from partially disrupted sporozoites collected by sucrose density centrifugation and viewed by negative staining (Petry and Harris 1999). Another striking feature of the isolated CB noted in situ was the structural similarity of the negatively stained vesicles to those of ferritin, a non-toxic iron storage molecule typical of many eukaryotes (Harris and Scheffler 2002; Johnson et al. 2005). A reasonable hypothesis that remains to be tested is whether the CB might be a compartmentalized reservoir of soluble iron for import into the relic mitochondrion where ISC are assembled (Miller et al. 2018b).

6.2 Putative Functions

To date, there are still no data to suggest the exact structure, function or origins of this enigmatic organelle. The three-dimensional structure of the CB in sporozoites visualized by tomographic reconstruction suggests that the closely packed vesicles observed by TEM might actually be a series of interconnected channels [see Fig. 20 in Keithly et al. (2005)], but this observation needs a more detailed examination by high-resolution cryo-electron tomography for all stages of *C. parvum* life cycle. This technique permits observations of *C. parvum* in its "native state" following rapid freezing and freeze substitution of cells (Lucic et al. 2013). The in vitro cultivation of *C. parvum* in COLO-680N cells (Miller et al. 2018a) combined with cryo-electron tomography should permit a greater understanding of the structure and function of this remarkable organelle.

To date, the most compelling evidence that the CB of *C. parvum* sporozoites may be an essential component of the life cycle is the observation that *Cp*PNO appears to localize within this organelle (Figs. 3a–c, 4; Ctrnacta et al. 2006). As mentioned previously, it is well known that oxygen-sensitive PFO is the key enzyme for anaerobic core energy metabolism in some parasitic protists and that it may be compartmentalized either into the cytosol (*Giardia*, *Entamoeba*) or within the hydrogenosome (*Trichomonas*, *Nyctotherus*) (Embley et al. 2003; Muller et al. 2012). Interestingly, the PNO of *C. parvum* sporozoites is localized not only in the cytosol, where substrate-level phosphorylation generates two molecules of ATP [reviewed in Katinka et al. (2001), Williams and Keeling (2003), Liu et al. (2016)], but apparently also within the *Cryptosporidium parvum*-specific CB (Ctrnacta et al. 2006). The significance of this observation is yet to be determined and needs to be expanded to include all life cycle stages. Recent advances in the in vitro cultivation of species of *Cryptosporidium* (Miller et al. 2018a) combined with genetic manipulation of cells (Vinayak et al. 2015) should help facilitate an understanding of its function.

7 Diversity of Mitochondria Within the Clade

Recently, additional genomes of *Cryptosporidium* species have been (or are being) sequenced; (Liu et al. 2016; www.cryptodb.com). These genomes reveal a wide spectrum of adaptations of these organelles within the *Cryptosporidium*/gregarine lineage. As mentioned previously, *C. ubiquitum*, an intestinal parasite and close relative of *C. parvum* and *C. hominis*, has a mitosome that is predicted to contain even fewer mitochondrial proteins than *C. parvum*, whereas the mitosomes of *C. andersoni* and *C. muris*, both gastric parasites, possess a broader spectrum of mitochondrial proteins including those of the ETC and TCA cycle (Liu et al. 2016; Fig. 4). These in silico predictions have raised questions regarding the evolution and adaptation of mitochondria within the clade. An elegant attempt to answer some of these is discussed by Liu et al. (2016) and depicted here in Fig. 4. Briefly, this figure graphically shows the reductive evolution of predicted mitochondrial-localizing proteins within class Gregarinomorphea (subclass Cryptogregaria) in which species of *Cryptosporidium* are thought to have the most reduced organelle-derived metabolism, i.e. total loss of the apicoplast and a remnant mitochondrial structure (mitosome) lacking a genome. The *Cryptosporidium*/gregarine lineage differs in energy metabolism from other Apicomplexa because they lack an apicoplast and possess an alternative oxidase. Gregarines are thought to be the most ancestral of the *Cryptosporidium*/gregarine lineage because they still possess a TCA cycle and ET chain (Liu et al. 2016). Within the genus *Cryptosporidium*, *C. andersoni* is thought to be an earlier member of the clade because it uses both a TCA cycle and glycolysis for energy metabolism and retains a nearly conventional oxidative type phosphorylation, whereas *C. ubiquitum* is thought to be a more recent member because it has completely lost the ETC and with it the capacity for ubiquinone and polyisoprenoid biosynthesis. All its mitosome retains is ISC biosynthesis (Liu et al. 2016; Fig. 4).

8 Concluding Remarks and Future Perspectives

The phylum Apicomplexa consists of highly adapted parasites, with a unique cell biology, the members of which possess remarkably specialized organelles including the apical complex (a structure assisting invasion) and the apicoplast (a plastid without photosynthetic functions). Within class Gregarinomorphea (subclass Cryptogregaria), the *Cryptosporidium*/gregarine lineage lacks a metabolically active apicoplast, depending instead on host cells for nutrition. Consequently, this early branching group has lost many metabolic capabilities during reductive evolution, which is especially reflected in a loss of mitochondrial structure and function. For example, the human parasites *C. parvum* and *C. hominis* possess mitosomes that lack major metabolic capacities, including a functional tricarboxylic acid cycle and a cytochrome-based electron transport chain. On the other hand, human cryptosporidia have gained other metabolic features, i.e. tolerance to oxygen stress (Abrahamsen 2001) and a functional alternative oxidase (Roberts et al. 2004; Suzuki et al. 2004; Fig. 4), characteristic of anaerobic eukaryotes.

Research in the last few years, including the completion of the genome projects for various *Cryptosporidium* species (Liu et al. 2016), has provided a wealth of in silico information on the structure and likely functions of the relict mitochondrion (mitosome), as well as its associated membranes and organelles. The development of a long-term in vitro system for cultivating species of *Cryptosporidium* in COLO-680N cells (Miller et al. 2018a) would permit comparisons of gene expression during the life cycle (Mauzy et al. 2012) and, combined with the new CRISPR-Cas9 system (Vinayak et al. 2015), could enable genetic manipulation of all life cycle stages of *Cryptosporidium* species. Subsequently, these techniques should permit experimental determination of the functions of the proteins and pathways for mitochondrial and cytosolic iron-sulphur cluster biosynthesis, protein import and export into the mitosome, adaptations to oxygen stress, as well as new tools for dissecting the mechanisms of core carbohydrate and energy metabolism. Intriguing clues about the structure and function of the crystalloid body are beginning to emerge as a result of cryo-electron microscopy and tomography (Lucic et al. 2013), and it is likely that in the near future, the isolation of the CB and mitosome, combined with proteomics, will finally unravel the mystery of an intimate association of these organelles with the RER and with one another. Further exploration of this unusual compartmentalization might also provide insights into the evolutionary position of the cryptosporidia among mitosome- and hydrogenosome-bearing anaerobic eukaryotes, as well as eventually leading to better strategies for the development of new drugs against human and animal cryptosporidiosis.

Acknowledgements This research was supported by BBSRC research grant (BB/M009971/1) to ADT. We would like to thank Miklos Müller for proofreading the manuscript and his constructive comments.

References

Abrahamsen MS (2001) *Cryptosporidium parvum* genome project. Comp Funct Genomics 2 (1):19–21. https://doi.org/10.1002/cfg.67

Abrahamsen MS, Templeton TJ, Enomoto S et al (2004) Complete genome sequence of the apicomplexan, *Cryptosporidium parvum*. Science 304(5669):441–445. https://doi.org/10.1126/science.1094786

Aji T, Flanigan T, Marshall R et al (1991) Ultrastructural study of asexual development of *Cryptosporidium parvum* in a human intestinal cell line. J Protozool 38(6):82S–84S

Alcock F, Webb CT, Dolezal P et al (2012) A small Tim homohexamer in the relict mitochondrion of *cryptosporidium*. Mol Biol Evol 29(1):113–122. https://doi.org/10.1093/molbev/msr165
Alvarez-Pellitero P, Quiroga MI, Sitja-Bobadilla A et al (2004) *Cryptosporidium scophthalmi* n. Sp. (Apicomplexa: Cryptosporidiidae) from cultured turbot *Scophthalmus maximus*. Light and electron microscope description and histopathological study. Dis Aquat Org 62 (1–2):133–145. https://doi.org/10.3354/dao062133
Beyer TV, Svezhova NV, Sidorenko NV et al (2000) *Cryptosporidium parvum* (Coccidia, apicomplexa): some new ultrastructural observations on its endogenous development. Eur J Protistol 36(2):151–159. https://doi.org/10.1016/S0932-4739(00)80034-6
Boxma B, de Graaf RM, van der Staay GW et al (2005) An anaerobic mitochondrion that produces hydrogen. Nature 434(7029):74–79
Braymer JJ, Lill R (2017) Iron-sulfur cluster biogenesis and trafficking in mitochondria. J Biol Chem 292(31):12754–12763. https://doi.org/10.1074/jbc.R117.787101
Brown JR, Doolittle WF (1997) Archaea and the prokaryote-to-eukaryote transition. Microbiol Mol Biol Rev 61(4):456–502
Cai X, Fuller AL, McDougald LR et al (2003) Apicoplast genome of the coccidian *Eimeria tenella*. Gene 321:39–46. https://doi.org/10.1016/j.gene.2003.08.008
Cai X, Herschap D, Zhu G (2005) Functional characterization of an evolutionarily distinct phosphopantetheinyl transferase in the apicomplexan *Cryptosporidium parvum*. Eukaryot Cell 4(7):1211–1220. https://doi.org/10.1128/EC.4.7.1211-1220.2005
Cavalier-Smith T (2014) Gregarine site-heterogeneous 18S rDNA trees, revision of gregarine higher classification, and the evolutionary diversification of Sporozoa. Eur J Protistol 50 (5):472–495. https://doi.org/10.1016/j.ejop.2014.07.002
Chan KW, Slotboom DJ, Cox S et al (2005) A novel ADP/ATP transporter in the Mitosome of the Microaerophilic human parasite *Entamoeba histolytica*. Curr Biol 15(8):737–742
Clark CG, Roger AJ (1995) Direct evidence for secondary loss of mitochondria in *Entamoeba histolytica*. Proc Natl Acad Sci U S A 92(14):6518–6521
Crawford MJ, Fraunholz MJ, Roos D (2003) Energy Metabolism in the Apicomplexa. In: Marr JJ, Nielsen TW, Kouiecki RW (eds) Molecular Medical parasitology, vol 7. Academic, New York, pp 154–169
Csordas G, Renken C, Varnai P et al (2006) Structural and functional features and significance of the physical linkage between ER and mitochondria. J Cell Biol 174(7):915–921. https://doi.org/10.1083/jcb.200604016
Ctrnacta V, Ault JG, Stejskal F et al (2006) Localization of pyruvate:NADP+ oxidoreductase in sporozoites of *Cryptosporidium parvum*. J Eukaryot Microbiol 53(4):225–231. https://doi.org/10.1111/j.1550-7408.2006.00099.x
Dean P, Sendra KM, Williams TA et al (2018) Transporter gene acquisition and innovation in the evolution of Microsporidia intracellular parasites. Nat Commun 9(1):1709. https://doi.org/10.1038/s41467-018-03923-4
Deng Y, Marko M, Buttle KF et al (1999) Cubic membrane structure in amoeba (*Chaos carolinensis*) mitochondria determined by electron microscopic tomography. J Struct Biol 127 (3):231–239. https://doi.org/10.1006/jsbi.1999.4147
Desportes I, Schrével J (2013) Treatise on zoology - anatomy, taxonomy, biology. The gregarines. The early branching Apicomplexa. BRILL, Leiden
Dolezal P, Smid O, Rada P et al (2005) Giardia mitosomes and trichomonad hydrogenosomes share a common mode of protein targeting. Proc Natl Acad Sci U S A 102(31):10924–10929
Dolezal P, Likic V, Tachezy J et al (2006) Evolution of the molecular machines for protein import into mitochondria. Science 313(5785):314–318
Dolezal P, Dagley MJ, Kono M et al (2010) The essentials of protein import in the degenerate mitochondrion of *Entamoeba histolytica*. PLoS Pathog 6(3):e1000812. https://doi.org/10.1371/journal.ppat.1000812
Ellis JE, Setchell KD, Kaneshiro ES (1994) Detection of ubiquinone in parasitic and free-living protozoa, including species devoid of mitochondria. Mol Biochem Parasitol 65(2):213–224. https://doi.org/10.1016/0166-6851(94)90073-6

Embley TM, van der Giezen M, Horner DS et al (2003) Mitochondria and hydrogenosomes are two forms of the same fundamental organelle. Philos Trans R Soc Lond Ser B Biol Sci 358 (1429):191–201; discussion 201–202. https://doi.org/10.1098/rstb.2002.1190

Entrala E, Mascaro C (1997) Glycolytic enzyme activities in *Cryptosporidium parvum* oocysts. FEMS Microbiol Lett 151(1):51–57. https://doi.org/10.1111/j.1574-6968.1997.tb10393.x

Fayer R, Xiao L (2007) Cryptosporidium and cryptosporidiosis. CRC Press

Fenchel T, Perry T, Thane A (1977) Anaerobiosis and symbiosis with bacteria in free-living ciliates. J Protozool 24(1):154–163

Freibert SA, Goldberg AV, Hacker C et al (2017) Evolutionary conservation and in vitro reconstitution of microsporidian iron-sulfur cluster biosynthesis. Nat Commun 8:13932. https://doi.org/10.1038/ncomms13932

Frey TG, Perkins GA, Ellisman MH (2006) Electron tomography of membrane-bound cellular organelles. Annu Rev Biophys Biomol Struct 35:199–224. https://doi.org/10.1146/annurev.biophys.35.040405.102039

Fry M, Beesley JE (1991) Mitochondria of mammalian Plasmodium spp. Parasitology 102 (Pt 1):17–26

Gentle IE, Perry AJ, Alcock FH, Likić VA, Dolezal P, Ng ET, Purcell AW, McConnville M, Naderer T, Chanez AL, Charrière F, Aschinger C, Schneider A, Tokatlidis K, Lithgow T (2007) Conserved motifs reveal details of ancestry and structure in the small TIM chaperones of the mitochondrial intermembrane space. Mol Biol Evol 24(5):1149–1160

Goldberg AV, Molik S, Tsaousis AD et al (2008) Localization and functionality of microsporidian iron-Sulphur cluster assembly proteins. Nature 452(7187):624–628. https://doi.org/10.1038/nature06606

Harris JR, Scheffler D (2002) Routine preparation of air-dried negatively stained and unstained specimens on holey carbon support films: a review of applications. Micron 33(5):461–480. https://doi.org/10.1016/S0968-4328(01)00039-7

Henriquez FL, Richards TA, Roberts F et al (2005) The unusual mitochondrial compartment of *Cryptosporidium parvum*. Trends Parasitol 21(2):68–74. https://doi.org/10.1016/j.pt.2004.11.010

Horner DS, Foster PG, Embley TM (2000) Iron hydrogenases and the evolution of anaerobic eukaryotes. Mol Biol Evol 17(11):1695–1709

Inui H, Ono K, Miyatake K et al (1987) Purification and characterization of pyruvate:NADP+ oxidoreductase in *Euglena gracilis*. J Biol Chem 262(19):9130–9135

Jerlstrom-Hultqvist J, Einarsson E, Xu F et al (2013) Hydrogenosomes in the diplomonad *Spironucleus salmonicida*. Nat Commun 4:2493. https://doi.org/10.1038/ncomms3493

Johnson E, Cascio D, Sawaya MR et al (2005) Crystal structures of a tetrahedral open pore ferritin from the hyperthermophilic archaeon *Archaeoglobus fulgidus*. Structure 13(4):637–648. https://doi.org/10.1016/j.str.2005.01.019

Katinka MD, Duprat S, Cornillot E et al (2001) Genome sequence and gene compaction of the eukaryote parasite *Encephalitozoon cuniculi*. Nature 414(6862):450–453

Kayser O, Waters WR, Woods KM et al (2002) Evaluation of in vitro and in vivo activity of benzindazole-4,9-quinones against *Cryptosporidium parvum*. J Antimicrob Chemother 50 (6):975–980

Keeling PJ (2004) Reduction and compaction in the genome of the apicomplexan parasite *Cryptosporidium parvum*. Dev Cell 6(5):614–616. https://doi.org/10.1016/S1534-5807(04)00135-2

Keeling PJ, Fast NM (2002) Microsporidia: biology and evolution of highly reduced intracellular parasites. Annu Rev Microbiol 56:93–116

Keithly JS, Langreth SG, Buttle KF et al (2005) Electron tomographic and ultrastructural analysis of the *Cryptosporidium parvum* relict mitochondrion, its associated membranes, and organelles. J Eukaryot Microbiol 52(2):132–140. https://doi.org/10.1111/j.1550-7408.2005.04-3317.x

Kita K, Hirawake H, Miyadera H et al (2002) Role of complex II in anaerobic respiration of the parasite mitochondria from *Ascaris suum* and *Plasmodium falciparum*. Biochim Biophys Acta 1553(1–2):123–139. https://doi.org/10.1016/S0005-2728(01)00237-7

Krungkrai J, Burat D, Kudan S et al (1999) Mitochondrial oxygen consumption in asexual and sexual blood stages of the human malarial parasite, *Plasmodium falciparum*. Southeast Asian J Trop Med Public Health 30(4):636–642

Krungkrai J, Prapunwattana P, Krungkrai SR (2000) Ultrastructure and function of mitochondria in gametocytic stage of *Plasmodium falciparum*. Parasite 7(1):19–26. https://doi.org/10.1051/parasite/2000071019

LaGier MJ, Tachezy J, Stejskal F et al (2003) Mitochondrial-type iron-sulfur cluster biosynthesis genes (IscS and IscU) in the apicomplexan *Cryptosporidium parvum*. Microbiology 149 (Pt 12):3519–3530. https://doi.org/10.1099/mic.0.26365-0

Lantsman Y, Tan KS, Morada M et al (2008) Biochemical characterization of a mitochondrial-like organelle from *Blastocystis* sp. subtype 7. Microbiology 154.(Pt 9:2757–2766. https://doi.org/10.1099/mic.0.2008/017897-0

Leon-Avila G, Tovar J (2004) Mitosomes of *Entamoeba histolytica* are abundant mitochondrion-related remnant organelles that lack a detectable organellar genome. Microbiology 150 (Pt 5):1245–1250

Lill R, Dutkiewicz R, Freibert SA et al (2015) The role of mitochondria and the CIA machinery in the maturation of cytosolic and nuclear iron-sulfur proteins. Eur J Cell Biol 94(7–9):280–291. https://doi.org/10.1016/j.ejcb.2015.05.002

Lithgow T, Schneider A (2010) Evolution of macromolecular import pathways in mitochondria, hydrogenosomes and mitosomes. Philos Trans R Soc Lond Ser B Biol Sci 365(1541):799–817. https://doi.org/10.1098/rstb.2009.0167

Liu S, Roellig DM, Guo Y et al (2016) Evolution of mitosome metabolism and invasion-related proteins in *Cryptosporidium*. BMC Genomics 17(1):1006–016-3343-5. https://doi.org/10.1186/s12864-016-3343-5

Lucic V, Forster F, Baumeister W (2005) Structural studies by electron tomography: from cells to molecules. Annu Rev Biochem 74:833–865. https://doi.org/10.1146/annurev.biochem.73.011303.074112

Lucic V, Rigort A, Baumeister W (2013) Cryo-electron tomography: the challenge of doing structural biology in situ. J Cell Biol 202(3):407–419. https://doi.org/10.1083/jcb.201304193

Lukes J (1992) Life cycle of *Goussia pannonica* (Molnar, 1989) (Apicomplexa, Eimeriorina), an Extracytoplasmic Coccidium from the white bream *Blicca bjoerkna*. J Protozool 39 (4):484–494. https://doi.org/10.1111/j.1550-7408.1992.tb04836.x

Madern D, Cai X, Abrahamsen MS et al (2004) Evolution of *Cryptosporidium parvum* lactate dehydrogenase from malate dehydrogenase by a very recent event of gene duplication. Mol Biol Evol 21(3):489–497. https://doi.org/10.1093/molbev/msh042

Mannella CA (2006) The relevance of mitochondrial membrane topology to mitochondrial function. Biochim Biophys Acta 1762(2):140–147. https://doi.org/10.1016/j.bbadis.2005.07.001

Mannella CA, Pfeiffer DR, Bradshaw PC et al (2001) Topology of the mitochondrial inner membrane: dynamics and bioenergetic implications. IUBMB Life 52(3–5):93–100. https://doi.org/10.1080/15216540152845885

Maralikova B, Ali V, Nakada-Tsukui K et al (2010) Bacterial-type oxygen detoxification and iron-sulfur cluster assembly in amoebal relict mitochondria. Cell Microbiol 12(3):331–342. https://doi.org/10.1111/j.1462-5822.2009.01397.x

Mauzy MJ, Enomoto S, Lancto CA et al (2012) The *Cryptosporidium parvum* transcriptome during in vitro development. PLoS One 7(3):e31715. https://doi.org/10.1371/journal.pone.0031715

Melo EJ, Attias M, De Souza W (2000) The single mitochondrion of tachyzoites of *Toxoplasma gondii*. J Struct Biol 130(1):27–33. https://doi.org/10.1006/jsbi.2000.4228

Mi-Ichi F, Takeo S, Takashima E et al (2003) Unique properties of respiratory chain in *Plasmodium falciparum* mitochondria. Adv Exp Med Biol 531:117–133

Miller CN, Josse L, Brown I et al (2018a) A cell culture platform for *Cryptosporidium* that enables long-term cultivation and new tools for the systematic investigation of its biology. Int J Parasitol 48(3–4):197–201. https://doi.org/10.1016/j.ijpara.2017.10.001

Miller CN, Josse L, Tsaousis AD (2018b) Localization of Fe-S biosynthesis machinery in *Cryptosporidium parvum* Mitosome. J Eukaryot Microbiol 65:913. https://doi.org/10.1111/jeu.12663

Mogi T, Kita K (2010) Diversity in mitochondrial metabolic pathways in parasitic protists *Plasmodium* and *Cryptosporidium*. Parasitol Int 59(3):305–312. https://doi.org/10.1016/j.parint.2010.04.005

Muller M, Mentel M, van Hellemond JJ et al (2012) Biochemistry and evolution of anaerobic energy metabolism in eukaryotes. Microbiol Mol Biol Rev 76(2):444–495. https://doi.org/10.1128/MMBR.05024-11

Nakazawa M, Inui H, Yamaji R et al (2000) The origin of pyruvate: NADP+ oxidoreductase in mitochondria of *Euglena gracilis*. FEBS Lett 479(3):155–156

Nasirudeen AM, Tan KS (2004) Isolation and characterization of the mitochondrion-like organelle from Blastocystis hominis. J Microbiol Methods 58(1):101–109

Ovciarikova J, Lemgruber L, Stilger KL et al (2017) Mitochondrial behaviour throughout the lytic cycle of *Toxoplasma gondii*. Sci Rep 7:42746. https://doi.org/10.1038/srep42746

Painter HJ, Morrisey JM, Mather MW et al (2007) Specific role of mitochondrial electron transport in blood-stage *Plasmodium falciparum*. Nature 446(7131):88–91. https://doi.org/10.1038/nature05572

Petry F, Harris JR (1999) Ultrastructure, fractionation and biochemical analysis of *Cryptosporidium parvum* sporozoites. Int J Parasitol 29(8):1249–1260. https://doi.org/10.1016/S0020-7519(99)00080-6

Putignani L, Tait A, Smith HV et al (2004) Characterization of a mitochondrion-like organelle in *Cryptosporidium parvum*. Parasitology 129(Pt 1):1–18

Richardson E, Zerr K, Tsaousis A et al (2015) Evolutionary cell biology: functional insight from "endless forms most beautiful". Mol Biol Cell 26(25):4532–4538. https://doi.org/10.1091/mbc.E14-10-1433

Riordan CE, Langreth SG, Sanchez LB et al (1999) Preliminary evidence for a mitochondrion in *Cryptosporidium parvum*: phylogenetic and therapeutic implications. J Eukaryot Microbiol 46(5):52S–55S

Riordan CE, Ault JG, Langreth SG et al (2003) *Cryptosporidium parvum* Cpn60 targets a relict organelle. Curr Genet 44(3):138–147. https://doi.org/10.1007/s00294-003-0432-1

Roberts CW, Roberts F, Henriquez FL et al (2004) Evidence for mitochondrial-derived alternative oxidase in the apicomplexan parasite *Cryptosporidium parvum*: a potential anti-microbial agent target. Int J Parasitol 34(3):297–308. https://doi.org/10.1016/j.ijpara.2003.11.002

Rotte C, Stejskal F, Zhu G et al (2001) Pyruvate: NADP+ oxidoreductase from the mitochondrion of *Euglena gracilis* and from the apicomplexan *Cryptosporidium parvum*: a biochemical relic linking pyruvate metabolism in mitochondriate and amitochondriate protists. Mol Biol Evol 18(5):710–720. https://doi.org/10.1093/oxfordjournals.molbev.a003853

Ryan U, Paparini A, Monis P et al (2016) It's official – *Cryptosporidium* is a gregarine: what are the implications for the water industry? Water Res 105:305–313. https://doi.org/10.1016/j.watres.2016.09.013

Schneider A, Bursac D, Lithgow T (2008) The direct route: a simplified pathway for protein import into the mitochondrion of trypanosomes. Trends Cell Biol 18(1):12–18. https://doi.org/10.1016/j.tcb.2007.09.009

Senkovich O, Speed H, Grigorian A et al (2005) Crystallization of three key glycolytic enzymes of the opportunistic pathogen *Cryptosporidium parvum*. Biochim Biophys Acta 1750(2):166–172. https://doi.org/10.1016/j.bbapap.2005.04.009

Siddall ME, Desser SS (1992) Ultrastructure of gametogenesis and Sporogony of Haemogregarina (sensu lato) myoxocephali (Apicomplexa: Adeleina) in the marine leech *Malmiana scorpii*. J Protozool 39(5):545–554. https://doi.org/10.1111/j.1550-7408.1992.tb04849.x

Slapeta J, Keithly JS (2004) *Cryptosporidium parvum* mitochondrial-type HSP70 targets homologous and heterologous mitochondria. Eukaryot Cell 3(2):483–494

Stejskal F, Slapeta J, Ctrnacta V et al (2003) A Narf-like gene from *Cryptosporidium parvum* resembles homologues observed in aerobic protists and higher eukaryotes. FEMS Microbiol Lett 229(1):91–96. https://doi.org/10.1016/S0378-1097(03)00794-8

Suzuki T, Hashimoto T, Yabu Y et al (2004) Direct evidence for cyanide-insensitive quinol oxidase (alternative oxidase) in apicomplexan parasite *Cryptosporidium parvum*: phylogenetic and therapeutic implications. Biochem Biophys Res Commun 313(4):1044–1052. https://doi.org/10.1016/j.bbrc.2003.12.038

Tan KS (2008) New insights on classification, identification, and clinical relevance of *Blastocystis* spp. Clin Microbiol Rev 21(4):639–665. https://doi.org/10.1128/CMR.00022-08
Templeton TJ, Iyer LM, Anantharaman V et al (2004) Comparative analysis of apicomplexa and genomic diversity in eukaryotes. Genome Res 14(9):1686–1695. https://doi.org/10.1101/gr.2615304
Templeton TJ, Enomoto S, Chen WJ et al (2010) A genome-sequence survey for *Ascogregarina taiwanensis* supports evolutionary affiliation but metabolic diversity between a gregarine and *Cryptosporidium*. Mol Biol Evol 27(2):235–248. https://doi.org/10.1093/molbev/msp226
Tetley L, Brown SM, McDonald V et al (1998) Ultrastructural analysis of the sporozoite of *Cryptosporidium parvum*. Microbiology 144. (Pt 12)(Pt 12:3249–3255. https://doi.org/10.1099/00221287-144-12-3249
Thompson RC, Olson ME, Zhu G et al (2005) *Cryptosporidium* and cryptosporidiosis. Adv Parasitol 59:77–158
Tovar J, Fischer A, Clark CG (1999) The mitosome, a novel organelle related to mitochondria in the amitochondrial parasite *Entamoeba histolytica*. Mol Microbiol 32(5):1013–1021
Tovar J, Leon-Avila G, Sanchez LB et al (2003) Mitochondrial remnant organelles of *Giardia* function in iron-Sulphur protein maturation. Nature 426(6963):172–176
Trefiak WD, Desser SS (1973) Crystalloid inclusions in species of *Leucocytozoon, Parahaemoproteus*, and *Plasmodium*. J Protozool 20(1):73–80
Tsaousis AD, Kunji ER, Goldberg AV et al (2008) A novel route for ATP acquisition by the remnant mitochondria of *Encephalitozoon cuniculi*. Nature 453(7194):553–556. https://doi.org/10.1038/nature06903
Tsaousis AD, Gaston D, Stechmann A et al (2011) A functional Tom70 in the human parasite *Blastocystis* sp.: implications for the evolution of the mitochondrial import apparatus. Mol Biol Evol 28(1):781–791. https://doi.org/10.1093/molbev/msq252
Uni S, Iseki M, Maekawa T et al (1987) Ultrastructure of *Cryptosporidium muris* (strain RN 66) parasitizing the murine stomach. Parasitol Res 74(2):123–132
van der Giezen M (2005) Endosymbiosis: past and present. Heredity
van der Giezen M, Tovar J (2005) Degenerate mitochondria. EMBO Rep 6(6):525–530
van der Giezen M, Tovar J, Clark CG (2005) Mitochondrion-derived organelles in protists and fungi. Int Rev Cytol 244:175–225. https://doi.org/10.1016/S0074-7696(05)44005-X
van Dooren GG, Marti M, Tonkin CJ et al (2005) Development of the endoplasmic reticulum, mitochondrion and apicoplast during the asexual life cycle of *Plasmodium falciparum*. Mol Microbiol 57(2):405–419. https://doi.org/10.1111/j.1365-2958.2005.04699.x
van Hoek AH, Akhmanova AS, Huynen MA et al (2000) A mitochondrial ancestry of the hydrogenosomes of *Nyctotherus ovalis*. Mol Biol Evol 17(1):202–206. https://doi.org/10.1093/oxfordjournals.molbev.a026234
Vinayak S, Pawlowic MC, Sateriale A et al (2015) Genetic modification of the diarrhoeal pathogen *Cryptosporidium parvum*. Nature 523(7561):477–480. https://doi.org/10.1038/nature14651
Waller RF, Jabbour C, Chan NC et al (2009) Evidence of a reduced and modified mitochondrial protein import apparatus in microsporidian mitosomes. Eukaryot Cell 8(1):19–26. https://doi.org/10.1128/EC.00313-08
Wiedemann N, Pfanner N (2017) Mitochondrial machineries for protein import and assembly. Annu Rev Biochem 86:685–714. https://doi.org/10.1146/annurev-biochem-060815-014352
Williams BA, Keeling PJ (2003) Cryptic organelles in parasitic protists and fungi. Adv Parasitol 54:9–68
Williams BA, Hirt RP, Lucocq JM et al (2002) A mitochondrial remnant in the microsporidian *Trachipleistophora hominis*. Nature 418(6900):865–869
Xu P, Widmer G, Wang Y et al (2004) The genome of *Cryptosporidium hominis*. Nature 431(7012):1107–1112. https://doi.org/10.1038/nature02977
Zhu G (2004) Current progress in the fatty acid metabolism in *Cryptosporidium parvum*. J Eukaryot Microbiol 51(4):381–388
Zhu G, LaGier MJ, Stejskal F et al (2002) *Cryptosporidium parvum*: the first protist known to encode a putative polyketide synthase. Gene 298(1):79–89. https://doi.org/10.1016/S0378-1119(02)00931-9

The Mitochondrion-Related Organelles of *Blastocystis*

Anastasios D. Tsaousis, Nigel Yarlett, and Kevin S. W. Tan

Contents

1 Introduction 268
2 *Blastocystis* Genome, Adaptations and Lateral Gene Transfer 269
3 *Blastocystis* Mitochondrial DNA 270
4 Cell Biology and Functions 272
4.1 Adaptations to Oxygen 273
4.2 Fe-S Cluster Assembly Biosynthesis 274
4.3 Mitochondrial Protein Import 274
4.4 Glycolysis 275
5 Energy Metabolism: Biochemistry 275
6 Morphology 276
7 Role of *Blastocystis* Mitochondria in Cell Death 280
8 Concluding Remarks and Future Perspectives 281
References 282

Abstract *Blastocystis* is an anaerobic microbial eukaryote that inhabits the gut of various animals. While it was considered to be a parasite, its pathogenicity is controversial, since recent estimations suggest that *Blastocystis* could be present in more than one billion humans. *Blastocystis* belongs to the stramenopiles (heterokonts), a diverse group of eukaryotes that also include unicellular diatoms and giant multicellular kelps. The *Blastocystis* cell lacks typical features of other

---

A. D. Tsaousis (✉)
Laboratory of Molecular and Evolutionary Parasitology, RAPID Group, School of Biosciences, University of Kent, Canterbury, Kent, UK
e-mail: a.tsaousis@kent.ac.uk

N. Yarlett
Haskins Laboratories, Department of Chemistry and Physical Science, Pace University, New York, NY, USA

K. S. W. Tan
Laboratory of Molecular and Cellular Parasitology, Department of Microbiology and Immunology, Yong Loo Lin School of Medicine, National University of Singapore, Singapore, Singapore

J. Tachezy (ed.), *Hydrogenosomes and Mitosomes: Mitochondria of Anaerobic Eukaryotes*, Microbiology Monographs 9,
https://doi.org/10.1007/978-3-030-17941-0_11

stramenopiles; its genome has laterally acquired many genes from other prokaryotes and eukaryotes, resulting in having exceptional biochemistry and unique adaptations of its mitochondria. For example, *Blastocystis* mitochondrion-related organelles (MROs) have characteristics of typical mitochondria, including Complexes I and II from electron transport chain, mitochondrial DNA, Fe-S cluster assembly and amino acid metabolism, but also proteins that are typically present in obligate anaerobes including FeFe-hydrogenase, pyruvate metabolism and alternative oxidase. Some of the pathways have been localized and biochemically characterized, providing a better understanding on the functions of these organelles. In this chapter, we will present and examine the *status quo* regarding the biology and morphology of *Blastocystis* MROs, and we will discuss future avenues on exploring even further adaptations of these organelles.

1 Introduction

The compartmentalization of energy generation provides a mechanism for the increased efficiency of high-energy bond transfer to form the ultimate cellular fuel ATP and has been suggested to be the driving force behind the evolution of the mitochondrion. This viewpoint is supported by studies of the mitochondrial proteome, which demonstrates that proteins of eubacterial origin predominantly have roles in bioenergetics, whereas the eukaryotic-specific mitochondrial proteins perform transport and regulatory functions (Gray et al. 2001). This is consistent with the proposed α-proteobacterial origin for this organelle and the switch to aerobic respiration (Kurland and Andersson 2000). The identification and characterization of organelles, which shuttle electrons to protons forming molecular hydrogen (hydrogenosome) under anaerobic conditions, have challenged the concept that the mitochondrion is the sole surviving example of the evolution of an energy-generating organelle (Lindmark and Muller 1973; Yarlett et al. 1981, 1986; Akhmanova et al. 1998; Muller et al. 2012). However, as molecular techniques improved, it became clear that many anaerobes previously considered to be amitochondriate in fact harbour membranes or highly reduced structures, which can be considered of mitochondrial origin (Tovar et al. 1999; Williams et al. 2002; Keithly et al. 2005; Tsaousis et al. 2008). Even the most characterized organelle, the hydrogenosome, which is also the most complex of the anaerobic structures, has been demonstrated to have mitochondrial features (Muller et al. 2012; Bui et al. 1996; Embley et al. 2003; Hrdy et al. 2004; Tachezy et al. 2001). These findings have coalesced into essentially three concepts for the origin and evolution of energy-generating organelles: (1) these organelles arose multiple times from multiple origins; (2) the various structures that are present in all extant organisms are of the same eubacterial origin that subsequently diverged; and (3) the variations observed in organelles is the result of multiple invasions which have evolved to utilize key functions as each wave occurred with the result that distinctions between each ancestor have blurred into the present-day collage. There is compelling evidence that the hydrogenosomes described from the trichomonads, rumen protists, rumen fungi and protists inhabiting the cockroach hindgut have many mitochondrial features including the recognition of specific targeting sequences (Dacks et al. 2006), ADP/ATP carriers (Dyall et al. 2000; Tjaden et al. 2004), the production of ATP (Hrdy et al. 2004), the iron-sulphur cluster assembly (Tachezy et al. 2001), etc. These studies prompted examination of other previously considered amitochondriate organisms which led to the finding that many of these organisms possessed mitochondrion-related organelles (MROs) including the mitosome of *Entamoeba* (Tovar et al. 1999), the cryptic organelle of *Giardia* (Tovar et al. 2003), the relict organelles in *Cryptosporidium parvum* (Keithly et al. 2005) and the microsporidians, *Trachipleistophora hominis* and *Encephalitozoon cuniculi* (Williams et al. 2002; Tsaousis et al. 2008). Despite little homology to the mitochondrion in many cases, these organelles have many features in common with each other and the mitochondrion. These studies have reinforced the conclusion that all of these organelles arose from a common origin.

In this chapter, we will focus on *Blastocystis*, an obligate anaerobic microbial eukaryote, also found in the guts of several metazoans, but its pathogenicity is a mystery since it varies among reports from different geographic areas (Stensvold and Clark 2016). *Blastocystis* is extremely diverse, having (at least) 17 genetic subtypes, some of which are either host specific or have defined worldwide distribution (Clark 1997; Betts et al. 2018; Stensvold 2013). *Blastocystis* can remain in the intestine for weeks, months, or even years and can withstand fluctuations of oxygen in the gut and symptoms that include diarrhoea, abdominal pain, weight loss, constipation and many others (Andersen and Stensvold 2016). Despite the prevalence of this parasite among humans, our knowledge on the genetic diversity, host specificity, geographic distribution and even cell biology still remains partial. Molecular phylogenetic studies have shown that it is related to the stramenopiles (heterokonts), a group of predominantly photosynthetic eukaryotes including diatoms, brown algae and oomycetes (Silberman et al. 1996). Regardless of this relationship, *Blastocystis* lacks typical stramenopile morphology (including flagella and tubular hairs) evidently as a result of secondary loss (Clark et al. 2013). *Blastocystis* is currently the only organism within this group to harbour reduced mitochondrion-related organelles (MROs), with some of their unique adaptations to be discussed in this chapter.

2 *Blastocystis* Genome, Adaptations and Lateral Gene Transfer

Over the past decade, several genome sequencing projects were generated for various *Blastocystis* subtypes (STs), including ST7 (Denoeud et al. 2011), ST4 (Wawrzyniak et al. 2015) and more recently ST1 (NandII) (Gentekaki et al. 2017).

The genome size of *Blastocystis* was predicted to be around 19 Mbp, with ST1 being the largest one encoding around 6544 protein-coding genes. Remarkably, there is some variability between the different subtypes, and the percentage of proteins unique to each ST ranges from 6.2% to 20.5% (Gentekaki et al. 2017). More importantly, these genomes have identified many cellular adaptations of *Blastocystis*, including complete membrane-trafficking system and a near-complete meiotic toolkit, indicating a potential sexual cycle. Interestingly, a genome-wide analysis revealed that *Blastocystis*, especially ST1, has a "mixed" genome: 2.5% of its genes originated by recent lateral gene transfer laterally acquired from diverse prokaryotes and other eukaryotes (Eme et al. 2017). These subsequently expressed proteins are associated with numerous pathways including carbohydrate scavenging and metabolism, anaerobic amino acid and nitrogen metabolism, oxygen stress resistance and pH homeostasis. Moreover, the analysis identified proteins linked with secretion that are potentially involved in infection, escaping host defence and even affect composition of the prokaryotic microbiome and inflammation of the gut (Gentekaki et al. 2017; Eme et al. 2017). In regard to the metabolic capacities of *Blastocystis* and its mitochondria, it seems that the anaerobic metabolism of their MROs was shaped by lateral gene transfer acquisitions, including proteins involved in pyruvate decarboxylation [pyruvate dehydrogenase complex (PDC), pyruvate-ferredoxin oxidoreductase (PFOR) and pyruvate NADP+ oxidoreductase (PNO)], biosynthesis of rhodoquinone and hydrogen production (Denoeud et al. 2011; Wawrzyniak et al. 2015; Eme et al. 2017), among others (further discussed below). In general, the most recent analysis in *Blastocystis* ST1 has identified 371 mitochondrial predicted proteins (Gentekaki et al. 2017) presenting elements of canonical mitochondria, with a highly reduced cytochrome-mediated electron transport chain with only complexes I and II being present along with an expanded repertoire of functions, including lipid, cofactor and vitamin biosynthesis, as well as proteins that could be involved in regulating mitochondrial morphology and MRO/endoplasmic reticulum (ER) interactions (see below and Fig. 1). Overall, *Blastocystis*' organelles contain (anaerobic) features characteristic to hydrogenosomes (Stechmann et al. 2008; Gentekaki et al. 2017), which are unique in *Blastocystis* and not in other stramenopiles, thus blurring the boundaries between typical mitochondria and hydrogenosomes (Stechmann et al. 2008).

3 *Blastocystis* Mitochondrial DNA

Unlike the nuclear genome sequencing projects, there have been numerous reports on sequencing the mitochondrial DNA (mtDNA) from various *Blastocystis* subtypes. To our knowledge, 11 genomes from ST1, ST2, ST3, ST4, ST6, ST7, ST8 and ST9 have been sequenced so far, and comparative analyses have shown that they are identical in both gene content and gene order (Jacob et al. 2016). *Blastocystis* MRO genomes are smaller than other stramenopiles' mtDNA and contain 27 protein-coding genes (35–44 in other stramenopiles), 2 ribosomal RNA genes and 16 transfer

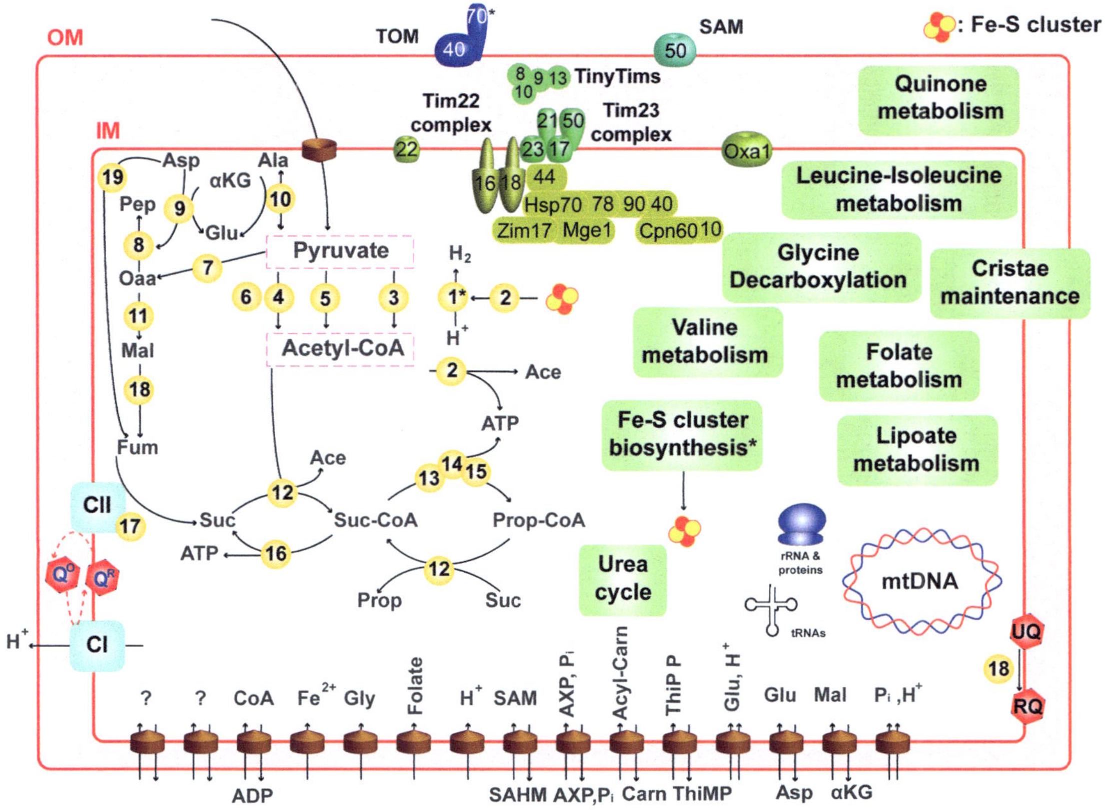

Fig. 1 Proposed metabolic map of *Blastocystis* mitochondrion-related organelles (MROs) based on the genome predictions [figure modified from Gentekaki et al. (2017)]. Various metabolic features of the MRO's role in energy generation and amino acid and lipid metabolism. Protein descriptions (numbers) are

RNA (tRNA) genes, but they also have shown a number of peculiarities. For example, in comparison with other stramenopile genomes, they have the lowest repertoire of tRNA genes seen, a longer than expected gene for ribosomal protein S4 (rps4) that lacks a start codon (Perez-Brocal and Clark 2008) and lastly, an unidentified open reading frame that has in frame stop codons in some of the subtypes. Because of the shorter genome, the gene repertoire of *Blastocystis* is reduced, mainly due to the absence of genes encoding for proteins involved in the electron transport chain including cytochrome oxidase subunits I, II and III (cox1–cox3), cytochrome b (cob) and all the F_0F_1-ATPase subunits (atp6, atp8, atp9 and in some species atp1) encoded by the mtDNA in other stramenopiles (Jacob et al. 2016; Perez-Brocal and Clark 2008; Stechmann et al. 2008). Thus, the reduction of the *Blastocystis* mtDNA size and subsequently of the electron transport could be one of the first, if not actually the first, pathways to be affected during the degenerative evolution of the mitochondrion into hydrogenosomal-related organelles (Perez-Brocal and Clark 2008).

4 Cell Biology and Functions

Over the past decade, there have been several attempts to elucidate the protein composition of the mitochondria of *Blastocystis*, using transcriptomics (Stechmann et al. 2008), genomics (Denoeud et al. 2011; Wawrzyniak et al. 2015; Gentekaki et al. 2017) and biochemistry (see below; Lantsman et al. 2008). *Blastocystis* NandII strain was predicted to have the highest composition of mitochondrial proteins

---

Fig. 1 (continued) outlined below: (1) FeFe-hydrogenase, flavodoxin; (2) iron-only hydrogenase maturation rSAM protein HydE; (3) por, nifJ; pyruvate-ferredoxin/flavodoxin oxidoreductase; (4) dihydrolipoamide succinyltransferase; (5) PDK2_3_4; pyruvate dehydrogenase kinase 2/3/4; (6) 2-oxoglutarate dehydrogenase; (7) pyruvate carboxylase, alpha subunit; (8) phosphoenolpyruvate carboxykinase (ATP); (9) aspartate aminotransferase; (10) alanine aminotransferase; (11) malate dehydrogenase; (12) acetate:succinate CoA-transferase; (13) methylmalonyl-CoA mutase; (14) methylmalonyl-CoA epimerase; (15) propionyl-CoA carboxylase alpha subunit; (16) succinyl-CoA synthetase; (17) succinate dehydrogenase subunit 5; (18) rhodoquinone biosynthesis enzyme RquA; (19) aspartate ammonia lyase. Proteins/pathways labelled with an asterisk were shown to be localized in *Blastocystis* MROs using immunofluorescence microscopy. Standard amino acid abbreviations are used: *Ace* acetate, *ACP* acyl carrier protein, *aKG* alpha-ketoglutarate, *BCD* branched-chain amino acid degradation, *CI* complex I, *CII* complex II, *Carn* carnitine, *CDP-DAG* cytidine diphosphate diacylglycerol, *CL* cardiolipin, *DHAP* dihydroxyacetone phosphate, *DHOro* dihydroorotate, *Fd* ferredoxin, *Fum* fumarate, *Gly3P* glycerol-3-phosphate, *Mal* malate, *MMC* methylmalonyl-CoA, *Nd(p)* NAD(P), *mtDNA* mitochondrial DNA, *Oaa* oxaloacetate, *Oro* orotate, *PA* phosphatidic acid, *PE* phosphatidylethanolamine, *Pep* phosphoenolpyruvate, *PI* phosphatidylinositol, *Prop* propionate, *PS* phosphatidylserine, *QO/R* quinone/quinol, oxidized or reduced, *RQ* rhodoquinone, *SAHC* *S*-adenosylhomocysteine, *SAM* *S*-adenosylmethionine, *Suc* succinate, *THF* tetrahydrofolate, *ThiMP* thiamine monophosphate, *ThiPP* thiamine pyrophosphate, and *UQ* ubiquinone among the other strains, based on the recent genomic (Gentekaki et al. 2017) and previous transcriptomic investigations (Stechmann et al. 2008). As such, several biochemically pathways were constructed based on a combination of diverse predictions methods, and it has been suggested that *Blastocystis* MROs could be involved in amino acid, cardiolipin and phospholipid metabolism, electron transport chain (presence of complex I and complex II), energy and pyruvate metabolism, glycolysis, quinone metabolism, Fe-S cluster biosynthesis (see below), ribosome translation and transcription, cofactor/vitamin metabolism (folate, B5, B12, steroid and lipoate), fatty acid biosynthesis, cristae maintenance and regulation (Stechmann et al. 2008; Gentekaki et al. 2017; Fig. 1). In addition, proteins involved in mitochondrial protein import, mitochondrial transport and adaptations to various oxygen conditions have also been identified (Eme et al. 2017; Stechmann et al. 2008; Gentekaki et al. 2017). Thus, *Blastocystis* MROs seem to be an intermediate stage between anaerobic mitochondria and hydrogenosomes, harbouring pathways (some of which incomplete) from both organelles (Muller et al. 2012). A fraction of these proteins and pathways were further characterized using a combination of bioinformatics and cell biological techniques, as discussed below.

4.1 Adaptations to Oxygen

The first protein to be localized in *Blastocystis* MROs was the FeFe-hydrogenase (Stechmann et al. 2008), an obligate anaerobic protein that functions under anaerobic conditions. *Blastocystis* encodes two homologues of this gene in ST1, which both have clear mitochondrial targeting peptides and also possess and additional electron-transferring domain (flavodoxin) (Gentekaki et al. 2017; Stechmann et al. 2008). This protein requires three maturases (HydE, HydF and HydG), which are typically necessary for the correct assembly of the mature hydrogenase enzyme, but *Blastocystis* genomes seems not to encode either HydF or HydG. Intriguingly, current reports are lacking any biochemical data to show that this protein is indeed functional or produce hydrogen (Muller et al. 2012), suggesting that there could be an alternative function of this enzyme in *Blastocystis*.

Alternative oxidase (AOX) is another protein that has been predicted and then shown to localize in *Blastocystis* MROs (Tsaousis et al. 2018). This protein was proposed to be involved in oxidative stress protective mechanisms and heat generation and to maintain tricarboxylic acid cycle turnover in high cytosolic phosphorylation potential in various organisms (Moore and Albury 2008), but its role in an obligate anaerobic organism such as *Blastocystis* is baffling. The study from the van der Giezen lab has not only shown that this protein is indeed functional both in vivo and in vitro and allow the organism to respire oxygen, but it could also act as a protective mechanism in a dysbiotic gut, where oxygen is abandoned (Tsaousis et al. 2018). As such, in addition to surviving under strict anaerobic conditions, *Blastocystis* also has the tools to survive under limited oxygenic conditions.

4.2 Fe-S Cluster Assembly Biosynthesis

One of the most eminent pathways in mitochondria and related organelles is the assembly of Fe-S clusters, which was considered to be the reason of the existence of mitochondria (Lill et al. 1999), and also all organelles investigated so far seem to harbour some version of this pathway. Thus, *Blastocystis* MROs could not deviate from this. Studies from the Roger laboratory demonstrated that *Blastocystis* MROs have a complete mitochondrial-type Fe-S cluster biosynthetic pathway (iron-sulphur cluster, ISC; Tsaousis et al. 2012), which is conserved, and its members can functionally replace homologues from other organisms, such as trypanosomes (Long et al. 2011; Tsaousis et al. 2012). This pathway functions to support the cytosolic iron-sulphur cluster assembly (CIA) machinery, which was also localized in *Blastocystis* (Tsaousis et al. 2014). In addition to these pathways, *Blastocystis* genomes also harbour a fused gene encoding for a SufCB protein that is typically found in methanogens (Tsaousis et al. 2012). This protein, which was later found and encoded by other microbial eukaryotes genomes, including *Pygsuia* (Stairs et al. 2014) and *Stygiella* (Leger et al. 2016), was further characterized in *Blastocystis* and was demonstrated to localized in its cytosol, having biochemical properties of its prokaryotic homologues and more intriguingly is upregulated under oxygen stress (Tsaousis et al. 2012).

4.3 Mitochondrial Protein Import

One of the most essential mitochondrial apparatuses, shared by mitochondria, hydrogenosomes and mitosomes, is the one of mitochondrial protein import, since it is responsible for the translocations of nuclear-encoded proteins in mitochondria. Overall, there are five major protein import pathways of mitochondria (Wiedemann and Pfanner 2017). Firstly, preproteins-carrying presequences are imported by the translocase of the outer mitochondrial membrane (TOM), followed by the presequence translocase of the inner membrane (TIM23). These proteins, especially the hydrophilic ones, are imported into the mitochondrial matrix, with the assistance of the presequence translocase-associated motor (PAM). Preproteins of metabolite carriers of the inner membrane are imported via the TOM pathway and subsequently inserted in the inner membrane through the carrier translocase TIM22. Finally, precursors of β-barrel proteins are inserted into the outer membrane by the sorting and assembly machinery (SAM). The *Blastocystis* genome encodes proteins from all the aforementioned pathways (Fig. 1), and studies using a combination of bioinformatics, genetic complementation and immunofluorescence microscopy analyses demonstrated that the translocase of the outer membrane 70 (Tom70) protein functions to import preproteins with internal targeting sequences into the MROs (Tsaousis et al. 2011). Interestingly, prior to this discovery, Tom70 has only been found in animals and fungi (Chan et al. 2006). The presence of Tom70 in

Blastocystis enabled the discovery of other homologues in stramenopiles and a haptophyte, suggesting an expansion of the evolutionary spectrum of eukaryotes that contain this protein and consequently proposing that it may have been part of the core mitochondrial protein import apparatus of the last common ancestral eukaryote (Tsaousis et al. 2011).

4.4 Glycolysis

Glycolysis is a major catabolic pathway for the subsequent production of ATP; in most eukaryotes, the pathway is localized in their cytosol, with the exception of trypanosomes where it is localized in compartments referred to as glycosomes. Exceptionally, in *Blastocystis* and other stramenopiles, the C3 part of glycolysis was predicted to localize in their mitochondria, based on the presence of mitochondrial targeting signals in many of these proteins (Kamikawa et al. 2017; Abrahamian et al. 2017). Follow-up studies using immunofluorescence microscopy and enzymatic assays (see below) confirmed the mitochondrial location of these glycolytic enzymes in *Blastocystis* (Río Bártulos et al. 2018). *Blastocystis* genome does not encode for a pyruvate transporter (Gentekaki et al. 2017) that is responsible for the import of the end product of glycolysis into the MROs, and thus potential novel mitochondrial transporters might have been recruited to assist with the transport of C3 products within the organelle. The reason behind the translocation of such pathway in the mitochondria needs to be explored (Río Bártulos et al. 2018), along with the potential implications to the evolution/adaptations of the organelle and the cell itself.

5 Energy Metabolism: Biochemistry

The lack of key energy coupling reactions typical of aerobic mitochondria by the *Blastocystis* MRO reflects the dependence on an anaerobic lifestyle by this organism. *Blastocystis* possess an oxygen-sensitive pyruvate:NADP+ oxidoreductase that is similar to the *Euglena gracilis* mitochondrial enzyme (Inui et al. 1987) and functions to convert pyruvate acetyl-CoA, trapping the high-energy thioester bond of coenzyme A. The high-energy bond is used to drive ATP formation in a two-step mechanism that couples acetate with succinate formation via an acetate:succinate CoA-transferase and a ATP-specific succinyl-CoA synthetase (SCS, aka succinate thiokinase; Hamblin et al. 2008), which is consistent with the proposed common origin of these organelles (Cavalier-Smith 1987). The literature is scattered with reports of some or all of the glycolytic enzymes relocating to various areas of the cell or specific organelles, including the mitochondrion, during energy depletion by restricting competition by intermediates (Plaxton 1996; Liaud et al. 2000; Giege et al. 2003; Graham et al. 2007; Nakayama et al. 2012; Abrahamian et al. 2017). In

Blastocystis the glycolytic enzymes after the split leading to the so-called payoff phase is localized to mitochondrion-related organelle (Río Bártulos et al. 2018). The terminal enzyme of the pathway pyruvate kinase is present indicating that the pathway can operate in the direction of pyruvate synthesis resulting in additional ATP yield. A pyruvate phosphate dikinase (PPDK) was also identified in the MROs; unlike pyruvate kinase, PPDK has a small ΔG and is reversible, potentially producing phosphoenolpyruvate from pyruvate. PPDK is a common adaptation to an anaerobic metabolism (Slamovits and Keeling 2006) and has been described in a range of anaerobic protists including *Entamoeba*, *Giardia*, *Streblomastix* and *Trichomonads* (Bruderer et al. 1996; Saavedra-Lira et al. 1998; Slamovits and Keeling 2006). The kinetoplastids, *Trypanosoma brucei*, *T. cruzi* and *Leishmania mexicana*, possess a functional PPDK that localizes to the glycosome and functions in the direction of ATP and pyruvate synthesis from glucose (Deramchia et al. 2014; Rodriguez-Contreras and Hamilton 2014). In the case of *L. mexicana*, it is proposed that PPDK is involved in the entry of alanine to gluconeogenesis (Rodriguez-Contreras and Hamilton 2014).

Localization of the glycolytic pathway after the split results in centralizing the four ATP produced from glycolysis to the MRO; when coupled with the additional two ATP per glucose resulting from SCS metabolism of the acetyl-CoA produced from pyruvate, a total of six ATP per glucose is produced in the MRO (Fig. 2). It is interesting to note that *Blastocystis* has a pyrophosphate-fructose 6-phosphate 1-phosphotransferase gene (AV274_1738, accession number A0A196SHQ8) that functions to reversibly catalyse the phosphorylation of fructose-6-phosphate using inorganic phosphate in place of ATP; if active, this enzyme would potentially further increase the efficiency of ATP production from glucose in this anaerobic protist.

6 Morphology

Within vacuolar forms, MROs are usually situated within the cytoplasmic rim, near the nucleus, and the configuration of a nuclei flanked on each side by one of more MROs is often observed in transmission electron microscopy (TEM) sections (Fig. 3) (Zierdt 1986; Dunn et al. 1989; Stenzel et al. 1994; Stenzel and Boreham 1996). However, there have been reports of cytoplasmic invagination resulting in organelles, including MROs, being deposited within the central vacuolar space (Pakandl 1999; Nasirudeen et al. 2001; Tan et al. 2001; Yin et al. 2010b). The exact role of this phenomenon is currently unclear although this feature, which has been described in *Blastocystis* undergoing cell death, was postulated to be part of an apoptotic body deposition process (Tan and Nasirudeen 2005). A later study by Yin et al. (2010a) revealed deposition, into the *Blastocystis* central vacuole, of membrane-bound cytoplasmic material harbouring electron-dense mitochondria, upon induction by classical autophagy inducers. This suggests that *Blastocystis* autophagy involves breakdown of cellular organelles within the parasite central vacuole, although such structures have been misconstrued as *Blastocystis* progeny

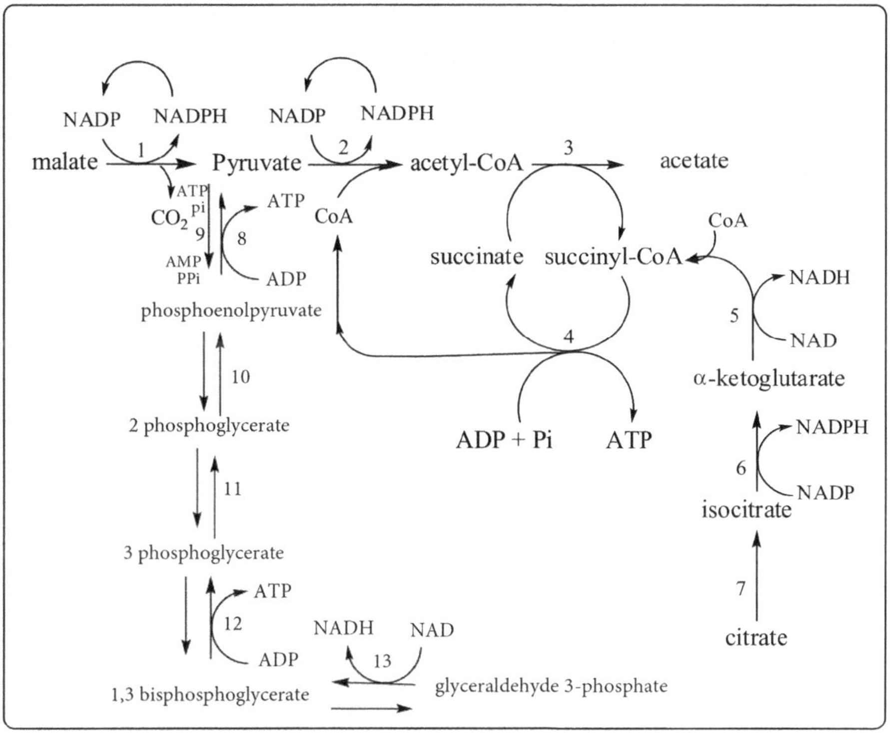

Fig. 2 Enzymes biochemically localized to *Blastocystis* mitochondrion-related organelle. The enzymes are (1) malic enzyme; (2) pyruvate:NADP- oxidoreductase; (3) acetate:succinate CoA-transferase; (4) succinyl-CoA synthetase succinate thiokinase; (5) α-ketoglutarate dehydrogenase; (6) isocitrate dehydrogenase; (7) aconitase; (8) pyruvate kinase; (9) pyruvate phosphate dikinase; (10) enolase; (11) phosphoglycerate mutase; (12) phosphoglycerate kinase; and (13) glyceraldehyde

(Raman et al. 2016). The *Blastocystis* MRO usually ranges from two to four per cell but can reach up to hundreds in giant and older cells (Zierdt et al. 1988). Dunn et al. (1989) performed an ultrastructural study of 10 different *Blastocystis* stocks and observed that the MROs of *Blastocystis* exhibited considerable variation in morphology, both between different isolates and within the same isolate. In fact, morphological heterogeneity could be observed between MROs of a single cell (Dunn et al. 1989).

Ultrastructurally, *Blastocystis* MRO possesses basic features archetypical of eukaryotic mitochondria. The organelles are round to ovoid in shape, are 0.5–2 μm in diameter and are bounded by an outer bilaminar membrane and an inner bilaminar membrane, from which the cristae arise. The MRO cristae are pleomorphic and have been described as bulbous, tubular and vesicular. Occasionally, cristae may take on a circular shape and appear as a ring-like structure within the MRO matrix (Dunn et al. 1989). The MRO matrix may exhibit variation in electron density (Silard et al. 1983; Stenzel et al. 1991), although MROs with

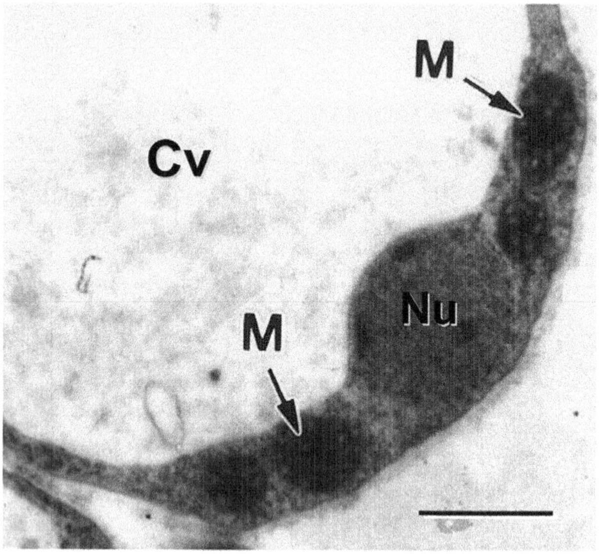

Fig. 3 Ultrastructure of *Blastocystis*. Transmission electron micrograph revealing organelle structures. In this cell, MROs (M) flank the nucleus (Nu) and are characterized by electron-dense matrix and distinct saccate cristae, Cv, central vacuole. Bar = 1 μm

electron-dense matrix are most often described (Zierdt 1986; Boreham and Stenzel 1993; Stenzel and Boreham 1996). An early freeze-fracture study (Yoshikawa et al. 1988) revealed that the MROs of *Blastocystis* were round and contained tubular cristae. Intramembranous particles were distributed heterogeneously on both the protoplasmic and exoplasmic surfaces of the MRO outer membrane. There has been a single report describing a technique for the isolation and purification of *Blastocystis* ST7 MROs (Nasirudeen and Tan 2004a). After density-gradient centrifugation, the authors observed, by negative staining and TEM, that the organelles were predominantly elongate and ranged approximately from 1.2 to 3.0 μm in length and approximately 0.7–1.0 μm in width. This is in general agreement with previous reports on MRO size (Dunn et al. 1989). The MRO membrane appeared wrinkled, and the organelle appeared morphologically intact (Fig. 4). This method will be useful for addressing the major metabolic functions of the organelle as well as characterizing the DNA contained within.

The MROs of *Blastocystis* are selectively stained by a number of dyes employed for staining mitochondria. Early studies reported the use of Janus green (LAVIER 1952; Zierdt 1986; Zierdt et al. 1988) and rhodamine 123 (Zierdt 1986, Zierdt et al. 1988), while more recent reports describe specific staining with MitoTracker Green™ (Fig. 5) (Nasirudeen and Tan 2004a), MitoTracker Orange™ (Stechmann et al. 2008), MitoTracker Deep Red™ (Tsaousis et al. 2011, 2012), MitoLight™ (Nasirudeen and Tan 2004b) and MitoCapture™ (Dhurga et al. 2016). Rhodamine 123, MitoTracker, MitoLight and MitoCapture accumulate in mitochondria due to the presence of a mitochondrial membrane potential. This indicates that the MROs of *Blastocystis* are functional and are able to generate energy to maintain this potential. Interestingly, the ability of the potential-sensitive fluorophore MitoTracker Green to accumulate in MROs was compromised in *Blastocystis* cells undergoing

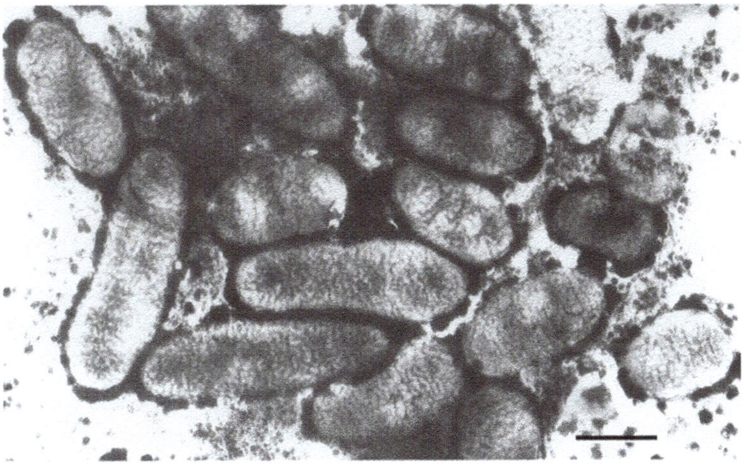

Fig. 4 Phosphate dehydrogenase. Negative staining of purified *Blastocystis* MROs. Note the distinct ovoid to elongate shapes, with apparent wrinkled surface texture. Bar = 1 μm

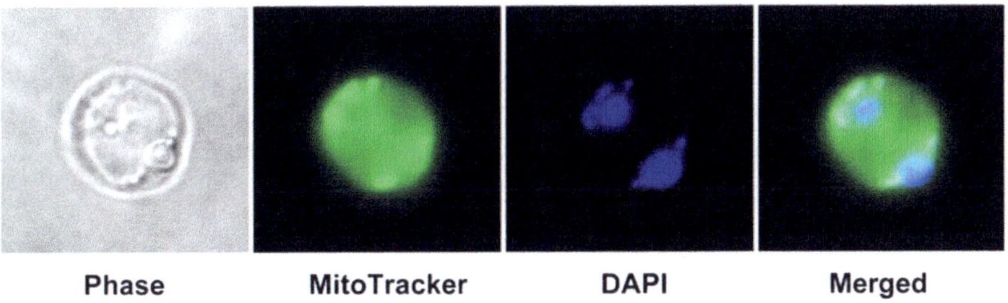

Fig. 5 Fluorescence microscopy of *Blastocystis* revealing MitoTracker Green uptake of MROs with DAPI colocalization. The MROs are situated very close to the nucleus (large DAPI-stained structures) and are clustered at opposite poles of the cell

programmed cell death (PCD) (Nasirudeen and Tan 2004a, 2005). This is a common feature of PCD in the metazoa and functions to release proapoptotic factors sequestered within the mitochondria into the cytoplasm. Pre-incubation of cells with cyclosporine A, a mitochondria permeability transition pore blocker, prevented a decrease in MitoTracker Green uptake in PCD-induced cells, with a concomitant decrease in DNA fragmentation (Nasirudeen and Tan 2005), while in another study (Yin et al. 2010a), cyclosporin pre-incubation rescued cells from staurosporine-induced phosphatidylserine externalization and DNA fragmentation. Both DNA fragmentation and PS externalization are key markers of programmed cell death (see next section). Interestingly, the MRO also stains positively for DAPI (4′,6-diamidino-2-phenylindole), indicating the presence of DNA within the organelle (Fig. 5) (Matsumoto et al. 1987; Tsaousis et al. 2011; Nasirudeen and Tan 2004b). New imaging platforms, such as the Amnis imaging flow cytometer (IFC), have

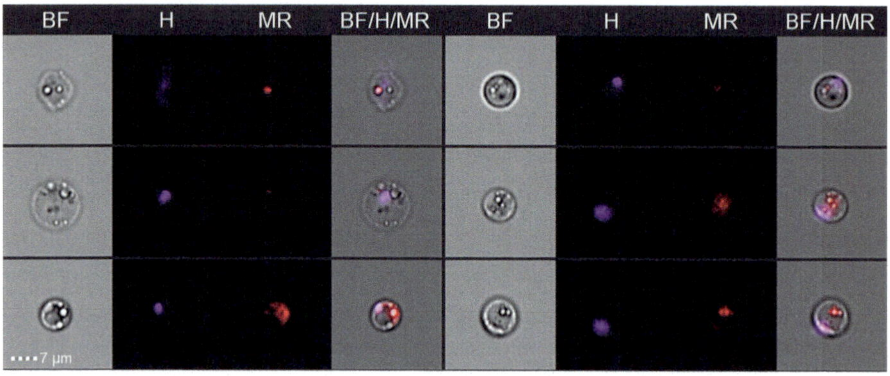

Fig. 6 Image gallery of *Blastocystis* cells captured by imaging flow cytometry. MitoTracker Red mitochondria (MR) can be observed in close proximity to Hoechst-stained nuclei (H). *BF* bright field

revolutionized high-content imaging by combining throughout flow cytometry with the imaging capabilities of high-resolution fluorescence microscopy (Yason and Tan 2015). The IFC has enormous potential for the study of *Blastocystis* mitochondria morphology and function in the context of the parasite cell (Fig. 6). An obvious application would be using an autosampler-capable IFC to screen drugs that perturb the parasite mitochondria via morphological alterations.

7 Role of *Blastocystis* Mitochondria in Cell Death

Programmed cell death (PCD) is a cell suicide mechanism that enables multicellular organisms to control cell number and eliminate undesirable cells. It has been established that PCD also exists among the unicellular eukaryotes and probably function similarly, when these microbes are viewed from a population-based perspective (DosReis and Barcinski 2001). *Blastocystis* has also been reported to undergo PCD, when exposed to a variety of undesirable stimuli (Nasirudeen and Tan 2005). As in higher eukaryotic mitochondria, the MROs are also involved in PCD. Mitochondrial outer membrane potential (MOMP), an indication of an actively respiring organelle, was compromised in cells that were exposed to a cytotoxic monoclonal antibody (Nasirudeen and Tan 2005). In mammalian cells, MOMP is directly governed by a subset of proapoptotic proteins, which induce disruptions in the outer mitochondrial membrane and subsequent release of death-promoting proteins like cytochrome *c*. Cytochrome *c* triggers a cascade of events that lead to the activation of caspases, enzymes that are intimately involved in dismantling the dying cell. It is currently unknown what proapoptotic factors are released by *Blastocystis* upon PCD induction, since only cytochrome oxidase has been identified in this organism and no other (Zierdt 1986; Gentekaki et al. 2017).

PCD involving mitochondria can be either caspase-dependent or caspase-independent. Interestingly, blocking caspases in PCD-induced cells led to a greater compromise in MOMP, suggesting that MROs trigger a compensatory caspase-independent pathway (Nasirudeen and Tan 2005). We postulate that one of the key roles of *Blastocystis* MRO is the regulation of PCD during times of cellular starvation and stress (Yin et al. 2010b), as has been reported for other cell types (Goyeneche et al. 2006; Yeo et al. 2006). Subsequent studies on *Blastocystis* cell death have implicated a major role for the parasite mitochondria in the PCD pathway (Yin et al. 2010a, b). Inhibiting MOMP loss with cyclosporin A resulted in amelioration of PCD features in autophagy- and staurosporine-induced cell death (Yin et al. 2010a, b). The identification of novel MRO-associated PCD mediators in *Blastocystis* may provide novel strategies for antiparasite therapies.

Recent studies on the role of *Blastocystis* mitochondria in metronidazole-induced PCD have suggested that drug-treated cells possess functional mitochondria that increase in number during cell death (Dhurga et al. 2016; Raman et al. 2016). The authors propose that the increased number and activity of mitochondria in these cells are important for transformation to the amoeboid form, as a stress response. These studies need to be read with caution as the conclusions were based on manual counts of fluorescence and electron micrographs, which are prone to subjective interpretation.

8 Concluding Remarks and Future Perspectives

Blastocystis MROs seem to accumulate unique morphological and biochemical characteristics enabling us to study the adaptations of mitochondria in various environments. Although there has been a huge progress on the studies of these organelles in *Blastocystis* over the past decade, it is necessary to explore them even further. For example, to accelerate understanding of the unique biology of these organelles, it will be necessary to perform localization of proteins and functional characterization of pathways that have been in silico predicted (Fig. 1). This could be either performed as previously using generated antibodies (Tsaousis et al. 2011, 2012, 2018) or with the development of a transfection system (or even a CRISPR genetic modification method) that will allow subcellular localization and depletion of proteins. In addition, it will be ideal to perform proteomic and biochemical studies to explore the composition of the mitochondrial proteome in various stages of the life cycle and under various environmental or biological fluctuations that could be present in the gut, such as immune responses, oxygen or drug stress, iron or other nutrient depletion, etc. The importance of the *Blastocystis* mitochondria in PCD has been established in earlier studies, despite a lack of PCD-related orthologues in the parasite genome. Proteomic, molecular and biochemical studies will help shed light on the identity of novel mitochondria-associated molecules involved in *Blastocystis* PCD. Lastly, it will be interesting to investigate and identify potential differences in the mitochondrial proteomes between the various *Blastocystis* strains and species but also examine the evolution of anaerobism in these mitochondria with follow-up studies in close relatives of *Blastocystis* (e.g. *Proteromonas*; Perez-Brocal et al. 2010).

Acknowledgements Dr. Anastasios D. Tsaousis' research on *Blastocystis* mitochondria was supported by BBSRC research grant (BB/M009971/1).

References

Abrahamian M, Kagda M, Ah-Fong AMV et al (2017) Rethinking the evolution of eukaryotic metabolism: novel cellular partitioning of enzymes in stramenopiles links serine biosynthesis to glycolysis in mitochondria. BMC Evol Biol 17(1):241. https://doi.org/10.1186/s12862-017-1087-8

Akhmanova A, Voncken F, van Alen T et al (1998) A hydrogenosome with a genome. Nature 396 (6711):527–528

Andersen LO, Stensvold CR (2016) *Blastocystis* in health and disease: are we moving from a clinical to a public health perspective? J Clin Microbiol 54(3):524–528. https://doi.org/10.1128/JCM.02520-15

Betts EL, Gentekaki E, Thomasz A et al (2018) Genetic diversity of *Blastocystis* in non-primate animals. Parasitology 145:1–7. https://doi.org/10.1017/S0031182017002347

Boreham PF, Stenzel DJ (1993) The current status of *Blastocystis hominis*. Parasitol Today 9 (7):251

Bruderer T, Wehrli C, Kohler P (1996) Cloning and characterization of the gene encoding pyruvate phosphate dikinase from *Giardia duodenalis*. Mol Biochem Parasitol 77(2):225–233

Bui ET, Bradley PJ, Johnson PJ (1996) A common evolutionary origin for mitochondria and hydrogenosomes. Proc Natl Acad Sci USA 93(18):9651–9656

Cavalier-Smith T (1987) Eukaryotes with no mitochondria. Nature 326(6111):332–333

Chan NC, Likic VA, Waller RF et al (2006) The C-terminal TPR domain of Tom70 defines a family of mitochondrial protein import receptors found only in animals and fungi. J Mol Biol 358 (4):1010–1022. https://doi.org/10.1016/j.jmb.2006.02.062

Clark CG (1997) Extensive genetic diversity in *Blastocystis hominis*. Mol Biochem Parasitol 87 (1):79–83. https://doi.org/10.1016/S0166-6851(97)00046-7

Clark CG, van der Giezen M, Alfellani MA et al (2013) Recent developments in *Blastocystis* research. Adv Parasitol 82:1–32. https://doi.org/10.1016/B978-0-12-407706-5.00001-0

Dacks JB, Dyal PL, Embley TM et al (2006) Hydrogenosomal succinyl-CoA synthetase from the rumen-dwelling fungus *Neocallimastix patriciarum*; an energy-producing enzyme of mitochondrial origin. Gene 373:75–82. https://doi.org/10.1016/j.gene.2006.01.012

Denoeud F, Roussel M, Noel B et al (2011) Genome sequence of the stramenopile *Blastocystis*, a human anaerobic parasite. Genome Biol 12(3):R29. https://doi.org/10.1186/gb-2011-12-3-r29

Deramchia K, Morand P, Biran M et al (2014) Contribution of pyruvate phosphate dikinase in the maintenance of the glycosomal ATP/ADP balance in the *Trypanosoma brucei* procyclic form. J Biol Chem 289(25):17365–17378. https://doi.org/10.1074/jbc.M114.567230

Dhurga DB, Suresh K, Tan TC (2016) Granular formation during apoptosis in *Blastocystis* sp. exposed to metronidazole (MTZ). PLoS One 11(7):e0155390. https://doi.org/10.1371/journal.pone.0155390

DosReis GA, Barcinski MA (2001) Apoptosis and parasitism: from the parasite to the host immune response. Adv Parasitol 49:133–161

Dunn LA, Boreham PF, Stenzel DJ (1989) Ultrastructural variation of *Blastocystis hominis* stocks in culture. Int J Parasitol 19(1):43–56. https://doi.org/10.1016/0020-7519(89)90020-9

Dyall SD, Koehler CM, Delgadillo-Correa MG et al (2000) Presence of a member of the mitochondrial carrier family in hydrogenosomes: conservation of membrane-targeting pathways between hydrogenosomes and mitochondria. Mol Cell Biol 20(7):2488–2497

Embley TM, van der Giezen M, Horner DS et al (2003) Hydrogenosomes, mitochondria and early eukaryotic evolution. IUBMB Life 55(7):387–395

Eme L, Gentekaki E, Curtis B et al (2017) Lateral gene transfer in the adaptation of the anaerobic parasite Blastocystis to the gut. Curr Biol 27(6):807–820. https://doi.org/10.1016/j.cub.2017.02.003

Gentekaki E, Curtis BA, Stairs CW et al (2017) Extreme genome diversity in the hyper-prevalent parasitic eukaryote Blastocystis. PLoS Biol 15(9):e2003769. https://doi.org/10.1371/journal.pbio.2003769

Giege P, Heazlewood JL, Roessner-Tunali U et al (2003) Enzymes of glycolysis are functionally associated with the mitochondrion in *Arabidopsis* cells. Plant Cell 15(9):2140–2151

Goyeneche AA, Harmon JM, Telleria CM (2006) Cell death induced by serum deprivation in luteal cells involves the intrinsic pathway of apoptosis. Reproduction 131(1):103–111. https://doi.org/10.1530/rep.1.00751

Graham JW, Williams TC, Morgan M et al (2007) Glycolytic enzymes associate dynamically with mitochondria in response to respiratory demand and support substrate channeling. Plant Cell 19(11):3723–3738. https://doi.org/10.1105/tpc.107.053371

Gray MW, Burger G, Lang BF (2001) The origin and early evolution of mitochondria. Genome Biol 2(6). https://doi.org/10.1186/gb-2001-2-6-reviews1018

Hamblin K, Standley DM, Rogers MB et al (2008) Localization and nucleotide specificity of Blastocystis succinyl-CoA synthetase. Mol Microbiol 68(6):1395–1405. https://doi.org/10.1111/j.1365-2958.2008.06228.x

Hrdy I, Hirt RP, Dolezal P et al (2004) *Trichomonas* hydrogenosomes contain the NADH dehydrogenase module of mitochondrial complex I. Nature 432(7017):618–622

Inui H, Ono K, Miyatake K et al (1987) Purification and characterization of pyruvate:NADP+ oxidoreductase in *Euglena gracilis*. J Biol Chem 262(19):9130–9135

Jacob AS, Andersen LO, Bitar PP et al (2016) Blastocystis mitochondrial genomes appear to show multiple independent gains and losses of start and stop codons. Genome Biol Evol 8(11):3340–3350. https://doi.org/10.1093/gbe/evw255

Kamikawa R, Moog D, Zauner S et al (2017) A non-photosynthetic diatom reveals early steps of reductive evolution in plastids. Mol Biol Evol 34(9):2355–2366. https://doi.org/10.1093/molbev/msx172

Keithly JS, Langreth SG, Buttle KF et al (2005) Electron tomographic and ultrastructural analysis of the *Cryptosporidium parvum* relict mitochondrion, its associated membranes, and organelles. J Eukaryot Microbiol 52(2):132–140. https://doi.org/10.1111/j.1550-7408.2005.04-3317.x

Kurland CG, Andersson SG (2000) Origin and evolution of the mitochondrial proteome. Microbiol Mol Biol Rev 64(4):786–820

Lantsman Y, Tan KS, Morada M et al (2008) Biochemical characterization of a mitochondrial-like organelle from *Blastocystis* sp. subtype 7. Microbiology 154(Pt 9):2757–2766. https://doi.org/10.1099/mic.0.2008/017897-0

Lavier G (1952) *Blastocystis* spp. Ann Parasitol Hum Comp 27(1–3):339–356

Leger MM, Eme L, Hug LA et al (2016) Novel Hydrogenosomes in the microaerophilic Jakobid *Stygiella incarcerata*. Mol Biol Evol 33(9):2318–2336. https://doi.org/10.1093/molbev/msw103

Liaud MF, Lichtle C, Apt K et al (2000) Compartment-specific isoforms of TPI and GAPDH are imported into diatom mitochondria as a fusion protein: evidence in favor of a mitochondrial origin of the eukaryotic glycolytic pathway. Mol Biol Evol 17(2):213–223. https://doi.org/10.1093/oxfordjournals.molbev.a026301

Lill R, Diekert K, Kaut A et al (1999) The essential role of mitochondria in the biogenesis of cellular iron-sulfur proteins. Biol Chem 380(10):1157–1166

Lindmark DG, Muller M (1973) Hydrogenosome, a cytoplasmic organelle of the anaerobic flagellate Tritrichomonas foetus, and its role in pyruvate metabolism. J Biol Chem 248 (22):7724–7728

Long S, Changmai P, Tsaousis AD et al (2011) Stage-specific requirement for Isa1 and Isa2 proteins in the mitochondrion of *Trypanosoma brucei* and heterologous rescue by human and *Blastocystis* orthologues. Mol Microbiol 81(6):1403–1418. https://doi.org/10.1111/j.1365-2958.2011.07769.x

Matsumoto Y, Yamada M, Yoshida Y (1987) Light-microscopical appearance and ultrastructure of *Blastocystis hominis*, an intestinal parasite of man. Zentralbl Bakteriol Mikrobiol Hyg A 264 (3–4):379–385

Moore AL, Albury MS (2008) Further insights into the structure of the alternative oxidase: from plants to parasites. Biochem Soc Trans 36.(Pt 5:1022–1026. https://doi.org/10.1042/BST0361022

Muller M, Mentel M, van Hellemond JJ et al (2012) Biochemistry and evolution of anaerobic energy metabolism in eukaryotes. Microbiol Mol Biol Rev 76(2):444–495. https://doi.org/10.1128/MMBR.05024-11

Nakayama T, Ishida K, Archibald JM (2012) Broad distribution of TPI-GAPDH fusion proteins among eukaryotes: evidence for glycolytic reactions in the mitochondrion? PLoS One 7(12): e52340. https://doi.org/10.1371/journal.pone.0052340

Nasirudeen AM, Tan KS (2004a) Caspase-3-like protease influences but is not essential for DNA fragmentation in *Blastocystis* undergoing apoptosis. Eur J Cell Biol 83(9):477–482. https://doi.org/10.1078/0171-9335-00411

Nasirudeen AM, Tan KS (2004b) Isolation and characterization of the mitochondrion-like organelle from *Blastocystis hominis*. J Microbiol Methods 58(1):101–109. https://doi.org/10.1016/j.mimet.2004.03.008

Nasirudeen AM, Tan KS (2005) Programmed cell death in *Blastocystis hominis* occurs independently of caspase and mitochondrial pathways. Biochimie 87(6):489–497. https://doi.org/10.1016/j.biochi.2005.03.003

Nasirudeen AM, Tan KS, Singh M et al (2001) Programmed cell death in a human intestinal parasite, *Blastocystis hominis*. Parasitology 123(Pt 3):235–246

Pakandl M (1999) *Blastocystis* sp. from pigs: ultrastructural changes occurring during polyxenic cultivation in Iscove's modified Dulbecco's medium. Parasitol Res 85(8–9):743–748

Perez-Brocal V, Clark CG (2008) Analysis of two genomes from the mitochondrion-like organelle of the intestinal parasite *Blastocystis*: complete sequences, gene content, and genome organization. Mol Biol Evol 25(11):2475–2482. https://doi.org/10.1093/molbev/msn193

Perez-Brocal V, Shahar-Golan R, Clark CG (2010) A linear molecule with two large inverted repeats: the mitochondrial genome of the stramenopile *Proteromonas lacertae*. Genome Biol Evol 2:257–266. https://doi.org/10.1093/gbe/evq015

Plaxton WC (1996) The organization and regulation of plant glycolysis. Annu Rev Plant Physiol Plant Mol Biol 47:185–214. https://doi.org/10.1146/annurev.arplant.47.1.185

Raman K, Kumar S, Chye TT (2016) Increase number of mitochondrion-like organelle in symptomatic *Blastocystis* subtype 3 due to metronidazole treatment. Parasitol Res 115(1):391–396. https://doi.org/10.1007/s00436-015-4760-0

Río Bártulos C, Rogers MB, Williams TA et al (2018) Mitochondrial targeting of glycolysis in a major lineage of eukaryotes. bioRxiv. https://doi.org/10.1101/257790

Rodriguez-Contreras D, Hamilton N (2014) Gluconeogenesis in Leishmania mexicana: contribution of glycerol kinase, phosphoenolpyruvate carboxykinase, and pyruvate phosphate dikinase. J Biol Chem 289(47):32989–33000. https://doi.org/10.1074/jbc.M114.569434

Saavedra-Lira E, Ramirez-Silva L, Perez-Montfort R (1998) Expression and characterization of recombinant pyruvate phosphate dikinase from *Entamoeba histolytica*. Biochim Biophys Acta 1382(1):47–54. https://doi.org/10.1016/S0167-4838(97)00139-8

Silard R, Panaitescu D, Burghelea B (1983) Ultrastructural aspects of *Blastocystis hominis*. Arch Roum Pathol Exp Microbiol 42(2–3):233–242

Silberman JD, Sogin ML, Leipe DD et al (1996) Human parasite finds taxonomic home. Nature 380 (6573):398. https://doi.org/10.1038/380398a0

Slamovits CH, Keeling PJ (2006) Pyruvate-phosphate dikinase of oxymonads and parabasalia and the evolution of pyrophosphate-dependent glycolysis in anaerobic eukaryotes. Eukaryot Cell 5 (1):148–154. https://doi.org/10.1128/EC.5.1.148-154.2006

Stairs CW, Eme L, Brown MW et al (2014) A SUF Fe-S cluster biogenesis system in the mitochondrion-related organelles of the anaerobic protist *Pygsuia*. Curr Biol 24 (11):1176–1186. https://doi.org/10.1016/j.cub.2014.04.033

Stechmann A, Hamblin K, Perez-Brocal V et al (2008) Organelles in *Blastocystis* that blur the distinction between mitochondria and hydrogenosomes. Curr Biol 18(8):580–585. https://doi.org/10.1016/j.cub.2008.03.037

Stensvold CR (2013) Blastocystis: genetic diversity and molecular methods for diagnosis and epidemiology. Trop Parasitol 3(1):26–34. https://doi.org/10.4103/2229-5070.113896

Stensvold CR, Clark CG (2016) Current status of *Blastocystis*: a personal view. Parasitol Int 65(6 Pt B):763–771. https://doi.org/10.1016/j.parint.2016.05.015

Stenzel DJ, Boreham PF (1996) *Blastocystis hominis* revisited. Clin Microbiol Rev 9(4):563–584

Stenzel DJ, Boreham PF, McDougall R (1991) Ultrastructure of *Blastocystis hominis* in human stool samples. Int J Parasitol 21(7):807–812. https://doi.org/10.1016/0020-7519(91)90149-2

Stenzel DJ, Cassidy MF, Boreham PF (1994) Morphology of *Blastocystis* sp. from domestic birds. Parasitol Res 80(2):131–137

Tachezy J, Sanchez LB, Muller M (2001) Mitochondrial type iron-sulfur cluster assembly in the amitochondriate eukaryotes *Trichomonas vaginalis* and *Giardia intestinalis*, as indicated by the phylogeny of IscS. Mol Biol Evol 18(10):1919–1928. https://doi.org/10.1093/oxfordjournals.molbev.a003732

Tan KS, Nasirudeen AM (2005) Protozoan programmed cell death--insights from *Blastocystis* deathstyles. Trends Parasitol 21(12):547–550. https://doi.org/10.1016/j.pt.2005.09.006

Tan KS, Howe J, Yap EH et al (2001) Do *Blastocystis hominis* colony forms undergo programmed cell death? Parasitol Res 87(5):362–367

Tjaden J, Haferkamp I, Boxma B et al (2004) A divergent ADP/ATP carrier in the hydrogenosomes of *Trichomonas gallinae* argues for an independent origin of these organelles. Mol Microbiol 51 (5):1439–1446

Tovar J, Fischer A, Clark CG (1999) The mitosome, a novel organelle related to mitochondria in the amitochondrial parasite *Entamoeba histolytica*. Mol Microbiol 32(5):1013–1021

Tovar J, Leon-Avila G, Sanchez LB et al (2003) Mitochondrial remnant organelles of *Giardia* function in iron-Sulphur protein maturation. Nature 426(6963):172–176

Tsaousis AD, Kunji ER, Goldberg AV et al (2008) A novel route for ATP acquisition by the remnant mitochondria of *Encephalitozoon cuniculi*. Nature 453(7194):553–556. https://doi.org/10.1038/nature06903

Tsaousis AD, Gaston D, Stechmann A et al (2011) A functional Tom70 in the human parasite *Blastocystis* sp.: implications for the evolution of the mitochondrial import apparatus. Mol Biol Evol 28(1):781–791. https://doi.org/10.1093/molbev/msq252

Tsaousis AD, Ollagnier de Choudens S, Gentekaki E et al (2012) Evolution of Fe/S cluster biogenesis in the anaerobic parasite *Blastocystis*. Proc Natl Acad Sci U S A 109 (26):10426–10431. https://doi.org/10.1073/pnas.1116067109

Tsaousis AD, Gentekaki E, Eme L et al (2014) Evolution of the cytosolic iron-sulfur cluster assembly machinery in *Blastocystis* species and other microbial eukaryotes. Eukaryot Cell 13 (1):143–153. https://doi.org/10.1128/EC.00158-13

Tsaousis AD, Hamblin K, Elliot CR et al (2018) The human gut colonizer *Blastocystis* respires using complex II and alternative oxidase to buffer transient oxygen fluctuations in the gut. Front Cell Infect Microbiol 8:371

Wawrzyniak I, Courtine D, Osman M et al (2015) Draft genome sequence of the intestinal parasite *Blastocystis* subtype 4-isolate WR1. Genom Data 4:22–23. https://doi.org/10.1016/j.gdata.2015.01.009

Wiedemann N, Pfanner N (2017) Mitochondrial machineries for protein import and assembly. Annu Rev Biochem 86:685–714. https://doi.org/10.1146/annurev-biochem-060815-014352

Williams BA, Hirt RP, Lucocq JM et al (2002) A mitochondrial remnant in the microsporidian *Trachipleistophora hominis*. Nature 418(6900):865–869

Yarlett N, Hann AC, Lloyd D et al (1981) Hydrogenosomes in the rumen protozoon *Dasytricha ruminantium* Schuberg. Biochem J 200(2):365–372

Yarlett N, Orpin CG, Munn EA et al (1986) Hydrogenosomes in the rumen fungus Neocallimastix patriciarum. Biochem J 236(3):729–739

Yason JA, Tan KS (2015) Seeing the whole elephant: imaging flow cytometry reveals extensive morphological diversity within *Blastocystis* isolates. PLoS One 10(11):e0143974. https://doi.org/10.1371/journal.pone.0143974

Yeo JH, Lo JC, Nissom PM et al (2006) Glutamine or glucose starvation in hybridoma cultures induces death receptor and mitochondrial apoptotic pathways. Biotechnol Lett 28 (18):1445–1452. https://doi.org/10.1007/s10529-006-9110-y

Yin J, Howe J, Tan KS (2010a) Staurosporine-induced programmed cell death in *Blastocystis* occurs independently of caspases and cathepsins and is augmented by calpain inhibition. Microbiology 156(Pt 5):1284–1293. https://doi.org/10.1099/mic.0.034025-0

Yin J, Ye AJ, Tan KS (2010b) Autophagy is involved in starvation response and cell death in *Blastocystis*. Microbiology 156(Pt 3):665–677. https://doi.org/10.1099/mic.0.033944-0

Yoshikawa H, Yamada M, Yoshida Y (1988) Freeze-fracture study of Blastocystis hominis. J Protozool 35(4):522–528

Zierdt CH (1986) Cytochrome-free mitochondria of an anaerobic protozoan--*Blastocystis hominis*. J Protozool 33(1):67–69

Zierdt CH, Donnolley CT, Muller J et al (1988) Biochemical and ultrastructural study of *Blastocystis hominis*. J Clin Microbiol 26(5):965–970

Mitochondrion-Related Organelles in Free-Living Protists

Michelle M. Leger, Martin Kolísko, Courtney W. Stairs, and Alastair G. B. Simpson

Contents

1 Introduction 288
2 Fornicata 289
2.1 Iron-Sulfur Cluster Assembly 289
2.2 Pyruvate Metabolism and Energy Generation 290
2.3 Protein Import 293
2.4 Glycine Cleavage System 293
2.5 Hydrogen Production Coupled with GCS 294
2.6 Overall Characteristics and Evolution of MROs in Ancestrally Free-Living Fornicata 294
3 *Trepomonas* sp. 296
4 Other Taxa 296
4.1 Jakobida 296
4.2 Heterolobosea 297
4.3 Stramenopila 297
4.4 Rhizaria 298
4.5 Alveolata 299
4.6 Breviatea 299
4.7 Amoebozoa 300

M. M. Leger (✉)
Institute of Evolutionary Biology (CSIC-UPF), Barcelona, Spain
e-mail: michelle.leger@upf.edu

M. Kolísko
Institute of Parasitology, Biology Centre, Czech Academy of Science, České Budějovice, Czech Republic
e-mail: kolisko@paru.cas.cz

C. W. Stairs
Department of Cell and Molecular Biology, Science for Life Laboratory, Uppsala University, Uppsala, Sweden
e-mail: courtney.stairs@icm.uu.se

A. G. B. Simpson
Department of Biology, Centre for Comparative Genomics and Evolutionary Bioinformatics, Dalhousie University, Halifax, Canada
e-mail: alastair.simpson@dal.ca

J. Tachezy (ed.), *Hydrogenosomes and Mitosomes: Mitochondria of Anaerobic Eukaryotes*, Microbiology Monographs 9,
https://doi.org/10.1007/978-3-030-17941-0_12

5 Conclusions ... 301
References ... 302

Abstract Mitochondrion-related organelles (MROs) are organelles that have independently evolved from mitochondria in eukaryotes that live in low-oxygen conditions. These organelles are functionally diverse, possessing a range of ancestrally mitochondrial or horizontally acquired biochemical pathways. Early studies of MROs focused mainly on parasitic organisms; however, the past decade has seen a growing body of work on the MROs of free-living eukaryotes based on comparative genomics, making it possible to tease apart adaptations to low-oxygen conditions from adaptations to parasitism. Here, we review current knowledge of MROs in free-living eukaryotes.

1 Introduction

"Mitochondrion-related organelles" (MROs) is a term that encompasses the functionally diverse set of organelles that are derived from canonical mitochondria, and which are found in various distantly related eukaryotes that are capable of surviving under low-oxygen conditions. Five classes of MROs have been proposed according to their role in ATP generation (Müller et al. 2012): (1) *aerobic mitochondria*, which produce ATP aerobically using the electron transport chain, with oxygen as the terminal electron acceptor; (2) *anaerobically functioning mitochondria*, which use alternative electron acceptors such as fumarate or succinate under low-oxygen conditions; (3) *hydrogen-producing mitochondria*, which possess both an electron transport chain and a [FeFe]-hydrogenase that, in combination with other enzymes, allows them to produce ATP under anaerobic conditions; (4) *hydrogenosomes*, which have lost the capacity for oxidative phosphorylation and instead couple ATP biosynthesis to hydrogen production using an [FeFe]-hydrogenase; and (5) *mitosomes*, which lack any role in ATP generation. Recent studies have blurred the distinctions between these categories, showing that the true functional diversity of MROs falls along a continuum (Roger et al. 2017; Gawryluk et al. 2016).

Mitochondrion-related organelles were first described in parabasalid parasites (Lindmark and Müller 1973) (see chapter "Introduction"), and most of the early work on MROs was carried out in a variety of other parasites or commensals [Yarlett et al. 1981, 1986; Mai et al. 1999; Tovar et al. 1999; reviewed in Santos et al. (2018)], including *Trichomonas vaginalis* (Lindmark et al. 1975) and *Giardia lamblia* (Tovar et al. 2003). As a result, uncertainty surrounded the adaptations that they displayed: these might have been primarily adaptations to a low-oxygen environment, but some might equally be attributed to a parasitic lifestyle. In the past 10 years, work focusing on MROs in a range of different free-living eukaryotes has broadened our understanding of these organelles. However, almost all of these organisms exist in culture with dense and diverse populations of bacteria, which complicates and limits possible experiments; as a result, the majority of these organelles have so far only been studied using in silico predictions of biochemical functions based on transcriptomic data. Therefore, any conclusions about the functions of the MROs of these organisms must be drawn with this limitation in mind. The picture that these studies paint is one of functional diversity among the MROs of free-living eukaryotes, with commonalities to the MROs of parasites at the level of ATP generation but with a larger complement of ancestrally mitochondrial pathways and proteins.

In the following, we first describe the range of processes inferred for the MROs of the free-living representatives of the Fornicata subclade of Metamonada, the most extensively studied group of anaerobes that includes a range of both free-living and endobiotic species. Then we survey free-living anaerobes from across the diversity of other eukaryotes, most of which are poorly understood, often with only one or two species examined per clade.

2 Fornicata

Fornicata is a diverse group of flagellated eukaryotes that includes parasitic, commensal, and free-living species. Ancestrally free-living representatives of Fornicata are crucial for understanding the evolution of MROs in these anaerobes: Fornicata is a sister lineage to the hydrogenosome-bearing Parabasalia, and the free-living members form the deepest branches of the lineage that includes the mostly parasitic Diplomonadida, which have either hydrogenosomes or mitosomes (Kolísko et al. 2010; Takishita et al. 2012; Yubuki et al. 2016). Several free-living members of Fornicata have been studied for the presence and function of MROs—*Carpediemonas membranifera*, *Ergobibamus cyprinoides*, *Aduncisulcus palustris*, *Chilomastix cuspidata*, *Kipferlia bialata*, and *Dysnectes brevis* (Leger et al. 2017; Tanifuji et al. 2018). Transmission electron microscopy has shown that all of these lineages contain organelles that have a similar appearance to the hydrogenosomes of *Trichomonas vaginalis* (Park et al. 2009; Kolísko et al. 2010; Park et al. 2010; Yubuki et al. 2016). Overall, the predicted mitochondrial metabolism of free-living fornicates is similar to that of *Trichomonas* hydrogenosomes; however, there are a few additional pathways that appear to be correlated with the free-living lifestyle.

2.1 Iron-Sulfur Cluster Assembly

The one mitochondrial pathway that appears almost universally present across eukaryote lineages (exceptions are discussed below) is the mitochondrial iron-sulfur cluster (Fe-S) (ISC) assembly system. This consists of several proteins, but the two signature proteins are IscU, which provides a scaffold for newly formed Fe-S clusters, and IscS, which functions as a cysteine desulfurase and transfers sulfur to a newly formed Fe-S cluster. Fe-S clusters are crucial parts of iron-sulfur proteins, which are essential for a variety of biological processes, including transcription, translation, and mitochondrial electron transport. In addition to its role in the biosynthesis of mitochondrial Fe-S enzymes, the mitochondrial ISC system is also essential for the cell because the mitochondrial ISC system provides the yet-unknown signal molecule that transmits the iron status of the mitochondria and is crucial for proper functioning of cytosolic Fe-S synthesis (the CIA machinery) (Lill et al. 2012). This molecule is transferred from the organelle to the cytosol through the *ATP-binding cassette transporter of the mitochondria* [ATM (Lill et al. 2012)]. It is hypothesized that most highly reduced type of MRO known in Fornicata, the mitosome of *Giardia*, still exists because the essential ISC pathway is localized to them (Tovar et al. 2003), even though the ATM has not yet been identified in the *Giardia* genome. All studied Fornicata, whether free-living, commensal, or parasitic, appear to possess proteins involved in the ISC system (Leger et al. 2017).

2.2 *Pyruvate Metabolism and Energy Generation*

All members of Fornicata lack the following: mitochondrial genomes; all components of the ETC, with the exception of the 24 kDa and 51 kDa subunits of complex I (CI); and all components of the TCA cycle, with the exception of succinyl-CoA synthetase (SCS), which is capable of functioning without oxygen in these organisms (see below; Fig. 1). Instead, most of the free-living Fornicata appear to possess typical anaerobic pyruvate fermentation and ATP generation pathways. Aerobic pyruvate decarboxylation in typical mitochondria is carried out by the pyruvate decarboxylation complex (PDC). In contrast, the pyruvate fermentation that is widespread in MROs involves the conversion of pyruvate to acetyl-CoA by one of three enzymes: a pyruvate:ferredoxin oxidoreductase (PFO), the related enzyme pyruvate:NADP oxidoreductase (PNO), or pyruvate formate-lyase (PFL) [reviewed in Stairs et al. (2015)]. Acetyl-CoA is subsequently metabolized to either acetate (while generating ATP) or to ethanol (without ATP generation). PFO enzymes are indeed found in free-living fornicates, but their localization is unclear, and therefore the conversion of pyruvate to acetyl-CoA may be localized to either the cytosol or MROs or both (Leger et al. 2017). There are two ways by which acetyl-CoA can be converted to acetate: (1) acetate:succinyl-CoA transferase (ASCT) may convert acetyl-CoA to acetate and succinate to succinyl-CoA, a reaction that is coupled to the generation of ATP through substrate-level phosphorylation catalyzed by SCS (Steinbuchel and Muller 1986; Müller et al. 2012); and (2) acetyl-CoA may be directly converted to acetate by acetyl-CoA synthetase (ACS), with concomitant

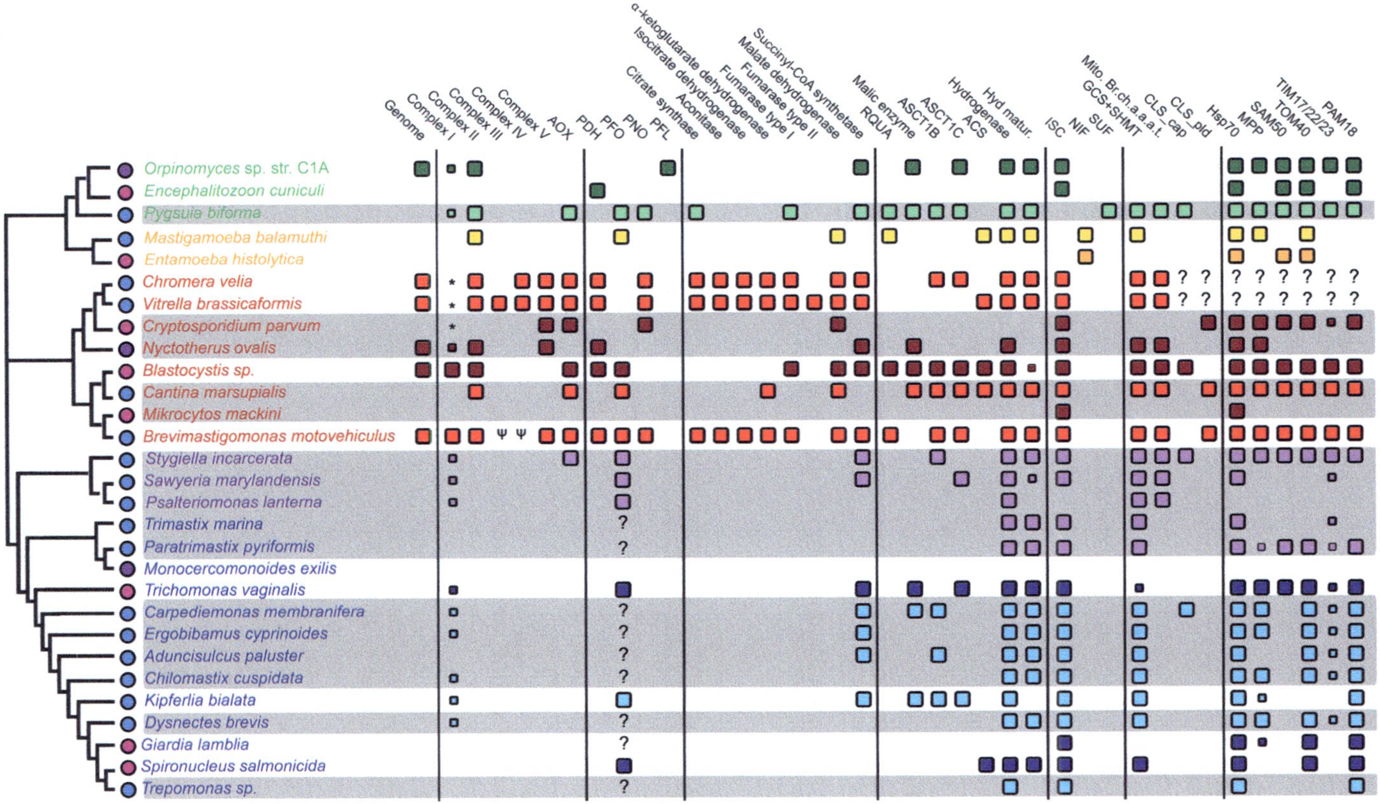

Fig. 1 Selected proteins predicted to localize to mitochondria and related organelles in free-living eukaryotes and in selected parasites and commensals. Green, opisthokonts; yellow, amoebozoa; red, stramenopiles, alveolates, and Rhizaria (SAR clade); purple, discobids; blue, fornicates. Lighter colors indicate the taxa

generation of ATP (Sanchez and Muller 1996). SCS and ASCT are both enzymes that localize exclusively to MROs, with SCS being part of the TCA cycle in aerobic mitochondria (Tielens et al. 2010). The ACS enzyme most likely originated through a horizontal gene transfer (HGT) event from bacteria and is usually localized to the cytosol (Sanchez and Muller 1996; Tielens et al. 2010). In deeper-branching free-living fornicates—*C. membranifera*, *Ergobibamus*, *Aduncisulcus*, and *Kipferlia*—genes encoding mitochondrially targeted ASCT and SCS, and cytosolic ACS, have been identified, suggesting a dual location for the conversion of acetyl-CoA to acetate in both the cytosol and in mitochondria. In contrast, only ACS has been identified in *Chilomastix* and *Dysnectes*, suggesting a metabolism highly similar to that of *Giardia*, in which ATP is generated only in the cytosol. Based on the phylogeny of ACS, this enzyme appears to have been acquired within Fornicata several times by independent HGT events (Leger et al. 2017). *Kipferlia*, *Chilomastix*, and *Dysnectes* contain ACS of the same evolutionary origin as *Giardia* (Leger et al. 2017), which has been experimentally shown to generate ATP while converting acetyl-CoA to acetate (Sanchez and Muller 1996). A second species of *Carpediemonas*, *C. frisia*, has been reported to possess MROs that lack any role in pyruvate metabolism or hydrogen production (Hamann et al. 2017), although this may have been based on incomplete sequence data.

Fig. 1 (continued) discussed in detail in this chapter. Squares indicate the presence in the genome or transcriptome and predicted localization to the organelle; small squares indicate that only a subset of the expected proteins or subunits are present and predicted to be present in the organelle. Blue circles, free-living; purple circles, commensals; pink circles, parasites. Asterisk, alternative NADH dehydrogenases. Psi symbol, pseudogenised. Question mark, insufficient literature. Gray, transcriptome only. *AOX* alternative oxidase, *PDH* pyruvate dehydrogenase, *PFO* pyruvate:ferredoxin oxidoreductase, *PNO* pyruvate:NADP+ oxidoreductase, *PFL* pyruvate formate-lyase, *RQUA* rhodoquinone biosynthesis methyltransferase-like protein RQUA, *ASCT* acetate:succinyl-CoA transferase, *ACS* acetyl-CoA synthetase, *Hyd. matur.* [FeFe]-hydrogenase maturases, *ISC* iron-sulfur cluster assembly proteins, *SUF* sulfur mobilization system of iron-sulfur cluster assembly proteins, *NIF* nitrogen fixation system of iron-sulfur cluster assembly proteins, *GCS + SHMT* glycine cleavage system and serine hydroxymethyltransferase, *Mito.Br.ch.a.a.a.t.* mitochondrial branched-chain amino acid aminotransferase, *CLS_cap* cytidine diphosphate-alcohol phosphatidyltransferase domain-containing cardiolipin synthase, *CLS_pld* phospholipase D domain-containing cardiolipin synthase, *MPP* mitochondrial processing peptidase, *SAM50* sorting and assembly machinery component 50, *TOM40* translocase of the outer membrane component 40, *TIM17/22/23* translocase of the inner membrane components 17, 22, and 23, *PAM18* presequence translocase-associated motor component 18. Data from Henriquez et al. (2005), Doležal et al. (2006, 2010), de Graaf et al. (2009, 2011), Barberà et al. (2010), Fritz-Laylin et al. (2010), Jedelský et al. (2011), Opperdoes et al. (2011), Stairs et al. (2011, 2014, 2015, 2018), Schneider et al. (2011), Tian et al. (2012), Burki et al. (2013), Jerlström-Hultqvist et al. (2013), Leger et al. (2013, 2016, 2017), Zubácová et al. (2013), Maguire and Richards (2014), Makiuchi and Nozaki (2014), Flegontov et al. (2015), Noguchi et al. (2015), Nývltová et al. (2015), Oborník and Lukeš (2015), Gawryluk et al. (2016), Karnkowska et al. (2016), Noguchi et al. (2016), Gentekaki et al. (2017), Santos et al. (2018)

2.3 Protein Import

The protein import machinery of MROs in free-living Fornicata is highly reduced to a "bare-bones" version of the canonical mitochondrial import system. Most free-living Fornicata possess the key components of each of the major complexes of the protein import machinery (Dudek et al. 2013; Leger et al. 2017): these include Tom40 and Sam50, which facilitate import across and into the outer membrane, as well as a set of Tim proteins facilitating import across and into the inner membrane (Dudek et al. 2013). Interestingly, all free-living Fornicates possess a single Tim17/22/23 pore-forming protein, like that recently discovered in *Giardia* (Pyrihova et al. 2018); meanwhile, most other eukaryotes have all three of this family of paralogs (Mokranjac and Neupert 2010). Accessory proteins for the import machinery have also been identified, including Pam18, Pam16, Tim44, and a single small Tim8/9/10/13 protein (Leger et al. 2017), as have TIM23 and mitochondrial HSP70, which is the molecular motor that pulls proteins across the inner membrane (Schneider et al. 1994) through the TIM23 complex. Similarly, both subunits of mitochondrial processing peptidase (MPP) are present in free-living members of Fornicata (Leger et al. 2017). This is an interesting contrast to the diplomonads, in which the alpha subunit of MPP has been secondarily lost (Šmid et al. 2008).

2.4 Glycine Cleavage System

The glycine cleavage system (GCS), a component of amino acid metabolism, is essential for the degradation of glycine. It is a set of four enzymes—T-protein, P-protein, L-protein, and H-protein—and is exclusively localized to mitochondria in aerobes (Kikuchi 1973). The predicted mitochondrial targeting signals found in many of the GCS sequences of free-living fornicates suggest that they are likewise localized to MROs (Leger et al. 2017). The presence and absence of GCS within Fornicata shows an interesting evolutionary pattern (Fig. 1): all free-living representatives most likely have a fully functioning GCS, with all four enzymes being present, suggesting that the presence of GCS is somehow essential for free-living species (though see *Trepomonas*, below). By contrast, the host-associated species appear to have lost the GCS completely or partially (Leger et al. 2017; Tanifuji et al. 2018). Concerning "partial losses," some host-associated species contain the L-protein; however, this is commonly involved in other metabolic processes (Nývltová et al. 2016), and therefore its presence cannot be taken as evidence for a functioning GCS.

2.5 Hydrogen Production Coupled with GCS

[FeFe]-hydrogenase, its maturases (HydE, HydF, and HydG) (Putz et al. 2006), and ferredoxin have been identified in all free-living fornicates (Fig. 1). The phylogeny of [FeFe]-hydrogenase weakly suggests multiple independent origins within free-living Fornicata, while maturases show strong monophyly with other eukaryotic homologs (Leger et al. 2017). The 24 kDa and 51 kDa subunits of complex I (CI), which are often the only CI subunits present in MROs (Müller et al. 2012; Stairs et al. 2015; Hrdý et al. 2004; Dyall et al. 2004), have been identified in *C. membranifera*, *Ergobibamus*, *Chilomastix*, *Kipferlia*, and *Dysnectes* (Leger et al. 2017; Tanifuji et al. 2018). These subunits form an electron bifurcating trimeric complex with [FeFe]-hydrogenase to reoxidize NADH to NAD+ and ferredoxin, ultimately generating hydrogen (Schut and Adams 2009). The presence of these NADH-oxidizing enzymes suggests an interesting metabolic link with the GCS, as their presence in the MROs will help to maintain the NAD+/NADH required for GCS function (Hampson et al. 1983). Altogether, this suggests that MROs of all primarily free-living Fornicata are capable of the following: NADH production (via the GCS), H_2 production (via [FeFe]-hydrogenase), NADH oxidation (24 kDa and 51 kDa subunits of CI), and electron transport through ferredoxin (Leger et al. 2017).

2.6 Overall Characteristics and Evolution of MROs in Ancestrally Free-Living Fornicata

Recent comparative transcriptomic studies of metamonad MROs have shed light on the evolutionary trajectory of primarily free-living lineages of Fornicata. Ancestrally, Fornicata possessed an organelle similar to the hydrogenosomes of parabasalids (Fig. 2), with several additional metabolic functions but with an already highly reduced protein import system. The ancestral lineages were capable of ATP generation by converting acetyl-CoA to acetate, in both MROs via coupled SCS and ASCT enzymes and in the cytosol via the ACS enzyme; they possessed a functioning GCS; and they were capable of hydrogen production. In addition, they likely possessed additional amino acid metabolism enzymes and cardiolipin synthase.

Major changes occurred within the lineage leading to *Dysnectes* and diplomonads (Fig. 2), in which the MRO lost its capacity for ATP generation but retained a functioning GCS as well as NADH dehydrogenase and [FeFe]-hydrogenase. This resulted in the so far unique MRO of *Dysnectes*, which is predicted to be capable of hydrogen production but which is unable to produce ATP (in contrast to the canonical hydrogenosomes of *Trichomonas*) (Leger et al. 2017). Further reduction of the organelle occurred in diplomonads, which are ancestrally host-associated and have completely lost the GCS. The pattern of MRO reduction continued in the lineage leading to *Giardia*, which resulted in the existence of mitosomes (Tovar et al. 2003; Jedelský et al. 2011). In contrast, *Spironucleus salmonicida*, another parasitic

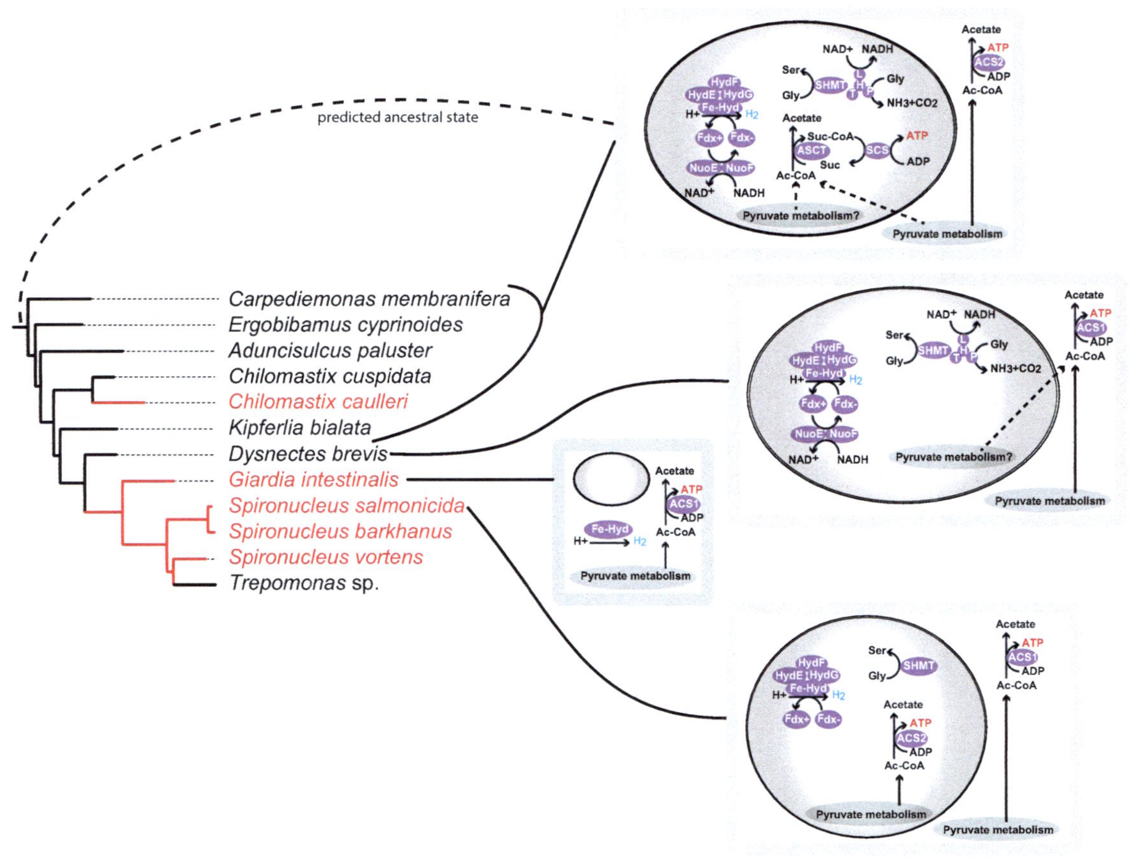

Fig. 2 Phylogenetic tree of Fornicata and the reconstruction of the putative metabolism of their MROs [adapted from Leger et al. (2017)]. Capacity for ATP synthesis in the organelle is lost on the branch leading to *Dysnectes* and diplomonads and is subsequently regained in *Spironucleus salmonicida*. The organelle of *Dysnectes* represents a rare organelle capable of hydrogen production but incapable of ATP synthesis. Free-living taxa are shown in black; parasitic or commensal taxa are shown in red

diplomonad, regained ATP generation capacity in its MRO by localizing the ACS enzyme, acquired through an independent HGT event, to the organelle (Jerlström-Hultqvist et al. 2013).

3 *Trepomonas* sp.

Trepomonas represents an interesting lineage within diplomonads, as it is likely a secondarily free-living species (Xu et al. 2016). Unfortunately, there are very few data available, and therefore it is hard to draw strong conclusions. However, *Trepomonas* possesses cytosol-localized ACS of the same evolutionary origin as that of *Giardia*, suggesting that the MROs of both organisms share characteristic features (Leger et al. 2017). It does not possess the GCS, which is intriguing as all other free-living taxa do. It is very likely that the GCS was lost from the genome of the *Trepomonas* lineage during its evolutionary history as parasites and has never been regained. To date, it remains unclear whether GCS-like functions in glycine decarboxylation and 5,10-methylenetetrahydrofolate production are carried out in *Trepomonas*.

4 Other Taxa

4.1 Jakobida

Stygiella incarcerata [formerly *Jakoba incarcerata* or *Andalucia incarcerata* (Pánek et al. 2015)] is a member of Jakobida, a group of free-living flagellates which together with Euglenozoa, Heterolobosea, and *Tsukubamonas* forms the clade known as Discoba (Hampl et al. 2009; Yabuki et al. 2011). Most jakobids are aerobes with canonical mitochondria, yet members of the family Stygiellidae (Pánek et al. 2015) appear to have secondarily adapted to an obligatory anaerobic lifestyle, possessing uniformly staining mitochondrion-related organelles that lack cristae (Simpson and Patterson 2001; Pánek et al. 2015).

As with the free-living fornicates, predictions of the biochemical features of these organelles are based on a transcriptomic survey from cells in culture with diverse prokaryotes (Leger et al. 2016). Based on this, *Stygiella* appears to lack all components of the ETC, save the two CI subunits described above, and all components of the TCA cycle. Searches for orthologs of known jakobid mitochondrial gene transcripts, PCR screens using mitochondrial 16S rRNA gene probes, and pulsed-field gel electrophoresis failed to uncover any evidence for a mitochondrial genome (Leger et al. 2016). *Stygiella* MROs possess pyruvate metabolism similar to that of *Trichomonas* hydrogenosomes. [FeFe]-hydrogenase and all three of its maturases, as well as PFO, SCS, and ASCT, are present (although the transferase present in *Stygiella*, ASCT1B, is a different subfamily (Tielens et al. 2010) from the ASCT1C found in

Trichomonas). Immunoelectron microscopy showed that [FeFe]-hydrogenase was primarily located in the MROs, suggesting that they are the site of pyruvate metabolism and ATP production. IscS, the cysteine desulfurase component of the ISC system, was similarly localized to the MROs. Overall, *Stygiella* appears to have retained a wider complement of mitochondrial amino acid metabolism genes than *Trichomonas*, including a complete branched-chain amino acid degradation pathway and cardiolipin synthase (Leger et al. 2016) (Fig. 1).

4.2 Heterolobosea

Heterolobosea is a diverse group of amoebae, flagellates, and amoeboflagellates that includes a large number of anaerobic species, almost all of which belong to one clade, Psalteriomonadidae (Pánek et al. 2017). *Psalteriomonas lanterna* and *Sawyeria marylandensis*, both free-living members of Psalteriomonadidae, possess uniformly staining MROs. In *Psalteriomonas*, these are elongated ["sausage- and dumbbell-shaped" (de Graaf et al. 2009)] and form clusters; in *Sawyeria*, they are cup-shaped (Barberà et al. 2010). The transcriptome data available for both organisms were derived from Sanger-sequenced expressed sequence tag clones and as a result are relatively sparse. Nevertheless, the transcriptomes of both organisms encode [FeFe]-hydrogenase and PFO with mitochondrial targeting signals as well as the 51 kDa subunit of CI, suggestive of the hydrogenosomal pyruvate fermentation pathway (de Graaf et al. 2009; Barberà et al. 2010). Both also encode ISC system enzymes. The larger *Sawyeria* transcriptome additionally encodes SCS, the 24 kDa subunit of CI; elements of the branched-chain amino acid degradation pathway; proline, serine, and ornithine metabolism pathways; and the GCS (Barberà et al. 2010) (Fig. 1).

4.3 Stramenopila

MROs with cristae have been described in four taxa of anaerobic stramenopile: the commensal flagellate *Proteromonas lacertae* (Perez-Brocal et al. 2010), the commensalic/parasitic *Blastocystis* (Zierdt et al. 1988; Zierdt 1991), and the free-living flagellate *Rictus lutensis* (Yubuki et al. 2010) and *Cantina marsupialis* (Yubuki et al. 2015). Despite their very different appearance, *Proteromonas* and *Blastocystis* are closely related and likely descend from a common ancestor that was anaerobic and endobiotic (Kostka 2017). The MROs of *Blastocystis* spp. and of *Cantina* have been studied at the molecular level (Lantsman et al. 2008; Stechmann et al. 2008; Denoeud et al. 2011; Tsaousis et al. 2011; Noguchi et al. 2015, 2016; Gentekaki et al. 2017). *Blastocystis* MROs retain a mitochondrial genome and CI as well as complex II (CII) of the ETC (Lantsman et al. 2008; Stechmann et al. 2008; Denoeud et al. 2011; Gentekaki et al. 2017) (see chapter "The Mitochondrion-

Related Organelles of Blastocystis"). By contrast, *Cantina* MROs appear to lack both the genome and all ETC complexes other than CII (Noguchi et al. 2015). The MROs of both organisms appear to possess anaerobic ATP generation enzymes—[FeFe]-hydrogenase and its associated maturases, PFO, ASCT1B, ASCT1C, and ACS. *Blastocystis* also possesses RQUA (Stairs et al. 2014), an enzyme that catalyzes the biosynthesis of rhodoquinone, which is essential for fumarate reduction by CII in some anaerobes (Van Hellemond et al. 1995; Lonjers et al. 2012; Stairs et al. 2018); however, RQUA was not identified in *Cantina* (Noguchi et al. 2015; Stairs et al. 2018). Both also possess the ISC system, a complete GCS, and the branched-chain amino acid degradation pathway (Noguchi et al. 2015; Gentekaki et al. 2017) (Fig. 1). Interestingly, while both organisms possess an MRO-targeted cardiolipin synthase, the specific type of enzyme is not the same, due to differential loss in the two lineages: the *Cantina* enzyme contains two phospholipase D domains and synthesizes cardiolipin from two phosphatidylglycerol molecules, while the *Blastocystis* enzyme contains an CDP-alcohol phosphatidyltransferase domain and synthesizes cardiolipin from one phosphatidylglycerol molecule and one cytidine diphosphate diacylglycerol molecule (Noguchi et al. 2016).

4.4 *Rhizaria*

As in stramenopiles, MROs have only been studied in two rhizarians that are not closely related to each other; the first, *Mikrocytos mackini*, is an obligate parasite with highly reduced MROs that have only retained the ISC system and elements of the mitochondrial protein import machinery. In stark contrast, the organelle of the free-living *Brevimastigomonas motovehiculus* appears to represent an early step in mitochondrial adaptation to low-oxygen conditions. *Brevimastigomonas* MROs retain a partial electron transport chain consisting of CI, CII, and complex V (CV); at least seven subunits of these complexes are encoded in the mitochondrial genome, as are the highly divergent and possibly nonfunctional *cob* and *cox1* genes. Also retained are PDH and at least seven, and possibly all, enzymes of the TCA cycle. The TCA cycle enzymes malate dehydrogenase, fumarase, and succinate dehydrogenase (CII) may also be capable of acting reductively under anaerobic conditions, as they are predicted to do in *Blastocystis*. Supporting this possibility, the *Brevimastigomonas* genome also encodes RQUA (Gawryluk et al. 2016). *Brevimastigomonas* MROs are also predicted to contain [FeFe]-hydrogenase and its maturases, PFO, PNO, ASCT1B, and SCS (Gawryluk et al. 2016). Consistent with this co-occurrence of mitochondrial and hydrogenosomal-like elements of pyruvate metabolism, *Brevimastigomonas* grows well under low-oxygen conditions but also tolerates aerobic conditions (Gawryluk et al. 2016). In addition, the ISC system, branched-chain amino acid degradation pathway, cardiolipin biosynthesis pathway, and GCS were identified as putatively present in the MROs (Fig. 1).

4.5 *Alveolata*

The majority of MROs in alveolates have been described in ciliates that are commensals of mammals (Yarlett et al. 1981, 1983, 1984; Paul et al. 1990); however hydrogenosomes have also been described in free-living members of at least six orders of **ciliates** [reviewed in Fenchel and Finlay (1991)]. These hydrogenosomes contain cristae and are generally associated with endosymbiotic methanogenic archaea, which consume the hydrogen that is produced as a metabolic by-product [van Bruggen et al. 1983; reviewed in Fenchel and Finlay (1991), Hackstein (2010)]. Apart from the presence of [FeFe]-hydrogenase, little is known about the metabolic potential of the MROs of free-living ciliates (but see also chapter "Hydrogenosomes of Anaerobic Ciliates").

MROs have also been described in two free-living chromerids, *Chromera velia* and *Vitrella brassicaformis*. Both are photosynthetic organisms that are found associated with corals; both belong to the out-group of the parasitic apicomplexan clade (which includes *Plasmodium*, *Toxoplasma*, and *Cryptosporidium*). The MROs of both organisms are capable of oxidative phosphorylation but lack CI; *Chromera* additionally lacks complex III (CIII) (Flegontov et al. 2015; Oborník and Lukeš 2015). Instead, their ETCs include an alternative NADH dehydrogenase and an alternative oxidase, features that are shared with apicomplexan parasites. The ETC is continuous in *Vitrella*, but it is uniquely split in *Chromera*, with complex IV (CIV) and CV acting independently of the alternate NADH dehydrogenase, the alternative oxidase, and CII (Flegontov et al. 2015). The genomes of both organisms encode MRO-targeted [FeFe]-hydrogenase and its maturases, PNO and SCS (Flegontov et al. 2015; Oborník and Lukeš 2015); both genomes also encode ACS, but only that of *Vitrella* possesses a mitochondrial targeting peptide, while *Chromera* also possesses mitochondrially targeted ASCT1B (Flegontov et al. 2015; Oborník and Lukeš 2015; Stairs et al. 2015) (Fig. 1).

4.6 *Breviatea*

Pygsuia biforma is a member of Breviatea, a group of free-living flagellates that, together with apusomonads, are the closest relatives to Opisthokonta (Brown et al. 2013). Superficially similar MROs are present in *Pygsuia*, *Breviata anathema*, and *Subulatomonas tetraspora*, although only partial transcriptome data are available for the latter two organisms (Stairs et al. 2014; Grant et al. 2012). The single MRO of *Pygsuia* is uniformly staining and elongated (Brown et al. 2013; Stairs et al. 2014). It appears to lack a mitochondrial genome and most components of the electron transport chain: only CII and the 24 kDa and 51 kDa subunits of CI are encoded in its transcriptome (Stairs et al. 2014). Of the TCA cycle, only fumarase and succinate dehydrogenase are present. Consistent with these enzymes acting reductively, *Pygsuia* also possesses RQUA for rhodoquinone (RQ) biosynthesis, like

Blastocystis and *Brevimastigomonas*; recently, RQ was detected in *Pygsuia* lipid extracts (Stairs et al. 2018 REF). [FeFe]-hydrogenase and its maturases, PFO, PNO, ASCT1B, ASCT1C, and SCS, are also predicted to be targeted to the MROs. The GCS, branched-chain amino acid degradation pathway, and cardiolipin biosynthesis pathway are also present (Stairs et al. 2014) (Fig. 1).

Interestingly, *Pygsuia* lacks all components of the ISC system. Instead, it possesses two genes encoding an enzyme, SufCB, that appears to be a fusion of the SufC and SufB components of the sulfur mobilization (SUF) system of iron-sulfur cluster biosynthesis (Stairs et al. 2014). SufCB is also encoded in the transcriptomes of *Blastocystis* (Tsaousis et al. 2012) and *Stygiella* (Leger et al. 2016), where it is likely located in the cytosol; in *Pygsuia*, however, one of the SufCB proteins possesses a mitochondrial targeting signal and has been localized to the MRO (Stairs et al. 2014). Phylogenies suggest that SufC and SufB were acquired from an archaeon, possibly as part of an operon, and became fused and then transferred in a handful of eukaryotes. An independent transfer event, this time from bacteria, appears to have led to the acquisition of cytosolic Suf genes in Preaxostyla (Karnkowska et al. 2016; Vacek et al. 2018; see chapter "Protists Without Mitochondria, How It May Happen?").

Genome, transcriptome, and proteome data have also been generated for a second free-living breviate, *Lenisia limosa* (Hamann et al. 2016). Unlike *Pygsuia*, *Lenisia* cells contain multiple, cristate MROs. Much like *Pygsuia* and other protists, *Lenisia* encodes MRO-targeted PFO, [FeFe]-hydrogenase, and its associated maturases, the 24 kDa and 51 kDa subunits of CI, SCS, and ASCT. However, this protist uses the typical mitochondrial ISC system for Fe-S cluster biosynthesis, rather than the SUF system discussed above (Stairs, *pers. comm.*). Furthermore, no evidence of RQUA was found in the genome (Stairs et al. 2018). Like anaerobic ciliates (see above), *Lenisia* forms a symbiotic association with prokaryotic symbionts—in this case, epibionts of the bacterial genus *Arcobacter*—which consume the hydrogen gas produced by its MROs (Hamann et al. 2016).

4.7 Amoebozoa

The first free-living organism shown to contain a MRO was the amoebozoan *Mastigamoeba balamuthi* (Gill et al. 2007), a close relative of the human pathogen *Entamoeba histolytica*. Both of these archaeamoebae have replaced the ISC system for the biosynthesis of Fe-S clusters with a horizontally acquired nitrogen fixation (NIF) system of iron-sulfur cluster assembly (Ali et al. 2004; Nývltová et al. 2013). While the pathway is exclusively cytoplasmic in *Entamoeba*, *Mastigamoeba* encodes two copies of each component, targeted to the cytoplasm and MRO, respectively (Nývltová et al. 2013). Extensive biochemical and molecular biological analyses have revealed that the *Mastigamoeba* MROs oxidize pyruvate and generate hydrogen via the so-called "extended glycolysis" pathway (PFO, [FeFe]-hydrogenase, and ACS), malate dehydrogenase (MDH), and an NAD +-dependent

D-lactate dehydrogenase (D-LDH) (Nývltová et al. 2015) (Fig. 1). Here, the authors propose that malate is converted to oxaloacetate which in turn is used by an unidentified dicarboxylic acid/lactate antiporter to import D-lactate into the MROs. This D-lactate serves as the substrate for D-LDH to generate the pyruvate necessary for extended glycolysis. Therefore, unlike other MROs, the *Mastigamoeba* MROs derive pyruvate from lactate, and not oxaloacetate or via the pyruvate transporter (Nývltová et al. 2015). Like *Pygsuia biforma*, *Mastigamoeba* encodes MRO-targeted quinone-dependent metabolism (e.g., CII, RQUA, electron-transferring flavoprotein system) despite lacking the remainder of the respiratory chain (i.e., no components of CI, CIII, CIV, or CV have been identified); however, the exact functions of these proteins have not been elucidated. Both *Entamoeba* and *Mastigamoeba* have MRO-localized sulfate activation (Mi-ichi et al. 2009; Nývltová et al. 2015). In *Entamoeba*, this pathway is essential for the generation of sulfated lipids necessary for cell proliferation and encystation (Mi-ichi et al. 2009, 2011, 2015); however, its role in *Mastigamoeba balamuthi* MROs remains to be investigated.

5 Conclusions

Both the categories of "mitochondrion-related organelles" and "free-living eukaryotes" are artificial and lie on a spectrum. Several eukaryotes possess mitochondrially targeted fermentation enzymes in addition to a complete mitochondrial electron transport chain (ETC) and tricarboxylic acid (TCA) cycle, resulting in mitochondria that may be capable of functioning anaerobically (Clarke et al. 2013; Fritz-Laylin et al. 2010; Hug et al. 2010; Leger et al. 2013; Ebenezer et al. 2017). Some of these are opportunistic pathogens that inhabit a variety of freshwater, marine, or soil environments in addition to host tissues (Clarke et al. 2013; Leger et al. 2013); organisms with this type of lifestyle may reveal a wider diversity of mitochondrial functions than previously described in most MROs of parasites or commensals. Other eukaryotes have lost CI of the ETC while retaining the capacity for oxidative phosphorylation and other mitochondrial functions (Gabaldón et al. 2005; Maclean et al. 2018; Senkler et al. 2018).

Microbial eukaryotes, particularly free-living ones, remain relatively poorly studied. Given the recent advent of low-cost sequencing technologies, and the discoveries of new major groups of eukaryotes that these have enabled, it is likely that much progress is still to be made in our understanding of eukaryote diversity and MROs. Nevertheless, the work summarized above has greatly expanded our understanding of these organelles and has enabled some themes to emerge. MROs in free-living eukaryotes may still retain the capacity for aerobic ATP generation or may have lost it entirely; most of MROs of free-living eukaryotes described to date are capable of pyruvate fermentation under low-oxygen conditions using any of several characteristic enzymes, although in some cases the capacity for ATP production may be absent under any conditions. Many have retained specific mitochondrial pathways not commonly found in the MROs of parasites or commensals: the glycine cleavage system, the branched-chain amino acid degradation pathway, and cardiolipin biosynthesis enzymes. However, evidence for other key mitochondrial enzymes, such as those involved in heme biosynthesis or the urea cycle, has not been described in any of these MROs. Comparative genomics based on high-quality, high-coverage sequence data has been instrumental in extending our predictions of the biology of MROs beyond a small number of key enzymes. Proteomic and biochemical analyses that focus on mitochondria and MROs in a wider range of eukaryotes may uncover new functions in these organelles, not previously identified in model eukaryotes. The interactions between MROs and the prokaryotic endosymbionts of their hosts, as in ciliates or *Lenisia*, are a particularly exciting avenue of future study.

Acknowledgments We wish to thank Rachel Kolísko for helpful input and editing of the manuscript.

References

Ali V, Shigeta Y, Tokumoto U, Takahashi Y, Nozaki T (2004) An intestinal parasitic protist, *Entamoeba histolytica*, possesses a non-redundant nitrogen fixation-like system for iron-sulfur cluster assembly under anaerobic conditions. J Biol Chem 279(16):16863–16874. https://doi.org/10.1074/jbc.M313314200

Barberà MJ, Ruiz-Trillo I, Tufts JY, Bery A, Silberman JD, Roger AJ (2010) *Sawyeria marylandensis* (Heterolobosea) has a hydrogenosome with novel metabolic properties. Eukaryot Cell 9(12):1913–1924. https://doi.org/10.1128/EC.00122-10

Brown MW, Sharpe SC, Silberman JD, Heiss AA, Lang BF, Simpson AG, Roger AJ (2013) Phylogenomics demonstrates that breviate flagellates are related to opisthokonts and apusomonads. Proc Biol Sci 280(1769):20131755. https://doi.org/10.1098/rspb.2013.1755

Burki F, Corradi N, Sierra R, Pawlowski J, Meyer GR, Abbott CL, Keeling PJ (2013) Phylogenomics of the intracellular parasite *Mikrocytos mackini* reveals evidence for a mitosome in rhizaria. Curr Biol 23(16):1541–1547. https://doi.org/10.1016/j.cub.2013.06.033

Clarke M, Lohan AJ, Liu B, Lagkouvardos I, Roy S, Zafar N, Bertelli C, Schilde C, Kianianmomeni A, Burglin TR, Frech C, Turcotte B, Kopec KO, Synnott JM, Choo C, Paponov I, Finkler A, Heng Tan CS, Hutchins AP, Weinmeier T, Rattei T, Chu JS, Gimenez G, Irimia M, Rigden DJ, Fitzpatrick DA, Lorenzo-Morales J, Bateman A, Chiu CH, Tang P, Hegemann P, Fromm H, Raoult D, Greub G, Miranda-Saavedra D, Chen N, Nash P, Ginger ML, Horn M, Schaap P, Caler L, Loftus BJ (2013) Genome of *Acanthamoeba castellanii* highlights extensive lateral gene transfer and early evolution of tyrosine kinase signaling. Genome Biol 14(2):R11. https://doi.org/10.1186/gb-2013-14-2-r11

de Graaf RM, Duarte I, van Alen TA, Kuiper JW, Schotanus K, Rosenberg J, Huynen MA, Hackstein JH (2009) The hydrogenosomes of *Psalteriomonas lanterna*. BMC Evol Biol 9:287. https://doi.org/10.1186/1471-2148-9-287

de Graaf RM, Ricard G, van Alen TA, Duarte I, Dutilh BE, Burgtorf C, Kuiper JW, van der Staay GW, Tielens AG, Huynen MA, Hackstein JH (2011) The organellar genome and metabolic potential of the hydrogen-producing mitochondrion of Nyctotherus ovalis. Mol Biol Evol 28 (8):2379–2391. https://doi.org/10.1093/molbev/msr059

Denoeud F, Roussel M, Noel B, Wawrzyniak I, Da Silva C, Diogon M, Viscogliosi E, Brochier-Armanet C, Couloux A, Poulain J, Segurens B, Anthouard V, Texier C, Blot N, Poirier P, Ng GC, Tan KS, Artiguenave F, Jaillon O, Aury JM, Delbac F, Wincker P, Vivares CP, El Alaoui H (2011) Genome sequence of the stramenopile *Blastocystis*, a human anaerobic parasite. Genome Biol 12(3):R29. https://doi.org/10.1186/gb-2011-12-3-r29

Doležal P, Likic V, Tachezy J, Lithgow T (2006) Evolution of the molecular machines for protein import into mitochondria. Science 313(5785):314–318. https://doi.org/10.1126/Science.1127895

Doležal P, Dagley MJ, Kono M, Wolynec P, Likic VA, Foo JH, Sedinova M, Tachezy J, Bachmann A, Bruchhaus I, Lithgow T (2010) The essentials of protein import in the degenerate mitochondrion of *Entamoeba histolytica*. PLoS Pathog 6(3):e1000812. https://doi.org/10.1371/journal.ppat.1000812

Dudek J, Rehling P, van der Laan M (2013) Mitochondrial protein import: common principles and physiological networks. BBA-Mol Cell Res 1833(2):274–285. https://doi.org/10.1016/j.bbamcr.2012.05.028

Dyall SD, Yan W, Delgadillo-Correa MG, Lunceford A, Loo JA, Clarke CF, Johnson PJ (2004) Non-mitochondrial complex I proteins in a hydrogenosomal oxidoreductase complex. Nature 431(7012):1103–1107. https://doi.org/10.1038/nature02990

Ebenezer TE, Carrington M, Lebert M, Kelly S, Field MC (2017) *Euglena gracilis* genome and transcriptome: organelles, nuclear genome assembly strategies and initial features. Adv Exp Med Biol 979:125–140. https://doi.org/10.1007/978-3-319-54910-1_7

Fenchel T, Finlay BJ (1991) The biology of free-living anaerobic ciliates. Eur J Protistol 26 (3–4):201–215. https://doi.org/10.1016/S0932-4739(11)80143-4

Flegontov P, Michalek J, Janouškovec J, Lai DH, Jirku M, Hajduskova E, Tomcala A, Otto TD, Keeling PJ, Pain A, Oborník M, Lukeš J (2015) Divergent mitochondrial respiratory chains in phototrophic relatives of apicomplexan parasites. Mol Biol Evol 32(5):1115–1131. https://doi.org/10.1093/molbev/msv021

Fritz-Laylin LK, Prochnik SE, Ginger ML, Dacks JB, Carpenter ML, Field MC, Kuo A, Paredez A, Chapman J, Pham J, Shu S, Neupane R, Cipriano M, Mancuso J, Tu H, Salamov A, Lindquist E, Shapiro H, Lucas S, Grigoriev IV, Cande WZ, Fulton C, Rokhsar DS, Dawson SC (2010) The genome of *Naegleria gruberi* illuminates early eukaryotic versatility. Cell 140(5):631–642. https://doi.org/10.1016/j.cell.2010.01.032

Gabaldón T, Rainey D, Huynen MA (2005) Tracing the evolution of a large protein complex in the eukaryotes, NADH:ubiquinone oxidoreductase (complex I). J Mol Biol 348(4):857–870. https://doi.org/10.1016/j.jmb.2005.02.067

Gawryluk RMR, Kamikawa R, Stairs CW, Silberman JD, Brown MW, Roger AJ (2016) The earliest stages of mitochondrial adaptation to low oxygen revealed in a novel rhizarian. Curr Biol 26(20):2729–2738. https://doi.org/10.1016/j.cub.2016.08.025

Gentekaki E, Curtis BA, Stairs CW, Klimeš V, Eliáš M, Salas-Leiva DE, Herman EK, Eme L, Arias MC, Henrissat B, Hilliou F, Klute MJ, Suga H, Malik SB, Pightling AW, Kolísko M, Rachubinski RA, Schlacht A, Soanes DM, Tsaousis AD, Archibald JM, Ball SG, Dacks JB, Clark CG, van der Giezen M, Roger AJ (2017) Extreme genome diversity in the hyper-prevalent parasitic eukaryote *Blastocystis*. PLoS Biol 15(9):e2003769. https://doi.org/10.1371/journal.pbio.2003769

Gill EE, Diaz-Trivino S, Barbera MJ, Silberman JD, Stechmann A, Gaston D, Tamas I, Roger AJ (2007) Novel mitochondrion-related organelles in the anaerobic amoeba *Mastigamoeba balamuthi*. Mol Microbiol 66(6):1306–1320. https://doi.org/10.1111/j.1365-2958.2007.05979.x

Grant JR, Lahr DJG, Rey FE, Burleigh JG, Knight R, Molestina RE, Katz LA (2012) Gene discovery from a pilot study of the transcriptomes from three diverse microbial eukaryotes: *Corallomyxa tenera*, *Chilodonella uncinata*, and *Subulatomonas tetraspora*. Protist Genomics 1:3–18

Hackstein JHP (2010) Anaerobic ciliates and their methanogenic endosymbionts. In: Hackstein JHP (ed) (Endo)symbiotic methanogenic archaea. Microbiology monographs. Springer, Heidelberg, pp 13–23

Hamann E, Gruber-Vodicka H, Kleiner M, Tegetmeyer HE, Riedel D, Littmann S, Chen J, Milucka J, Viehweger B, Becker KW, Dong X, Stairs CW, Hinrichs KU, Brown MW, Roger AJ, Strous M (2016) Environmental Breviatea harbour mutualistic Arcobacter epibionts. Nature 534(7606):254–258. https://doi.org/10.1038/nature18297

Hamann E, Tegetmeyer HE, Riedel D, Littmann S, Ahmerkamp S, Chen J, Hach PF, Strous M (2017) Syntrophic linkage between predatory *Carpediemonas* and specific prokaryotic populations. ISME J 11(5):1205–1217. https://doi.org/10.1038/ismej.2016.197

Hampl V, Hug L, Leigh JW, Dacks JB, Lang BF, Simpson AG, Roger AJ (2009) Phylogenomic analyses support the monophyly of Excavata and resolve relationships among eukaryotic "supergroups". Proc Natl Acad Sci U S A 106(10):3859–3864. https://doi.org/10.1073/pnas.0807880106

Hampson RK, Barron LL, Olson MS (1983) Regulation of the glycine cleavage system in isolated rat-liver mitochondria. J Biol Chem 258(5):2993–2999

Henriquez FL, Richards TA, Roberts F, McLeod R, Roberts CW (2005) The unusual mitochondrial compartment of *Cryptosporidium parvum*. Trends Parasitol 21(2):68–74. https://doi.org/10.1016/j.pt.2004.11.010

Hrdý I, Hirt RP, Doležal P, Bardonova L, Foster PG, Tachezy J, Embley TM (2004) *Trichomonas* hydrogenosomes contain the NADH dehydrogenase module of mitochondrial complex I. Nature 432(7017):618–622. https://doi.org/10.1038/nature03149

Hug LA, Stechmann A, Roger AJ (2010) Phylogenetic distributions and histories of proteins involved in anaerobic pyruvate metabolism in eukaryotes. Mol Biol Evol 27(2):311–324. https://doi.org/10.1093/molbev/msp237

Jedelský PL, Doležal P, Rada P, Pyrih J, Šmid O, Hrdý I, Sedinova M, Marcincikova M, Voleman L, Perry AJ, Beltran NC, Lithgow T, Tachezy J (2011) The minimal proteome in the reduced mitochondrion of the parasitic protist *Giardia intestinalis*. PLoS One 6(2):e17285. https://doi.org/10.1371/journal.pone.0017285

Jerlström-Hultqvist J, Einarsson E, Xu FF, Hjort K, Ek B, Steinhauf D, Hultenby K, Bergquist J, Andersson JO, Svärd SG (2013) Hydrogenosomes in the diplomonad *Spironucleus salmonicida*. Nat Commun 4:2493. https://doi.org/10.1038/ncomms3493

Karnkowska A, Vacek V, Zubácová Z, Treitli SC, Petrzelkova R, Eme L, Novak L, Žárský V, Barlow LD, Herman EK, Soukal P, Hroudova M, Doležal P, Stairs CW, Roger AJ, Eliáš M, Dacks JB, Vlček C, Hampl V (2016) A eukaryote without a mitochondrial organelle. Curr Biol 26(10):1274–1284. https://doi.org/10.1016/j.cub.2016.03.053

Kikuchi G (1973) Glycine cleavage system - composition, reaction mechanism, and physiological significance. Mol Cell Biochem 1(2):169–187. https://doi.org/10.1007/bf01659328

Kolísko M, Silberman J, Čepička I, Yubuki N, Takishita K, Yabuki A, Leander B, Inouye I, Inagaki Y, Roger A, Simpson A (2010) A wide diversity of previously undetected free-living relatives of diplomonads isolated from marine/saline habitats. Environ Microbiol 12(10):2700–2710. https://doi.org/10.1111/j.1462-2920.2010.02239.x

Kostka M (2017) Opalinata. In: Archibald JM, Slamovits CH, Simpson AGB (eds) Handbook of the protists. Springer, Cham

Lantsman Y, Tan KS, Morada M, Yarlett N (2008) Biochemical characterization of a mitochondrial-like organelle from *Blastocystis* sp. subtype 7. Microbiology 154(Pt 9):2757–2766. https://doi.org/10.1099/mic.0.2008/017897-0

Leger MM, Gawryluk RM, Gray MW, Roger AJ (2013) Evidence for a hydrogenosomal-type anaerobic ATP generation pathway in *Acanthamoeba castellanii*. PLoS One 8(9):e69532. https://doi.org/10.1371/journal.pone.0069532

Leger MM, Eme L, Hug LA, Roger AJ (2016) Novel hydrogenosomes in the microaerophilic jakobid *Stygiella incarcerata*. Mol Biol Evol 33(9):2318–2336. https://doi.org/10.1093/molbev/msw103

Leger MM, Kolísko M, Kamikawa R, Stairs CW, Kume K, Čepička I, Silberman JD, Andersson JO, Xu F, Yabuki A, Eme L, Zhang Q, Takishita K, Inagaki Y, Simpson AGB, Hashimoto T, Roger AJ (2017) Organelles that illuminate the origins of *Trichomonas* hydrogenosomes and *Giardia* mitosomes. Nature Ecol Evol 1(4):0092. https://doi.org/10.1038/s41559-017-0092

Lill R, Hoffmann B, Molik S, Pierik AJ, Rietzschel N, Stehling O, Uzarska MA, Webert H, Wilbrecht C, Muhlenhoff U (2012) The role of mitochondria in cellular iron-sulfur protein biogenesis and iron metabolism. Biochim Biophys Acta 1823(9):1491–1508. https://doi.org/10.1016/j.bbamcr.2012.05.009

Lindmark DG, Müller M (1973) Hydrogenosome, a cytoplasmic organelle of the anaerobic flagellate *Tritrichomonas foetus*, and its role in pyruvate metabolism. J Biol Chem 248 (22):7724–7728

Lindmark DG, Müller M, Shio H (1975) Hydrogenosomes in *Trichomonas vaginalis*. J Parasitol 61 (3):552–554

Lonjers ZT, Dickson EL, Chu TP, Kreutz JE, Neacsu FA, Anders KR, Shepherd JN (2012) Identification of a new gene required for the biosynthesis of rhodoquinone in Rhodospirillum rubrum. J Bacteriol 194(5):965–971. https://doi.org/10.1128/JB.06319-11

Maclean AE, Hertle AP, Ligas J, Bock R, Balk J, Meyer EH (2018) Absence of complex I is associated with diminished respiratory chain function in European mistletoe. Curr Biol 28 (10):1614–1619 e1613. https://doi.org/10.1016/j.cub.2018.03.036

Maguire F, Richards TA (2014) Organelle evolution: a mosaic of 'mitochondrial' functions. Curr Biol 24(11):R518–R520. https://doi.org/10.1016/j.cub.2014.03.075

Mai Z, Ghosh S, Frisardi M, Rosenthal B, Rogers R, Samuelson J (1999) Hsp60 is targeted to a cryptic mitochondrion-derived organelle ("crypton") in the microaerophilic protozoan parasite *Entamoeba histolytica*. Mol Cell Biol 19(3):2198–2205

Makiuchi T, Nozaki T (2014) Highly divergent mitochondrion-related organelles in anaerobic parasitic protozoa. Biochimie 100:3–17. https://doi.org/10.1016/j.biochi.2013.11.018

Mi-ichi F, Abu Yousuf M, Nakada-Tsukui K, Nozaki T (2009) Mitosomes in *Entamoeba histolytica* contain a sulfate activation pathway. Proc Natl Acad Sci U S A 106 (51):21731–21736. https://doi.org/10.1073/pnas.0907106106

Mi-ichi F, Makiuchi T, Furukawa A, Sato D, Nozaki T (2011) Sulfate activation in mitosomes plays an important role in the proliferation of *Entamoeba histolytica*. PLoS Negl Trop Dis 5(8):e1263. https://doi.org/10.1371/journal.pntd.0001263

Mi-ichi F, Miyamoto T, Takao S, Jeelani G, Hashimoto T, Hara H, Nozaki T, Yoshida H (2015) *Entamoeba* mitosomes play an important role in encystation by association with cholesteryl sulfate synthesis. Proc Natl Acad Sci U S A 112(22):E2884–E2890. https://doi.org/10.1073/pnas.1423718112

Mokranjac D, Neupert W (2010) The many faces of the mitochondrial TIM23 complex. BBA-Bioenergetics 1797(6–7):1045–1054. https://doi.org/10.1016/j.bbabio.2010.01.026

Müller M, Mentel M, van Hellemond JJ, Henze K, Woehle C, Gould SB, Yu R-Y, van der Giezen M, Tielens AGM, Martin WF (2012) Biochemistry and evolution of anaerobic energy metabolism in eukaryotes. Microbiol Mol Biol Rev 76(2):444–495. https://doi.org/10.1128/mmbr.05024-11

Noguchi F, Shimamura S, Nakayama T, Yazaki E, Yabuki A, Hashimoto T, Inagaki Y, Fujikura K, Takishita K (2015) Metabolic capacity of mitochondrion-related organelles in the free-living anaerobic stramenopile *Cantina marsupialis*. Protist 166(5):534–550. https://doi.org/10.1016/j.protis.2015.08.002

Noguchi F, Tanifuji G, Brown MW, Fujikura K, Takishita K (2016) Complex evolution of two types of cardiolipin synthase in the eukaryotic lineage stramenopiles. Mol Phylogenet Evol 101:133–141. https://doi.org/10.1016/j.ympev.2016.05.011

Nývltová E, Šuták R, Harant K, Šedinová M, Hrdý I, Pačes J, Vlček C, Tachezy J (2013) NIF-type iron-sulfur cluster assembly system is duplicated and distributed in the mitochondria and cytosol of *Mastigamoeba balamuthi*. Proc Natl Acad Sci U S A 110(18):7371–7376. https://doi.org/10.1073/pnas.1219590110

Nývltová E, Stairs CW, Hrdý I, Ridl J, Mach J, Pačes J, Roger AJ, Tachezy J (2015) Lateral gene transfer and gene duplication played a key role in the evolution of *Mastigamoeba balamuthi* hydrogenosomes. Mol Biol Evol 32(4):1039–1055. https://doi.org/10.1093/molbev/msu408

Nývltová E, Smutna T, Tachezy J, Hrdý I (2016) OsmC and incomplete glycine decarboxylase complex mediate reductive detoxification of peroxides in hydrogenosomes of *Trichomonas vaginalis*. Mol Biochem Parasitol 206(1–2):29–38. https://doi.org/10.1016/j.molbiopara.2016.01.006

Oborník M, Lukeš J (2015) The organellar genomes of *Chromera* and *Vitrella*, the phototrophic relatives of apicomplexan parasites. Annu Rev Microbiol 69:129–144. https://doi.org/10.1146/annurev-micro-091014-104449

Opperdoes FR, De Jonckheere JF, Tielens AG (2011) *Naegleria gruberi* metabolism. Int J Parasitol 41(9):915–924. https://doi.org/10.1016/j.ijpara.2011.04.004

Pánek T, Taborský P, Pachiadaki MG, Hroudová M, Vlček Č, Edgcomb VP, Čepička I (2015) Combined culture-based and culture-independent approaches provide insights into diversity of jakobids, an extremely plesiomorphic eukaryotic lineage. Front Microbiol 6:1288. https://doi.org/10.3389/fmicb.2015.01288

Pánek T, Simpson AGB, Brown MW, Dyer BD (2017) Heterolobosea. In: Archibald JM, Slamovits CH, Simpson AGB (eds) Handbook of the Protists. Springer

Park J, Kolísko M, Heiss A, Simpson A (2009) Light microscopic observations, ultrastructure, and molecular phylogeny of *Hicanonectes teleskopos* n. G., n. Sp., a deep-branching relative of Diplomonads. J Eukaryot Microbiol 56(4):373–384. https://doi.org/10.1111/j.1550-7408.2009.00412.x

Park J, Kolísko M, Simpson A (2010) Cell morphology and formal description of *Ergobibamus cyprinoides* n. G., n. Sp., another *Carpediemonas*-like relative of Diplomonads. J Eukaryot Microbiol 57(6):520–528. https://doi.org/10.1111/j.1550-7408.2010.00506.x

Paul RG, Williams AG, Butler RD (1990) Hydrogenosomes in the rumen entodiniomorphid ciliate *Polyplastron multivesiculatum*. J Gen Microbiol 136(10):1981–1989. https://doi.org/10.1099/00221287-136-10-1981

Perez-Brocal V, Shahar-Golan R, Clark CG (2010) A linear molecule with two large inverted repeats: the mitochondrial genome of the stramenopile *Proteromonas lacertae*. Genome Biol Evol 2:257–266. https://doi.org/10.1093/gbe/evq015

Putz S, Doležal P, Gelius-Dietrich G, Boháčová L, Tachezy J, Henze K (2006) Fe-hydrogenase maturases in the hydrogenosomes of *Trichomonas vaginalis*. Eukaryot Cell 5(3):579–586. https://doi.org/10.1128/ec.5.3.579-586.2006

Pyrihova E, Motyckova A, Voleman L, Wandyszewska N, Fiser R, Seydlova G, Roger A, Kolísko M, Doležal P (2018) A single Tim translocase in the mitosomes of *Giardia intestinalis* illustrates convergence of protein import machines in anaerobic eukaryotes. Genome Biol Evol 10:2813. https://doi.org/10.1093/gbe/evy215

Roger AJ, Munoz-Gomez SA, Kamikawa R (2017) The origin and diversification of mitochondria. Curr Biol 27(21):R1177–R1192. https://doi.org/10.1016/j.cub.2017.09.015

Sanchez LB, Muller M (1996) Purification and characterization of the acetate forming enzyme, acetyl-CoA synthetase (ADP-forming) from the amitochondriate protist, *Giardia lamblia*. FEBS Lett 378(3):240–244. https://doi.org/10.1016/0014-5793(95)01463-2

Santos HJ, Makiuchi T, Nozaki T (2018) Reinventing an organelle: the reduced mitochondrion in parasitic protists. Trends Parasitol 34:1038. https://doi.org/10.1016/j.pt.2018.08.008

Schneider HC, Berthold J, Bauer MF, Dietmeier K, Guiard B, Brunner M, Neupert W (1994) Mitochondrial Hsp70/MIM44 complex facilitates protein import. Nature 371(6500):768–774. https://doi.org/10.1038/371768a0

Schneider RE, Brown MT, Shiflett AM, Dyall SD, Hayes RD, Xie Y, Loo JA, Johnson PJ (2011) The *Trichomonas vaginalis* hydrogenosome proteome is highly reduced relative to mitochondria, yet complex compared with mitosomes. Int J Parasitol 41(13–14):1421–1434. https://doi.org/10.1016/j.ijpara.2011.10.001

Schut GJ, Adams MW (2009) The iron-hydrogenase of *Thermotoga maritima* utilizes ferredoxin and NADH synergistically: a new perspective on anaerobic hydrogen production. J Bacteriol 191(13):4451–4457. https://doi.org/10.1128/JB.01582-08

Senkler J, Rugen N, Eubel H, Hegermann J, Braun HP (2018) Absence of complex I implicates rearrangement of the respiratory chain in European mistletoe. Curr Biol 28(10):1606–1613 e1604. https://doi.org/10.1016/j.cub.2018.03.050
Simpson AG, Patterson DJ (2001) On core jakobids and excavate taxa: the ultrastructure of *Jakoba incarcerata*. J Eukaryot Microbiol 48(4):480–492
Šmid O, Matuskova A, Harris SR, Kucera T, Novotny M, Horvathova L, Hrdý I, Kutejova E, Hirt RP, Embley TM, Janata J, Tachezy J (2008) Reductive evolution of the mitochondrial processing peptidases of the unicellular parasites *Trichomonas vaginalis* and *Giardia intestinalis*. PLoS Pathog 4(12):e1000243. https://doi.org/10.1371/journal.ppat.1000243
Stairs CW, Roger AJ, Hampl V (2011) Eukaryotic pyruvate formate lyase and its activating enzyme were acquired laterally from a firmicute. Mol Biol Evol 28(7):2087–2099. https://doi.org/10.1093/Molbev/Msr032
Stairs CW, Eme L, Brown MW, Mutsaers C, Susko E, Dellaire G, Soanes DM, van der Giezen M, Roger AJ (2014) A SUF Fe-S cluster biogenesis system in the mitochondrion-related organelles of the anaerobic protist *Pygsuia*. Curr Biol 24(11):1176–1186. https://doi.org/10.1016/j.cub.2014.04.033
Stairs CW, Leger MM, Roger AJ (2015) Diversity and origins of anaerobic metabolism in mitochondria and related organelles. Philos Trans R Soc Lond Ser B Biol Sci 370 (1678):20140326. https://doi.org/10.1098/rstb.2014.0326
Stairs CW, Eme L, Muñoz-Gómez SA, Cohen A, Dellaire G, Shepherd JN, Fawcett JP, Roger AJ (2018) Microbial eukaryotes have adapted to hypoxia by horizontal acquisitions of a gene involved in rhodoquinone biosynthesis. eLife 7. https://doi.org/10.7554/eLife.34292
Stechmann A, Hamblin K, Perez-Brocal V, Gaston D, Richmond GS, van der Giezen M, Clark CG, Roger AJ (2008) Organelles in *Blastocystis* that blur the distinction between mitochondria and hydrogenosomes. Curr Biol 18(8):580–585. https://doi.org/10.1016/j.cub.2008.03.037
Steinbuchel A, Muller M (1986) Anaerobic pyruvate metabolism of *Tritrichomonas fetus* and *Trichomonas vaginalis* hydrogenosomes. Mol Biochem Parasitol 20(1):57–65. https://doi.org/10.1016/0166-6851(86)90142-8
Takishita K, Kolísko M, Komatsuzaki H, Yabuki A, Inagaki Y, Čepička I, Smejkalova P, Silberman J, Hashimoto T, Roger A, Simpson A (2012) Multigene phylogenies of diverse *Carpediemonas*-like organisms identify the closest relatives of 'amitochondriate' diplomonads and retortamonads. Protist 163(3):344–355. https://doi.org/10.1016/j.protis.2011.12.007
Tanifuji G, Takabayashi S, Kume K, Takagi M, Nakayama T, Kamikawa R, Inagaki Y, Hashimoto T (2018) The draft genome of *Kipferlia bialata* reveals reductive genome evolution in fornicate parasites. PLoS One 13(3):e0194487. https://doi.org/10.1371/journal.pone.0194487
Tian HF, Feng JM, Wen JF (2012) The evolution of cardiolipin biosynthesis and maturation pathways and its implications for the evolution of eukaryotes. BMC Evol Biol 12:32. https://doi.org/10.1186/1471-2148-12-32
Tielens AG, van Grinsven KW, Henze K, van Hellemond JJ, Martin W (2010) Acetate formation in the energy metabolism of parasitic helminths and protists. Int J Parasitol 40(4):387–397. https://doi.org/10.1016/j.ijpara.2009.12.006
Tovar J, Fischer A, Clark CG (1999) The mitosome, a novel organelle related to mitochondria in the amitochondrial parasite *Entamoeba histolytica*. Mol Microbiol 32(5):1013–1021
Tovar J, León-Avila G, Sanchez LB, Sutak R, Tachezy J, van der Giezen M, Hernandez M, Muller M, Lucocq JM (2003) Mitochondrial remnant organelles of *Giardia* function in iron-sulphur protein maturation. Nature 426(6963):172–176. https://doi.org/10.1038/nature01945
Tsaousis AD, Gaston D, Stechmann A, Walker PB, Lithgow T, Roger AJ (2011) A functional Tom70 in the human parasite *Blastocystis* sp.: implications for the evolution of the mitochondrial import apparatus. Mol Biol Evol 28(1):781–791. https://doi.org/10.1093/molbev/msq252
Tsaousis AD, Ollagnier de Choudens S, Gentekaki E, Long S, Gaston D, Stechmann A, Vinella D, Py B, Fontecave M, Barras F, Lukeš J, Roger AJ (2012) Evolution of Fe/S cluster biogenesis in the anaerobic parasite *Blastocystis*. Proc Natl Acad Sci U S A 109(26):10426–10431. https://doi.org/10.1073/pnas.1116067109

Vacek V, Novák LVF, Treitli SC, Táborský P, Čepička I, Kolísko M, Keeling PJ, Hampl V (2018) Fe-S cluster assembly in oxymonads and related protists. Mol Biol Evol 35(11):2712–2718. https://doi.org/10.1093/molbev/msy168

van Bruggen JJA, Stumm CK, Vogels GD (1983) Symbiosis of methanogenic bacteria and sapropelic protozoa. Arch Microbiol 136(2):89–95

Van Hellemond JJ, Klockiewicz M, Gaasenbeek CP, Roos MH, Tielens AG (1995) Rhodoquinone and complex II of the electron transport chain in anaerobically functioning eukaryotes. J Biol Chem 270(52):31065–31070

Xu F, Jerlstrom-Hultqvist J, Kolísko M, Simpson AGB, Roger AJ, Svard SG, Andersson JO (2016) On the reversibility of parasitism: adaptation to a free-living lifestyle via gene acquisitions in the diplomonad *Trepomonas* sp. PC1. BMC Biol 14:62. https://doi.org/10.1186/s12915-016-0284-z

Yabuki A, Nakayama T, Yubuki N, Hashimoto T, Ishida K, Inagaki Y (2011) Tsukubamonas globosa n. Gen., n. Sp., a novel excavate flagellate possibly holding a key for the early evolution in "Discoba". J Eukaryot Microbiol 58(4):319–331. https://doi.org/10.1111/j.1550-7408.2011.00552.x

Yarlett N, Hann AC, Lloyd D, Williams A (1981) Hydrogenosomes in the rumen protozoon *Dasytricha ruminantium* Schuberg. Biochem J 200(2):365–372

Yarlett N, Hann AC, Lloyd D, Williams AG (1983) Hydrogenosomes in a mixed isolate of *Isotricha prostoma* and *Isotricha intestinalis* from ovine rumen contents. Comp Biochem Physiol B 74 (2):357–364

Yarlett N, Coleman GS, Williams AG, Lloyd D (1984) Hydrogenosomes in known species of rumen entodiniomorphid protozoa. FEMS Microbiol Lett 21(1):15–19

Yarlett N, Orpin CG, Munn EA, Yarlett NC, Greenwood CA (1986) Hydrogenosomes in the rumen fungus *Neocallimastix patriciarum*. Biochem J 236(3):729–739

Yubuki N, Leander BS, Silberman JD (2010) Ultrastructure and molecular phylogenetic position of a novel phagotrophic stramenopile from low oxygen environments: *Rictus lutensis* gen. Et sp. nov. (Bicosoecida, incertae sedis). Protist 161(2):264–278. https://doi.org/10.1016/j.protis.2009.10.004

Yubuki N, Pánek T, Yabuki A, Čepička I, Takishita K, Inagaki Y, Leander BS (2015) Morphological identities of two different marine stramenopile environmental sequence clades: *Bicosoeca kenaiensis* (Hilliard, 1971) and *Cantina marsupialis* (Larsen and Patterson, 1990) gen. Nov., comb. nov. J Eukaryot Microbiol 62(4):532–542. https://doi.org/10.1111/jeu.12207

Yubuki N, Huang S, Leander B (2016) Comparative ultrastructure of fornicate excavates, including a novel free-living relative of diplomonads: *Aduncisulcus paluster* gen. Et sp nov. Protist 167 (6):584–596. https://doi.org/10.1016/j.protis.2016.10.001

Zierdt CH (1991) *Blastocystis hominis* – past and future. Clin Microbiol Rev 4(1):61–79

Zierdt CH, Donnolley CT, Muller J, Constantopoulos G (1988) Biochemical and ultrastructural study of Blastocystis hominis. J Clin Microbiol 26(5):965–970

Zubácová Z, Novak L, Bublikova J, Vacek V, Fousek J, Ridl J, Tachezy J, Doležal P, Vlček C, Hampl V (2013) The mitochondrion-like organelle of *Trimastix pyriformis* contains the complete glycine cleavage system. PLoS One 8(3):e55417. https://doi.org/10.1371/journal.pone.0055417

Organisms Without Mitochondria, How It May Happen?

Vladimír Hampl

Contents

1 Mitochondria in Anaerobes Are Reduced but Typically Not Lost 310
2 Oxymonads: Protists Without Mitochondria 311
3 Prerequisites and Consequences of Mitochondrial Loss 313
4 Why Should We Be Interested in Amitochondriate Protists? 314
References 316

Abstract Decades of investigations have clearly shown that protists living in low-oxygen environments possess mitochondria despite their textbook function, oxidative phosphorylation, is usually absent. The presence of these, in some cases, very rudimental mitochondria has been ascribed to their irreplaceable role in the synthesis of FeS clusters, prosthetic groups of several essential proteins. The deep investigation of the oxymonad *Monocercomonoides exilis* (Preaxostyla, Metamonada) revealed that this organism very likely represents a notable exception, in which the synthesis of FeS clusters runs in the cytosol and mitochondrion is absent. Investigation of a broader spectrum of oxymonads and their relatives provided evidence that the profound reorganisation of FeS cluster synthesis was initiated by a HGT of the bacterial pathway SUF and a loss of the mitochondrial pathway ISC already before the last common ancestor of this clade. This innovation was very likely a preadaptation for (and not a consequence of) the mitochondrial loss, which happened much later and only in the oxymonad lineage. *M. exilis* and other oxymonads are being further studied because they represent valuable examples relevant to our understanding of the reductive evolution of organelles and to the origin of the eukaryotic cell.

V. Hampl (✉)
Faculty of Science, Department of Parasitology, Charles University, BIOCEV, Vestec, Czech Republic
e-mail: vlada@natur.cuni.cz

J. Tachezy (ed.), *Hydrogenosomes and Mitosomes: Mitochondria of Anaerobic Eukaryotes*, Microbiology Monographs 9,
https://doi.org/10.1007/978-3-030-17941-0_13

1 Mitochondria in Anaerobes Are Reduced but Typically Not Lost

Endosymbiosis greatly contributed to the formation of eukaryotic complexity, because it is the only known mechanism how membrane-bound and genome-containing cell compartments evolve (Sagan 1967; Lane 2011). Its most important products, mitochondria and plastids, had changed the biosphere completely and arguably are behind the evolutionary success of eukaryotes. At the same time, while sampling the diverse eukaryotic lineages, it has become apparent that these organelles are not being dispensed at occasions, when their most pronounced functions, oxidative phosphorylation in the case of mitochondria and photosynthesis in the case of plastids, are not needed or are disabled by environmental conditions. So then, vast majority of secondary non-photosynthetic algae and plants still contain colourless plastids (Hadariová et al. 2017), and eukaryotes inhabiting environments with low-oxygen concentrations still contain mitochondria (Roger et al. 2017), the examples of which have been in details described in this book.

Careful examinations provided *raison d'etre* of these sometimes very rudimental mitochondria in the synthesis of FeS clusters (Williams et al. 2002). FeS clusters serve as prosthetic groups of some proteins and are essential in every cell, and their synthesis is initiated in the mitochondrion by the iron-sulphur cluster assembly pathway of alpha-proteobacterial origin and continued in the cytosol by the cytosolic iron-sulphur protein assembly (CIA) pathway (Kispal et al. 1999; Lill et al. 2015; Braymer and Lill 2017). Interesting exception is represented by Archamoebae (*Entamoeba*, *Mastigamoeba*), which lack ISC pathway, and their FeS clusters are apparently synthesised by the combination of nitrogen fixation (NIF) and CIA pathway; the former was acquired by horizontal gene transfer (HGT) from *Epsilonproteobacteria*. In the human pathogen *Entamoeba histolytica*, the NIF pathway is probably not localised in its mitochondrion, although controversies remain (Mi-ichi et al. 2009; Maralikova et al. 2010; Nývltová et al. 2013), and so it may be the first known mitochondrion not involved in the iron-sulphur cluster assembly. If true, the case of *E. histolytica* demonstrates that under some circumstances, like acquisition of new upstream biosynthetic pathway via horizontal gene transfer, the process of FeS cluster assembly can be performed entirely in the cytosol. Also, it seems that FeS cluster assembly is not the reason why *E. histolytica* keeps its mitosomes and another essential process, sulphate activation, was proposed to localise into this organelle (Mi-ichi et al. 2009).

The presence of some form of mitochondrion in every studied eukaryote and the fact that even the most reduced mitosomes harbour essential pathways strengthened the paradigm about the ubiquity of mitochondria (Roger and Silberman 2002; Embley et al. 2003; Hackstein et al. 2006). On the other hand, the example of *E. histolytica* has shown that the seeming essentiality of mitochondria is not provided by the same function in all eukaryotes, and so the question remained, whether it is possible for a mitochondrion to dispense all functions and disappear completely.

Investigation of additional eukaryotic lineages has been instrumental, as many times before, and this question was answered when it came to oxymonads.

2 Oxymonads: Protists Without Mitochondria

Oxymonads are relatively small collection of ~140 described species (Hampl 2016), which live in the intestine of some insect and mammalian hosts, and together with the genera *Trimastix* and *Paratrimastix*, they represent the third principal lineage of Metamonada (Hampl et al. 2009; Adl et al. 2019). In wood-eating insects, certain termites and cockroaches, oxymonads seem to be beneficial part of the intestinal microflora; however, no solid data on their metabolic involvement is available. Majority of studies involving oxymonads focused on description of new species, phylogeny and ultrastructure (e.g. Nie 1950; Radek 1994; Brugerolle et al. 1997; Heiss and Keeling 2006) and rarely on their other features. Exceptions are few studies investigating the movement of axostyle, pronounced microtubular part of their cytoskeleton (McIntosh 1973), relationships with ectosymbiotic bacteria (Leander and Keeling 2004; Noda et al. 2006; Utami et al. 2018), modifications of glycolysis (Slamovits and Keeling 2006a; Liapounova et al. 2006), introns (Slamovits and Keeling 2006b) and noncanonical genetic codes (Keeling and Leander 2003). Importantly for this text, ultrastructural investigations have not brought clear evidence for the presence of neither mitochondrion nor Golgi body. Notable is the study of Carpenter et al. (2008), in which was reported the presence of relatively large electron dense and potentially double-membrane-bounded vesicles more than 1 μm in diameter, putative mitochondria or peroxisomes. Similar vesicles are visible also in earlier TEM images of (McIntosh 1973) and might be homologous to cytoplasmic granules reported from light microscopy by Cleveland in *Saccinobaculus lata* (1950). They were not investigated further.

In 2016 we have published a study (Karnkowska et al. 2016) in which the traces of mitochondrion were carefully investigated in the genomic and transcriptomic datasets of an oxymonad now known as *Monocercomonoides exilis* (Treitli et al. 2018; Fig. 1). The generated draft genome is ~75 MB in size and contains 16,629 predicted protein-coding loci, and according to the available measure, it is reasonably complete albeit divergent. Core Eukaryotic Genes Mapping Approach (CEGMA) recovered 63.3% of core eukaryotic genes, and after manual curation of gene models, and not considering mitochondrion-related genes in the CEGMA dataset, the conserved gene recovery increased to 90%. The predicted proteins as well as the six-frame translation of the genome have been searched for mitochondrial hallmark proteins (e.g. components of TIM/TOM complexes, mitochondrial chaperones and ISC pathway enzymes); furthermore, the data were exhaustively searched using homology-based and targeting signal-based approaches [summarised in Fig. 2 in Karnkowska et al. (2016)]. No approach has resulted in recovering a reliable candidate(s) for mitochondrial proteins, and this led to the conclusion that the mitochondrion is not present. This hypothesis still holds, while further investigation

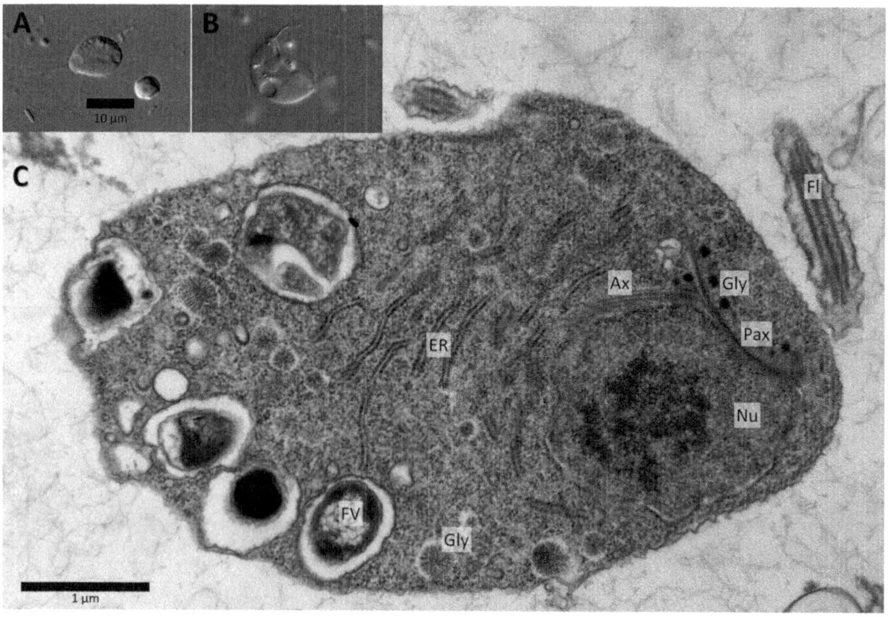

Fig. 1 *Monocercomonoides exilis* and *Paratrimastix pyriformis*. DIC images of (**a**) *Monocercomonoides exilis* and (**b**) *Paratrimastix pyriformis* from cultures. Cells are shown in the same scale. (**c**) TEM micrograph of *Monocercomonoides exilis* (photo by Naoji Yubuki), *Ax* axostyle, *ER* endoplasmic reticulum, *Fl* flagellum, *FV* food vacuoles, *Gly* glycogen granules, *Nu* nucleus, *Pax* preaxostyle

of *M. exilis* genome and genomic and transcriptomic data of other oxymonads is in progress (Karnkowska et al. under review).

If we admit the absence of mitochondrion in *M. exilis*, we should provide an explanation how the cell substitutes the essential functions typically provided by this compartment. As in the mitosome-containing anaerobes, the cellular ATP requirements must be met by the cytosolic ATP production associated with the catabolism of organic compounds. Indeed, *M. exilis* contains full set of glycolytic enzymes including ATP saving analogues PPi-dependent phosphofructokinase (PFK) and pyruvate-orthophosphate dikinase (PPDK), which are known to be present in other eukaryotic anaerobes (Liapounova et al. 2006; Karnkowska et al. 2016). The breakdown of one molecule of glucose to two molecules of pyruvate should theoretically provide the *M. exilis* cell with three ATPs, and because two of these ATP molecules were formed from AMP and PPi, *M. exilis* gains five new high-energy phosphate bonds on ATP molecules. Enzymes of extended glycolysis, pyruvate-ferredoxin oxidoreductase (PFOR), [FeFe]hydrogenase and acetyl-CoA synthetase (ADP forming) are also present, suggesting the yield of additional one ATP per pyruvate molecule. Enzymes for the arginine dihydrolase pathway, known as significant contributor to ATP pool in *Giardia intestinalis* and *Trichomonas vaginalis* (Schofield et al. 1992; Yarlett et al. 1996), have been detected in *M. exilis* genome (Novák et al. 2016), and other ATP-producing pathways were and will be revealed by the detailed analysis of the *M. exilis* genome (Karnkowska et al. under review) and by follow-up metabolic studies.

Less usual explanation was proposed for the way how FeS clusters are synthetized without the mitochondrial ISC system. Like in the case of *E. histolytica*, HGT apparently played the key role. In the genome and transcriptome of *M. exilis*, a set of genes for sulphur mobilisation system (SUF) was found. It consists of a fusion gene SufDSU, and genes for SufB and SufC, which branch from within eubacteria and have no close bacterial sister lineage (Karnkowska et al. 2016). SUF pathway is frequently present in prokaryotes including *E. coli*, in all plastids, which inherited the pathway from cyanobacteria, and it rarely occurs also in eukaryotic anaerobes (*Blastocystis*, *Pygsuia* and *Stygiella*) (Tsaousis et al. 2012; Roche et al. 2013; Stairs et al. 2014; Leger et al. 2016). The SUF pathway of *M. exilis* is simple but theoretically functional with SufDSU extracting sulphur from cysteine and SufB and SufC creating a scaffold for the cluster formation. Besides the SUF pathway, *M. exilis* contains basic inventory of genes for the CIA pathway, namely, Nbp35, Nar1, Cia1, Cia2a and Cia2b. As there are no functional data regarding the SUF and CIA pathways, it is unclear whether and how this unique combination of pathway cooperates during the FeS cluster formation.

3 Prerequisites and Consequences of Mitochondrial Loss

With such an "extreme" product of mitochondrial reduction in hand, it would be interesting to learn more about the mitochondria of *M. exilis* ancestors. Luckily, the mitochondrion-bearing relatives of oxymonads, genera *Trimastix* and *Paratrimastix*, split from two points of the lineage leading to *M. exilis*, and so they represent independent descendants of two such intermediate stages (Zhang et al. 2015). Bits of knowledge on the function of the mitochondrion in *Trimastix marina* acquired thanks to transcriptomic project on this species suggest that this organelle possesses [FeFe]hydrogenase and a complete glycine cleavage system (GCS) (Leger et al. 2017). GCS is a complex of four enzymes with strictly mitochondrial localisation catabolising glycine to CO_2 and NH_3 with concomitant production of cofactors NADH and 5,10-methylene-H_4folate. ATP is probably not produced in the *T. marina* mitochondrion, and pyruvate is oxidatively decarboxylated in the cytosol by PFO and ATP is produced by substrate-level phosphorylation by acetyl-CoA synthetase (ADP forming). The mitochondrion of *Paratrimastix* (formerly *Trimastix*) *pyriformis* has been scrutinised more thoroughly supporting the picture sketched for *T. marina* (Hampl et al. 2008; Zubáčová et al. 2013). GCS has been proven to be localised in the organelle by specific antibodies, and the ongoing genomic project on this organism promises to provide more complete picture of its metabolism. Importantly, the mitochondrial ISC pathway is absent in *P. pyriformis*, and it is substituted by the closely related set of the three SUF pathway enzymes orthologous to those in *M. exilis*. The localisation of these enzymes is unknown, and it is of interest for understanding the functioning of *P. pyriformis* mitochondrion, as well as the loss of mitochondrion in *M. exilis*.

Taxon-wide investigation of the presence/absence of FeS cluster assembly pathways using genomic and transcriptomic datasets of seven species of oxymonads and three species of (para)trimastigids has confirmed the observation from *P. pyriformis* and *M. exilis* that in this whole clade, the ISC was lost and substituted by SUF composed of the same three enzymes found in *M. exilis* (Vacek et al. 2018). This indicates that the switch from ISC to SUF happened relatively early in the common ancestor of the Preaxostyla clade, but the loss of mitochondrion, which is well-documented for *M. exilis* but expected to be common for all oxymonads, happened much later. So, ISC to SUF transition was likely a preadaptation for the mitochondrial loss, but it was definitely not the direct and immediate cause of it, as these protists have undergone millions of years of evolution with mitochondrion along with the SUF system before losing the organelle in oxymonads and (para) trimastigids have preserved this status until present. Dating the protist evolution is a tricky task, but in this particular case, we have a solid hint from 100-MYA-old amber fossils of termites, which demonstrate that at this time, oxymonads in their hindguts were already present and diversified (Poinar 2009). This leads to conclusion that the loss of mitochondria is at least 100-MYA-old event and the substitution of ISC by SUF is an event considerably older.

In the situation of having the organism that lives millions of years without such a canonical part of the eukaryotic cell as the mitochondrion, it is fair to ask how the other systems of such cell have been affected by this unique loss. With this question in mind, we have performed careful analysis of the genomic and transcriptomic data focusing on many cellular systems and the answer will hopefully be published soon (Karnkowska et al. under review).

4 Why Should We Be Interested in Amitochondriate Protists?

The case of *M. exilis*, which is given all currently available data a mitochondrion-free eukaryote, provides example relevant to several ongoing debates related to evolution of eukaryotes. Firstly, it indicates that mitochondrion, despite being essential and evolutionarily very stable part of the cell, can be disposed under circumstances, when the lineage modifies its FeS cluster synthetic pathway into a cytosolic one (Karnkowska and Hampl 2016). This is apparently plausible but very improbable process that has never been achieved with the ISC pathway, and in both known cases (*Monocercomonoides exilis* and *Entamoeba histolytica*), it was initiated by a rare event of HGT. From this perspective, the cellular setup of mitosome-containing eukaryotes (*Giardia, Entamoeba*, microsporidia, *Microcytos*) is not in the "global optimum". Maintenance of mitosomes represents a non-zero cost, and it is also not error-prone. Imagine synthesising costly lipids, setting up proteins and metabolite transporters, performing specific protein import, dividing vesicles in synchrony with cell cycle, and this everything only for one or few pathways that are in principle operable in the cytosol as the example of *M. exilis*demonstrates. I expect that many if not all mitosome-containing eukaryotes keep mitosomes not because the setup of their cells without mitosomes is impossible but because the step-by-step evolutionary journey to a mitosome-free cell is so narrow and involves so improbable (but possible) steps that the random walk of evolution has not made it through yet.

The second topic for which I propose *M. exilis* relevancy is the hotly debated origin of eukaryotes; see Zachar and Szathmáry (2017) for the recent summary. In a nutshell, two major schools of thought have formed in this debate, which disagree on the role of mitochondria in the whole process. While some (Yutin et al. 2009; Martijn and Ettema 2013; Cavalier-Smith 2014) consider mitochondrion as an important innovation that was acquired during the formation of the eukaryotic cell but not at the very beginning of this evolutionary transformation, others (Lane and Martin 2010) argue that acquisition of mitochondrial symbiont was the starter of eukaryogenesis. The latter claim that without mitochondrial ATP, evolution of many other eukaryotic novelties would be energetically just implausible. Here comes the example case of *M. exilis* but also other eukaryotic anaerobes that thrive without mitochondrial ATP but maintain their eukaryotic features. Existence of these organisms suggests, as we have recently argued in detail (Hampl et al. 2018), that life forms of phagotrophic, nucleus-containing eukaryotes existing prior to the mitochondrial endosymbiosis are plausible. Hypotheses, which include amitochondriate (Archezoa) stages, are therefore not logically incorrect or energetically unbearable, which at the same time does not mean that evolution must have proceeded through this path.

There are several reasons why *Monocercomonoides exilis* and other representatives of Preaxostyla should be deeply explored. The main is to confirm or reject our hypothesis that mitochondrion is absent, which still needs to be treated with caution and critically scrutinised. Further studies of cell biology and biochemistry, currently hindered by the absence of laboratory tools including axenic cultivation, should follow to depict in more details how the amitochondriate cells function. The highest priority remains to elucidate the process of the FeS cluster synthesis in such cells. Finally, other oxymonads and Preaxostyla should be investigated for the presence of mitochondria and, if present, the functions of these organelles should be understood as much as possible, because their evolution has gone in such an interesting direction within this group of protists.

Acknowledgement The salary of VH was funded from the European Research Council (ERC) under the European Union's Horizon 2020 research and innovation programme (grant agreement No 771592), from the Centre for research of pathogenicity and virulence of parasites reg. nr.: CZ.02.1.01/0.0/0.0/16_019/0000759, from the Ministry of Education, Youth and Sports of CR within the National Sustainability Program II (Project BIOCEV-FAR) LQ1604 and from the project 'BIOCEV' (CZ.1.05/1.1.00/02.0109).

References

Adl SM, Bass D, Lane CE et al (2019) Revisions to the classification, nomenclature, and diversity of eukaryotes. J Eukaryot Microbiol 66:4–119

Braymer JJ, Lill R (2017) Iron–sulfur cluster biogenesis and trafficking in mitochondria. J Biol Chem 292:12754–12763. https://doi.org/10.1074/jbc.R117.787101

Brugerolle G, Koenig H, Konig H (1997) Ultrastructure and organization of the cytoskeleton in *Oxymonas*, an intestinal flagellate of termites. J Eukaryot Microbiol 44:305–313

Carpenter KJ, Waller RF, Keeling PJ (2008) Surface morphology of *Saccinobaculus* (Oxymonadida): implications for character evolution and function in oxymonads. Protist 159:209–221. https://doi.org/10.1016/j.protis.2007.09.002

Cavalier-Smith T (2014) The neomuran revolution and phagotrophic origin of eukaryotes and cilia in the light of intracellular coevolution and a revised tree of life. Cold Spring Harb Perspect Biol 6:a016006. https://doi.org/10.1101/cshperspect.a016006

Cleveland LR (1950) Hormone-induced sexual cycles of flagellates. IV. Meiosis after syngamy and before nuclear fusion in Notila. J Morphol 87(2):317–347

Embley TM, Van Der Giezen M, Horner DS et al (2003) Hydrogenosomes, mitochondria and early eukaryotic evolution. IUBMB Life 55:387–395. https://doi.org/10.1080/15216540310001592834

Hackstein JHP, Tjaden J, Huynen M (2006) Mitochondria, hydrogenosomes and mitosomes: products of evolutionary tinkering! Curr Genet 50:225–245. https://doi.org/10.1007/s00294-006-0088-8

Hadariová L, Vesteg M, Hampl V, Krajčovič J (2017) Reductive evolution of chloroplasts in non-photosynthetic plants, algae and protists. Curr Genet 64:365. https://doi.org/10.1007/s00294-017-0761-0

Hampl V (2016) Preaxostyla. In: Handbook of the protists. Springer, Cham, pp 1–36

Hampl V, Silberman JD, Stechmann A et al (2008) Genetic evidence for a mitochondriate ancestry in the 'amitochondriate' flagellate *Trimastix pyriformis*. PLoS One 3(9):e1383. https://doi.org/10.1371/journal.pone.0001383

Hampl V, Hug L, Leigh JW et al (2009) Phylogenomic analyses support the monophyly of Excavata and resolve relationships among eukaryotic "supergroups". Proc Natl Acad Sci U S A 106:3859–3864. https://doi.org/10.1073/pnas.0807880106

Hampl V, Čepička I, Eliáš M (2018) Was the mitochondrion necessary to start eukaryogenesis? Trends Microbiol 27:1–9. https://doi.org/10.1016/j.tim.2018.10.005

Heiss A, Keeling PJ (2006) The phylogenetic position of the oxymonad *Saccinobaculus* based on SSU rRNA. Protist 157:335–344. https://doi.org/10.1016/j.protis.2006.05.007

Karnkowska A, Hampl V (2016) The curious case of vanishing mitochondria. Microb Cell 3:491–494. https://doi.org/10.15698/mic2016.10.531

Karnkowska A, Vacek V, Zubáčová Z et al (2016) A eukaryote without a mitochondrial organelle. Curr Biol 26:1274–1284. https://doi.org/10.1016/j.cub.2016.03.053

Keeling PJ, Leander BS (2003) Characterisation of a non-canonical genetic code in the oxymonad *Streblomastix strix*. J Mol Biol 326:1337–1349

Kispal G, Csere P, Prohl C, Lill R (1999) The mitochondrial proteins Atm1p and Nfs1p are essential for biogenesis of cytosolic Fe/S proteins. EMBO J 18:3981–3989. https://doi.org/10.1093/emboj/18.14.3981

Lane N (2011) Energetics and genetics across the prokaryote-eukaryote divide. Biol Direct 6:35. https://doi.org/10.1186/1745-6150-6-35

Lane N, Martin W (2010) The energetics of genome complexity. Nature 467:929–934. https://doi.org/10.1038/nature09486

Leander BS, Keeling PJ (2004) Symbiotic innovation in the oxymonad *Streblomastix strix*. J Eukaryot Microbiol 51:291–300. https://doi.org/10.1111/j.1550-7408.2004.tb00569.x

Leger MM, Eme L, Hug LA, Roger AJ (2016) Novel hydrogenosomes in the microaerophilic jakobid *Stygiella incarcerata*. Mol Biol Evol 33:2318–2336. https://doi.org/10.1093/molbev/msw103

Leger MM, Kolisko M, Kamikawa R et al (2017) Organelles that illuminate the origins of *Trichomonas* hydrogenosomes and *Giardia* mitosomes. Nat Ecol Evol 1:0092. https://doi.org/10.1038/s41559-017-0092

Liapounova NA, Hampl V, Gordon PMK et al (2006) Reconstructing the mosaic glycolytic pathway of the anaerobic eukaryote *Monocercomonoides*. Eukaryot Cell 5:2138–2146. https://doi.org/10.1128/EC.00258-06

Lill R, Dutkiewicz R, Freibert SA et al (2015) The role of mitochondria and the CIA machinery in the maturation of cytosolic and nuclear iron-sulfur proteins. Eur J Cell Biol 94:280–291. https://doi.org/10.1016/j.ejcb.2015.05.002

Maralikova B, Ali V, Nakada-Tsukui K et al (2010) Bacterial-type oxygen detoxification and iron-sulfur cluster assembly in amoebal relict mitochondria. Cell Microbiol 12:331–342. https://doi.org/10.1111/j.1462-5822.2009.01397.x

Martijn J, Ettema TJG (2013) From archaeon to eukaryote: the evolutionary dark ages of the eukaryotic cell. Biochem Soc Trans 41:451–457. https://doi.org/10.1042/BST20120292

McIntosh JR (1973) The axostyle of *Saccinobaculus*. II. Motion of the microtubule bundle and a structural comparison of straight and bent axostyles. J Cell Biol 56:324–339

Mi-ichi F, Abu Yousuf M, Nakada-Tsukui K, Nozaki T (2009) Mitosomes in *Entamoeba histolytica* contain a sulfate activation pathway. Proc Natl Acad Sci U S A 106:21731–21736. https://doi.org/10.1073/pnas.0907106106

Nie D (1950) Morphology and taxonomy of the intestinal Protozoa of the Guinea-pig, *Cavia porcella*. J Morphol 86:381–494. https://doi.org/10.1002/jmor.1050860302

Noda S, Inoue T, Hongoh Y et al (2006) Identification and characterization of ectosymbionts of distinct lineages in Bacteroidales attached to flagellated protists in the gut of termites and a wood-feeding cockroach. Environ Microbiol 8:11–20. https://doi.org/10.1111/j.1462-2920.2005.00860.x

Novák L, Zubáčová Z, Karnkowska A et al (2016) Arginine deiminase pathway enzymes: evolutionary history in metamonads and other eukaryotes. BMC Evol Biol 16:197. https://doi.org/10.1186/s12862-016-0771-4

Nývltová E, Šuták R, Harant K et al (2013) NIF-type iron-sulfur cluster assembly system is duplicated and distributed in the mitochondria and cytosol of *Mastigamoeba balamuthi*. Proc Natl Acad Sci U S A 110:7371–7376. https://doi.org/10.1073/pnas.1219590110

Poinar G Jr (2009) Early cretaceous protist flagellates (Parabasalia: Hypermastigia: Oxymonada) of cockroaches (Insecta: Blattaria) in Burmese amber. Cretac Res 30:1066–1072. https://doi.org/10.1016/j.cretres.2009.03.008

Radek R (1994) *Monocercomonides termitis* n. sp., an oxymonad from the lower termite *Kalotermes sinaicus*. Arch Protistenkd 144:373–382. https://doi.org/10.1016/S0003-9365(11)80240-X

Roche B, Aussel L, Ezraty B et al (2013) Iron/sulfur proteins biogenesis in prokaryotes: formation, regulation and diversity. Biochim Biophys Acta 1827:455–469. https://doi.org/10.1016/j.bbabio.2012.12.010

Roger AJ, Silberman JD (2002) Cell evolution: mitochondria in hiding. Nature 418:827–829

Roger AJ, Muñoz-Gómez SA, Kamikawa R (2017) The origin and diversification of mitochondria. Curr Biol 27:R1177–R1192. https://doi.org/10.1016/j.cub.2017.09.015

Sagan L (1967) On the origin of mitosing cells. J Theor Biol 14:255–274

Schofield PJ, Edwards MR, Matthews J, Wilson JR (1992) The pathway of arginine catabolism in *Giardia intestinalis*. Mol Biochem Parasitol 51:29–36

Slamovits CH, Keeling PJ (2006a) Pyruvate-phosphate dikinase of oxymonads and Parabasalia and the evolution of pyrophosphate-dependent glycolysis in anaerobic eukaryotes. Eukaryot Cell 5:148–154. https://doi.org/10.1128/EC.5.1.148

Slamovits CH, Keeling PJ (2006b) A high density of ancient spliceosomal introns in oxymonad excavates. BMC Evol Biol 6(34):34. https://doi.org/10.1186/1471-2148-6-34

Stairs CW, Eme L, Brown MW et al (2014) A SUF Fe-S cluster biogenesis system in the mitochondrion-related organelles of the anaerobic protist *Pygsuia*. Curr Biol 24:1176–1186. https://doi.org/10.1016/j.cub.2014.04.033

Treitli SC, Kotyk M, Yubuki N, Jirounková E, Vlasáková J, Smejkalová P, Šípek P, Čepička I, Hampl V (2018) Molecular and morphological diversity of the oxymonad genera *Monocercomonoides* and *Blattamonas* gen. nov. Protist 169(5):744–783. https://doi.org/10.1016/j.protis.2018.06.005

Tsaousis AD, Ollagnier de Choudens S, Gentekaki E et al (2012) Evolution of Fe/S cluster biogenesis in the anaerobic parasite *Blastocystis*. Proc Natl Acad Sci U S A 109:10426–10431. https://doi.org/10.1073/pnas.1116067109

Utami YD, Kuwahara H, Igai K et al (2018) Genome analyses of uncultured TG2/ZB3 bacteria in "Margulisbacteria" specifically attached to ectosymbiotic spirochetes of protists in the termite gut. ISME J 13:455. https://doi.org/10.1038/s41396-018-0297-4

Vacek V, Novák LVF, Treitli SC et al (2018) Fe-S cluster assembly in oxymonads and related protists. Mol Biol Evol. https://doi.org/10.1093/molbev/msy168

Williams BAP, Hirt RP, Lucocq JM, Embley TM (2002) A mitochondrial remnant in the microsporidian *Trachipleistophora hominis*. Nature 418:865–869

Yarlett N, Martinez MP, Moharrami MA, Tachezy J (1996) The contribution of the arginine dihydrolase pathway to energy metabolism by *Trichomonas vaginalis*. Mol Biochem Parasitol 78:117–125. https://doi.org/10.1016/S0166-6851(96)02616-3

Yutin N, Wolf MY, Wolf YI, Koonin EV (2009) The origins of phagocytosis and eukaryogenesis. Biol Direct 4:9. https://doi.org/10.1186/1745-6150-4-9

Zachar I, Szathmáry E (2017) Breath-giving cooperation: critical review of origin of mitochondria hypotheses. Biol Direct 12:19. https://doi.org/10.1186/s13062-017-0190-5

Zhang Q, Táborský P, Silberman JD et al (2015) Marine isolates of *Trimastix marina* form a plesiomorphic deep-branching lineage within Preaxostyla, separate from other known trimastigids (*Paratrimastix* n. Gen.). Protist 166:468–491. https://doi.org/10.1016/j.protis.2015.07.003

Zubáčová Z, Novák L, Bublíková J et al (2013) The mitochondrion-like organelle of *Trimastix pyriformis* contains the complete glycine cleavage system. PLoS One 8:e55417. https://doi.org/10.1371/journal.pone.0055417

Index

A

Abnormal hydrogenosomes, 95
Acetate/succinate CoA-transferase (ASCT), 117, 290, 292, 294, 296, 300
Adenylate kinase, 141, 142
ADP/ATP carrier (AAC), 45, 54–56, 60, 61
Aerobic mitochondria, 288, 292
Alanine, 190
Alcohol dehydrogenase E (ADHE), 164, 168, 169
Alveolates, 291, 299
Amitochondriate protists, 314, 315
Amitochondriates, 208, 209
Amoebozoa, 291, 300, 301
Anaerobic ciliates
- double membrane, 113
- evolution, 122–123
- methanogenic endosymbionts, 121

Anaerobic energy metabolism, 10, 15, 17–21
Anaerobic enzymes, 122
Anaerobic fungi
- bacterial mixed acid fermentations, 164
- cellulosomes, 161, 164
- diversity, 161
- fibre-degrading microorganisms, 161
- genera, 160
- genome sequencing projects, 162
- hydrolysis, cellulose, 164
- lignocellulosic biomass, 161
- methanogenic bacteria, 161
- mitochondria *vs*. hydrogenosomes, 162–164
- monocentric, 160
- *N. frontalis*, 160
- phylum *Neocallimastigomycota*, 160
- *Piromyces* and *Neocallimastix*, hydrogenosomal metabolism, 165–169
- *Piromyces* sp. E2, energy metabolism, 169–170
- zoospores, 160

Anaerobic metabolism, 3
Anoxic oceans, 14, 15, 17
Apicoplast, 244–246, 258, 260
Archaebacteria, 13
Archaic translocase of the outer membrane (ATOM), 58
Archamoebae, 207
Arginine deiminase (ADI1), 190
Aspartate aminotransferase (AAT), 190
ATP-binding cassette transporter of the mitochondria (ATM), 290
Autophagic vacuole, 104
Autophagy, 96, 104

B

BLAST, 43, 60
Blastocystis species
- adaptations and lateral gene transfer, 269–271
- AOX, 273
- cell biology and functions, 272, 273
- energy metabolism, 275–277
- energy-generating organelles, 268
- Fe-S clusters, 274
- glycolysis, 275
- heterokonts, 269
- hydrogenosomes, 268, 269

J. Tachezy (ed.), *Hydrogenosomes and Mitosomes: Mitochondria of Anaerobic Eukaryotes*, Microbiology Monographs 9,
https://doi.org/10.1007/978-3-030-17941-0

Blastocystis species (*cont.*)
metazoans, 269
mitochondrial DNA, 270, 272
mitochondrial protein, 274, 275
morphology, 276–280
MROs, 269, 273, 281
PCD, 280, 281
stramenopile morphology, 269
Brain-eating amoeba, 21
Breviatea, 299, 300

C
Carbohydrate metabolism, 134, 151
core metabolism, 254
glycolysis, 255
PNO, 254, 255
Carbohydrates, 95, 99
Cardiolipin, 129
Cell fractionation analysis, 105
Cellulosomes, 161, 164
Chlamydomonas, 17, 18
Chromatic granules, 2, 3
Chytrid hydrogenosomes, 17, 18
Ciliate hydrogenosomes, 93
Ciliates
description, 112
free-living anaerobic, 113
hydrogenosomes, 113
orders, 112
See also Anaerobic ciliates
Core Eukaryotic Genes Mapping Approach (CEGMA), 311
Cortical cilia, 112
CRISPR/Cas9-based strategy, 42
CRISPRi, 41
Cristae, 248, 249
Crypton, 8
Cryptosporidium mitosomes, 51
Cryptosporidium parvum
apicomplexan, 212, 220
ATP/ADP transporter, 229
and *Encephalitozoon cuniculi*, 231
and *Entamoeba histolytica*, 230
Fe-S cluster assembly, 225
intermembrane space, 216
IscS, 228
mitosome, 216
NADH dehydrogenase, 232
relict mitochondrion, 209
Cryptosporidium species
apicoplast, 260
carbohydrate metabolism (*see* Carbohydrate metabolism)
CB (*see* Crystalloid body (CB))
in *C. parvum*, 245
CRISPR-Cas9 system, 261
cristae, 248, 249
cytochrome c oxidase, 246
diverse environments, 244
energy metabolism, 256, 257
ETC, 246
human cryptosporidia, 261
hydrogenosomes, 245
in silico information, 261
intracellular location, 246, 247
ISC, 250–253
medical and veterinary importance, 244
mitochondrial protein, 253, 254
mitosome, 261
RER, 250
subcellular (internal) organization, 249
subclass, 244
TCA cycle and glycolysis, 246
ultrastructural and phylogenetic analyses, 245
ultrastructure morphology, 246
Crystalloid body (CB)
C. parvum, 259
mitochondrial IM, 259
non-toxic iron, 259
RER, 257
sporozoites, 259
ultrastructural data, 257
C-tail-anchored proteins (CTAs), 193
Cyanobacteria, 13, 14, 20
5-Cyano-2,3-ditolyl tetrazolium chloride (CTC), 230
Cysteine, 189
Cytosol, 10, 17–19
Cytosolic iron-sulphur cluster assembly (CIA), 274, 310, 313

D
1,4-Diamino-2-butanone (DAB), 150
Diplomonads, 207, 226
Discobids, 291
DnaK, 63
Dynamin-related proteins (DRPs), 90, 105

E
Earth history, 9–11
Ectosymbiotic bacteria, 311
Electron tomography, 245, 250, 259
Electron transport chain (ETC), 290, 296–299, 301
Embden-Meyerhof pathway, 12
Endomembrane system, 33
Endoplasmic reticulum (ER), 103–104

nascent luminal proteins, 34
signal peptidase, 34
translation and translocation, 34
Endosymbiotic theory, 12
Energetic benefit, 11
Entamoeba mitosomes, 52
ER-mitochondria encounter structures (ERMES), 103, 104
ER signal peptidase, 34
Ethanol, 17
Eukaryotes
LGT theories, 10
oxygen role, 10
Eukaryotic anaerobes, 8, 10, 13, 14, 18, 19
Experimental models, 32

F
[FeFe] hydrogenase, 133, 138, 139
Ferredoxins (Fdx), 130, 136, 138
Fe-S clusters
in bacteria, 221
biosynthesis, 274
Cryptosporidium spp., 227, 228
Entamoeba histolytica, 228, 229
in eukaryotes, 221
Giardia intestinalis, 223, 224
maturation of extramitochondrial, 223
microsporidia, 226, 227
Mikrocytos mackini, 228
mitochondrial model, 222
NIF system, 222
physiological function, 223
SUF system, 222
Fe-S proteins, 187, 195
Field emission scanning electron microscopy (FESEM), 92, 95
Flavodiiron (FDP), 145
Fornicata, 289, 290, 293–295
Diplomonadida, 289
energy generation, 290–292
Fe-S clusters, 289, 290
flagellated eukaryotes, 289
GCS, 293, 294
mitochondrial metabolism, 289
protein import machinery, 293
pyruvate metabolism, 290–292
Frataxin, 147
Free-living protists
ATP generation, 288
Fornicata (*see* Fornicata)
free-living and endobiotic species, 289
parabasalid parasites, 288
Fumarate reductase (FRD), 16, 17, 117
Fungal hydrogenosomes, 89, 102
Fungi, 160
anaerobic (*see also* Anaerobic fungi)
mycorrhiza, 160

G
Genetic transformation techniques, 41
Giardia intestinalis, 207–210, 214, 215, 218–220, 230, 232
Giardia mitosomes, 52–53
Giardia processing peptidase (GPP), 215
Global oxygen history, 13, 14
Glutamate dehydrogenase (GDH), 190
Glycine cleavage system (GCS), 190, 292–294, 296–298, 300, 313
Glycine decarboxylase complex (GDC), 148, 149, 190
Glycolysis, 275

H
Heart-shaped process, 105
Heterokonts, 269
Heterolobosea, 297
Heteroloboseans, 207
HHpred, 43
Hidden Markov model (HMM), 43, 58, 60, 69
HMMER software, 43
Horizontal gene transfer (HGT), 292, 296, 310, 313, 314
Hybrid cluster protein (HCP), 189
Hydrogenase, 138–139
Hydrogenosomal and mitosomal protein import
laboratory techniques and tools, 40–42
mining genome sequence data, 43
Hydrogenosomal malic enzyme, 139
Hydrogenosomal targeting sequence (HTS), 50
Hydrogenosome components
envelope, 90–93
matrix, 99–101
peripheral vesicles, 93–99
Hydrogenosome matrix
ammoniacal silver and PTA, 101
calcium deposits, 101
description, 99
electron-dense core, 99
finely granular and homogeneous, 101
granular structure, 101
Hydrogenosomes, 207, 208, 210, 213, 214, 217, 221, 231, 268–270, 273, 274, 288, 289, 294, 296, 299

Hydrogenosomes (*cont.*)
anaerobic ciliates (*see* Anaerobic ciliates)
anaerobic fungi (*see also* Anaerobic fungi)
autophagy, 96, 104
biological nomenclature, 2
carbohydrates, 95
cell cycle, 106
chytrid, 17, 18
definition, 128
discovery, 8, 178
division
dynamin, 105
furrowing process, 104
growth, 104, 105
heart-shaped process, 105
microscopic and cell fractionation analysis, 105
mitochondrial process, 105
segmentation process, 104
enzymatic activity, 88
and ER, 103–104
history, 1
identification of proteins, 88
immunolabeling, 107
and microtubules, 106
mitotic process, 106
morphology, 88
new structures and organelles, 86, 87
organelles, 4
rumen ciliates, 118–120
shape, 90
sizes, 90
trichomonad, 86, 128 (*see also* Trichomonad hydrogenosomes)
trichomonad flagellates, 3
Trichomonadida order, 88
types, 86
Hydrogen-producing mitochondria, 288

I
Immunocytochemistry, 102
Immunolabeling, 107
Inorganic pyrophosphatase (IPP), 231
Iron-only hydrogenase ([Fe]-Hyd), 18
Iron-sulfur cluster (ISC), 139, 146, 221–224, 226–228, 231–233, 250, 252, 253, 258–260

J
Jakobida, 296, 297
Jakobids, 207

K
Karyokinesis, 106

L
Lactate dehydrogenase (LDH), 255
Large-scale projects, 178
Lateral gene transfer (LGT), 9, 10, 15, 19, 222
Lignocellulosic biomass, 161
Lysosomes, 2

M
Malate, 133
Malic enzyme, 97, 107, 139–140, 150
Margulis's theory, 12, 13
Marine invertebrates, 16, 17, 20
Methanogenic bacteria, 161
Methanogenic endosymbionts, 121
Microfibrillar structures, 104
Microsporidia, 226, 227
Microtubules, 106
Mitochondria, 1
in anaerobes, 310
eukaryote evolution, 8
origin, 11
oxygen and ATP, 13
Mitochondrial ATP synthesis, 16
Mitochondrial carrier family (MCF) proteins, 37, 54, 61, 191, 193
Mitochondrial division process, 105
Mitochondrial import (MIM), 36, 41
Mitochondrial intermediate peptidase (MIP), 37, 64–66
Mitochondrial loss, 313, 314
Mitochondrial metabolism, 3
Mitochondrial outer membrane potential (MOMP), 280, 281
Mitochondrial processing peptidase (MPP), 35, 37, 44, 50, 53, 64–67, 214, 292, 293
Mitochondrial protein import machinery
endosymbiont transferring, 38
endosymbiotic theory, 38
mitochondria, 38
N-terminal presequence, 39
protein import machines, 38
in *S. cerevisiae*, 36
transferred genes, 39
Mitochondrial proteins
description, 34
IMP, 36
MCF proteins, 37
MPP, 35

N-terminal cleavable presequences, 35
PAM complex, 37
protein import, 34
SAM complex, 35
TIM23, 37
TOM complex, 35
Mitochondrial tail-anchored (TA) proteins, 55
Mitochondrial targeting signals, 44–45
Mitochondria-related organelles (MROs), 86, 170, 178, 199, 269–271, 273–281
Mitosomes, 1, 288, 289, 294
aerobic/anaerobic respiration, 206
amitochondrial groups, 208
anaerobes, 312
Archezoa concept, 210
ATP, membrane potential and electron transport, 229–231
biochemical capabilities, 206
biogenesis
genome, 213
protein targeting, translocation and maturation, 214–217
canonical mitochondrial pathways, 208
CPN60, 209
discovery, 8
E. histolytica, 310
Encephalitozoon cuniculi genome sequence, 209
endosymbiotic ancestor, 208
eukaryotes, 314
evolutionary relationship, 9
F_0F_1-ATPase, 206
Fe-S clusters, 207, 209, 210 (*see also* Fe-S clusters)
Giardia intestinalis, 211
history, 2
hydrogenosomes, 207
metabolic pathways, 207
mitochondrial origin, 209
mitochondrion-harboring protists, 208
morphology, 207, 210–212
oxygen-independent ATP synthesis, 8
parasitic protists, 208
perspectives, 225, 232, 233
physiological functions, 221
relict mitochondrion, 209
replication
division, 217–219
structures, 217
segregation, 219, 220
sulfate activation, 231, 232
TCA, 206
trypanosome, 206
Vavraia culicis, 212
Molecular hydrogen, 3
Monocentric anaerobic fungi, 160
Monocercomonoides exilis, 311–315

N

NADH dehydrogenase, 140
Neocallimastix frontalis, 160, 168
Neocallimastix hydrogenosomes, 51
[NiFe] hydrogenase, 138
Nitrogen-fixing (NIF) system, 221, 222, 228, 231, 232
Noncanonical genetic codes, 311
N-terminal targeting sequences (NTS), 88, 180, 181
Nucleoid, 99, 102
Nucleotide transporters (NTT), 229
Nucleus, 33
Nyctotherus ovalis
anaerobic mitochondrion, 113
description, 113
energy metabolism, 116–117
in silico reconstruction, 117–118
mitochondrial genome, 113
MROs, 115

O

Opisthokonts, 291
Organellar membranes
inner membrane
TIM22 complex, 61
TIM23 complex, 62
outer membrane
SAM complex, 59–60
TOM complex, 56–59
small TIM chaperones, 60–61
Organellar targeting signals, *see* Targeting signals
Origin of eukaryotes, 12, 13
OsmC system, 188, 189
Oxic environments, 11, 16, 19
Oxidative phosphorylation, 8, 11, 12
Oxygen (O_2)
accumulation, 14
atmospheric levels, 10
biochemical, 16
endosymbiotic theory, 14
eukaryotes, 11
global oxygen history, 13
O_2-producing algae, 21
oxidative phosphorylation, 11
Oxygen accumulation, earth history, 9, 11

Oxygen late, 15, 18
Oxygen/ATP, 12, 13
Oxymonads, 311–313

P
Parasitic protists, 207, 208, 218
Paratrimastix pyriformis, 312, 314
Percoll-sucrose density centrifugation, 93
Periodic acid-thiosemicarbazide-silver proteinate technique, 95
Peripheral vesicles, 93, 95, 97, 99
Peroxiredoxin, 144
Peroxisomes, 33, 55
  chromatic granules, 2
  marker enzyme, 3
  origin, 2
Petites, 162
Phosphoenolpyruvate (PEP), 132
Phosphofructokinase (PFK), 194, 312
Phosphotungstic acid (PTA), 101, 102
Piromyces sp. E2, 162, 163, 165, 170
  ADHE activity, 168
  energy metabolism, 166, 169–170
  epifluorescence micrograph, 163
  growth, 165
Plasmodium falciparum, 19
Polyamines, 150
Preaxostyla, 314, 315
Preproteins
  hydrogenosomal, 58
  hydrophobic, 56
  inner membrane, 66–67
  mitochondrial inner membrane, 61
  MPP, 65–66
  soluble matrix-destined, 37
  TIM23 complex, 37
Presequence translocase-associated motor (PAM), 37, 63, 214, 253
Programmed cell death (PCD), 279–281
Prokaryote-to-eukaryote transition, 13
Protein, 107
  DRP, 90
  dynamin, 105
  ERMES, 103, 104
  hydrogenosome matrix, 101
  malic enzyme, 107
  SDS-PAGE, 97
  T. vaginalis hydrogenosome, 88
Protein family (PFAM), 43
Protein folding, 63, 67, 68
Protein import motor, 63–64
Protein precursors, 46
Protein trafficking
  fungi and mammals, 32
  in eukaryotes
    ER, 34
    mitochondrion, 34
    nucleus, 33
    proteins travelling, 33
Proteome, *Pentatrichomonas hominis* hydrogenosome, 198
Proteome, *T. vaginalis* hydrogenosomes
  amino acid metabolism, 189–190
  analysis, 179
  carriers, hydrogenosomal inner membrane, 193
  contaminants/surface-associated proteins, 194
  CTAs, 193–194
  energy metabolism, 182–187
  Fe-S cluster assembly, 187–188
  oxygen metabolism, 181
  protein import, 190–192
  protein sequences, 179
  proteins, 181
  putative pore-forming Hmp35, 181
  quantitative analysis, iron-induced changes, 195
  ROS defense, 188–189
  silico identification, genes coding, 180
Proteomic studies, 179, 181, 183, 191, 194
Protists, 88, 89, 105
Proton motive force (PMF), 118
Protons, 117
Pyridine-nucleotide transhydrogenase (PNT), 209, 230, 256
Pyroantimonate, 101
Pyruvate dehydrogenase complex (PDH), 117, 118, 121, 135, 270
Pyruvate ferredoxin oxidoreductase (PFO), 18, 19, 117, 119, 121, 122, 168, 169, 171, 181, 182, 187, 188, 195, 198, 290, 292, 296–298, 300
Pyruvate formate-lyase (PFL), 17–19, 117, 121, 290, 292
Pyruvate NADP+ oxidoreductase (PNO), 254, 255, 270, 290, 292, 298–300
Pyruvate phosphate dikinase (PPDK), 276

R
Reactive oxygen species (ROS)
  enzymes and molecules, 188
  peroxide, 188
  SOD, 188
  in *Trichomonas* hydrogenosomes, 188, 189

Relict mitochondrion, 209, 212
Revived endosymbiotic theory, 12
Rhizaria, 298
RNA silencing, 41
Rough endoplasmic reticulum (RER), 245
Rubrerythrin, 144, 188, 189
Rumen ciliates, 118–120, 122

S
SAM-like pathway, 55
SecYEG translocon, 34
Segmentation process, 100, 104, 105
Serine hydroxymethyltransferase (SHMT), 148, 149
Small translocase of the inner membrane (small TIMs) chaperone, 36, 55, 60–61
Sorting and assembly machinery (SAM), 35, 38, 56, 57, 59–60, 69, 214, 274
Sporozoites, 245–247, 249, 250, 252, 254, 257, 259
Standard model, 13
Stramenopila, 297, 298
Stramenopiles, 291, 298
Structured illumination microscopy (SIM), 219
Succinate, 17
Succinate thiokinase (STK), 18, 133, 141, 181
Succinyl-CoA synthetase (SCS), 50, 141, 181, 290, 292, 294, 296–300
Sulfur-mobilization (SUF) system, 221, 222, 228, 232

T
Targeting signals, 38, 39, 41, 44
  Cryptosporidium mitosomes, 51
  Entamoeba mitosomes, 52
  Giardia mitosomes, 52–53
  membrane proteins, hydrogenosomal and mitosomal, 54–55
  microsporidia mitosomes, 53–54
  mitochondrial, 44–45
  Neocallimastix hydrogenosomes, 51
  Trichomonas hydrogenosomes, 45–50
3′-Phosphoadenosine-5′-phosphosulfate (PAPS), 231
Tom40-like proteins, 58
Trachipleistophora hominis, 207, 209–211, 226, 227, 232
Transcriptional silencing techniques, 41
Translocase of the outer membrane (TOM) complex, 35, 37, 56–60
Transmission electron microscopy (TEM), 245, 247, 250, 252, 254, 259, 276, 289
Trepomonas sp.
  ACS, 296
  GCS, 296
Tricarboxylic acid (TCA), 206, 227, 290, 292, 296, 298, 299, 301
Trichomonad ferredoxin, 136
Trichomonad flagellates, 2, 3
Trichomonad hydrogenosomes
  amino acid metabolism, 148
  carbohydrate metabolism, 134
  cardiolipin, 129
  energy metabolism
    ADP to ATP, 129
    glycolysis, 132
    SDS-polyacrylamide gels, 132
  GDC, 148, 149
  ISC assembly machinery, 146–148
  malate, 133
  membranes, 128–129
  NADH:ferredoxin, 134
  oxygen and reactive oxygen species, 142–146
  polyamine metabolism, 150
  proteins, core catabolic pathway
    adenylate kinase, 141, 142
    ferredoxin, 136–138
    hydrogenase, 138–139
    malic enzyme, 139–140
    NADH dehydrogenase, 140–141
    PFOR, 135–136
    SCS, 141
    succinyl-CoA:acetate CoA transferase, 141
  pyruvate, 133
  serine, 149
  SHMT, 148, 149
Trichomonadida, 86, 88
Trichomonads, 128, 207, 208, 269
Trichomonas hydrogenosomes, 45–50
Trichomonas vaginalis, 178, 207, 208, 215, 217
  See also Proteome, T. vaginalis hydrogenosomes
Trichomonas vaginals, 2
Trichomoniasis, 2
Tritrichomonas foetus
  autophagy, 104
  division, 103
  FESEM, 92
  freeze-fracture images, 98
  interphase, 94
  peripheral vesicle, 95

Tritrichomonas foetus (*cont.*)
  peripheral vesicles, 93
  plasma membrane, 87
  segmentation process, 105
Trypanosoma brucei
  Excavata group, 220
  kinetoplastid protist, 218
  mitochondrial Fe-S cluster, 223
  mitochondrion, 206
  tubular cristae, 207

U
Ubiquinone, 129, 134, 140

W
WGA, 95, 99

Z
Zoospores, 160